T0266821

Introduction to Imprecise Probabilities

WILEY SERIES IN PROBABILITY AND STATISTICS

ESTABLISHED BY WALTER A. SHEWHART AND SAMUEL S. WILKS

A complete list of the titles in this series appears at the end of this volume.

Introduction to Imprecise Probabilities

Edited by

Thomas Augustin
Department of Statistics, LMU Munich, Germany

Frank P. A. Coolen
Department of Mathematical Sciences, Durham University, UK

Gert de Cooman
SYSTeMS Research Group, Ghent University, Belgium

Matthias C. M. Troffaes
Department of Mathematical Sciences, Durham University, UK

This edition first published 2014

Registered office
John Wiley & Sons Ltd, The Atrium, Southern Gate, Chichester, West Sussex, PO19 8SQ, United Kingdom

For details of our global editorial offices, for customer services and for information about how to apply for permission to reuse the copyright material in this book please see our website at www.wiley.com.

Library of Congress Cataloging-in-Publication Data

Introduction to imprecise probabilities / edited by Thomas Augustin, Frank Coolen, Gert de Cooman,
Matthias Troffaes.
pages cm
Includes bibliographical references and index.
ISBN 978-0-470-97381-3 (hardback)
1. Probabilities. I. Augustin, Thomas, editor of compilation.
QA273.I56 2014
519.2–dc23

2013041146

A catalogue record for this book is available from the British Library.

ISBN: 978-0-470-97381-3

Typeset in 10/12pt TimesLTStd by Laserwords Private Limited, Chennai, India

1 2014

Contents

Introduction

One of the big challenges for science is coping with uncertainty, omnipresent in modern societies and of ever increasing complexity. Quantitative modelling of uncertainty is traditionally based on the use of precise probabilities: for each event A, a single (classical, precise) probability $P(A)$ is assigned, typically implicitly assumed to satisfy Kolmogorov's axioms. Although there have been many successful applications of this concept, an increasing number of researchers in different areas keep warning that the concept of classical probability has severe limitations. The mathematical formalism of classical probability indispensably requires, and implicitly presupposes, an often unrealistically high level of precision and internal consistency of the information modelled, and thus relying on classical probability under complex uncertainty may lead to unjustified, and possibly deceptive, conclusions.

Against this background, a novel, more flexible theory of uncertainty has evolved: *imprecise probabilities*. Imprecise probabilities have proven a powerful and elegant framework for quantifying, as well as making inferences and decisions, under uncertainty. They encompass and extend the traditional concepts and methods of probability and statistics by allowing for incompleteness, imprecision and indecision, and provide new modelling opportunities where reliability of conclusions from incomplete information is important.

Given that it is a more flexible, and indeed a more honest description of complex uncertainty, the term 'imprecise probabilities' taken literally is an unfortunate misnomer, which, regrettably, has provoked misunderstandings. Nevertheless 'imprecise probabilities' has established itself as an umbrella term for related theories of generalized uncertainty quantification, most notably including *lower and upper probabilities* and *lower* and *upper previsions*. There a classical (σ-) additive probability or a classical, linear prevision (expectation) $P(\cdot)$ is replaced by an interval-valued set-function or functional $[\underline{P}(\cdot), \overline{P}(\cdot)]$. For every gamble (random variable) X, the lower prevision $\underline{P}(X)$ and the upper prevision $\overline{P}(X)$ have a clear behavioural meaning as (typically different) supremum acceptable buying price and infimum acceptable selling price for X, respectively. For events A, the lower probability $\underline{P}(A)$ can informally be interpreted as reflecting the evidence certainly in favour of event A, while the upper probability $\overline{P}(A)$ reflects all evidence possibly in favour of A. Imprecise probabilities live in the continuum between two extreme cases: the case of classical probabilities and linear previsions where $\underline{P}(\cdot)$ and $\overline{P}(\cdot)$ coincide, appropriate only if there is complete information on the stochastic behaviour of a perfect precise random process, and on the other extreme, the case of complete probabilistic ignorance, expressed by $\underline{P}(A) = 0$ and $\overline{P}(A) = 1$ for all non-trivial events A.

Using imprecise probabilities or related concepts seems very natural, and indeed it has a long tradition (see, in particular, [343] for a review). The first formal treatment dates back at least to Boole's book [81] in the middle of the nineteenth century. Further milestones include among others the work of Keynes [394] on incomplete probability orderings; Smith's [602] introduction of lower and upper betting odds and Williams' [702] generalization of coherence; Ellsberg's [264] experiments demonstrating the constitutive role of complex uncertainty in rational decision making; Dempster's [222] concept of multivalued mappings in inference, and its powerful reinterpretation by Shafer [576], that became particularly popular in artificial intelligence ('Dempster-Shafer theory' of belief functions); Huber's [367] reflection on the power of capacity-based neighbourhoods models in robust statistics; and Walley and Fine's [678] work on a frequentist theory of lower and upper probabilities.

Walley's book [672] had a boosting impact, it provided a comprehensive foundation for the theory of lower and upper previsions and coined the term 'imprecise probabilities'. Further influential books include the monographs by Kuznetsov [414], Weichselberger [692] on an introduction of interval probability in generalization of Kolmogorov's axioms, and Shafer and Vovk [582] developing game-theoretic probability.

Based on this solid foundation, work on imprecise probabilities has been gathering strong momentum, and their high potential should be clear in any area of application where the reliability of conclusions drawn from reasoning under uncertainty is of major concern. As a consequence, the literature is spread widely over different fields, each with its own journals and conferences. The Society for Imprecise Probability: Theories and Applications (SIPTA, www.sipta.org) provides an umbrella over these activities, also organizing biennial summer schools and the biennial ISIPTA symposia [1], which since 1999 have been fostering the exchange of the cutting-edge research on imprecise probabilities.

Given the increasing interest in imprecise probabilities, the time is right to present an overview of the main aspects of imprecise probabilities theory and applications. In graduate teaching, research and consultancy, we felt there was a need for an introductory book on imprecise probabilities. The present book aims at filling this gap, and it attempts to achieve this by a providing a synthesis between a book as a holistic entity and a collection of contributions on the current state of the art by leading researchers. We have first established a common structure and a concise framework including a unique notation, and then, in order to cover the most important areas of the vastly developing field, we have asked specialists for specific contributions within this framework, emphasizing continued exchange with the other chapter authors and the editors. The book shows the current state of the art, it sets out research challenges on theory and applications of imprecise probabilities and it provides guidance for further reading.

[1] International Symposium on Imprecise Probability: Theories and Applications, see www.sipta.org, where also the electronic versions of the proceedings are available.

A brief outline of this book

This book consists of 16 closely interrelated chapters. It starts by introducing the basic concepts of the theory (Chapters 1 to 6), then turns to general fields of application (Chapters 7 to 11), provides an insight into the use of imprecise probabilities in engineering, financial risk measurement and reliability analysis (Chapters 12 to 14) and concludes with aspects of elicitation and computation (Chapters 15 and 16).

The notion of desirability, as described in Chapter 1, lies at the core of the behavioural approach to imprecise probabilities, and the discussion there provides a foundation and justification for the account of lower previsions presented in later chapters. It deals with coherence, and with the associated notion of inference using natural extension, marginalization and conditioning.

Chapter 2 discusses (conditional) lower previsions and the notions of coherence associated with them. It also shows how the notion of natural extension provides the theory with a powerful method of conservative (or least committal) probabilistic inference.

Chapter 3 explains how to deal with structural assessments of independence and symmetry in the theory of coherent lower previsions. In particular, the different concepts of independence used in later chapters are explained in detail.

Chapter 4 investigates simplified representations by less general model classes, which are often useful as they are easier for computation and elicitation, and because the full generality of the theory of lower previsions is not always required.

Chapter 5 discusses, within the context of imprecise probabilities, a variety of ways of looking at uncertainty modelling in general, including common constructive elements as well as discrepancies and differences.

Chapter 6 offers a look at lower and upper previsions (or prices) with an interpretation that is rather different from the behavioural one introduced in Chapter 2. The authors discuss their game-theoretic approach to probability, which has its origins in the seminal work of Jean Ville on martingales.

Chapter 7 gives an introduction to the powerful use of imprecise probabilities in statistical inference. In particular it discusses generalized Bayesian inference, extensions of frequentist testing and estimation theory, including their relation to robust statistical methods, as well as nonparametric predictive inference, and provides some insight into modelling of data imprecision.

Chapter 8 discusses nonsequential and sequential decision making under imprecise probability. In particular, the main criteria generalizing expected utility are introduced.

Chapter 9 provides an overview of the powerful use of imprecise probabilities in probabilistic graphical modelling, where complex models over a possible large set of variables are built by combining several submodels. The construction of such credal networks and their updating is discussed.

Chapter 10 describes imprecise probability methods for classification, i.e. predicting the class of an object on the basis of its attributes. This is an application area where the paradigmatic shift to imprecise probability has proven to be particularly powerful, and its success has also been corroborated by large scale comparison studies. Imprecise probabilities allow to overcome major difficulties of traditional methods, notably prior dependency of traditional Bayesian classifiers and overfitting and instability of classification trees.

Chapter 11 gives a first introduction to stochastic processes under imprecise probabilities. The chapter discusses the event-driven approach to stochastic processes and its generalization. It then focuses on the practically very important case of imprecise Markov chains. It compares models based on different independence concepts and investigates the limit behaviour of imprecise Markov chains.

Chapter 12 presents strong arguments for the need to use imprecise previsions in finance. It is shown that theory of imprecise probabilities provides many tools that are closely linked to popular concepts in financial risk measurement, and it enables modelling based on fewer and simpler assumptions than the standard approaches.

Chapter 13 illustrates several engineering applications in which the opportunity to use weaker modelling assumptions underlying structural models fits neatly with the lack of perfect information about certain aspects in reality. Links to sensitivity analysis and practical decision making in engineering are also discussed.

Chapter 14 provides examples where the theory of imprecise probabilities provides natural problem formulations in reliability and risk analysis. Problems of stress-strength reliability are formulated as constrained optimization problems, with the constraints reflecting the limited information available. Some statistical methods for reliability and risk problems are also discussed, including inference about unobserved or even unknown failure modes.

Chapter 15 gives an introduction to the important but relatively unexplored topic of elicitation of imprecise probability models, looking at methods for evaluating judged probabilities and influence factors on probability judgements.

Chapter 16 briefly discusses a number of computational aspects of imprecise probabilities, focusing on natural extension and decision making in particular.

Guide to the reader

An overview of the logical structure of the book is given in the diagram below (Figure 1). A full arrow from one chapter to another indicates that the former is intended to be read first, as the discussion in the latter relies on ideas and notions discussed and introduced there. A dashed arrow identifies a weaker dependence, and merely indicates that one chapter can provide more context for, a better understanding of, or even a different way to motivate the discussion in another chapter.

Although the chapters are closely interrelated, we also have tried to make the book suitable for more selective reading, trying to accommodate readers with different focus and aims. When designing the book, we had in particular three types of readers in mind, namely:

1. Mathematicians mainly interested in foundations of uncertainty (see Figure 2),
2. readers mainly interested in statistics, data analysis and machine learning (see Figure 3), and
3. applied probabilistic modellers (see Figure 4).

In the figures below, we give, for each of these, a number of suggested chapters for reading, along with a diagram depicting their mutual dependencies and recommended reading order. We have also classified these suggested chapters into three categories: a *must read* category (black rectangles), a *nice to have* category (black ellipses), and an *optional* category (grey ellipses).

The material selected should also be suitable for corresponding courses at master or PhD level.

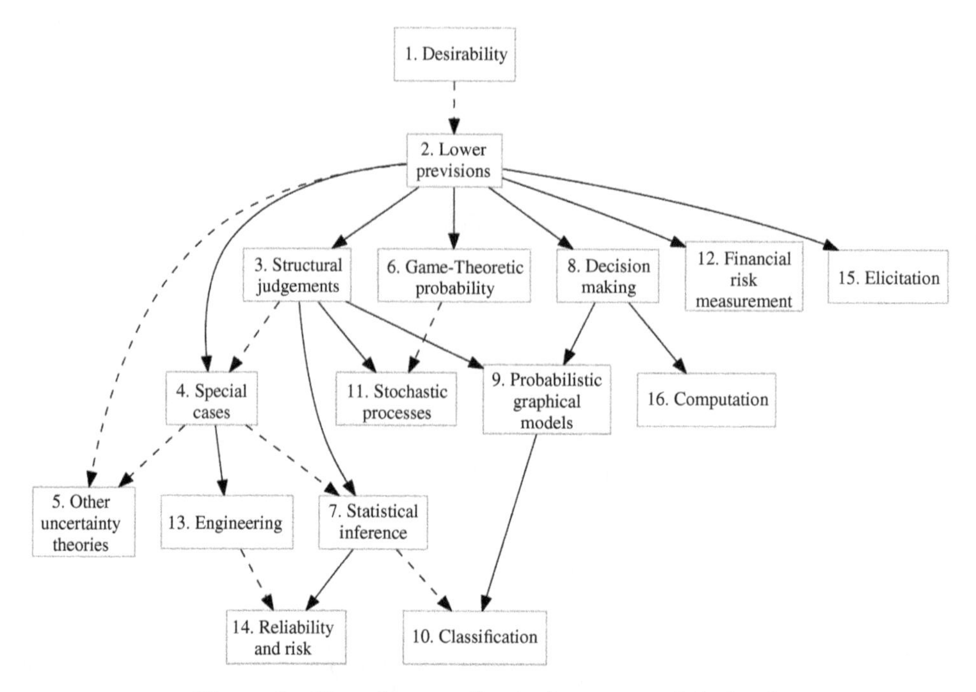

Figure 1 Flow diagram: Logical structure of the book.

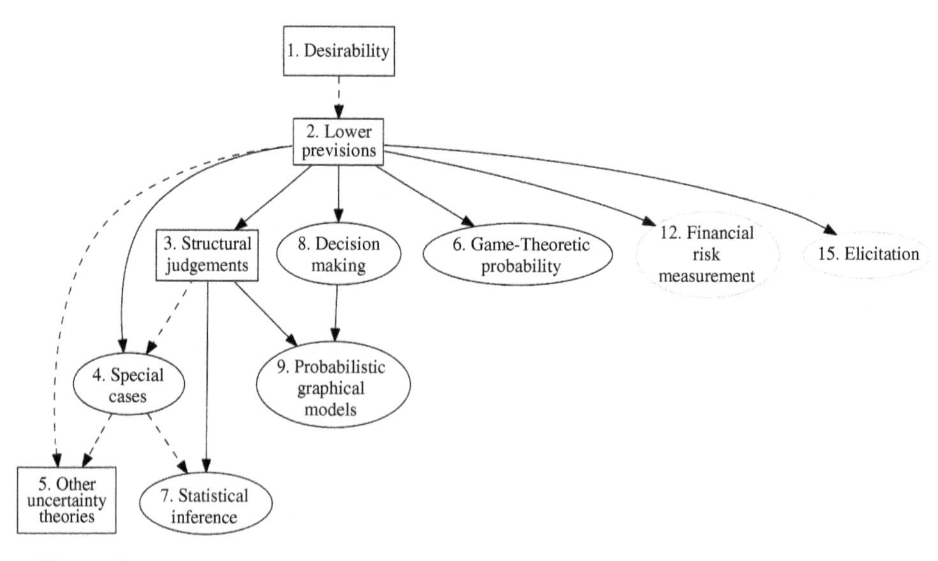

Figure 2 Flow diagram: Mathematicians interested in foundations of uncertainty.

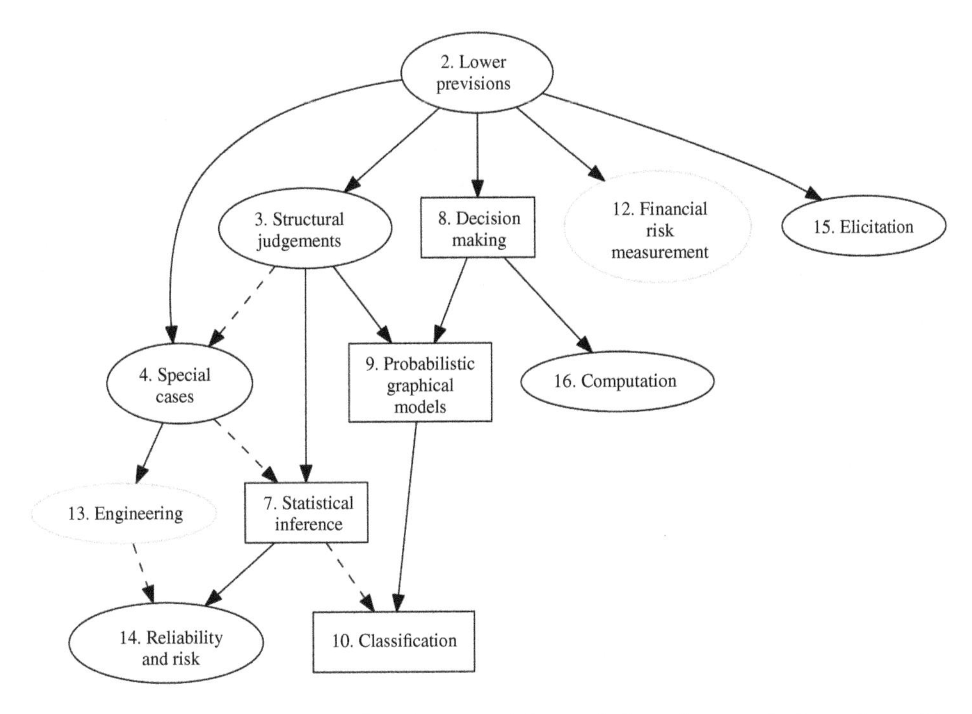

Figure 3 Flow diagram: Readers interested in statistics, data analysis and machine learning.

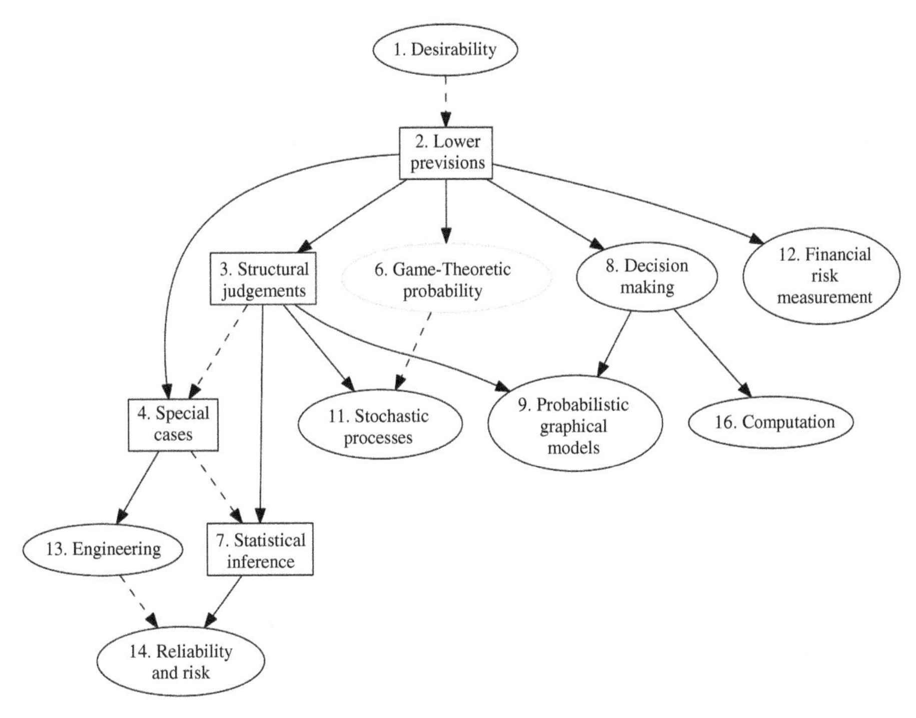

Figure 4 Flow diagram: Applied probabilistic modellers.

Contributors

Joaquín Abellán co-authored Chapter 10 (Classification). He is an associate professor of Computer Science and Artificial Intelligence and a member of the research group on Uncertainty Treatment on Artificial Intelligence (UTAI) at the University of Granada. His current research interests are representation of the information via imprecise probabilities; definition and analysis of information/uncertainty measures on imprecise probabilities; and their applications, principally in the data mining area. For further information, see http://decsai.ugr.es/~jabellan/

Alessandro Antonucci (PhD in computer science, '08) co-authored Chapter 9 (Probabilistic graphical models). He is a Researcher at IDSIA, in Switzerland, with the Imprecise Probability Group (http://ipg.idsia.ch). He also teaches computational mathematics at SUPSI, in Switzerland. He is the Treasurer and an former Executive Editor of SIPTA, the Society for Imprecise Probability: Theory and Applications. His interests include probabilistic graphical models, classification, and the development of knowledge-based expert systems. For further information, see http://www.idsia.ch/~alessandro

Thomas Augustin co-authored Chapter 7 (Statistical inference). He is Professor at the Department of Statistics, Ludwig-Maximilians University (LMU) in Munich, and head of the group 'Foundations of Statistics and Their Applications' there. He received a PhD at LMU Munich in 1997 and worked as postdoc in Munich and Bielefeld. His research aims at a statistical methodology where data quality can be taken into account critically, resulting in generalized uncertainty modelling of complex, non-idealized data in empirical social research and biometrics. He has mainly published on imprecise probabilities in statistical inference and decision making as well as on measurement error modelling. For further information, see www.statistik.lmu.de/~thomas/

Frank P. A. Coolen co-authored Chapters 7 (Statistical inference) and 14 (Reliability and risk). He is Professor of Statistics at the Department of Mathematical Sciences, Durham University, UK. He holds a PhD from Eindhoven University of Technology (The Netherlands, 1994). He has contributed to theory of statistics using imprecise probability for about two decades, most noticeably by developing nonparametric predictive inference, with applications in statistics, reliability, risk, operations research and related topic areas. He has served as joint guest-editor to several special journal issues related to imprecise probabilities, and is on the editorial board of several journals in Statistics and Reliability. For further information, see http://maths.dur.ac.uk/stats/people/fc/fc.html

Gert de Cooman co-authored Chapters 2 (Lower previsions) and 3 (Structural judgements). He is Full Professor in Uncertainty Modelling and Systems Science at the SYSTeMS Research Group, Ghent University, Belgium. He holds an MSc in Physics Engineering (1987) and a PhD in Engineering (1993) from Ghent University. He has been working in imprecise probabilities for more than twenty years, is a founding member and former President of SIPTA, the Society for Imprecise Probability: Theories and Applications, and has helped organise many of the ISIPTA conferences and SIPTA Schools. In the context of imprecise probability theory, he has worked on mathematical aspects of coherence for lower previsions, notions of independence and their consequences, desirability, symmetry and exchangeability, theoretical and algorithmic aspects of credal networks, and stochastic processes. For further information, see http://users.ugent.be/~gdcooma/

Giorgio Corani co-authored Chapter 10 (Classification). He obtained an MSc degree in Engineering in 1999 and a PhD degree in Engineering in 2005, both at Politecnico di Milano (Italy). He is a researcher at IDSIA, the 'Dalle Molle' Institute for Artificial Intelligence, in Switzerland. He is author of about 50 international publications in the area of probabilistic graphical models, data mining and applied statistics. For further information, see http://www.idsia.ch/~giorgio/

Cassio de Campos (PhD in engineering, '05) co-authored Chapter 9 (Probabilistic graphical models). He is a Researcher at IDSIA, in Switzerland. He works with the Imprecise Probability Group (http://ipg.idsia.ch), currently appointed as at-large member of SIPTA, the Society for Imprecise Probability: Theories and Applications. Formerly, he has been an Assistant Professor at the University of Sao Paulo. His interests include probabilistic graphical models, bioinformatics, computational complexity, and imprecise probability. For further information, see http://www.idsia.ch/~cassio

Sébastien Destercke co-authored Chapters 4 (Special cases) and 5 (Other uncertainty theories based on capacities). He graduated from the Faculté Polytechnique de Mons as an Engineer in 2004 and obtained a PhD degree in computer science from the University of Paul-Sabatier, Toulouse. Since 2011, he has been a researcher at the National Centre for Scientific Research (CNRS) and member of the Heuristics and Diagnostics for Complex Systems (Heudiasyc) laboratory. Between 2011 and 2013 he has served as Executive Editor of SIPTA, the Society for Imprecise Probability: Theories and Applications. His topics of interest include imprecise probabilistic approaches (lower previsions theory, Dempster-Shafer theory, possibility theory) and their applications in the fields of risk and reliability analyses and machine learning. For further information, see https://www.hds.utc.fr/~sdesterc/

Didier Dubois co-authored Chapters 4 (Special cases) and 5 (Other uncertainty theories based on capacities). He received the Dr. Eng. degree from the Ecole Nationale Supérieure de l'Aéronautique et de l'Espace, Toulouse, France, in 1977, the Doctorat d'Etat degree from Grenoble University, Grenoble, France, in 1983, and the Honorary Doctorate degree from the Faculté Polytechnique de Mons, Mons, Belgium, in 1997. He is currently a Research Director with the Institute of Research in Informatics of Toulouse, Department of Computer Science, Paul Sabatier University, Toulouse, and belongs to the French National Centre for Scientific Research. He is the co-author of more than 15 edited volumes on uncertain reasoning and fuzzy sets. He has authored or co-authored about 200 technical journal papers on uncertainty theories and applications. He is the Editor-in-Chief of the journal *Fuzzy Sets and Systems*. His current research interests include artificial intelligence, operations research, and decision sciences, with emphasis on the modelling, representation, and processing of imprecise and uncertain information in reasoning and problem-solving tasks. For further information, see http://www.irit.fr/~Didier.Dubois/

Robert Hable co-authored Chapters 8 (Decision making) and 16 (Computation). He was born in Landshut, Germany, in 1981 and received the Diploma in Mathematics from the University of Bayreuth in 2006, the PhD degree in Statistics from the Ludwig Maximilian University of Munich in 2009, and the Habilitation in Mathematics from the University of Bayreuth in 2012. Currently, he is Privatdozent at the University of Bayreuth. His research interests include decision theory and statistical methods for imprecise probabilities, robust statistics, and nonparametric regression. For further information, see http://www.stoch.uni-bayreuth.de/de/team/J_Robert_Hable/

Filip Hermans co-authored Chapter 11 (Stochastic processes). He started his PhD under supervision of prof. Gert de Cooman in 2006 at the SYSTeMS research group in Ghent (Belgium). In 2012 he left the group as a postdoctoral researcher. During his stay, he focused on the gambling interpretation of imprecise probabilities and the reinterpretation of directed graphical networks in the light of epistemic irrelevance.

Nathan Huntley co-authored Chapter 8 (Decision making). He received his PhD in 2011 from Durham University, UK, for work on imprecise decision theory. He has since worked as a researcher at Ghent University, Belgium, and Durham University. His main research interest is sequential decision making with very weak assumptions, in particular the relationship between local and global solutions and the circumstances under which local solutions can be used to find global solutions. The principal results provide simple and easily-checked conditions on choice functions for particular interesting behaviours such as backward induction to apply. He has also worked on imprecise linear programming, and on uncertainty analysis for complex computer models.

Andrés Masegosa co-authored Chapter 10 (Classification). He obtained MSc and PhD degrees in computer sciences from the University of Granada, Granada, Spain, in 2003 and 2009, respectively. He is currently with the Department of Computer Science and A.I., University of Granada, and a member of the research group on Uncertainty Treatment on Artificial Intelligence (UTAI). His topics of interests include supervised classification models (with both precise and imprecise probabilities), Bayesian networks (automatic and interactive learning), credal networks (learning from data) and probabilistic graphical models, in general. For further information, see http://decsai.ugr.es/~andrew/

Enrique Miranda co-authored Chapters 2 (Lower previsions) and 3 (Structural judgements). He is Associate Professor at the University of Oviedo (Spain), where he earned his PhD in 2003. Between 2003 and 2008, he was an Assistant Professor at Rey Juan Carlos University, in Madrid (Spain). He has made research stays at Ghent University (Belgium) and at IDSIA (Lugano, Switzerland). He served SIPTA, the Society for Imprecise Probability: Theories and Applications, as Treasurer between 2009 and 2011 and Secretary between 2011 and 2013. His research interests are the foundations of the theory of coherent lower previsions and some of the alternative models of imprecise probabilities, such as possibility measures, 2-monotone lower previsions, and p-boxes. For further information, see http://grupos.uniovi.es/web/mirandaenrique

Serafín Moral co-authored Chapter 10 (Classification). He obtained a Master Degree in Sciences (Mathematics) in 1981 and a PhD Degree in Sciences (Mathematics) in 1985, both at the University of Granada. Now, he is Full Professor at the Department of Computer Science at the University of Granada. Topics of interest include imprecise probability (conditioning, independence, uncertainty measurement, and algorithms in graphical structures), Bayesian networks (propagation algorithms, most probable explanation, learning and classification), data mining, and foundations of uncertain reasoning. For further information, see http://decsai.ugr.es/~smc/

Michael Oberguggenberger authored Chapter 13 (Engineering). He is Full Professor and Chair of the Unit of Engineering Mathematics, University of Innsbruck, Austria. He received the MSc degree in Mathematics and Physics from the University of Innsbruck in 1979 and the PhD degree in Mathematics from Duke University, Durham NC, in 1981. His research interests are partial differential equations, generalized functions, stochastic analysis, engineering mathematics, risk and uncertainty modelling in engineering applications, especially imprecise probability and hybrid stochastic models in engineering. For further information, see http://www.uibk.ac.at/techmath/michael/index.html.en

Erik Quaeghebeur authored Chapter 1 (Desirability). He is a researcher at the SYSTeMS research group of Ghent University (Belgium), where he also earned his PhD in 2009. He is currently (Septemper 2013) a visiting scholar at the Decision Support Systems group of the Department of Informatics of Utrecht University (the Netherlands). In 2009–2010, he was a visiting scholar at the Department of Philosophy of Carnegie Mellon University (USA). From 2009–2011, he was Secretary of SIPTA, the Society for Imprecise Probability: Theories and Applications. His past and current research interests in imprecise probability theory include learning from samples, applications of polytope and linear programming theory, modelling structural assessments, representations based on sets of gambles, graphical models, and optimization under uncertainty. For further information, see http://ac.erikquaeghebeur.name

Glenn Shafer co-authored Chapter 6 (Game-theoretic probability). He is Professor and Dean of the Rutgers Business School – Newark and New Brunswick. He was awarded a PhD in Statistics in 1973 by Princeton University. He is well known for his work in the 1970s and 1980s on the Dempster-Shafer theory, especially for his path-breaking book, *A Mathematical Theory of Evidence*. His most important work is *Probability and Finance: It's Only a Game!*, co-authored by Vladimir Vovk and published in 2001. He is a fellow of both the Institute of Mathematical Statistics and the American Association for Artificial Intelligence. For further information, see www.glennshafer.com

Damjan Škulj co-authored Chapter 11 (Stochastic processes). He received his PhD in Mathematics from University of Ljubljana in 2004. Since 2006 he has been Assistant Professor of Mathematics at the Faculty of Social Sciences, University of Ljubljana. His research interests are mostly in the area of imprecise probabilities. At the moment his focus is mainly in stochastic processes and modelling of risk and uncertainty. For further information, see http://www.fdv.uni-lj.si/en/news-and-information/contacts/teachers/info/damjan-skulj

Michael Smithson authored Chapter 15 (Elicitation). He is a Professor in the Research School of Psychology at The Australian National University in Canberra, and received his PhD from the University of Oregon. He is the author of four books, co-author of two books, and co-editor of two books. His other publications include more than 140 refereed journal articles and book chapters. His primary research interests are in judgement and decision making under uncertainty, statistical methods for the social sciences, and applications of fuzzy set theory to the social sciences. For further information, see https://researchers.anu.edu.au/researchers/smithson-mj

Matthias C. M. Troffaes co-authored Chapters 8 (Decision making) and 16 (Computation). He is currently Senior Lecturer in Statistics at the Mathematical Sciences Department, Durham University (UK). He earned his PhD in 2005 at Ghent University (Belgium), and visited Carnegie Mellon University (USA) for one year as postdoctoral researcher, before coming to Durham. He has authored numerous peer-reviewed papers in the field of imprecise probability theory and decision making, and edited special issues for various journals. Since 2009, he has been a member of the executive committee of SIPTA, the Society for Imprecise Probability: Theories and Applications. For further information, see http://www.maths.dur.ac.uk/users/matthias.troffaes/

Lev V. Utkin co-authored Chapter 14 (Reliability and risk). He is currently the Vice-Rector for Research and a Head of the Department of Control, Automation and System Analysis, Saint-Petersburg State Forest Technical University, Saint-Petersburg, Russia. He holds a PhD in Information Processing and Control Systems (1989) from Saint-Petersburg Electrotechnical University and a DSc in Mathematical Modelling (2001) from Saint-Petersburg State Institute of Technology, Russia. His research interests are focused on imprecise probability theory, reliability analysis, decision making, risk analysis and machine learning theory. For further information, see http://levvu.narod.ru/

Paolo Vicig authored Chapter 12 (Financial risk measurement). He is Professor of Mathematics for Economics at the University of Trieste (Italy), Department D.E.A.M.S. 'B. de Finetti'. After his degree in 1986, he worked in the Research Office of the insurance company Lloyd Adriatico (now: Allianz Group) until 1991, when he initiated his academic career as Assistant Professor in Probability Theory. His main research interests focus on foundations of imprecise probability theory (imprecise previsions, models for consistent imprecise assessments, conditioning) and on their applications to risk measurement. He has been member of the executive committee of SIPTA, the Society for Imprecise Probability: Theories and Applications (2006 to 2011). For further information, see http://www.units.it/persone/index.php/from/abook/persona/4539

Vladimir Vovk co-authored Chapter 6 (Game-theoretic probability). He is Professor of Computer Science at Royal Holloway, University of London. He received a PhD in the Department of Mathematical Logic at Moscow State University in 1988. Among his research interests are Kolmogorov complexity, machine learning and the foundations of probability and statistics. He co-authored two books, *Probability and Finance: It's only a Game!* (with Glenn Shafer, 2001) and *Algorithmic Learning in a Random World* (with Alex Gammerman and Glenn Shafer, 2005). For further information, see http://www.vovk.net/

Gero Walter co-authored Chapter 7 (Statistical inference). He graduated from the Department of Statistics at Ludwig-Maximilians-University (LMU) Munich as Dipl.-Statistician in 2007. He is currently a research assistant in the group 'Foundations of Statistics and their Applications' at the same department, obtained his PhD degree in Statistics, on generalized Bayesian inference with sets of conjugate priors, in 2013. He was invited as a visiting researcher to the Department of Mathematical Sciences, Durham University, in 2010 and

2013, and is a member of the WPMSIIP working group, helping to organize several workshops in the series. For further information, see http://www.statistik.lmu.de/~walter/

Marco Zaffalon (PhD in Applied Mathematics, '97) co-authored Chapters 9 (Probabilistic graphical models) and 10 (Classification). He is a Professor at IDSIA, in Switzerland, where he has founded and heads the Imprecise Probability Group (http://ipg.idsia.ch). Formerly, he has been the President of SIPTA, the Society for Imprecise Probability: Theories and Applications, co-chaired the conferences ISIPTA '03 and '07, and started the SIPTA schools in imprecise probability. He is currently the Area Editor for Imprecise Probability for the *International Journal of Approximate Reasoning*. His interests include probabilistic graphical models (such as Bayesian and credal networks, or influence diagrams), data mining and credal classification, statistics and the foundations of imprecise probability. For further information, see http://www.idsia.ch/~zaffalon

Acknowledgements

This book is based on the work of the many experts contributing to the sixteen chapters. We thank the chapter authors for their dedication in writing these chapters, and for their willingness to embed their contribution into the overall structure, as well as their openness to discussion.

We are also grateful to Teddy Seidenfeld for expert advice on this project. During the final stages towards completion of this book, we were greatly assisted by Eva Endres, Julia Plaß, Katharina Stöhr, Nina Scharrer and Martin Viehrig, whom we thank most warmly. We are also very grateful to Radhika Sivalingam and her team from Laserwords. Staff at Wiley have been very helpful and understanding, we want to thank in particular Richard Davies, Heather Kay, Jo Taylor and Prachi Sinha Sahay. The development of this book was a process that took much longer than we had initially hoped, and we appreciate the patience of the chapter authors and staff at Wiley.

Thomas Augustin, Frank Coolen, Gert de Cooman, Matthias Troffaes

1

Desirability

Erik Quaeghebeur
SYSTeMS Research Group, Ghent University, Belgium

1.1 Introduction

There are many ways to model uncertainty. The most widely used type of model in the literature is a function that maps something we are uncertain about to a value that expresses what we know or believe to know about it. Examples are probabilities, which may specify a degree of belief that an event will occur, and previsions, which specify acceptable prices for gambles (cf. Section 1.6 and Chapter 2).

In this chapter, we show that other types of models that are conceptually and intuitively attractive can be built and used as well. The focus lies on the notion of desirability and the theory of sets of desirable gambles. Next to introducing its concepts and structure, we also use it as a nexus for clarifying the relationships between many of the equivalent or almost equivalent models for uncertainty appearing in the imprecise-probability literature: partial preference orders, credal sets, and lower previsions.

We formulate desirability in the context of an abstract betting framework:

We – short for 'an intentional system' – are uncertain about the outcome of an experiment. A *possibility space* for the experiment is a finite or infinite set of *elementary events* – i.e., mutually exclusive outcomes – that is exhaustive in the sense that other outcomes are deemed practically or pragmatically impossible. A bounded real-valued function on a possibility space is called a *gamble* and interpreted as an uncertain payoff. The set of all gambles combined – as we do – with pointwise addition of gambles, pointwise multiplication with real numbers, and the supremum norm (topology), forms a normed real vector space, the bounded function (Banach) space to be precise [256].

Consider a possibility space $\mathcal{X}$, the associated set of all gambles $\mathcal{L}(\mathcal{X})$, and one of its elements, f. After the experiment's outcome x is determined, the gamble's owner receives the (possibly negative) payoff $f(x)$.

Introduction to Imprecise Probabilities, First Edition.
Edited by Thomas Augustin, Frank P. A. Coolen, Gert de Cooman and Matthias C. M. Troffaes.

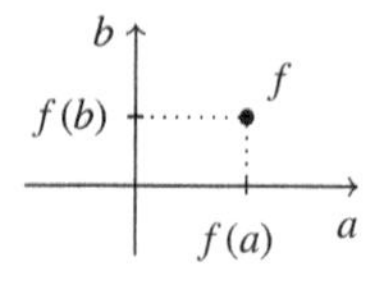

Figure 1.1 A depiction of the set of gambles on the possibility space $\{a, b\}$: The plane depicts the two-dimensional vector space $\mathcal{L}(\{a, b\})$; a gamble f and its component payoffs are shown.

We will supplement the formal exposition by intuition-building toy examples or illustrations, as in Figure 1.1.

A special class of gambles are *indicators* of events $A \subseteq \mathcal{X}$, denoted I_A: they are one on A and zero elsewhere. Indicators of elementary events $x \in \mathcal{X}$ are written $I_x := I_{\{x\}}$.

Payoffs have a certain *utility*. We assume that the gambles' payoffs are expressed in units of utility that are linear for us. So if we, e.g., receive double the payoff, we consider this to be twice as good, or bad.

A gamble is *desirable* for us if we accept ownership of it when offered to us. We model our uncertainty about the experiment's outcome using a set of desirable gambles.

1.2 Reasoning about and with sets of desirable gambles

We first need to establish what constitutes a reasonable set of desirable gambles. This forms the foundation of the theory and provides us with the basic rules that allow us to reason with sets of desirable gambles: make decisions and draw conclusions. This is the topic of this section.

1.2.1 Rationality criteria

So we use a set of desirable gambles $\mathcal{D} \subseteq \mathcal{L}(\mathcal{X})$ to model our uncertainty about the experiment's outcome. Do we need to specify for each gamble individually whether we consider it desirable? Or can we argue that if some gambles are desirable, then others should be as well, i.e., can we automatically extend a partial specification? And must we consider some specific set of gambles to be desirable or not desirable? If we adopt some *rationality criteria*, the answer to the second and third questions is yes, and thus no to the first.

The first two, constructive rationality criteria express that we are working with a linear utility scale and say that a gamble's desirability should be independent of the stake and that the combination of two desirable gambles should also be desirable:

$$\text{Positive scaling:} \quad \lambda > 0 \wedge f \in \mathcal{D} \Rightarrow \lambda f \in \mathcal{D} \quad \text{or} \quad \lambda > 0 \Rightarrow \lambda\mathcal{D} = \mathcal{D}, \tag{1.1}$$

$$\text{Addition:} \quad f, g \in \mathcal{D} \Rightarrow f + g \in \mathcal{D} \quad \text{or} \quad \mathcal{D} + \mathcal{D} \subseteq \mathcal{D}, \tag{1.2}$$

where $\lambda\mathcal{D} := \{\lambda f : f \in \mathcal{D}\}$ and $\mathcal{D} + \mathcal{D} := \{f + g : f, g \in \mathcal{D}\}$ define the elementwise operations used. These two criteria enable the positive answer to question two.

So, given a partial desirability specification, or *assessment* $\mathcal{A} \subseteq \mathcal{L}(\mathcal{X})$, these criteria extend desirability to all gambles in its so-called *positive hull*:

$$\text{posi}\,\mathcal{A} := \left\{ \sum_{k=1}^{n} \lambda_k f_k : \lambda_k > 0 \wedge f_k \in \mathcal{A} \wedge n \in \mathbb{N} \right\}. \tag{1.3}$$

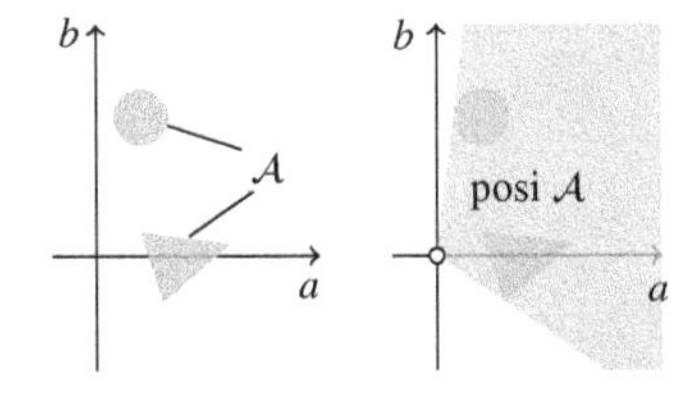

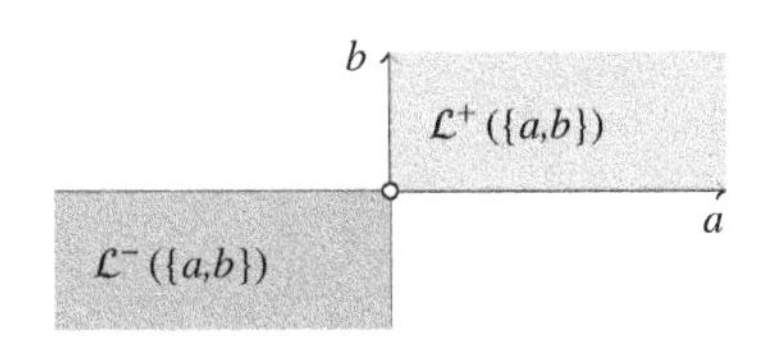

Figure 1.2 An illustration of the positive hull operation (left) and the positive and negative orthants (right).

The constructive rationality criteria define a convex cone and therefore the positive hull operator generates cones of gambles. Its effect is illustrated in Figure 1.2; note the exclusion of the zero gamble in the extension. The positive hull operator moreover allows us to express the constructive rationality criteria succinctly: posi $\mathcal{D} = \mathcal{D}$.

The third question effectively asks whether there are compelling reasons to insist that some gambles should be desirable or not desirable, independent of the information available to us about the experiment's outcome. And indeed there are: any gamble that might give a positive payoff without ever giving a negative one is certainly desirable, and any gamble that might give a negative payoff without ever giving a positive one is not. This results in a second pair of rationality criteria:

$$\text{Accepting partial gain:} \quad f > 0 \Rightarrow f \in \mathcal{D} \quad \text{or} \quad \mathcal{L}^+(\mathcal{X}) \subseteq \mathcal{D}, \tag{1.4}$$

$$\text{Avoiding partial loss:} \quad f < 0 \Rightarrow f \notin \mathcal{D} \quad \text{or} \quad \mathcal{D} \cap \mathcal{L}^-(\mathcal{X}) = \emptyset, \tag{1.5}$$

where we have used gamble (vector) inequalities and the zero gamble; i.e., $f \geq g$ if and only if $\inf(f - g) \geq 0$ and $f > g$ if moreover $\sup(f - g) > 0$, or $f \neq g$.

We see that the so-called *positive orthant* $\mathcal{L}^+(\mathcal{X})$ and *negative orthant* $\mathcal{L}^-(\mathcal{X})$ must, respectively, be included in and excluded from the set of desirable gambles. Their definition is implicit in Criteria (1.4) and (1.5). A toy illustration of them is shown in Figure 1.2. By replacing the positive and negative orthant in Criteria (1.4) and (1.5) by their interior, we obtain a weaker pair of criteria, used only in Section 1.6.4 on simplified variants of desirability:

$$\text{Accepting sure gain:} \quad \inf f > 0 \Rightarrow f \in \mathcal{D} \quad \text{or} \quad \text{int}(\mathcal{L}^+(\mathcal{X})) \subseteq \mathcal{D}, \tag{1.6}$$

$$\text{Avoiding sure loss:} \quad \sup f < 0 \Rightarrow f \notin \mathcal{D} \quad \text{or} \quad \mathcal{D} \cap \text{int}(\mathcal{L}^-(\mathcal{X})) = \emptyset. \tag{1.7}$$

1.2.2 Assessments avoiding partial or sure loss

We say that an assessment $\mathcal{A}$ *incurs partial loss* if any desirable gamble, including those implied by the constructive rationality criteria, incurs a partial loss, or formally, if posi $\mathcal{A} \cap \mathcal{L}^-(\mathcal{X}) \neq \emptyset$. Conversely, an assessment $\mathcal{A}$ *avoids partial loss* if

$$\text{posi } \mathcal{A} \cap \mathcal{L}^-(\mathcal{X}) = \emptyset. \tag{1.8}$$

Again, we talk about a *sure loss* when $\mathcal{L}^-(\mathcal{X})$ is replaced by its interior $\text{int}(\mathcal{L}^-(\mathcal{X}))$. In the illustration on the left in Figure 1.3, the assessment $\mathcal{A}$ incurs a sure loss because the combination of two of its elements, f and g, always has a strictly negative outcome.

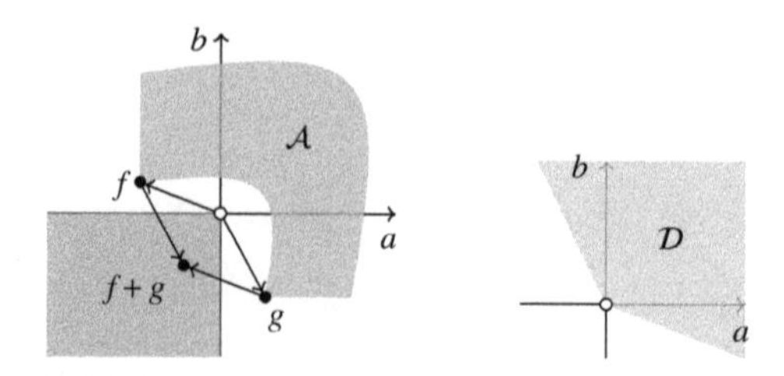

Figure 1.3 An illustration of an assessment $\mathcal{A}$ incurring sure loss (left) and an illustration of a coherent set of desirable gambles $\mathcal{D}$ (right).

1.2.3 Coherent sets of desirable gambles

A set of desirable gambles $\mathcal{D}$ is *coherent* if it satisfies the four rationality criteria (1.1), (1.2), (1.4), and (1.5). So a coherent set of desirable gambles is a convex cone excluding $\mathcal{L}^-(\mathcal{X})$ and including $\mathcal{L}^+(\mathcal{X})$, which is the smallest coherent set of desirable gambles. One is drawn as an illustration on the right in Figure 1.3. The set of all coherent sets of desirable gambles on the possibility space $\mathcal{X}$ is denoted $\mathbb{D}(\mathcal{X})$.

No coherent set of desirable gambles includes an assessment that incurs partial loss. However, an assessment that avoids partial loss can, in general, be included in an infinity of coherent sets of desirable gambles. This is illustrated for an assessment $\mathcal{A}$ in Figure 1.4: six of its coherent *extensions* are shown. (Ignore the dashed lines for now.) The set $\mathbb{D}_{\mathcal{A}} := \{\mathcal{D} \in \mathbb{D}(\mathcal{X}) : \mathcal{A} \subseteq \mathcal{D}\}$ contains all the coherent desirable gambles that include an assessment $\mathcal{A}$.

The coherent sets of desirable gambles can be ordered according to set inclusion $\subseteq$. A set of desirable gambles that includes another one is called more committal than the other, because it models a state in which we are committed to accepting more gambles. We know that this partial order has a least element, $\mathcal{L}^+(\mathcal{X})$.

The set of coherent extensions of an assessment that avoids partial loss inherit this order. The Hasse diagram for the selection of extensions given in Figure 1.4 is given in Figure 1.5.

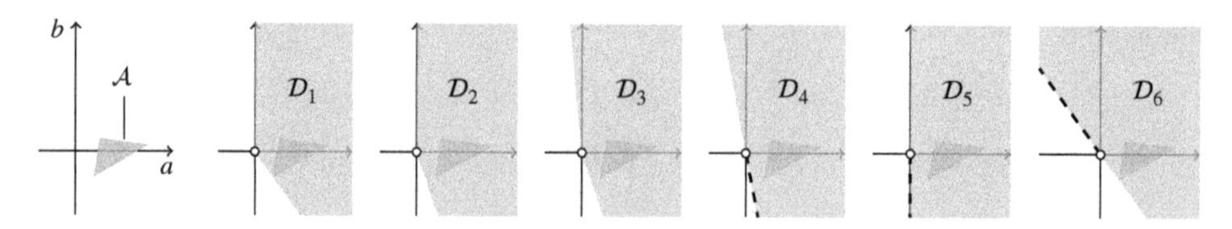

Figure 1.4 Illustration of the multiplicity of the coherent extensions of an assessment $\mathcal{A}$.

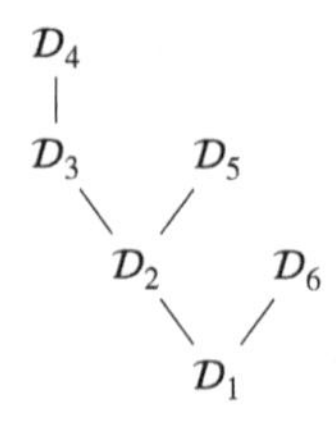

Figure 1.5 Illustration of the partial order of coherent sets of desirable gambles induced by the inclusion relation.

Based on the sets shown there, our intuition tells us that $\mathcal{D}_1$ is the *least committal* coherent extension of the assessment $\mathcal{A}$ shown, i.e., the intersection $\bigcap \mathbb{D}_{\mathcal{A}}$ of all coherent extensions. The intuition that there is such an extension is correct and we will show that this is a general feature in the next section. Later on in Section 1.5, we will discuss the class of *maximally committal* coherent extensions.

1.2.4 Natural extension

Given an assessment $\mathcal{A}$, the fact that all gambles in $\mathcal{L}^+(\mathcal{X})$ must be desirable, and the constructive rationality criteria, there is a *natural extension*, defined as

$$\mathcal{E}(\mathcal{A}) := \text{posi}\,(\mathcal{A} \cup \mathcal{L}^+(\mathcal{X})) = \text{posi}\,\mathcal{A} \cup \mathcal{L}^+(\mathcal{X}) \cup (\text{posi}\,\mathcal{A} + \mathcal{L}^+(\mathcal{X})). \tag{1.9}$$

The rightmost expression follows from Equation (1.3) and the fact that $\mathcal{L}^+(\mathcal{X})$ is already a convex cone.

An important result links the natural and the least committal coherent extensions:

Theorem 1.1 *The natural extension $\mathcal{E}(\mathcal{A})$ of $\mathcal{A} \subseteq \mathcal{L}(\mathcal{X})$ coincides with its least committal coherent extension $\bigcap \mathbb{D}_{\mathcal{A}}$ if and only if $\mathcal{A}$ avoids partial loss.*

Proof. By construction, the natural extension $\mathcal{E}(\mathcal{A})$ must be included in any coherent extension, if they exist, as they must satisfy Criteria (1.1), (1.2), and (1.4): it is therefore the least committal one if it is coherent itself. This is the case if and only if it also satisfies (1.5). From Equation (1.9) we see that $\mathcal{E}(\mathcal{A})$'s pointwise smallest gambles lie in posi $\mathcal{A}$ or $\mathcal{L}^+(\mathcal{X})$, which proves the necessary equivalence of $\mathcal{A}$ avoiding partial loss, i.e., posi $\mathcal{A} \cap \mathcal{L}^-(\mathcal{X}) = \emptyset$, and $\mathcal{E}(\mathcal{A}) \cap \mathcal{L}^-(\mathcal{X}) = \emptyset$. □

Natural extension is the prime tool for *deductive inference* in desirability: given an initial assessment, it allows us to straightforwardly deduce which gambles must also be desirable in order to satisfy coherence, but makes no further commitments.

1.2.5 Desirability relative to subspaces with arbitrary vector orderings

In this section, we have considered desirability relative to the linear space of all gambles $\mathcal{L}(\mathcal{X})$ with the ordinary vector ordering: $f \geq g \Leftrightarrow f - g \in \mathcal{L}_0^+(\mathcal{X})$ and $f > g \Leftrightarrow f - g \in \mathcal{L}^+(\mathcal{X})$ for all gambles f and g on $\mathcal{X}$. The theory can be developed, mutatis mutandis, for linear subspaces $\mathcal{K} \subseteq \mathcal{L}(\mathcal{X})$ with a vector ordering determined by an arbitrary cone $\mathcal{C} \subset \mathcal{K}$ instead of $\mathcal{L}^+(\mathcal{X})$ [213]: for all gambles f and g in $\mathcal{K}$, $f \geq g \Leftrightarrow f - g \in \mathcal{C}_0 := \mathcal{C} \cup \{0\}$ and $f > g \Leftrightarrow f - g \in \mathcal{C}$. (We let the cone be a strict subset of the subspace to prevent triviality of the vector ordering.)

Restricting attention to linear subspaces can be especially useful for infinite possibility spaces. An example is the set of polynomials on the unit simplex [213]. Considering vector orderings different from the ordinary can be useful when looking at transformations between different vector spaces that do not map the positive orthant to the positive orthant, but to some other cone.

We do not pursue a treatment of desirability in this generality here, not to overly burden the notation. Nevertheless, it is good to keep this possibility in mind, especially when dealing with transformations, as in the upcoming section; it allows us to deal straightforwardly with a much wider class of them.

1.3 Deriving and combining sets of desirable gambles

Sometimes, we want to be able to look at the experiment from a different viewpoint, focus on some specific aspect, or combine different aspects. We will then also want to carry over the information we have formalized as a set of desirable gambles on one possibility space to another. In this section we are first going to get acquainted with gamble space transformations, the basic mathematical tool to deal with this kind of problem, then look at how one coherent set of desirable gambles can be derived from another and investigate two special instances, conditioning and marginalization, and finally explain how coherent sets of desirable gambles can be combined.

1.3.1 Gamble space transformations

Given two possibility spaces $\mathcal{X}$ and $\mathcal{Z}$ and the corresponding sets of gambles, relationships between them can be expressed using transformations Γ from $\mathcal{L}(\mathcal{Z})$ to $\mathcal{L}(\mathcal{X})$ that – to preserve topological structure – are continuous. To work with such transformations, we first introduce some concepts:

$$\text{Linearity:} \qquad \lambda \in \mathbb{R} \Rightarrow \Gamma(\lambda f + g) = \lambda \Gamma f + \Gamma g, \tag{1.10}$$

$$\text{Increasing:} \qquad f > g \Rightarrow \Gamma f > \Gamma g. \tag{1.11}$$

The direct image of a set of gambles $\mathcal{A} \subseteq \mathcal{L}(\mathcal{Z})$ is $\Gamma\mathcal{A} := \{\Gamma f : f \in \mathcal{A}\}$; in particular, the vector (sub)space $\mathrm{im}\Gamma := \Gamma(\mathcal{L}(\mathcal{Z}))$ is the range. The inverse image of a set of gambles $\mathcal{A} \subseteq \mathrm{im}\Gamma$ is $\Gamma^{-1}\mathcal{A} := \{f \in \mathcal{L}(\mathcal{Z}) : \Gamma f \in \mathcal{A}\}$; in particular, $\ker\Gamma := \Gamma^{-1}\{0\}$ is the kernel. Whenever Γ is injective, its inverse Γ^{-1} is well-defined on $\mathrm{im}\Gamma$ by $\{\Gamma^{-1}f\} = \Gamma^{-1}\{f\}$. The inverse of a linear increasing injective transformation is also linear, but need not be increasing ([213, Example 9]). (This may lead one to use non-ordinary vector orderings.)

While other transformations may also be interesting, we here restrict attention to a class of transformations that are guaranteed to leave the partial-loss criterion invariant and commute with natural extension. For brevity's sake, we call linear increasing injective transformations with an increasing inverse *liiwii transformations*.

Lemma 1.1 *Given a liiwii transformation Γ from $\mathcal{L}(\mathcal{Z})$ to $\mathcal{L}(\mathcal{X})$, then (i) Γ and* posi *commute, i.e.,* $\Gamma(\mathrm{posi}\mathcal{A}) = \mathrm{posi}(\Gamma\mathcal{A}) = \mathrm{posi}(\Gamma\mathcal{A}) \cap \mathrm{im}\Gamma$, *and (ii)* $\Gamma(\mathcal{L}^+(\mathcal{Z})) = \mathcal{L}^+(\mathcal{X}) \cap \mathrm{im}\Gamma$ *and* $\Gamma(\mathcal{L}^-(\mathcal{Z})) = \mathcal{L}^-(\mathcal{X}) \cap \mathrm{im}\Gamma$.

Proof. Claim (i) is a consequence of linearity. Claim (ii) follows from both Γ and Γ^{-1} being increasing. □

Proposition 1.1 *Given a liiwii transformation Γ from $\mathcal{L}(\mathcal{Z})$ to $\mathcal{L}(\mathcal{X})$, then: (i) for any $\mathcal{A} \subseteq \mathcal{L}(\mathcal{Z})$, $\Gamma(\mathcal{E}(\mathcal{A})) = \mathcal{E}(\Gamma\mathcal{A}) \cap \mathrm{im}\Gamma$; (ii) $\Gamma\mathcal{A}$ avoids partial loss if and only if $\mathcal{A}$ does.*

Proof. To prove Claim (i), we transform the right-hand side of Equation (1.9). Given Lemma 1.1, we need to prove that $\Gamma(\mathrm{posi}\ \mathcal{A} + \mathcal{L}^+(\mathcal{Z})) = (\mathrm{posi}\ (\Gamma\mathcal{A}) + \mathcal{L}^+(\mathcal{X})) \cap \mathrm{im}\Gamma$. Its left-hand side can be rewritten as $\Gamma(\mathrm{posi}\ \mathcal{A}) + \Gamma(\mathcal{L}^+(\mathcal{Z}))$ because of linearity; the right-hand side can be put in this form by first applying commutativity of Γ and posi and then realizing the first intersection factor can only contain elements outside of $\mathrm{im}\Gamma$ due to elements in $\mathcal{L}^+(\mathcal{X})$ outside of $\mathcal{L}^+(\mathcal{X}) \cap \mathrm{im}\Gamma = \Gamma(\mathcal{L}^+(\mathcal{Z}))$.

For Claim (ii), we show that Equation (1.8) remains invariant: $\Gamma(\mathrm{posi}\ \mathcal{A} \cap \mathcal{L}^-(\mathcal{Z})) \subseteq \Gamma(\mathrm{posi}\ \mathcal{A}) \cap \Gamma(\mathcal{L}^-(\mathcal{Z})) = \mathrm{posi}\ (\Gamma\mathcal{A}) \cap \mathcal{L}^-(\mathcal{X})$, where the first step is immediate and the second follows from Lemma 1.1; similarly $\Gamma^{-1}(\mathrm{posi}\ (\Gamma\mathcal{A}) \cap \mathcal{L}^-(\mathcal{X})) \subseteq \mathrm{posi}\ \mathcal{A} \cap \mathcal{L}^-(\mathcal{Z})$. □

Liiwii transformations appear when combining coherent sets of desirable gambles; their inverse is used when deriving one coherent set of desirable gambles from another.

1.3.2 Derived coherent sets of desirable gambles

Now, given a set of desirable gambles $\mathcal{D} \subset \mathcal{L}(\mathcal{X})$, we can use a liiwii transformation Γ from $\mathcal{L}(\mathcal{Z})$ to $\mathcal{L}(\mathcal{X})$ to derive a set of desirable gambles $\mathcal{D}_\Gamma := \Gamma^{-1}\mathcal{D}$ on $\mathcal{Z}$.

Proposition 1.2 *Given a liiwii transformation Γ from $\mathcal{L}(\mathcal{Z})$ to $\mathcal{L}(\mathcal{X})$, then $\mathcal{D}_\Gamma$ is coherent if $\mathcal{D} \subset \mathcal{L}(\mathcal{X})$ is coherent.*

Proof. The transformation Γ^{-1} from $\operatorname{im}\Gamma$ to $\mathcal{L}(\mathcal{Z})$ is also liiwii. Let $\mathcal{A} := \mathcal{D} \cap \operatorname{im}\Gamma$, which avoids partial loss, then Proposition 1.1 tells us $\mathcal{D}_\Gamma = \Gamma^{-1}\mathcal{A}$ avoids partial loss. Lemma 1.1 allows us to conclude $\mathcal{D}_\Gamma$ is a convex cone because $\mathcal{D} \cap \operatorname{im}\Gamma$ is. We know $\mathcal{L}^+(\mathcal{X}) \subseteq \mathcal{D}$; applying Γ^{-1} allows us to conclude, using Lemma 1.1, that $\mathcal{L}^+(\mathcal{Z}) \subseteq \mathcal{D}_\Gamma$. So $\mathcal{D}_\Gamma$ satisfies Criteria (1.1), (1.2), (1.4), and (1.5) and is therefore coherent. □

In the illustration on the left in Figure 1.6, Γ maps from $\mathcal{L}(\{d,b\})$ to $\mathcal{L}(\{a,b\})$ and is defined for any gamble h on $\{d,b\}$: $(\Gamma h)(a) = \frac{1}{2}h(d)$ and $(\Gamma h)(b) = h(b)$.

Linear transformations map linear vector spaces to linear vector spaces. So the range $\operatorname{im}\Gamma$ will either coincide with $\mathcal{L}(\mathcal{X})$ or be a strict subspace of it, that, because Γ is increasing, includes part of the positive and negative orthants. In the latter case, one can conceptually visualize an important part of the definition of $\mathcal{D}_\Gamma$ as taking a *slice* of $\mathcal{D}$ by intersecting it with the linear subspace $\operatorname{im}\Gamma$.

Let us give a more involved illustration on the right in Figure 1.6 that clarifies this. First, look at the left figure on the right. To understand it, we first note that the intersection of many a convex cone with an appropriate hyperplane results in a convex polytope. In light grey, we have depicted the intersection of a coherent set of desirable gambles $\mathcal{D} \subset \mathcal{L}(\{a,b,c\})$ and the hyperplane consisting of those gambles with components that sum to one. The triangle that is the convex hull of the indicators of elementary events indicates the intersection with the positive orthant. Dotted lines indicate open parts of the border and undecorated ones indicate closed parts. Darkly-filled dots are closure points of interest belonging to $\mathcal{D}$; white-filled ones do not belong to $\mathcal{D}$. For each, we indicate which gamble f it depicts, explicitly or as a vector $(f(a), f(b), f(c))$.

Rightmost, we have indicated two transformations that slice our cone $\mathcal{D}$, with resulting derived sets of coherent desirable gambles depicted: Γ_1 from $\mathcal{L}(\{c,d\})$ to $\mathcal{L}(\{a,b,c\})$

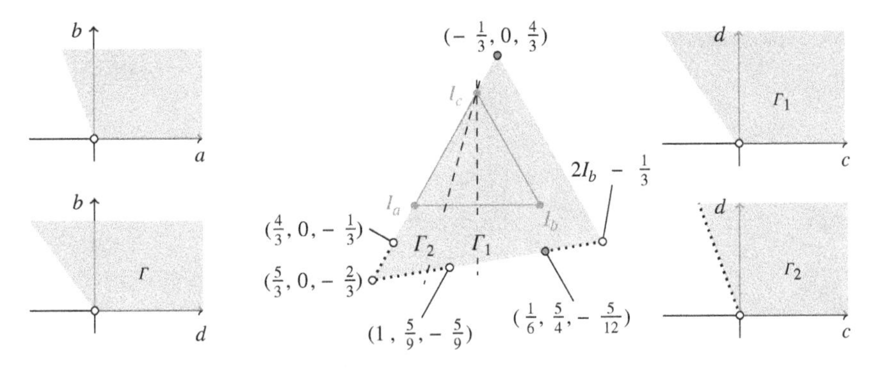

Figure 1.6 Illustrations of derived coherent sets of desirable gambles.

is defined by $(\Gamma_1 h)(a) = (\Gamma_1 h)(b) = h(d)$ and $(\Gamma_1 h)(c) = h(c)$; Γ_2 from $\mathcal{L}(\{c,d\})$ to $\mathcal{L}(\{a,b,c\})$ is defined by $(\Gamma_2 h)(a) = \frac{3}{4}h(d)$, $(\Gamma_2 h)(b) = \frac{1}{4}h(d)$ and $(\Gamma_2 h)(c) = h(c)$.

1.3.3 Conditional sets of desirable gambles

We may be interested in obtaining an uncertainty model for the situation in which the experiment's outcome belongs to a *conditioning event* $B \subseteq \mathcal{X}$. In that case, we can focus on gambles that are *contingent* on B occurring: these are gambles such that if B does not occur, no payoff is received or due – *status quo* is maintained.

This idea can be formalized by introducing the liiwii transformation $\lceil_{B^c}$ that maps gambles on any conditioning event B to contingent gambles on $\mathcal{X}$, defined for every gamble h on B by

$$(\lceil_{B^c} h)(x) := \begin{cases} h(x), & x \in B, \\ 0, & x \in B^c, \end{cases} \tag{1.12}$$

where $B^c := \mathcal{X} \backslash B$ is B's complement. Given a set of desirable gambles $\mathcal{D} \subseteq \mathcal{L}(\mathcal{X})$, we call

$$\mathcal{D}\rfloor B := \mathcal{D}_{\lceil_{B^c}} = (\lceil_{B^c})^{-1}\mathcal{D} = \{h \in \mathcal{L}(B) : \lceil_{B^c} h \in \mathcal{D}\}$$

its set of desirable gambles conditional on B. Often, this name is also given to $\lceil_{B^c}(\mathcal{D}\rfloor B) = \{f \in \mathcal{D} : f = fI_B\}$ and

$$\mathcal{D}|B := \lceil_{B^c}(\mathcal{D}\rfloor B) + \lceil_B(\mathcal{L}(B^c)) = \{f \in \mathcal{L}(\mathcal{X}) : fI_B \in \mathcal{D}\},$$

which are equivalent to $\mathcal{D}\rfloor B$ and each other as uncertainty models.

As an illustration, we depict, in Figure 1.7, the three conditional sets of desirable gambles that can be associated to the set of desirable gambles $\mathcal{D}$ from the illustration on the right in Figure 1.6. These are obtained by slicing along subspaces corresponding to the span of subsets of axes. We have explicitly included the gambles that correspond to the contingent border gambles of the full uncertainty model.

A conditional set of desirable gambles expresses current commitments contingent on the occurrence of the conditioning event. Although it is not required by the rationality criteria we have adopted, conditional models are also often assumed to express current or future commitments after (nothing but) the conditioning event is learned to have occurred: they can be used as updated sets of desirable gambles.

1.3.4 Marginal sets of desirable gambles

When the experiment's possibility space has a Cartesian product structure, so if $\mathcal{X} = \mathcal{Y} \times \mathcal{Z}$, we may wish to focus on one of the component possibility spaces, meaning that we can

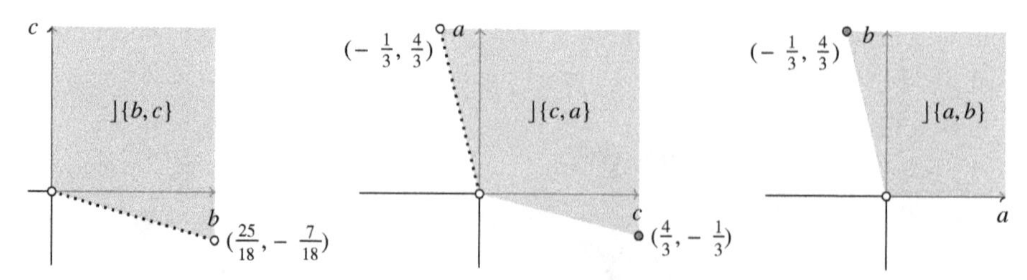

Figure 1.7 An illustration of conditional sets of desirable gambles.

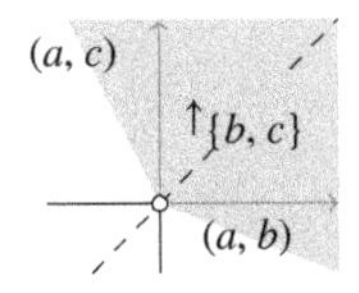

Figure 1.8 Illustration of marginal sets of desirable gambles.

ignore one aspect of the experiment's outcome. So, assume we focus on $\mathcal{Y}$ – or ignore $\mathcal{Z}$. Then the interesting gambles are those whose payoffs do not depend on the $\mathcal{Z}$-component.

This idea can be formalized by introducing the liiwii transformation $\uparrow_{\mathcal{Z}}$ that performs the so-called *cylindrical extension* of a gamble on any possibility space $\mathcal{Y}$ to its Cartesian product with $\mathcal{Z}$; it is defined for any gamble h on $\mathcal{Y}$ and outcome (y, z) in $\mathcal{Y} \times \mathcal{Z}$ by

$$(\uparrow_{\mathcal{Z}} h)(y, z) := h(y). \tag{1.13}$$

Given a set of desirable gambles $\mathcal{D} \subseteq \mathcal{L}\mathcal{Y} \times \mathcal{Z}$, we call

$$\mathcal{D}{\downarrow}\mathcal{Y} := \mathcal{D}_{\uparrow_{\mathcal{Z}}} = (\uparrow_{\mathcal{Z}})^{-1}\mathcal{D} = \{h \in \mathcal{L}(Y) : \uparrow_{\mathcal{Z}} h \in \mathcal{D}\}$$

its *$\mathcal{Y}$-marginal.*

To get a feel for what the effect is of cylindrical extension, visualization limitations force us to look, in Figure 1.8, at the trivial possibility space $\{a\} \times \{b, c\}$. Gambles on $\{a\}$ are constants, and $\uparrow_{\{b,c\}}$ maps them to the corresponding slice, the main diagonal of $\mathcal{L}(\{a\} \times \{b, c\})$.

Cylindrical extension $\uparrow_{\mathcal{Z}}$ works between gamble spaces, but it can be expressed using a surjective coordinate projection $\gamma_{\downarrow\mathcal{Y}}$ that maps elements of the Cartesian product possibility space $\mathcal{Y} \times \mathcal{Z}$ to $\mathcal{Y}$; it is defined for any outcome pair (y, z) by $\gamma_{\downarrow\mathcal{Y}}(y, z) = y$. Namely, $\uparrow_{\mathcal{Z}} h = h \circ \gamma_{\downarrow\mathcal{Y}}$, where $\circ$ denotes function composition. This projection generates the partition $\mathcal{B}_{\gamma_{\downarrow\mathcal{Y}}} := \{\gamma_{\downarrow\mathcal{Y}}^{-1}(y) : y \in \mathcal{Y}\} = \{\{y\} \times \mathcal{Z} : y \in \mathcal{Y}\}$, so the $\mathcal{Y}$-marginal can therefore also be seen as an uncertainty model with this partition as the possibility space.

Based on these observations, we can generalize the concept of marginalization to general surjective maps γ from $\mathcal{X}$ to $\mathcal{Y}$, where $\mathcal{X}$ does not need to have a Cartesian product space structure. The associated transformation γ^t defined for any gamble h on $\mathcal{Y}$ is obtained by the so-called lifting procedure:

$$\gamma^t h = h \circ \gamma. \tag{1.14}$$

Now, given a set of desirable gambles $\mathcal{D} \subseteq \mathcal{L}(\mathcal{X})$, the derived set of desirable gambles $\mathcal{D}_\gamma := \mathcal{D}_{\gamma^t} = (\gamma^t)^{-1}\mathcal{D} = \{h \in \mathcal{L}(Y) : \gamma^t h \in \mathcal{D}\}$ can be called its γ-marginal. And $\mathcal{Y}$ and the partition $\mathcal{B}_\gamma := \{\gamma^{-1}(y) : y \in \mathcal{Y}\}$ can be interchangeably used as its possibility space.

Given that surjective maps and partitions encode the same type of information, we can again translate the above concepts when a partition $\mathcal{B}$ of the possibility space $\mathcal{X}$ is given: $\Gamma_{\mathcal{B}}$ denotes the associated transformation and $\mathcal{D}_{\mathcal{B}}$ the derived $\mathcal{B}$-marginal. Return to the involved illustration on the right in Figure 1.6 for a moment: Γ_1 actually coincides with $\Gamma_{\{\{c\},\{a,b\}\}}$ under the identification $c := \{c\}$ and $d := \{a, b\}$.

1.3.5 Combining sets of desirable gambles

Now that we have gained some insight into how a set of desirable gambles and its derived sets of desirable gambles are related, *combining* sets of desirable gambles is straightforward: we view them as being derived from the unknown, sought for *joint* set of desirable gambles. This means that we must first make explicit the transformations between the gamble spaces of the sets of desirable gambles to be combined and the joint one, i.e., we must provide an

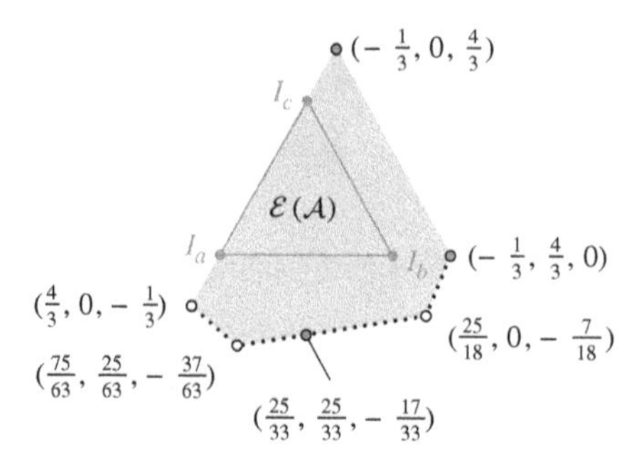

Figure 1.9 Illustration of combining sets of desirable gambles.

interpretation of their relationships. The union of the transformed sets of desirable gambles is then taken as an assessment. If this assessment avoids partial loss in the joint possibility space, the sets of desirable gambles are *compatible* and its natural extension is the joint.

Let us clarify this with two examples.

Example 1.1 For the first, consider three sets of desirable gambles, $\mathcal{E}(\{(-2,1)\}) \subset \mathcal{L}(\{a,b\})$, $\mathcal{E}(\{(-2,1)\}) \subset \mathcal{L}(\{b,c\})$, and $\mathcal{E}(\{(-2,1)\}) \subset \mathcal{L}(\{c,a\})$, interpreted as conditional sets of desirable gambles of a joint on $\{a,b,c\}$. Respective resulting contingent desirable gambles are $(-2,1,0)$, $(0,-2,1)$, and $(1,0,-2)$; their sum, desirable by additivity, is $(-1,-1,-1)$, so the coherent sets of desirable gambles are incompatible in the sense that their joint incurs sure loss. ♦

Example 1.2 For our second example, illustrated in Figure 1.9, we return to the illustration on the right in Figure 1.6 and the associated illustration in Figure 1.7. There, we started from a coherent set of desirable gambles $\mathcal{D} \subset \mathcal{L}(\{a,b,c\})$, and obtained the derived sets of desirable gambles $\mathcal{D}_{\Gamma_1}$ and $\mathcal{D}_{\Gamma_2}$, and conditional ones $\mathcal{D}\rfloor\{a,b\}$, $\mathcal{D}\rfloor\{b,c\}$, and $\mathcal{D}\rfloor\{c,a\}$. Here, we work back from those as the sets of desirable gambles to be combined: let

$$\mathcal{A} := \Gamma_1(\mathcal{D}_{\Gamma_1}) \cup \Gamma_2(\mathcal{D}_{\Gamma_2}) \cup \lceil_{\{c\}}(\mathcal{D}\rfloor\{a,b\}) \cup \lceil_{\{a\}}(\mathcal{D}\rfloor\{b,c\}) \cup \lceil_{\{b\}}(\mathcal{D}\rfloor\{c,a\})$$

be the assessment, which avoids partial loss, given that we are only putting slices back in place. The resulting joint $\mathcal{E}(\mathcal{A})$ is depicted in Figure 1.9 we see that it is strictly less committal than the original $\mathcal{D}$ because the particular sets of desirable gambles we derived do not contain all the information present in $\mathcal{D}$. ♦

Let us make our remark about 'putting slices back in place' precise with the following immediate result:

Proposition 1.3 *If $\mathcal{D} \subset \mathcal{L}(\mathcal{X})$ is coherent and $\mathcal{A} \subseteq \mathcal{D}$, then $\mathcal{E}(\mathcal{A})$ is coherent.*

Our first example showed that *individual coherence* – i.e., coherence of the individual sets of desirable gambles to be combined – does not imply coherence of the joint. However, there is an often-occurring specific situation where this does hold. Individual conditional sets of desirable gambles are *separately specified* if their conditioning events are disjoint. If they are moreover also coherent, then they are called *separately coherent*.

Theorem 1.2 *Given a partition $\mathcal{B}$ of $\mathcal{X}$, a coherent $\mathcal{B}$-marginal $\mathcal{D}_{\mathcal{B}} \subset \mathcal{L}(\mathcal{B})$, and separately coherent conditional sets of desirable gambles $\mathcal{D}\rfloor B \subset \mathcal{L}(B)$, $B \in \mathcal{B}' \subseteq \mathcal{B}$, then their combination $\mathcal{D} := \mathcal{E}(\mathcal{A}) \subseteq \mathcal{L}(\mathcal{X})$, with $\mathcal{A} := \Gamma_{\mathcal{B}}(\mathcal{D}_{\mathcal{B}}) \cup \bigcup_{B\in\mathcal{B}'} \lceil_{B^c}(\mathcal{D}\rfloor B)$, is coherent as well.*

Proof. We need to prove that $\mathcal{A}$ avoids partial loss. Assume it does not, then there will be a finite subset $\mathcal{B}''$ of $\mathcal{B}'$, a nonzero gamble $h \in \mathcal{D}_{\mathcal{B}}$, and nonzero gambles $g_B \in \mathcal{D}\rfloor B, B \in \mathcal{B}''$, such that $\Gamma_{\mathcal{B}}h + \sum_{B\in\mathcal{B}''} \lceil_{B^c} g_B = 0$. Coming from the $\mathcal{B}$-marginal $\mathcal{D}_{\mathcal{B}}$, $\Gamma_{\mathcal{B}}h$ is constant on elements of $\mathcal{B}$; because $\mathcal{D}_{\mathcal{B}}$ is coherent, it will be positive on at least one event $C \in \mathcal{B}''$. Being contingent gambles, the $\lceil_{B^c} g_B, B \in \mathcal{B}''$, have disjoint support. This means that $g_C < 0$, contradicting separate coherence. □

Combining separately coherent conditional uncertainty models and a coherent marginal one is called *marginal extension*.

In the proof above, we use the additivity criterion to combine a finite number of contingent gambles. Some authors specifically strengthen this [672, § 6.8, App. F]:

$\mathcal{B}$-Conglomerability:
$$((\forall B \in \mathcal{B}) fI_B \in \mathcal{D} \cup \{0\}) \Rightarrow fI_{\bigcup \mathcal{B}} \in \mathcal{D} \cup \{0\}, \tag{1.15}$$

where $\mathcal{B}$ is an (infinite) class of disjont events of $\mathcal{X}$. If this extra criterion is required for all possible partitions of $\mathcal{X}$, then $\mathcal{D}$ is called *fully conglomerable*.

Let us look at an example to illustrate the issue targeted with conglomerability.

Example 1.3 Consider $-\mathbb{N} \cup \mathbb{N}$, the set of nonzero integers, as the possibility space. Let the gamble f be defined by $f(-\mathbb{N}) = \{-1\}$ and $f(\mathbb{N}) = \{1\}$ and define $f_k := fI_{\{-k,k\}}$ for all $k \in \mathbb{N}$. Then $-f - \frac{1}{2}$ is compatible with the assessment $\{f_k : k \in \mathbb{N}\}$, but not with f itself. ♦

Apart from marginal extension, other types of extensions can be found in the literature, such as the epistemic irrelevance extension [482] and the exchangeable natural extension [214, 212, 213]. These extensions correspond to combinations under an additional structural assessment. Such structural assessments are often based on statements of indifference between pairs of gambles.

To make our understanding of desirability more precise, and e.g., to properly introduce the concept of indifference, we next look into the relationship with partial preference orders.

1.4 Partial preference orders

For purposes of decision making, it is useful to be able to compare gambles; e.g., each may correspond to a certain action and we wish to determine which should be preferred.

Desirability as presented in Section 1.2 is compatible with two types of *partial preference orders*, which specify pairwise comparisons of gambles. Each casts a different light on the phrase 'if we accept ownership of it when offered' in the definition of a desirable gamble. And in each case, this added clarification translates into a slight strengthening of one of the coherence criteria. So after this section, we end up with two variants of desirability, each of which describes coherent sets of desirable gambles that are, as uncertainty models, equivalent to a partial preference order of the corresponding type. Only one is used further on.

Much of the material in this chapter remains mathematically unaffected in the essentials by the strengthened criteria. We point out any minor modification necessary to preceding material.

1.4.1 Strict preference

The first partial preference order we look at, the strict one, is also the most straightforward one. A gamble f is (*strictly*) *preferred* to g – formally, $f \succ g$ – if we are eager to exchange g for f. This concept is linked to desirability by positing that a gamble is desirable if and only if it is preferred to status quo, i.e., the zero gamble. So, formally, the relationship between the strict preference relation $\succ$ on $\mathcal{L}(\mathcal{X}) \times \mathcal{L}(\mathcal{X})$ and a set of desirable gambles $\mathcal{D} \subset \mathcal{L}(\mathcal{X})$ becomes:

$$f \succ g \Leftrightarrow f - g \succ 0 \Leftrightarrow f - g \in \mathcal{D}. \tag{1.16}$$

A strict preference relation $\succ$ is coherent when the associated set of desirable gambles is. This can be expressed using the following rationality criteria:

$$\text{Irreflexivity:} \quad f \not\succ f, \tag{1.17}$$

$$\text{Transitivity:} \quad f \succ g \wedge g \succ h \Rightarrow f \succ h, \tag{1.18}$$

$$\text{Mix-independence:} \quad 0 < \mu \leq 1 \Rightarrow (f \succ g \Leftrightarrow \mu f + (1 - \mu)h \succ \mu g + (1 - \mu)h), \tag{1.19}$$

$$\text{Monotonicity:} \quad f > g \Rightarrow f \succ g, \tag{1.20}$$

where f, g, and h can be any gamble on $\mathcal{X}$. Irreflexivity makes the partial preference strict; adding transitivity, which is the equivalent of additivity for sets of desirable gambles, makes it a *preorder*. Mixture independence encompasses both positive scaling (take $h = 0$) and the first equivalence of Equation (1.16), the so-called cancellation property (take $(1 - \mu)h = -\mu g$). Monotonicity corresponds to accepting partial gains and together with irreflexivity and transitivity entails avoiding partial loss.

Irreflexivity precludes the zero gamble from belonging to the associated set of desirable gambles. Strengthening the rationality criteria of Section 1.2.1 accordingly is done by modifying avoiding partial loss, Criterion (1.5), to

$$\text{Avoiding nonpositivity:} \quad f \leq 0 \Rightarrow f \notin \mathcal{D} \quad \text{or} \quad \mathcal{D} \cap \mathcal{L}_0^-(\mathcal{X}) = \emptyset \quad \text{or} \quad 0 \notin \mathcal{D}, \tag{1.21}$$

where the last version can be used because together with additivity it entails the first two. Also, any coherent set of desirable gambles containing part of the kernel of a gamble space transformation is incompatible with that transformation.

1.4.2 Nonstrict preference

A gamble f is *nonstrictly preferred* to g – formally, $f \trianglerighteq g$ – if we are willing, i.e., not adverse, to exchange g for f. This concept is linked to desirability by positing that a gamble is desirable if and only if it is nonstrictly preferred to status quo, i.e., the zero gamble. So, formally, the relationship between the nonstrict preference relation $\trianglerighteq$ on $\mathcal{L}(\mathcal{X}) \times \mathcal{L}(\mathcal{X})$ and a set of desirable gambles $\mathcal{D} \subset \mathcal{L}(\mathcal{X})$ becomes:

$$f \trianglerighteq g \Leftrightarrow f - g \trianglerighteq 0 \Leftrightarrow f - g \in \mathcal{D}. \tag{1.22}$$

Analogously to what we saw before, a nonstrict preference relation $\trianglerighteq$ is coherent when the associated set of desirable gambles is. This can be expressed using the following rationality criteria:

$$\text{Reflexivity:} \qquad f \trianglerighteq f, \tag{1.23}$$

$$\text{Transitivity:} \qquad g \trianglerighteq h \wedge f \trianglerighteq g \Rightarrow f \trianglerighteq h, \tag{1.24}$$

$$\text{Mix-independence:} \qquad 0 < \mu \leq 1 \Rightarrow (f \trianglerighteq g \Leftrightarrow \mu f + (1-\mu)h \trianglerighteq \mu g + (1-\mu)h), \tag{1.25}$$

$$\text{Monotonicity:} \qquad f > g \Rightarrow f \trianglerighteq g \wedge g \ntrianglerighteq f, \tag{1.26}$$

where f, g, and h can be any gamble on $\mathcal{X}$. Reflexivity makes the partial preference nonstrict. Transitivity, mixture independence, and monotonicity have a similar function as for the strict variant.

Now, reflexivity forces the zero gamble to belong to the associated set of desirable gambles. Strengthening the rationality criteria of Section 1.2.1 accordingly is now done by modifying accepting partial gain, Criterion (1.4), to

$$\text{Accepting nonnegativity:} \qquad f \geq 0 \Rightarrow f \in \mathcal{D} \qquad \text{or} \qquad \mathcal{L}_0^+(\mathcal{X}) \subseteq \mathcal{D}. \tag{1.27}$$

Formula (1.9) for natural extension has to be modified accordingly, i.e., $\mathcal{L}^+(\mathcal{X})$ needs to be replaced by $\mathcal{L}_0^+(\mathcal{X})$.

There are two interesting symmetric relations that can be derived from a nonstrict preference order. The first is the equivalence relation called indifference: we are *indifferent* between two gambles f and g if we both nonstrictly prefer f over g and vice-versa:

$$f \equiv g \Leftrightarrow f \trianglerighteq g \wedge g \trianglerighteq f. \tag{1.28}$$

A nonstrict preference relation is compatible with a gamble space transformation if it entails indifference between the kernel gambles and the zero gamble. The second symmetric relation is *incomparability*: two gambles f and g are incomparable if we nonstrictly prefer neither over the other:

$$f \bowtie g \Leftrightarrow f \ntrianglerighteq g \wedge g \ntrianglerighteq f. \tag{1.29}$$

In keeping with the intuition provided by its name, incomparability is not an equivalence relation, as it is not reflexive.

We illustrate these two concepts in Figure 1.10: the zero gamble and three equivalence classes of the indifference relation are ordered according to the relative nonstrict preference of their elements. Although the gambles in $\mathcal{K}_1$ are nonstrictly preferred to those in $\mathcal{K}_2$, the gambles in both are incomparable with those in $\mathcal{K}_3$; this highlights that incomparability is often intransitive.

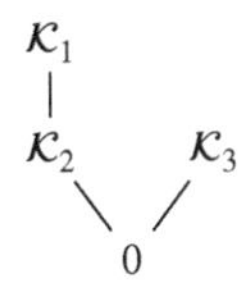

Figure 1.10 An illustration of nonstrict preference, indifference, and incomparability.

1.4.3 Nonstrict preferences implied by strict ones

Incomparability and indifference are very useful concepts: the former allows us to highlight where the information available is lacking and the latter allows us to straightforwardly specify many common structural – e.g., symmetry – assessments. However, to define them, we need a reflexive and thus nonstrict preference relation. Therefore, we give an example of how we can associate a nonstrict preference relation $\succcurlyeq$ to a strict one $\succ$, or equivalently, associate a set of nonstrictly desirable gambles $\mathcal{D}_{\succcurlyeq}$ to a set of strictly desirable gambles $\mathcal{D}_{\succ}$.

In general, there is no unique way of doing this and one must realize that it means adding commitments of some kind. Of course the resulting relation must satisfy the coherence criteria for nonstrict preference orders – reflexivity, specifically. That could be achieved simply by letting $\mathcal{D}_{\succcurlyeq}$ be equal to $\mathcal{D}_{\succ} \cup \{0\}$, but this approach would make indifference a vacuous concept: $f \succ g$ implies $g \not\succ f$ and therefore the same would hold for $\succcurlyeq$ whenever $f \neq g$, so there could only be the trivial indifference between a gamble and itself.

A nontrivial approach needs to be justified by giving an interpretation to the difference between strict and nonstrict preference that can be made mathematically precise. We draw inspiration from the difference between strict and nonstrict vector orderings (relative to $\mathcal{L}^+(\mathcal{X})$; cf. Section 1.2.5): $f \geq g$ if and only if $f + h > g$ for all gambles $h > 0$. For our preference relations, this becomes: a gamble f is nonstrictly preferred to a gamble g if and only if adding any gamble $h \succ 0$ to f makes $f + h$ strictly preferred to g. Formally:

$$f \succcurlyeq g \Leftrightarrow f - g \succcurlyeq 0 \Leftrightarrow (f - g) + \mathcal{D}_{\succ} \subseteq \mathcal{D}_{\succ}. \tag{1.30}$$

If adopted, this rule is essentially an additional rationality axiom.

Proof that $\succcurlyeq$ *satisfies* (1.23)–(1.26). Reflexivity is immediate, because $(f - f) + \mathcal{D}_{\succ} = \mathcal{D}_{\succ}$. Given $(g - h) + \mathcal{D}_{\succ} \subseteq \mathcal{D}_{\succ}$ and $(f - g) + \mathcal{D}_{\succ} \subseteq \mathcal{D}_{\succ}$, additivity implies $(f - h) + \mathcal{D}_{\succ} \subseteq \mathcal{D}_{\succ}$, so transitivity holds. To prove that mix-independence holds, i.e., $(f - g) + \mathcal{D}_{\succ} \subseteq \mathcal{D}_{\succ} \Leftrightarrow \mu(f - g) + \mathcal{D}_{\succ} \subseteq \mathcal{D}_{\succ}$, realize that positive scaling implies $\mathcal{D}_{\succ} = \mu\mathcal{D}_{\succ}$; then the equivalence's right-hand side can be written as $\mu(f - g) + \mu\mathcal{D}_{\succ} \subseteq \mu\mathcal{D}_{\succ}$, so that we see both sides only formally differ, in their arbitrary scaling factor. Monotonicity's first right-hand conjunct follows from $f - g \in \mathcal{L}^+(\mathcal{X}) \subseteq \mathcal{D}_{\succ}$ and additivity; the second, $(g - f) + \mathcal{D}_{\succ} \nsubseteq \mathcal{D}_{\succ}$, follows from the fact that $\frac{1}{2}(f - g) \in \mathcal{D}_{\succ}$, but $\frac{1}{2}(g - f) \notin \mathcal{D}_{\succ}$. □

An immediate consequence of Definition (1.30) is

$$f \succ g \Rightarrow f \succcurlyeq g \text{ and } g \not\succcurlyeq f. \tag{1.31}$$

The incomparability and indifference relations associated to $\succcurlyeq$ are respectively denoted by $\asymp$ and $\approx$. (De Cooman & Quaeghebeur [213] use an indifference relation thus defined.)

Let us clarify the material in this section by giving a somewhat more extensive illustration in Figure 1.11. We show a number of cases, i.e., sets of strictly desirable gambles (in grey). These are convex cones and are completely determined by a number of desirable and nondesirable extreme rays, i.e., cone lines starting in the zero gamble whose elements' (non)desirability is not imposed by convexity. Extreme rays whose elements are neither desirable in the strict preference sense nor in the associated nonstrict preference sense, are drawn dashed. Those whose elements are only nonstrictly desirable are drawn dotted. The ones with elements that are strictly – and thus also nonstrictly – desirable remain undecorated.

In the top row, the strongest relation between the gambles f and g shown in each of the cases, as implied by the given set of strictly desirable gambles is: $f \asymp g$, $f \succ g$, and $f \succcurlyeq g$.

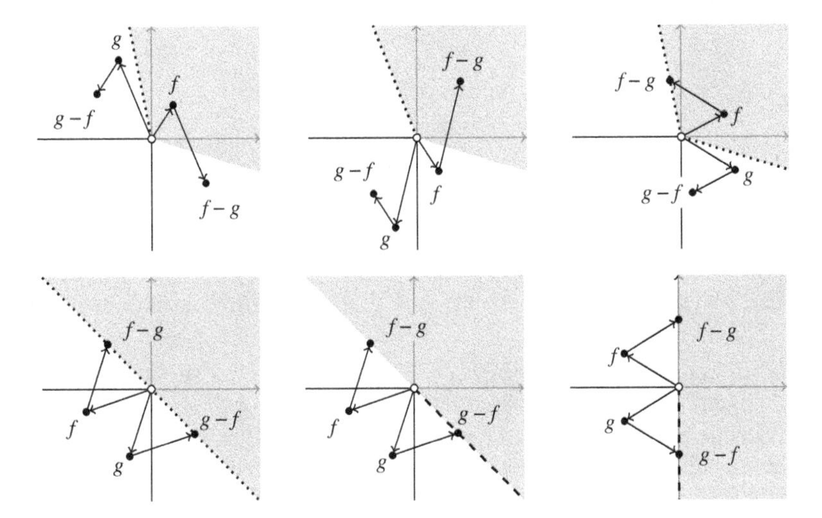

Figure 1.11 An illustration of the possible different preference relations between gambles.

For the first two cases on the bottom row, we have $f \approx g$ and $f \succ g$; we see that the cone's border structure can have an important effect. For the bottom rightmost case, we again have $f \succ g$; monotonicity leaves us no other choice.

In the first four cases of the illustration, the set $\mathcal{D}_{\succcurlyeq}$ is the closure of $\mathcal{D}_{\succ}$. The last two cases, however, remind us that this is not always so: the former because of Equation (1.31), the latter also, but more directly because of monotonicity.

1.4.4 Strict preferences implied by nonstrict ones

In decision making, strict preferences are especially useful, as they allow us to really choose one gamble over another. So it can be useful to be able to associate a strict preference relation $\rhd$ to a nonstrict one $\unrhd$, or equivalently, associate a set of strictly desirable gambles $\mathcal{D}_{\rhd}$ to a set of nonstrictly desirable gambles $\mathcal{D}_{\unrhd}$.

Also here we want to use the distinction between strict and nonstrict preference established in Equation (1.30). This essentially means that we need to find the set $\mathcal{D}_{\rhd}$ which has $\mathcal{D}_{\unrhd}$ as its associated set of nonstrictly desirable gambles.

However, as can be seen in the first and last first-row cases in the illustration of Figure 1.11, there is not always a unique such $\mathcal{D}_{\rhd}$. Moreover, within the current framework, not all sets of nonstrictly desirable gambles allowed by Criteria (1.23)–(1.26) even have such an associated set of strictly desirable gambles (cf. example below).

Although this is a topic about which little research has been done, a sketch of how the framework might be extended may be interesting for some readers. The leftmost drawing in Figure 1.12 gives an example of a situation incompatible with Equation (1.30): here both the set of strict and nonstrictly desirable gambles have open faces.

One way to make sense of it is by interpreting it as a partial view of a more complex uncertainty model in which the payoffs of gambles are defined with *infinitesimal* precision. For finite possibility spaces this can be implemented using – two-tier, for the illustration of Figure 1.12 – *lexicographic* utility: In such a context, a lexicographic gamble h can be written as $h_r + \epsilon h_i$, where ϵ is an infinitesimal quantity and h_r and h_i are, respectively, real-valued gambles for tier zero and one. The uncertainty model consists of a set of

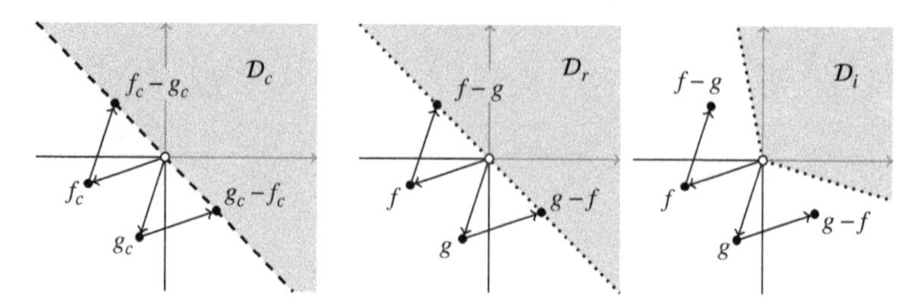

Figure 1.12 An illustration of an extension of desirability involving lexicographic utility.

desirable lexicographical gambles $\mathcal{D} := \mathcal{D}_r + \epsilon\mathcal{D}_i$ and the original situation is a view of lexicographic gambles that are constant over the tiers, i.e., of the type $f_c := f + \epsilon f$, with f some real-valued gamble. In this wider context, compatibility with (1.30) is preserved.

Another way could consist of defining separate but connected strict and nonstrict preference orders, and not derive one from the other.

1.5 Maximally committal sets of strictly desirable gambles

At the end of Section 1.2.3, we saw that coherent sets of desirable gambles form a partial order under inclusion. This carries over to coherent extensions of assessments that avoid partial loss. The least committal extension is their intersection and can be calculated using the natural extension (cf. Theorem 1.1).

In this short section (based on [213, § 2.4]), we are going to investigate the other extreme and look at the maximally committal – or *maximal* – coherent sets of desirable gambles, at maximally committal – or maximal – extensions of assessments, and their relationship with the least committal one. We from now on work with a strict preference interpretation of desirability, augmented with Rule (1.30). This means that avoiding partial loss is replaced by avoiding nonpositivity: the zero gamble is not desirable. This is crucial to obtaining interesting results; a similar analysis would not be as straightforward or even possible under a nonstrict preference interpretation.

A maximal coherent set of desirable gambles $\mathcal{D}$ is one that is not included in any other coherent set of desirable gambles. In other words, when adding any gamble f in $\mathcal{L}(\mathcal{X})\backslash\mathcal{D}$ to $\mathcal{D}$, this would result in an assessment that incurs nonpositivity. The set of maximal coherent sets of desirable gambles on $\mathcal{X}$ is denoted by $\hat{\mathbb{D}}(\mathcal{X})$.

What do these maximal coherent sets of desirable gambles look like? The following proposition provides a nice characterization:

Proposition 1.4 *The set $\mathcal{D}$ in $\mathbb{D}(\mathcal{X})$ is maximal if and only if $f \in \mathcal{D} \Leftrightarrow -f \notin \mathcal{D}$ for all nonzero gambles f on $\mathcal{X}$.*

Proof. Coherence, avoiding nonpositivity to be precise, makes $f \in \mathcal{D} \Rightarrow -f \notin \mathcal{D}$ a necessity for all nonzero gambles f on $\mathcal{X}$. So we have to prove that maximality of $\mathcal{D}$ is equivalent to $-f \notin \mathcal{D} \Rightarrow f \in \mathcal{D}$ for all nonzero gambles f.

First necessity; assume $\mathcal{D}$ is maximal but nevertheless $\{f, -f\} \cap \mathcal{D} = \emptyset$ for some nonzero gamble f. Then both $\{f\} \cup \mathcal{D}$ and $\{-f\} \cup \mathcal{D}$ incur nonpositivity, which, because

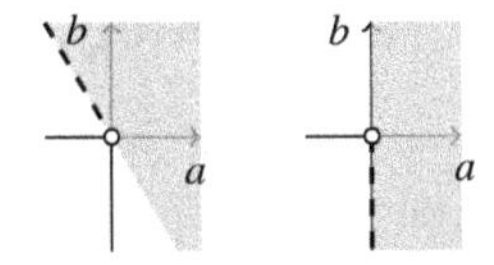

Figure 1.13 Illustration of maximal sets of desirable gambles.

$\mathcal{D}$ is coherent and thus a cone excluding the zero gamble, means both $0 \in \text{posi}\{f\} + \mathcal{D}$ and $0 \in \text{posi}\{-f\} + \mathcal{D}$. Or, in other words, $\{f, -f\} \subset \mathcal{D}$, a contradiction.

Now sufficiency; if $|\{f, -f\} \cap \mathcal{D}| = 1$ for all nonzero gambles f, then any set that strictly includes $\mathcal{D}$ incurs nonpositivity. □

We see that maximal sets of desirable gambles are halfspaces in a very concrete sense. They are neither open nor closed. An illustration with two cases is given in Figure 1.13.

Now, how can we use these maximal sets of desirable gambles to our advantage? Given an assessment $\mathcal{A}$, both nonpositivity avoidance and the least committal extension can be characterized using the set $\hat{\mathbb{D}}_{\mathcal{A}} := \{\mathcal{D} \in \hat{\mathbb{D}}(\mathcal{X}) : \mathcal{A} \subseteq \mathcal{D}\}$ of maximal coherent sets of desirable gambles. This is done in the following two results:

Theorem 1.3 *An assessment $\mathcal{A} \subseteq \mathcal{L}(\mathcal{X})$ avoids nonpositivity if and only if $\hat{\mathbb{D}}_{\mathcal{A}} \neq \emptyset$.*

Proof sketch. First sufficiency; assume there is a $\mathcal{D}$ in $\hat{\mathbb{D}}(\mathcal{X})$ such that $\mathcal{A} \subseteq \mathcal{D}$. Then $\text{posi}\,\mathcal{A} \subseteq \mathcal{D}$ and thus $0 \notin \text{posi}\,\mathcal{A}$.

Now necessity; assume $\mathcal{A} \subseteq \mathcal{L}(\mathcal{X})$ avoids nonpositivity. In case $\mathcal{X}$ is finite, one can then construct a maximal set of desirable gambles that includes $\mathcal{A}$ in a finite number of steps by each time enlarging the assessment with a gamble from outside its natural extension and the negation thereof [156, Theorem 12]. In case $\mathcal{X}$ is infinite, a nonconstructive approach can be applied in which infinite such chains of sets of desirable gambles that include $\mathcal{A}$ are employed [213, Theorem 3]. □

Corollary 1.1 *The least committal extension of an assessment $\mathcal{A} \subseteq \mathcal{L}(\mathcal{X})$ that avoids nonpositivity, i.e., its natural extension $\mathcal{E}(\mathcal{A})$, is the intersection $\bigcap \hat{\mathbb{D}}_{\mathcal{A}}$ of the maximal sets of desirable gambles that include $\mathcal{A}$.*

Proof. That the least committal extension is a subset follows from avoiding nonpositivity. Assume it is a strict subset and take f in $\bigcap \hat{\mathbb{D}}_{\mathcal{A}} \setminus \mathcal{E}(\mathcal{A})$. Then $\mathcal{E}(\mathcal{A}) \cup \{-f\}$ avoids nonpositivity and thus $\mathcal{E}(\mathcal{A} \cup \{-f\})$ is a coherent extension of $\mathcal{A}$, but it is not included in any element of $\hat{\mathbb{D}}_{\mathcal{A}}$, a contradiction. □

To close off this section, we present a result that shows derived – e.g., conditional or marginal – sets of desirable gambles of maximal models are maximal as well.

Proposition 1.5 *If $\mathcal{D} \subset \mathcal{L}(\mathcal{X})$ is maximal, then $\mathcal{D}_\Gamma$ is maximal for a linear increasing injective transformation Γ from $\mathcal{L}(\mathcal{Z})$ to $\mathcal{L}(\mathcal{X})$ with increasing inverse.*

Proof. Proposition 1.2 ensures that $\mathcal{D}_\Gamma$ is coherent. Now assume that the derived set of desirable gambles $\mathcal{D}_\Gamma$ is nonmaximal, i.e., that there is some gamble h on $\mathcal{Z}$ such that $\{h, -h\} \cap \mathcal{D}_\Gamma = \emptyset$, or $\{h, -h\} \subseteq (\Gamma^{-1}\mathcal{D})^c = \Gamma^{-1}\mathcal{D}^c$. Then, because of Γ's self-conjugacy, $\{\Gamma h, -\Gamma h\} \subseteq \mathcal{D}^c$, or $\{\Gamma h, -\Gamma h\} \cap \mathcal{D} = \emptyset$, a contradiction. □

1.6 Relationships with other, nonequivalent models

We have studied sets of desirable gambles and investigated their connection to partial preference orders, which gave rise to strict and nonstrict variants. In the preceding section we settled on the strict variant. Sets of desirable gambles and partial preference orders are equivalent uncertainty models: one can be expressed in terms of the other and vice versa. In this section, we investigate their connection with other, commonly used, but nonequivalent models.

Given an assessment denoted $\mathcal{A}_* \subset \mathcal{L}(\mathcal{X})$, where $*$ is a dummy variable that stands for an uncertainty model generating the assessment, we will use the notational conventions $\mathbb{D}_* := \mathbb{D}_{\mathcal{A}_*}$ and $\hat{\mathbb{D}}_* := \hat{\mathbb{D}}_{\mathcal{A}_*}$.

1.6.1 Linear previsions

Linear previsions are positive, linear, normed functionals and popular uncertainty models in classical probability theory. A linear prevision provides fair prices for gambles, i.e., it is a real-valued function on $\mathcal{L}(\mathcal{X})$. (We imbue the set of real-valued functionals on $\mathcal{L}(\mathcal{X})$ with the topology of pointwise convergence.) They are mathematically equivalent, as uncertainty models, to (finitely additive) probability measures and, on finite possibility spaces, to probability mass functions (cf. Section 2.2.2).

The set of all linear previsions is denoted by $\mathbb{P}(\mathcal{X})$. It is a closed convex set with extreme points that correspond to the $\{0, 1\}$-valued finitely additive probabilities [672, § 3.6.7]. For finite $\mathcal{X}$ it is a simplex, with the *degenerate previsions* as its extreme points. There is a degenerate prevision P_x for each elementary event x of $\mathcal{X}$. Its defining property is that $P_x(f) = f(x)$ for every gamble f.

Fair prices of indicators are called *probabilities*, and the following notational shorthand is used later on for every event $A \subseteq \mathcal{X}$: $P(A) := P(I_A)$.

Given a linear prevision $P \in \mathbb{P}(\mathcal{X})$ as an assessment, what gambles are desirable? We take a gamble to be strictly desirable whenever its fair price is positive. This results in the following rule for going from linear previsions to sets of desirable gambles:

$$\mathcal{D}_P := \mathcal{E}(\mathcal{A}_P), \quad \text{with} \quad \mathcal{A}_P := \{f \in \mathcal{L}(\mathcal{X}) : P(f) > 0\}. \tag{1.32}$$

Due to linearity, the space of gambles with fair price zero will be a linear subspace of $\mathcal{L}(\mathcal{X})$ and therefore the corresponding set of desirable gambles will be a halfspace. Therefore the assessment $\mathcal{A}_P$ is an open halfspace, which because of linearity and positivity avoids partial loss. When $\mathcal{D}_P$ coincides with $\mathcal{A}_P$, the border gambles on the delimiting hyperplane are nonstrictly, but not strictly desirable. This is illustrated on the left in Figure 1.14. Due to coherence, enforced by the natural extension, however, this is not always the case; for example, the partial loss incurring gamble that is negative everywhere except in the elementary event x, where it is zero, will have zero as its fair price under P_x. The illustration on the right in Figure 1.14 depicts $\mathcal{D}_{P_a}$.

Linear previsions are continuous functions, which explains why the associated assessment, the inverse image of $\mathbb{R}^+$, is an open set. Coherent sets of desirable gambles need not be open (or closed), and we know that in particular the maximal ones in $\hat{\mathbb{D}}_P \subseteq \mathbb{D}_P$, also halfspaces, are not. So linear previsions cannot express preferences as finely as coherent halfspaces of desirable gambles can. If we wish to define a correspondence between the latter and the former, we will end up with an uncertainty model that is less committal,

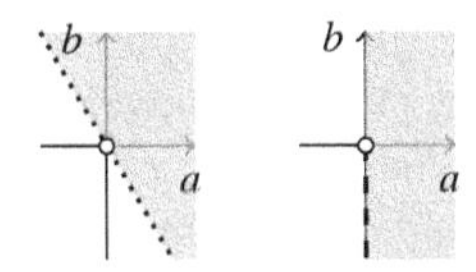

Figure 1.14 Illustration of sets of desirable gambles corresponding to linear previsions.

effectively replacing the possibly nonopen halfspace with the usually open natural extension of its interior.

1.6.2 Credal sets

Very popular imprecise-probabilistic models are *credal sets*, or sets of linear previsions. Credal sets can, for example, arise in situations when a linear prevision cannot be precisely elicited in the sense that only bounds on fair prices for gambles can be obtained, or also when differing linear previsions, each provided by a different expert, are pooled.

Given a credal set $\mathcal{M} \subseteq \mathbb{P}(\mathcal{X})$ as an assessment, what gambles are desirable? We take a gamble to be strictly desirable when it is so under any linear prevision in the credal set. This results in the following rule for going from credal sets to sets of desirable gambles:

$$\mathcal{D}_{\mathcal{M}} := \mathcal{E}(\mathcal{A}_{\mathcal{M}}), \quad \text{with} \quad \mathcal{A}_{\mathcal{M}} := \{f \in \mathcal{L}(\mathcal{X}) : (\forall P \in \mathcal{M}) P(f) > 0\} = \bigcap_{P \in \mathcal{M}} \mathcal{A}_P. \tag{1.33}$$

Each element in the credal set results in a linear constraint, i.e., restricts the desirable gambles to lie in a specific open halfspace. Because linear previsions are convex combinations of degenerate previsions, this rule will give the same result whether $\mathcal{M}$ or its convex hull is used; so the convex hull's border structure is uniquely important.

Let us look at an illustration of this rule for a possibility space $\{a, b, c\}$ in Figure 1.15.

On the left, we show a convex credal set $\mathcal{M}$, the light grey filled polytope, within the simplex spanned by the degenerate previsions, the grey triangle. The border structure is shown as follows: Open border segments are indicated by dashed lines; closed ones are undecorated. The endpoints of border segments are indicated by dots, filled darkly when part of the credal set and white otherwise. For each, we give the corresponding linear prevision's probability mass function: i.e., the coefficient vector (μ_a, μ_b, μ_c) of its convex combination in terms of the respective degenerate previsions P_a, P_b, and P_c.

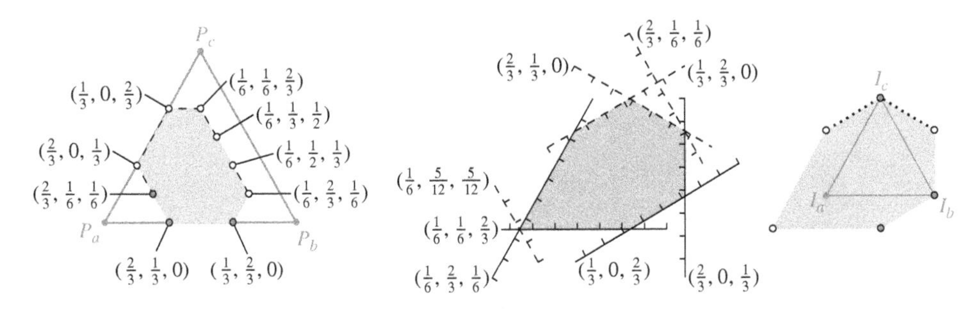

Figure 1.15 An illustration of the rule for obtaining the set of desirable gambles $\mathcal{D}_{\mathcal{M}}$ corresponding to a credal set $\mathcal{M}$.

In the middle, we show, in the plane of gambles with sum one, a sufficient subset of the constraints corresponding to $\mathcal{M}$; the stubs indicate the delimited halfspace. All but one correspond to border segment endpoints: open halfspaces $\sum_{x\in\mathcal{X}}\mu_x f_x > 0$ for endpoints in the credal set – indicated with dashed lines – and closed ones $\sum_{x\in\mathcal{X}}\mu_x f_x \geq 0$ otherwise – indicated with full lines. The constraint $(\frac{1}{6}, \frac{5}{12}, \frac{5}{12})$ does not correspond to such an endpoint. All nondepicted constraints are redundant, as well as $(\frac{2}{3}, \frac{1}{6}, \frac{1}{6})$, which is made redundant by $(\frac{2}{3}, \frac{1}{3}, 0)$ and $(\frac{2}{3}, 0, \frac{1}{3})$. Together, the constraints define the assessment $\mathcal{A}_{\mathcal{M}}$.

On the right, the resulting set of desirable gambles $\mathcal{D}_{\mathcal{M}} = \mathcal{A}_{\mathcal{M}} \cup \{I_c\}$ is shown. Note how included (excluded) exposed border segment endpoints of $\mathcal{M}$ map to open (closed) faces of $\mathcal{A}_{\mathcal{M}}$ and how nonopen (open) faces of $\mathcal{M}$ map to excluded (included) border rays of $\mathcal{A}_{\mathcal{M}}$. The nonopen to excluded mapping makes it clear that not all of $\mathcal{M}$'s border structure can be preserved: the resulting model is less committal.

Now, given a coherent set of strictly desirable gambles $\mathcal{D} \subset \mathcal{L}(\mathcal{X})$, what is the credal set that we should associate with it? The natural counterpart to the idea expressed in Equation (1.33) would be the set of linear previsions P in $\mathbb{P}(\mathcal{X})$ such that $P(f) > 0$ for all gambles f in $\mathcal{D}$. However, given a set $\mathcal{D}$ in $\hat{\mathbb{D}}_Q$ for some Q in $\mathbb{P}(\mathcal{X})$, then we know no linear prevision P in $\mathbb{P}(\mathcal{X})$ will satisfy the requirement, not even Q. Namely, $\mathcal{D}$ then includes border gambles f of the open set $\mathcal{D}_Q$ such that $Q(f) = 0$, and except for 0, all border gambles of $\mathcal{D}_Q$ are included in some element of $\hat{\mathbb{D}}_Q$. Therefore, to deal with the fact that previsions cannot express such border structure and to allow us to associate a credal set with any coherent set of desirable gambles, we must replace the strict inequality with a nonstrict one:

$$\mathcal{M}_{\mathcal{D}} := \bigcap_{f\in\mathcal{D}} \{P \in \mathbb{P}(\mathcal{X}) : P(f) \geq 0\}. \tag{1.34}$$

The resulting credal set has a simple structure:

Proposition 1.6 *The credal set $\mathcal{M}_{\mathcal{D}} \subseteq \mathbb{P}(\mathcal{X})$ associated to a coherent set of desirable gambles $\mathcal{D} \subset \mathcal{L}(\mathcal{X})$ is closed and convex.*

Proof. The set $\mathbb{P}(\mathcal{X})$ of all linear previsions is closed and convex. Considering for a moment previsions just as linear functionals on $\mathcal{L}(\mathcal{X})$, i.e., not restricted to $\mathbb{P}(\mathcal{X})$, then for any f in $\mathcal{L}(\mathcal{X})$, $P(f) \geq 0$ is a linear constraint, i.e., it specifies a closed half-space. So $\mathcal{M}_{\mathcal{D}}$ is an intersection of closed convex sets, which is closed and convex. □

Let us look at an illustration of this rule in Figure 1.16, again for a possibility space $\{a, b, c\}$. The meaning of the picture elements is the same as before. On the left, we show

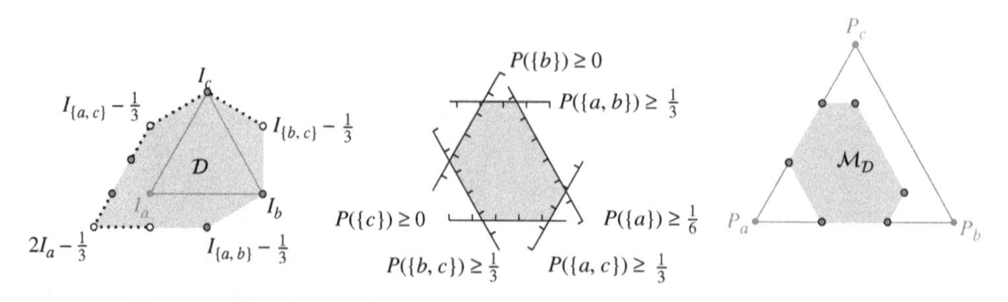

Figure 1.16 An illustration of the rule for obtaining the credal set $\mathcal{M}_{\mathcal{D}}$ corresponding to a set of desirable gambles $\mathcal{D}$.

a set of desirable gambles $\mathcal{D}$, or rather, its intersection with the sum-one plane. In the middle, we show the necessary and sufficient set of linear constraints in prevision-space corresponding to this set of desirable gambles: each of the constraints corresponds to an extreme point of the closure of $\mathcal{D}$; their intersection is shaded grey. On the right, we show the resulting credal set $\mathcal{M}_{\mathcal{D}}$.

1.6.3 To lower and upper previsions

We have seen that linear previsions correspond to a very specific type of sets of desirable gambles, halfspaces, although they cannot determine the border structure of the halfspace. Linear previsions specify fair prices for gambles. We now show how this idea can be generalized, i.e., how we can associate a pair of super and sublinear previsions to any set of desirable gambles, although again they will not determine border structure. These are respectively called lower and upper previsions, and defined in terms of, respectively, buying and selling prices for gambles.

The association is made by making comparisons with constant gambles, which we denote by their constant value. An illustration with a constant gamble α is given on the left in Figure 1.17. The reason is that the set of constant gambles can be linearly ordered – trivially so. So they facilitate the connection with the price scale.

Consider a coherent set of strictly desirable gambles $\mathcal{D} \subset \mathcal{L}(\mathcal{X})$ and the corresponding strict partial preference relation $\succ$. The corresponding *lower prevision* $\underline{P}_{\mathcal{D}}$ – interpreted to specify supremum acceptable buying prices – and *upper prevision* $\overline{P}_{\mathcal{D}}$ – specifying infimum acceptable selling prices – are defined as follows for every gamble f on $\mathcal{X}$:

$$\underline{P}_{\mathcal{D}}(f) := \sup\{\alpha \in \mathbb{R} : f \succ \alpha\} = \sup\{\alpha \in \mathbb{R} : f - \alpha \in \mathcal{D}\}, \tag{1.35}$$

$$\overline{P}_{\mathcal{D}} f := \inf\{\beta \in \mathbb{R} : \beta \succ f\} = \inf\{\beta \in \mathbb{R} : \beta - f \in \mathcal{D}\}. \tag{1.36}$$

An illustration is given in the middle in Figure 1.17. These definitions make lower and upper previsions conjugate: $\overline{P}_{\mathcal{D}}(f) = -\underline{P}_{\mathcal{D}}(-f)$. This means that both previsions are equivalent uncertainty models and either can be used on its own; we work with lower previsions.

Linear previsions are self-conjugate lower – and thus upper – previsions: coinciding acceptable buying and selling prices are fair prices.

Similarly as for linear previsions, lower and upper previsions of indicators are called *lower and upper probabilities*. The following notational shorthands are used:

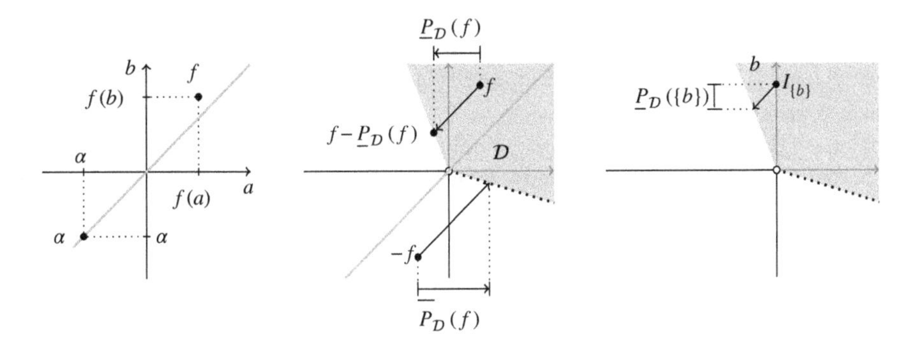

Figure 1.17 An illustration of the association of lower and upper previsions and probabilities $\underline{P}_{\mathcal{D}}$ and $\overline{P}_{\mathcal{D}}$ with sets of desirable gambles.

$\underline{P}_{\mathcal{D}}(A) := \underline{P}_{\mathcal{D}}(I_A)$ and $\overline{P}_{\mathcal{D}}(A) := \overline{P}_{\mathcal{D}}(I_A) = 1 - \underline{P}_{\mathcal{D}}(A^c)$. An illustration is given on the right in Figure 1.17.

Conceptually, desirability is relatively simple: generating convex cones from initial assessments and slicing those to obtain conditional and marginal models. This conceptual framework can also be exploited to derive results in the theory of lower previsions. For that, we need to be able to associate a set of desirable gambles with a given lower prevision. To facilitate this, we first look at a number of simplified variants of sets of desirable gambles.

1.6.4 Simplified variants of desirability

The previous sections have made it clear that the complicated border structures coherent sets of desirable gambles can possess cannot be expected to be preserved when translating them to other uncertainty models. In this section, we are going to have a look at variants of desirability where these issues are simplified away.

The first variant comes as a pair, with one member of the pair corresponding to a strict partial preference order, and the second to a nonstrict one: A coherent set of *almost desirable* gambles $\mathcal{D}_{\sqsupseteq} \subseteq \mathcal{L}(\mathcal{X})$ must satisfy positive scaling, additivity, accepting sure gain, and avoiding sure loss – i.e., Criteria (1.1), (1.2), (1.6), and (1.7) – and moreover be closed [672, §§ 3.7.3–6]. So it is a closed convex cone including the positive orthant, the zero gamble, but no gamble in the uniformly negative orthant. A coherent set of *surely desirable* gambles $\mathcal{D}_{\sqsupset} \subset \mathcal{L}(\mathcal{X})$ must satisfy the same criteria and moreover be open. So it is an open convex cone containing the uniformly positive orthant and no gamble of the negative orthant. Both models are equally expressive, but respectively correspond to a nonstrict and strict partial preference order (with modified monotonicity conditions): almost preference $\sqsupseteq$ and sure preference $\sqsupset$. In case one is given, the other can be defined in terms of it using the following correspondences: $\mathcal{D}_{\sqsupset} = \mathrm{int}(\mathcal{D}_{\sqsupseteq})$ and $\mathcal{D}_{\sqsupseteq} = \mathrm{cl}(\mathcal{D}_{\sqsupset})$, so their relationship is different from Equation (1.30).

Working with fully open or closed convex cones simplifies matters greatly. However, in order to be able to do so we have had to replace avoiding partial loss and accepting partial gains by their sure variants. In case one does not wish to do this and remain within the class of coherent sets of strictly and nonstrictly desirable gambles – i.e., those that satisfy Criteria (1.1), (1.2), (1.4), and (1.5) – the following variant can be used: A *simple* coherent set of strictly desirable gambles $\mathcal{D}_{\succ} \subseteq \mathcal{L}(\mathcal{X})$ is a coherent set of strictly desirable gambles such that $\mathcal{D}_{\succ} = \mathrm{int}(\mathcal{D}_{\succ}) \cup \mathcal{L}^{+}(\mathcal{X})$ [672, §§ 3.7.7–9]. As before, the associated set of nonstrictly desirable gambles obtained using Equation (1.30) is denoted $\mathcal{D}_{\succeq}$. Simple coherent sets of strictly desirable gambles are equally expressive as coherent sets of almost or surely desirable gambles. The correspondences are as follows: $\mathcal{D}_{\sqsupseteq} = \mathrm{cl}(\mathcal{D}_{\succ})$ and $\mathcal{D}_{\succ} = \mathcal{D}_{\sqsupset} \cup \mathcal{L}^{+}(\mathcal{X})$; the first also holds for nonsimple sets of strictly desirable gambles.

The last variant we look at are the *marginally desirable* gambles, those which are almost, but not surely desirable. In other words, a coherent set of marginally desirable gambles $\mathcal{G}$ is the border of a coherent set of (strictly, nonstrictly, almost, or surely) desirable gambles. Again, it is as expressive as the preceding variants; the correspondences are: $\mathcal{G} = \mathcal{D}_{\sqsupseteq} \setminus \mathcal{D}_{\sqsupset}$ and $\mathcal{D}_{\sqsupset} = \mathcal{G} + \mathbb{R}^{+}$. Sets of marginally desirable gambles usually are not as handy to work with mathematically, because they are not convex in general, but they can be useful for moving between representations.

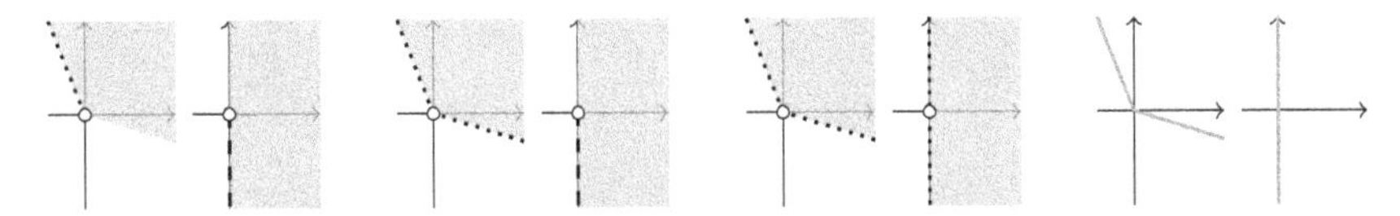

Figure 1.18 An illustration of the simplified variants of desirability and their relationship with standard desirability. Leftmost, we give a pair of sets of desirable gambles, each depicting, in the way we did before, both the set of strictly desirable gambles and the associated set of nonstrictly desirable gambles. Next to that, we do the same, but now for the derived simple sets of strictly desirable gambles and their associated sets of nonstrictly desirable gambles. To the right of that, we repeat things for the pair of sets of surely and almost desirable gambles. Finally, rightmost, we give the corresponding sets of marginally desirable gambles.

In Figure 1.18, we give an illustration of the variants of desirability discussed above and their relationship with standard desirability, which is purely a question of border structure distinctions.

What are the relationships between these simplified variants of desirability and the uncertainty models we have previously discussed in this section? We already know that the variants themselves are equivalent uncertainty models, so we only need to make the correspondence with one of them.

The set of marginally desirable gambles $\mathcal{G}_{\underline{P}}$ corresponding to a lower prevision $\underline{P}$ on $\mathcal{L}(\mathcal{X})$ is defined by:

$$\mathcal{G}_{\underline{P}} := \{f \in \mathcal{L}(\mathcal{X}) : \underline{P}(f) = 0\}. \tag{1.37}$$

The other direction is described by Equations (1.35) and (1.36), starting from a simple set of strictly desirable gambles. A linear prevision P is a special type of coherent lower prevision, and it is clear from Section 1.6.1 that the corresponding set of marginally desirable gambles $\mathcal{G}_P$ is a hyperplane.

1.6.5 From lower previsions

In the previous section, we have given the correspondence between simple coherent sets of strictly desirable gambles and lower previsions defined on the set of all gambles $\mathcal{L}(\mathcal{X})$. However, lower previsions are usually elicited on a gamble-by-gamble basis. This means that a lower prevision $\underline{P}$ is usually initially defined only on a subset $\mathcal{K}$ of gambles in $\mathcal{L}(\mathcal{X})$ and Equation (1.37) is not applicable.

How are we going to deal with this case? A first naive idea would be to obtain an assessment by taking those gambles in $\mathcal{K}$ to be desirable that have a positive lower prevision. This is not a good approach, because $\mathcal{K}$ may not contain such gambles, or only ones in the positive orthant.

For a better approach, we first need to realize that coherent lower previsions are *constant additive*: Equation (1.35) immediately implies $\underline{P}_{\mathcal{D}}(f + \alpha) = \underline{P}_{\mathcal{D}}(f) + \alpha$ for any gamble f on $\mathcal{X}$ and real number α. Second, we posit that this consequence of coherence can be used to extend $\underline{P}$ from $\mathcal{K}$ to a larger set of gambles; this is a reasonable thing to do if we want

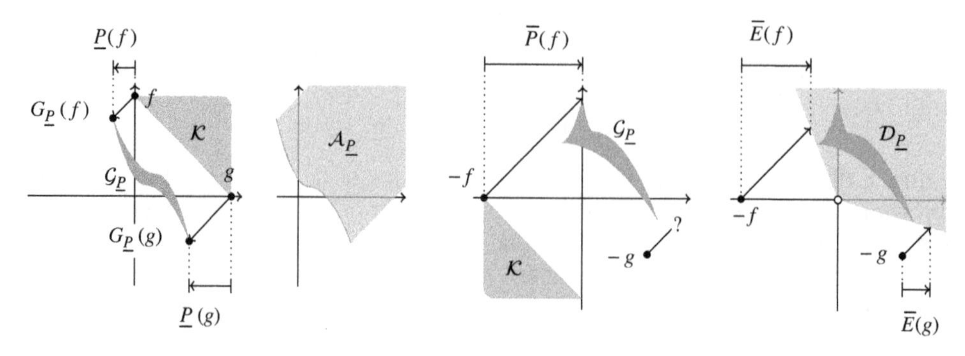

Figure 1.19 An illustration of the procedures to derive an assessment $\mathcal{A}_{\underline{P}}$ from a lower prevision $\underline{P}$ and how to use that assessment to derive the natural extension $\underline{E}$ of $\underline{P}$ and to check whether $\underline{P}$ is coherent.

the rationality criteria for lower previsions to essentially coincide with those for sets of desirable gambles. It allows us to generalize Equation (1.37):

$$\mathcal{G}_{\underline{P}} := G_{\underline{P}}(\mathcal{K}),$$

where for every f in $\mathcal{K}$, $G_{\underline{P}}(f) := f - \underline{P}(f)$ is the *marginal gamble* of f under $\underline{P}$. Then we can use the corresponding surely – and thus strictly – desirable gambles as an assessment:

$$\mathcal{D}_{\underline{P}} := \mathcal{E}(\mathcal{A}_{\underline{P}}), \quad \text{with} \quad \mathcal{A}_{\underline{P}} := \mathcal{G}_{\underline{P}} + \mathbb{R}^+. \tag{1.38}$$

On the left in Figure 1.19, we give an illustration of going from a lower prevision $\underline{P}$ to $\mathcal{A}_{\underline{P}}$. In this example, the form of $\mathcal{G}_{\underline{P}}$ betrays that $\underline{P}$ is not constant additive; this fact is corrected when going to $\mathcal{A}_{\underline{P}}$.

Now that we can go from lower previsions to sets of desirable gambles and back, it is possible to translate concepts derived for one uncertainty model to the other. As an example, let us here briefly look at the most basic ones. (A thorough treatment of coherent lower previsions can be found in Chapter 2.)

Example 1.4 A lower prevision $\underline{P}$ on $\mathcal{K} \subseteq \mathcal{L}(\mathcal{X})$ *avoids sure loss* if and only if $\mathcal{A}_{\underline{P}}$ avoids sure loss, or, equivalently due to its construction, partial loss. After some manipulation, one finds the following criterion:

$$(\forall g \in \operatorname{posi} \mathcal{G}_{\underline{P}}) \sup g \geq 0. \tag{1.39}$$

The *natural extension* $\underline{E}$ of a lower prevision $\underline{P}$ on $\mathcal{K} \subseteq \mathcal{L}(\mathcal{X})$ to the set of all gambles $\mathcal{L}(\mathcal{X})$ can be found by first moving to a representation in terms of desirable gambles, next performing natural extension as defined by Equation (1.9), and then going back, i.e., deriving supremum acceptable buying prices for gambles outside of $\mathcal{K}$ and possibly correcting those for gambles in $\mathcal{K}$. Working this out formally results in the following expression, valid for any gamble f on $\mathcal{X}$:

$$\underline{E}(f) = \sup\{\alpha \in \mathbb{R} : (\exists g \in \operatorname{posi} \mathcal{G}_{\underline{P}}) f - \alpha \geq g\} = \sup\{\inf(f - g) : g \in \operatorname{posi} \mathcal{G}_{\underline{P}}\}. \tag{1.40}$$

This procedure is illustrated on the right in Figure 1.19; note how $\underline{P}(f)$ is corrected to $\underline{E}(f)$.

A lower prevision $\underline{P}$ on $\mathcal{K} \subseteq \mathcal{L}(\mathcal{X})$ is *coherent* if and only if it coincides with its natural extension, which also automatically implies it avoids sure loss. This results in the following formal criterion:

$$(\forall f \in \mathcal{G}_{\underline{P}})(\forall g \in \text{posi}\, \mathcal{G}_{\underline{P}}) \sup(g - f) \geq 0. \tag{1.41}$$

So, in the illustration on the right in Figure 1.19, $\underline{P}$ avoids sure loss, but is nevertheless incoherent. ♦

1.6.6 Conditional lower previsions

After the previous sections, one might wonder why one would want to bother with nonsimple sets of strictly desirable gambles, i.e., why is it useful to have control over the border structure? In short, the answer is that it allows commitments to be specified more precisely. We make this more tangible with an example involving conditional lower previsions.

Consider the coherent lower prevision defined for every gamble f on a possibility space $\{a, b, c\}$ by $\underline{P}f := \min\{\frac{3}{4}f(a) + \frac{1}{4}f(b), \frac{1}{3}f(a) + \frac{2}{3}f(b), f(c)\}$. On the left in Figure 1.20, we give the associated closed convex credal set $\mathcal{M}_{\underline{P}}$. Next to that, in the upper row, we first give the associated simple set of strictly desirable gambles $\mathcal{D}_{\underline{P}}$ – by depicting its intersection with the sum-one plane – and then $\mathcal{D}_{\underline{P}}\rfloor\{a, b\}$, the corresponding set of strictly desirable gambles conditional on $\{a, b\}$. This allows us to derive the conditional lower prevision $\underline{P}(\cdot|\{a, b\}) := \underline{P}_{\mathcal{D}_{\underline{P}}\rfloor\{a,b\}} = \inf$, which is *vacuous*: it implies no commitments whatsoever. This is a general feature in that lower previsions conditional on events of lower probability zero are always vacuous when using this approach.

Another approach, *regular extension*, is sometimes used in the theory of coherent lower previsions to obtain nonvacuous lower previsions conditional on events of lower probability zero. It essentially corresponds with a rule for associating a set of desirable gambles with a lower prevision that generates sets that are in most cases strictly larger than the usual, simple one: $\mathcal{R}_{\underline{P}} := \mathcal{D}_{\underline{P}} \cup \{f \in \text{cl}(\mathcal{D}_{\underline{P}}) : \overline{P}f > 0\}$. The difference with the usual approach is illustrated in the second row. The conditional lower prevision we now obtain is

$$R(\cdot|\{a, b\}) := \underline{P}_{\mathcal{R}_{\underline{P}}\rfloor\{a,b\}} = \min\{\frac{3}{4}f(a) + \frac{1}{4}f(b), \frac{1}{3}f(a) + \frac{2}{3}f(b)\}.$$

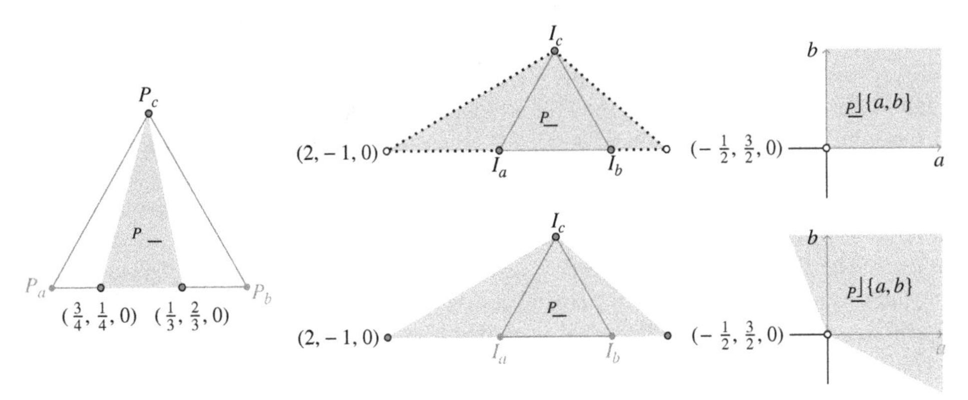

Figure 1.20 An illustration of the importance of control over the border structure: the difference between natural and regular extension.

So we see that being able to finely specify border structure can have a big effect on conditional uncertainty models.

1.7 Further reading

As far as we have been able to unearth, Smith [602] was the first to work with a desirability-type of model: he talks about 'open cones of exchange vectors', which would correspond to our surely desirable gambles (cf. Section 1.6.4). After that, Williams [701, 703, 702] put most of the essential ideas on paper: he talks about 'acceptable bets' and works with nonstrict desirability, producing many interesting results on conditional (lower and upper) previsions [703].

Many authors use partial preference orders as the primitive notion in their studies of decision making, utility, and probability; some also mention the equivalent desirability representation. Aumann [38, 39] does this while generalizing von Neumann–Morgenstern utility theory [660]. Girón & Rios [315] justify and axiomatize quasi-Bayesian probability theory – a theory of closed convex credal sets – mostly using partial preference orders, but they also talk about 'closed convex cones in the space of decisions' – our almost desirability. Fishburn [292] provides an accessible survey of what by now can be called earlier work. Seidenfeld et al. [571] take strict partial preference orders as primitive and mainly study the connection with credal sets, but do talk about 'favourable bets', which corresponds to strict desirability. In later work [572], to ensure a closer connection with credal set representations, they add an Archimedean (continuity) criterion. Another important point in which their work differs from the material in this chapter is by relaxing the linear (state independent) utility assumption. Other work in this direction, now with nonstrict partial preference, is done by Nau [490].

What can be seen as Williams's line has been continued and extended by Walley in his earlier work [672]. Because lower previsions are his workhorse uncertainty model, he focuses mostly on what in this chapter have been called simplified models [672, § 3.7–8], namely almost desirability and simple strict desirability – which he calls strict desirability. However, he also points out the usefulness of nonstrict desirability for conditional lower previsions [672, App F]. In later work [674], Walley advocates strict desirability.

Although Wilson & Moral [705] base a logical view of (imprecise) probability on nonstrict desirability, authors drawing inspiration from Walley's work mostly used almost desirability: Couso et al. [158] discuss independence concepts in imprecise probability and also mention some of their formalizations for almost desirability. De Cooman & Miranda [206] do the same for symmetry concepts. De Cooman [201] places (almost) desirability in a wider context of the order-theoretic structure shared by most belief models.

More recent work, all using strict desirability, has been done by the following authors: Moral [482] studies epistemic irrelevance with an eye on the usefulness of desirability for the generalization of Bayesian networks. For finite possibility spaces, Couso & Moral [156] discuss the relationship with closed convex credal sets, propose representations and algorithms, and introduce the concept of maximal sets of desirable gambles. De Cooman & Quaeghebeur [211, 212, 213] study exchangeability, but also mention maximal sets of desirable gambles for infinite possibility spaces and desirability relative to linear subspaces and with alternative vector inequalities.

The rule for associating a set of nonstrictly desirable gambles with a set of strictly desirable gambles espoused in this chapter is relatively new, and the explicit framing in terms of partial preference orders has not appeared before. Based on a specific definition of exchangeability by Quaeghebeur [519], De Cooman & Quaeghebeur [211, 212, 213] used it to introduce a 'set of weakly desirable gambles' – our associated set of nonstrictly desirable gambles.

Halpern [336] is a good starting reference for lexicographic uncertainty models.

Acknowledgements

This work was supported by a Francqui Fellowship of the Belgian American Educational Foundation. The whole chapter benefited greatly from discussions with Teddy Seidenfeld, who also suggested the lexicographic model example. Most of it was written while the author enjoyed the hospitality of Carnegie Mellon University's Department of Philosophy. Matthias Troffaes suggested the illustration for conglomerability.

2

Lower previsions

Enrique Miranda[1] and Gert de Cooman[2]

[1]*Department of Statistics and Operations Research, University of Oviedo, Spain*
[2]*SYSTeMS Research Group, Ghent University, Belgium*

2.1 Introduction

The material on desirability in the previous chapter constitutes an excellent foundation on which to build a general theory of imprecise probabilities. One of its drawbacks, however, is that it uses a language and mathematical approach that few experts in probability will be familiar with, or even recognise: most of the existing theory of probability rather uses the language of events, (conditional) probabilities and expectations.

The present chapter aims at bridging the gap between the material on desirability and the more traditional language of probability theory, by focusing on *lower* and *upper previsions*, which have already made an appearance in Section 1.6.3, and which are direct generalizations of the probabilities and expectations we encounter in the classical theory. We intend to present the most salient facts about the theory of coherent lower previsions, which falls squarely within the subjective approach to probability. Subjective probabilities can be given a number of different interpretations, but the one we consider here is *behavioural*: as we shall see, a subject's probability for an event reflects his willingness to take certain actions whose outcome depends on its occurrence, such as accepting bets on or against the event at certain betting rates.

In 1975, Williams [703] made a first attempt at a detailed study of imprecise subjective probabilities, based on the work that de Finetti had done on subjective probability [216, 218] and considering lower and upper previsions instead of de Finetti's previsions (or expectations). This was addressed in much more detail by Walley [672], who developed the arguably more mature *behavioural* theory of coherent lower previsions that we will focus on

Introduction to Imprecise Probabilities, First Edition.
Edited by Thomas Augustin, Frank P. A. Coolen, Gert de Cooman and Matthias C. M. Troffaes.

here. We do that mainly for two reasons: (i) from a mathematical point of view, it subsumes most of the other models in the literature as particular cases, and therefore has a unifying character; and (ii) it has a clear behavioural interpretation in terms of acceptable transactions. This chapter deals mainly with the mathematical groundwork for this theory. Many special instances of coherent lower previsions that occur in the literature will be discussed in detail in Chapter 4.

2.2 Coherent lower previsions

We consider a nonempty space $\mathcal{X}$, which could be interpreted as representing the set of possible outcomes for an experiment. As we have seen in the previous chapter, the behavioural theory of imprecise probabilities provides a mathematical framework for representing a subject's beliefs about the actual outcome of the experiment, in terms of accepting transactions (gambles) whose outcome depends on that of the experiment.

Recall that a gamble f on $\mathcal{X}$ is a bounded real-valued map on $\mathcal{X}$. It represents an uncertain reward, whose value is $f(x)$ if the outcome of the experiment is $x \in \mathcal{X}$.[1] This reward is expressed in units of some linear utility scale; see [672, § 2.2] for details.[2]

The theory of lower previsions considers two types of transactions involving a gamble f: accepting to buy f for a price μ, which amounts to accepting the gamble $f - \mu$; and accepting to sell f for a price λ, which comes down to accepting the gamble $\lambda - f$.

A subject's *lower prevision* $\underline{P}(f)$ for a gamble f represents his supremum acceptable buying price for f: it is the highest price s such that the subject accepts to buy f for all prices $\mu < s$, or in other words (see also Equation (1.35)):

$$\underline{P}(f) := \sup\{\mu \in \mathbb{R} : f - \mu \in \mathcal{D}\},$$

where $\mathcal{D}$ is the set of gambles that our subject finds desirable. This implies that he accepts to pay $\underline{P}(f) - \epsilon$ for the uncertain reward f, or in other words, that the transaction $f - \underline{P}(f) + \epsilon$, is desirable to him for every $\epsilon > 0$. Nothing is said about whether he would actually buy f for the price $\underline{P}(f)$.

We can also consider our subject's *upper prevision*, or infimum acceptable selling price, for the gamble f, which we denote by $\overline{P}(f)$: it is the lowest price t such that the subject accepts to buy f for all prices $\lambda > t$:

$$\overline{P}(f) := \inf\{\lambda \in \mathbb{R} : \lambda - f \in \mathcal{D}\}.$$

This implies that the subject accepts to get the price $\overline{P}(f) + \epsilon$ for selling f, for every $\epsilon > 0$, but, again, nothing is said about the price $\overline{P}(f)$ itself.

Example 2.1 Suppose a ball is drawn from an urn with green, red and black balls, and a subject is offered a reward depending on the colour of the ball drawn: he gets 10 euros if

[1] On Walley's approach, gambles are bounded, but the theory has been generalized to deal with unbounded rewards in [629, 632].

[2] The assumption of the linearity of the utility scale may sometimes not be reasonable in practice. We refer to [489, 509] for interesting extensions of the theory that may be useful for dealing with certain types of nonlinear utility.

the ball is green, 5 euros if it is a red ball, and nothing if he draws a black ball.[3] The gamble f representing these rewards is defined by

$$f(green) = 10, \quad f(red) = 5, \quad f(black) = 0.$$

It should be desirable to our subject, because it is never going to diminish his wealth, and can even increase it in some instances. But he may also be willing to pay some fixed amount of money, say x euros, in order to obtain this uncertain reward. In that case, the increase in his wealth would be $f - x$ euros: $10 - x$ euros if he draws a green ball, $5 - x$ euros if he draws a red ball, and $-x$ euros if he draws a black one (meaning that he loses x euros in that case). The supremum amount of money x that he is thus disposed to pay will be his lower prevision $\underline{P}(f)$ for the gamble f. If, for instance, he is completely certain that there are no black balls in the urn, he should be disposed to pay up to 5 euros, because he is certain his wealth is not going to decrease as a result of this. And if he has no additional information at all about the composition of the urn and wants to be cautious, he will not pay more than 5 euros, because it could very well be that all the balls in the urn are red, and in that case by paying more than 5 euros he would end up losing money for sure.

On the other hand, he can also consider selling the gamble f for a fixed price y. His increase in wealth will then be $y - 10$ euros if the ball is green, $y - 5$ euros if it is red, and y euros if it is black. The infimum amount of money y that he requires in order to sell f will be his upper prevision $\overline{P}(f)$ for this gamble. If he knows for sure that there are only red and green balls in the urn, and nothing more, then he should accept to sell f for any price higher than 10 euros, because he is certain not to lose money by this. But if he wants to be cautious he should not sell it for less: for all he knows all the balls might be green and then by selling it for less than 10 euros he would end up losing money for certain. ♦

In the theory of lower previsions, we are not paying attention to our subject's set of desirable gambles $\mathcal{D}$ – or at least not explicitly – but rather try to model our subject's beliefs by looking directly at his lower (and upper) previsions for a number of gambles.

Suppose our subject assesses a lower prevision $\underline{P}(f)$ for all gambles f on some subset $\mathcal{K}$ of the set $\mathcal{L}$ of all gambles on $\mathcal{X}$. This defines a real functional $\underline{P} : \mathcal{K} \to \mathbb{R}$, called a *lower prevision* with domain $\mathcal{K}$.

Since selling a gamble f for a price μ is the same thing as buying the gamble $-f$ for the price $-\mu$, he should disposed to accept these transactions under the same conditions. Hence, his infimum acceptable selling price for f should agree with minus his supremum acceptable buying price for $-f$:

$$\begin{aligned}
\overline{P}(f) &= \inf\{\lambda \in \mathbb{R} : \lambda - f \in \mathcal{D}\} \\
&= \inf\{\lambda \in \mathbb{R} : -f - (-\lambda) \in \mathcal{D}\} \\
&= \inf\{-\mu \in \mathbb{R} : -f - \mu \in \mathcal{D}\} \\
&= -\sup\{\mu \in \mathbb{R} : -f - \mu \in \mathcal{D}\} = -\underline{P}(-f).
\end{aligned}$$

[3] We are assuming, for the sake of simplicity, that these rewards are small compared to our subjects capital, which makes sure that the utility of such monetary rewards is linear to a good approximation.

As a consequence, given a lower prevision $\underline{P}$ on a set of gambles $\mathcal{K}$, we can define the so-called *conjugate* upper prevision $\overline{P}$ on $-\mathcal{K} := \{-f : f \in \mathcal{K}\}$, by $\overline{P}(-f) := -\underline{P}(f)$, and vice versa. Taking this into account, as we can freely go from one concept to the other, it suffices to discuss only one of them. We will concentrate here on lower previsions.

2.2.1 Avoiding sure loss and coherence

A subject's lower prevision $\underline{P}$ represents commitments to accept certain gambles, and is therefore subject to certain requirements of rationality, which can be derived from the rationality requirements for desirability discussed in Section 1.2. We recall from the discussion there that:

I. A transaction that makes our subject lose utiles, no matter the outcome of the experiment, should not be acceptable to him; [this follows from Condition (1.5)].

II. If he considers a transaction acceptable, he should also consider as acceptable any other transaction that gives him a reward that is at least as high, no matter the outcome of the experiment; [this follows from Conditions (1.4) and (1.2)].

III. A positive linear combination of acceptable transactions should also be acceptable; [this follows from Conditions (1.4) and (1.1)].

The first, and most basic requirement that we can derive from these is the following. A lower prevision $\underline{P}$ should *avoid sure loss*, meaning that:

$$\sup_{x\in\mathcal{X}} \sum_{i=1}^{n} [f_i(x) - \underline{P}(f_i)] \geq 0 \text{ for all natural } n \geq 0 \text{ and all } f_1, \dots, f_n \text{ in } \mathcal{K}. \tag{2.1}$$

To motivate this condition, assume that it is not satisfied. Then there are $n > 0, f_1, \dots, f_n$ in $\mathcal{K}$ and $\delta > 0$ such that $\sum_{k=1}^{n} [f_k - (\underline{P}(f_k) - \delta)] \leq -\delta$, meaning that the sum of the desirable transactions $f_k - (\underline{P}(f_k) - \delta)$ results in a loss of at least δ, no matter the outcome of the experiment. Since the sum is desirable by III, this violates I.

Example 2.2 Our subject decides to pay up to 5 euros for the gamble f considered in Example 2.1. Then considers another gamble g, which gives him a reward of 9 euros if he draws a black ball, 5 if he draws a red ball and nothing if he draws a green one. If he decides to pay up to 6 euros to get the reward g, he is paying a total of $11 = 5 + 6$ euros, and total reward he is going to get is $f + g$: 10 euros for a green or a red ball, and 9 euros for a black ball. He is therefore certain to lose at least one euro in specifying the lower previsions $\underline{P}(f) = 5$ and $\underline{P}(g) = 6$: these assessments incur a sure loss. ♦

There is a stronger rationality condition: *coherence*. It requires that our subject's supremum acceptable buying price for a gamble f cannot be raised by considering a positive linear combination of a finite number of other acceptable gambles. Formally:

$$\sup_{x\in\mathcal{X}} \sum_{i=1}^{n} [f_i(x) - \underline{P}(f_i)] - m[f_0(x) - \underline{P}(f_0)] \geq 0$$

$$\text{for all natural } n, m \geq 0 \text{ and all } f_0, f_1, \dots, f_n \text{ in } \mathcal{K}. \tag{2.2}$$

A lower prevision satisfying this condition will in particular avoid sure loss: consider the particular case that $m = 0$. Suppose that Equation (2.2) does not hold for some non-negative integers n, m and some $f_0, f_1, \ldots, f_n$ in $\mathcal{K}$. If $m = 0$, this means that $\underline{P}$ incurs a sure loss, which we have already argued is an inconsistency. Assume therefore that $m > 0$. Then there is some $\delta > 0$ such that

$$\sum_{i=1}^{n}[f_i - (\underline{P}(f_i) - \delta)] \leq m[f_0 - (\underline{P}(f_0) + \delta)].$$

The left-hand side is a sum of desirable transactions, and should therefore be desirable, by III. The dominating right-hand side should therefore be desirable as well, by II. But this means that our subject should accept to buy the gamble f_0 for the price $\underline{P}(f_0) + \delta$, which is strictly higher than the supremum buying price he has specified for it. This is an inconsistency that, admittedly, is not as bad as incurring a sure loss, but should nevertheless be avoided.

Example 2.3 After some reflection, our subject decides to pay up to 5 euros (but not more) for the first gamble f, and up to 4 for the second gamble g. He also decides that he will sell g for any price higher than 6 euros, but not for less. In the language of lower and upper previsions, this means that $\underline{P}(f) = 5$, $\underline{P}(g) = 4$ and $\overline{P}(g) = 6$, or equivalently, $\underline{P}(-g) = -6$. The corresponding changes to his wealth are:

	green	red	black
f	10	5	0
g	0	5	9
buy f for 5	5	0	−5
buy g for 4	−4	1	5
sell g for 6	6	1	−3

These assessments avoid sure loss. However, since our subject accepts to buy f for up to 5 euros, he should also be willing to sell g for any price higher than 5 euros, leading to an increase in his wealth of 5 euros for a green ball, 0 for a red, and −4 for a black ball. This increase is higher than the one associated with buying f for 5 euros, and should therefore also be desirable, by II. This leads to an inconsistency with his assessment of 6 euros as his infimum acceptable selling price for g. ♦

Coherent lower previsions satisfy a number of properties that often prove very useful when reasoning with them. We have gathered them in the following two propositions, which can be inferred directly and quite easily from the definition of coherence.

Proposition 2.1 [672, §§ 2.6.3–2.6.5]. *Let Γ be an arbitrary nonempty index set. For every $\gamma \in \Gamma$, let $\underline{P}_\gamma$ be a coherent lower prevision with domain $\mathcal{K}$.*

(i) *The lower prevision $\underline{Q}$ on $\mathcal{K}$ defined by $\underline{Q}(f) := \inf_{\gamma\in\Gamma}\underline{P}_\gamma(f)$ is coherent.*

(ii) *Let $(\underline{P}_n)_n$ be a sequence of coherent lower previsions from $\{\underline{P}_\gamma : \gamma \in \Gamma\}$ that converges point-wise to another lower prevision $\underline{P}$, meaning that $\underline{P}(f) := \lim_n \underline{P}_n(f)$ for every $f \in \mathcal{K}$. Then $\underline{P}$ is a coherent lower prevision on $\mathcal{K}$.*

(iii) *Given $\underline{P}_1, \underline{P}_2 \in \{\underline{P}_\gamma : \gamma \in \Gamma\}$ and $\alpha \in (0, 1)$, the lower prevision $\alpha\underline{P}_1 + (1 - \alpha)\underline{P}_2$ is coherent.*

Let us define, for any two gambles f and g, their 'meet' $f \wedge g$ as their point-wise minimum and their 'join' $f \vee g$ as their point-wise maximum:

$$(f \wedge g)(x) := \min\{f(x), g(x)\} \text{ and } (f \vee g)(x) := \max\{f(x), g(x)\} \text{ for all } x \in \mathcal{X}.$$

Proposition 2.2 [672, § 2.6.1]. *Let $\underline{P}$ be a coherent lower prevision, and $\overline{P}$ its conjugate upper prevision. Let f and g be gambles, $a \in \mathbb{R}$, λ and κ be real numbers with $\lambda \geq 0$ and $0 \leq \kappa \leq 1$. Let f_n be a sequence of gambles. Then the following statements hold whenever the gambles that appear in the arguments of $\underline{P}$ and $\overline{P}$ belong to their respective domains $\mathcal{K}$ and $-\mathcal{K}$.*

(i) $\inf f \leq \underline{P}(f) \leq \overline{P}(f) \leq \sup f$.

(ii) $\underline{P}(a) = \overline{P}(a) = a$.

(iii) $\underline{P}(f + a) = \underline{P}(f) + a$ *and* $\overline{P}(f + a) = \overline{P}(f) + a$.

(iv) *If* $f \leq g + a$ *then* $\underline{P}(f) \leq \underline{P}(g) + a$ *and* $\overline{P}(f) \leq \overline{P}(g) + a$.

(v) $\underline{P}(f) + \underline{P}(g) \leq \underline{P}(f + g) \leq \underline{P}(f) + \overline{P}(g) \leq \overline{P}(f + g) \leq \overline{P}(f) + \overline{P}(g)$.

(vi) $\underline{P}(\lambda f) = \lambda\underline{P}(f)$ *and* $\overline{P}(\lambda f) = \lambda\overline{P}(f)$.

(vii) $\kappa\underline{P}(f) + (1 - \kappa)\underline{P}(g) \leq \underline{P}(\kappa f + (1 - \kappa)g) \leq \kappa\underline{P}(f) + (1 - \kappa)\overline{P}(g) \leq \overline{P}(\kappa f + (1 - \kappa)g) \leq \kappa\overline{P}(f) + (1 - \kappa)\overline{P}(g)$.

(viii) $\underline{P}(|f|) \geq \underline{P}(f)$ *and* $\overline{P}(|f|) \geq \overline{P}(f)$.

(ix) $|\underline{P}(f)| \leq \overline{P}(|f|)$ *and* $|\overline{P}(f)| \leq \overline{P}(|f|)$.

(x) $|\underline{P}(f) - \underline{P}(g)| \leq \overline{P}(|f - g|)$ *and* $|\overline{P}(f) - \overline{P}(g)| \leq \overline{P}(|f - g|)$.

(xi) $\underline{P}(|f + g|) \leq \underline{P}(|f|) + \overline{P}(|g|)$ *and* $\overline{P}(|f + g|) \leq \overline{P}(|f|) + \overline{P}(|g|)$.

(xii) $\underline{P}(f \vee g) + \underline{P}(f \wedge g) \leq \underline{P}(f) + \overline{P}(g) \leq \overline{P}(f \vee g) + \overline{P}(f \wedge g)$, *and* $\underline{P}(f) + \underline{P}(g) \leq \underline{P}(f \vee g) + \overline{P}(f \wedge g) \leq \overline{P}(f) + \overline{P}(g)$, *and* $\underline{P}(f) + \underline{P}(g) \leq \overline{P}(f \vee g) + \underline{P}(f \wedge g) \leq \overline{P}(f) + \overline{P}(g)$.

(xiii) $\underline{P}$ *and* $\overline{P}$ *are continuous with respect to the topology of uniform convergence on their respective domains: for any* $\epsilon > 0$ *and any* f *and* g *in* $\mathcal{K}$, *if* $\sup|f - g| < \epsilon$, *then* $|\underline{P}(f) - \underline{P}(g)| < \epsilon$.

(xiv) *If* $\lim_{n\to\infty}\overline{P}(|f_n - f|) = 0$ *then* $\lim_{n\to\infty}\underline{P}(f_n) = \underline{P}(f)$ *and* $\lim_{n\to\infty}\overline{P}(f_n) = \overline{P}(f)$.

When the domain $\mathcal{K}$ of the lower prevision assessment $\underline{P}$ is a *linear space*, i.e. closed under taking linear combinations of gambles, coherence takes a much simpler mathematical form. It can be checked that in that case $\underline{P}$ is coherent if and only if the following three requirements are satisfied:

C1. $\underline{P}(f) \geq \inf f$ for all $f \in \mathcal{K}$ [accepting sure gains];

C2. $\underline{P}(f + g) \geq \underline{P}(f) + \underline{P}(g)$ for all $f, g \in \mathcal{K}$ [super-linearity];

C3. $\underline{P}(\lambda f) = \lambda \underline{P}(f)$ for all $f \in \mathcal{K}$ and all real $\lambda > 0$ [positive homogeneity].

In fact, a lower prevision on an arbitrary domain is coherent, or in other words, satisfies Condition (2.2), if and only if it can be extended to a lower prevision on a linear space (such as $\mathcal{L}$) that satisfies C1–C3. We will come back to this in Section 2.2.4.

A coherent lower prevision defined only on indicators I_A of events A is called a *coherent lower probability*, and we use the simplifying notation $\underline{P}(A) := \underline{P}(I_A)$. A subject's lower probability $\underline{P}(A)$ for an event A can also be seen as his supremum rate for betting *on* the event A, where betting on A at rate r means that he gets $1 - r$ if A occurs and $-r$ if it does not. Similarly, his upper probability $\overline{P}(A)$ for an event A can be interpreted as one minus his supremum rate for betting *against* the event A. Here, in contradistinction with the more classical approaches to probability theory, we start from (lower and upper) previsions of gambles and deduce the (lower and upper) probabilities of events as special cases, instead of going from probabilities to previsions using some expectation operator. See also [218, 700] for a similar approach involving precise probabilities.

As Walley himself mentions [672, § 2.11], the notion of coherence could be considered too weak to fully characterize the rationality of probabilistic reasoning. Indeed, other additional requirements could be added whenever they are deemed appropriate. In this sense, most of the specialized notions of upper and lower probabilities considered in the literature, and discussed in detail in Chapter 4, such as 2- and n-monotone capacities, belief functions, possibility measures, can be seen as particular cases of coherent lower previsions (probabilities), that satisfy additional requirements. It is debatable whether these additional requirements should be considered as rationality requirements, or rather as additional mathematical properties that simplify making inferences based on them. Whatever the answer may be, our main point here is that coherence is a necessary condition for rationality, and that we should at least require our subject's assessments to satisfy it. Other possible rationality axioms such as conglomerability, which come to the fore when conditioning is taken into account, will be discussed later in this chapter.

Assessments obtained by specifying the lower and upper previsions for certain gambles, could be called *local assessments*. There are other, so-called *structural assessments*, which instead of, say, assigning values to lower previsions of certain gambles, state that certain relations should exist between the lower previsions of a number of gambles. We may for instance require that these lower previsions satisfy certain symmetry requirements, or that they express irrelevance or independence, or satisfy the Principle of Direct Inference discussed in [672, § 1.7.2]. It will be explained in Chapter 3 how such structural assessments can be combined with local assessments under conditions of coherence.

The assessments expressed by means of a lower prevision can also be given two alternative mathematical representations: using sets of linear previsions, and sets of almost desirable gambles. This is what we turn to in the next two sections.

2.2.2 Linear previsions

One particular case of coherent lower previsions is the *vacuous* lower prevision, which is given by $\underline{P}(f) := \inf f$ for every gamble $f \in \mathcal{L}$. It corresponds to a situation where our subject only considers a gamble desirable when it cannot produce a loss, and thus his supremum acceptable buying price for a gamble is the infimum reward it can give. With a vacuous lower prevision a subject is being as uncommittal as possible, or, in other words, he is being maximally imprecise.

The other side of the spectrum corresponds to the case of maximal precision: when a subject's supremum buying price and the infimum selling price for a gamble f coincide, then the common value $P(f) := \underline{P}(f) = \overline{P}(f)$ is called a *fair price* or *prevision* for f. More generally, then, if a subject specifies a prevision P on a set of gambles $\mathcal{K}$, he is implicitly defining a lower prevision $\underline{P}$ and an upper prevision $\overline{P}$ on this domain $\mathcal{K}$, and stating that $\underline{P}$ and $\overline{P}$ coincide (with P) on $\mathcal{K}$.

Now, let us denote the conjugate upper prevision of $\underline{P}$ by $\overline{Q}$, and the conjugate lower prevision of $\overline{P}$ by $\underline{Q}$:

$$\overline{Q}(f) := -\underline{P}(-f) = -P(-f) \text{ and } \underline{Q}(f) := -\overline{P}(-f) = -P(-f) \text{ for all } f \in -\mathcal{K}.$$

If we consider any gamble f in $-\mathcal{K}$, we see that our subject effectively has a supremum lower price $\underline{Q}(f) := -P(-f)$ and an infimum selling price $\overline{Q}(f) := -P(-f)$ that coincide: the assessment of a fair price for all gambles in $\mathcal{K}$ implies an assessment of a fair price for gambles in $-\mathcal{K}$, so we can more or less automatically extend P to a prevision Q on the negation-invariant domain $\mathcal{K} \cup -\mathcal{K}$ that is self-conjugate in the sense that $Q(f) = -Q(-f)$ for all $f \in \mathcal{K} \cup -\mathcal{K}$. Such a prevision, defined on a negation-invariant domain, is called a *linear* or *coherent prevision* if it is coherent both when interpreted as a lower prevision, and as an upper prevision, on this domain.

More generally, Walley [672, § 2.8.1] calls a real-valued functional P defined on a (not necessarily negation-invariant) set of gambles $\mathcal{K}$ a *linear prevision* if for all natural numbers $m, n \geq 0$ and all gambles $f_1, \dots, f_m, g_1, \dots, g_n$ in the domain $\mathcal{K}$,

$$\sup_{x \in \mathcal{X}} \left[\sum_{i=1}^{m} [f_i(x) - P(f_i)] - \sum_{j=1}^{n} [g_j(x) - P(g_j)] \right] \geq 0. \tag{2.3}$$

This definition has the following justification as a rationality criterion in terms of the behaviour of a subject. Assume that the condition does not hold for certain $m, n \geq 0$ and certain gambles $f_1, \dots, f_m, g_1, \dots, g_n$ in $\mathcal{K}$. Then there is some $\delta > 0$ such that

$$\sum_{i=1}^{m} [f_i - P(f_i) + \delta] + \sum_{j=1}^{n} [P(g_j) - g_j + \delta] \leq -\delta. \tag{2.4}$$

Since $P(f_i)$ can in particular be interpreted as our subject's supremum acceptable buying price for the gamble f_i, he is willing to pay $P(f_i) - \delta$ for it, so the transaction $f_i - P(f_i) + \delta$ is desirable for him. On the other hand, since $P(g_j)$ can in particular be interpreted as his infimum acceptable selling price for g_j, he is willing to sell g_j for the price $P(g_j) + \delta$, so the transaction $P(g_j) - g_j + \delta$ is desirable to him. We then infer from III that the sum on the left-hand side of Equation (2.4) represents a desirable transaction, which produces a loss of at least δ, no matter the outcome of the experiment! This contradicts I.

Not every functional that is coherent both as a lower and as an upper prevision is a linear prevision, if its domain is not negation-invariant. This is because coherence as a lower prevision only guarantees that Equation (2.3) holds for $n \leq 1$, and coherence as an upper prevision only guarantees that the same equation holds for the case where $m \leq 1$. An example showing that these two properties do not imply Equation (2.3) can be found in [460, Example 2].

As is the case with the coherence condition for lower previsions, when the domain $\mathcal{K}$ satisfies some additional conditions, then the condition (2.3) for being a linear prevision can be simplified [672, §§ 2.8.2–2.8.9]. For instance, if the domain $\mathcal{K}$ is *negation-invariant*, meaning that $-\mathcal{K} = \mathcal{K}$, then P is a linear prevision if and only if it avoids sure loss (as a lower prevision) and satisfies the *self-conjugacy* condition $P(f) = -P(-f)$ for all $f \in \mathcal{K}$. In particular, when $\mathcal{K}$ is a linear space of gambles, then P is a linear prevision if and only if

P1. $P(f) \geq \inf f$ for all $f \in \mathcal{K}$;

P2. $P(f + g) = P(f) + P(g)$ for all $f, g \in \mathcal{K}$.

P then automatically also satisfies the homogeneity condition:

P3. $P(\lambda f) = \lambda P(f)$ for all $f \in \mathcal{K}$ and all real λ.

So, when their domain is a linear space, linear previsions are coherent lower previsions that are additive instead of only super-additive.

A linear prevision P on $\mathcal{L}$ can also be characterized as a *linear functional* that is *positive*, meaning that $P(f) \geq 0$ for all gambles $f \geq 0$, and *normalized*, meaning that $P(1) = P(\mathcal{X}) = 1$. A functional on a domain $\mathcal{K}$ is a linear prevision, or in other words satisfies condition (2.3), if and only if it can be extended to a linear prevision on $\mathcal{L}$. We will denote by $\mathbb{P}(\mathcal{X})$ the set of all linear previsions on $\mathcal{L}$.

A linear prevision P whose domain consists of the indicators of the events in some class $\mathcal{F}$ is called an *additive probability* on $\mathcal{F}$. If in particular $\mathcal{F}$ is a field of events, then P is a *finitely additive probability*, or *probability charge* [74], in the usual sense, as Condition (2.3) simplifies to the usual axioms for probability charges:

PC1. $P(A) \geq 0$ for all A in $\mathcal{F}$.

PC2. $P(\mathcal{X}) = 1$.

PC3. $P(A \cup B) = P(A) + P(B)$ whenever $A \cap B = \emptyset$.

Example 2.4 Assume our subject knows that there are only 10 balls in the urn, and that the drawing is fair, so that the probability of each colour is proportional to the number of balls of that colour. If he knows the composition of the urn, for instance that there are 5 green balls, 4 red balls and 1 black ball, then his expected gain for the gamble f is $10 \cdot \frac{5}{10} + 5 \cdot \frac{4}{10} - 0 \cdot \frac{1}{10} = 7$ euros, and this should be his fair price for f. Any linear prevision will be determined by its restriction to events via the expectation operator. This restriction to events corresponds to some particular composition of the urn: if he knows that there are 4 red balls out of 10 in the urn, then his fair betting rate on drawing a red ball (that is, his fair price for a gamble with reward 1 if he draws a red ball and 0 if he does not) should be $\frac{2}{5}$. ♦

Given a linear prevision P on all gambles – an element of $\mathbb{P}(\mathcal{X})$ – we can consider its restriction Q to the set of indicators of events. This restriction can also be seen as a set function defined on the set $\wp(\mathcal{X})$ of all subsets of $\mathcal{X}$, using the identification $Q(A) := P(I_A)$. This set function is a finitely additive probability, or probability charge [74]. Interestingly, P is the unique linear prevision on $\mathcal{L}$ that coincides with Q on events, and its the expectation functional with respect to Q that can be found in general using Dunford integration [74]. Hence, for linear previsions there is no difference in expressive power between representing a subject's beliefs in terms of fair betting rates for events, or in terms of fair prices for gambles: the restriction to events (the probability charge) determines the values on gambles (the linear prevision) uniquely, and vice versa. This is no longer true for lower previsions in general: there usually are infinitely many extensions of a coherent lower probability to a coherent lower prevision on all gambles [672, § 2.7.3], and this is why the theory of coherent lower previsions is formulated using the language of gambles, rather than events. We can characterize the coherence of a lower prevision $\underline{P}$ with domain $\mathcal{K}$ by means of its set of *dominating* linear previsions:

$$\mathcal{M}(\underline{P}) := \{P \in \mathbb{P}(\mathcal{X}) : (\forall f \in \mathcal{K}) P(f) \geq \underline{P}(f)\}. \tag{2.5}$$

$\mathcal{M}(\underline{P})$ is sometimes also called the *credal set* associated with $\underline{P}$.

Proposition 2.3 (Lower envelope theorem, [672, Proposition 3.3.3]). *$\underline{P}$ is coherent if and only if it is the lower envelope of $\mathcal{M}(\underline{P})$, that is, if and only if $\underline{P}(f) = \min\{P(f) : P \in \mathcal{M}(\underline{P})\}$ for all f in $\mathcal{K}$.*

Moreover, the set of dominating linear previsions allows us to establish a one-to-one correspondence between coherent lower previsions $\underline{P}$ and weak*-compact[4] and convex sets of linear previsions:

Proposition 2.4 [672, Proposition 3.6.1]. *Given a coherent lower prevision $\underline{P}$, the set of linear previsions $\mathcal{M}(\underline{P})$ is weak*-compact and convex. Conversely, every weak*-compact and convex set $\mathcal{M}$ of linear previsions determines uniquely a coherent lower prevision $\underline{P}$ by taking lower envelopes.*

Besides, it can be checked that these two operations (taking lower envelopes of compact convex sets of linear previsions and considering the linear previsions that dominate a given coherent lower prevision) commute.

We want to warn the reader at this point that the mathematical equivalence between coherent lower previsions and sets of linear previsions mentioned above does not hold in general for the *conditional* lower previsions that we will introduce in Section 2.3, although there exists an envelope result for the alternative approach by Williams, which we shall come to in Section 2.4.1.

[4] Weak*-compact means compact in the weak* topology, which is the weakest topology that makes all the evaluation functionals continuous [672, Appendix D].

As we have introduced them here, lower previsions have a direct behavioural interpretation in terms of a subject's supremum buying prices for gambles. But the representation of coherent lower previsions in terms of sets of linear previsions allows us to give them a *sensitivity analysis* representation as well: we might assume that the subject ideally has some fair price $P(f)$ for every gamble f on $\mathcal{X}$, leading to a specific linear prevision $P \in \mathbb{P}(\mathcal{X})$. Given that our subject has limited information, time and resources, he can only place P among a set $\mathcal{M}$ of possible candidates. The inferences he can make using this set $\mathcal{M}$ turn out to be equivalent to the ones he can make using the lower envelope $\underline{P}$ of this set – a coherent lower prevision. All the inferences made with (unconditional) coherent lower previsions can also be made with closed convex sets of finitely additive probabilities, or equivalently, with the sets of their associated expectation operators – which are linear previsions. In this sense, there is a link between the theory of coherent lower previsions and robust Bayesian analysis [538]. We shall see in Section 2.3 that, roughly speaking, this connection can only be extended to include conditioning provided we update with finite partitions.

Example 2.5 The lower previsions $\underline{P}(f) = 5$ and $\underline{P}(g) = 4$ that our subject has established in previous examples for the gambles f and g are coherent: this $\underline{P}$ constitutes a coherent lower prevision on the domain $\mathcal{K} = \{f, g\}$. The information embedded in $\underline{P}$ is equivalent to that of its set of dominating linear previsions:

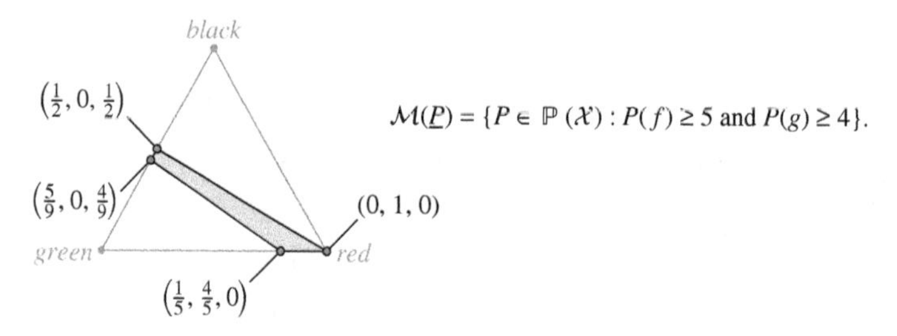

Any linear prevision P is completely determined by its mass function (p_1, p_2, p_3), where $p_1 := P(\{green\})$, $p_2 := P(\{red\})$ and $p_3 := P(\{black\})$. We see that $P \in \mathcal{M}(\underline{P})$ if and only if

$$\begin{cases} 10p_1 + 5p_2 \geq 5 \\ 5p_2 + 9p_3 \geq 4. \end{cases}$$

The convex closed set $\mathcal{M}(\underline{P})$ is completely determined by its set of extreme points

$$\text{ext}(\mathcal{M}(\underline{P})) = \left\{(0,1,0), \left(\frac{1}{5}, \frac{4}{5}, 0\right), \left(\frac{1}{2}, 0, \frac{1}{2}\right), \left(\frac{5}{9}, 0, \frac{4}{9}\right)\right\},$$

and the lower prevision $\underline{P}$ is also the lower envelope of the set of linear previsions associated with these extreme points, which therefore provides the same behavioural information as its convex hull $\mathcal{M}(\underline{P})$. ♦

A coherent lower prevision is determined by the extreme points of its set of dominating linear previsions:

Proposition 2.5 [672, Proposition 3.6.2]. *Let $\underline{P}$ be a coherent lower prevision with domain $\mathcal{K}$, and let $\mathcal{M}(\underline{P})$ be its associated set of dominating linear previsions.*

(i) *The set* $\text{ext}(\mathcal{M}(\underline{P}))$ *of extreme points of* $\mathcal{M}(\underline{P})$ *is nonempty.*

(ii) $\mathcal{M}(\underline{P})$ *is the weak* closure of the convex hull of* $\text{ext}(\mathcal{M}(\underline{P}))$.

(iii) *For every gamble* $f \in \mathcal{K}$ *there is a linear prevision* $P \in \text{ext}(\mathcal{M}(\underline{P}))$ *such that* $P(f) = \underline{P}(f)$, *i.e.,* $\underline{P}$ *is the lower envelope of* $\text{ext}(\mathcal{M}(\underline{P}))$.

2.2.3 Sets of desirable gambles

We have seen above that in order to introduce lower previsions as supremum acceptable buying prices, we need a notion of desirability for gambles: we need to know which gambles of the type $f - \alpha$ are desirable to our subject. This idea allows us to establish a correspondence between coherent lower previsions and the coherent sets of desirable gambles introduced in Chapter 1 (see Sections 1.6.3 and 1.6.5, [470] and [672, §§ 3.7–3.8] for more information).

If we define a lower prevision given a coherent set of desirable gambles $\mathcal{D}$ by letting

$$\underline{P}(f) := \sup\{\alpha : f - \alpha \in \mathcal{D}\} \text{ for all gambles } f \text{ on } \mathcal{X},$$

then $\underline{P}$ is a coherent lower prevision: it satisfies the coherence conditions C1–C3.

Conversely, if a subject specifies or assesses a lower prevision $\underline{P}$ on a set of gambles $\mathcal{K}$, he effectively states that he is willing to buy each gamble f in $\mathcal{K}$ for any price strictly lower than $\underline{P}(f)$, so all gambles $f - \underline{P}(f) + \alpha$ with $\alpha > 0$ are desirable to him. This means that if we define the sets of gambles

$$\mathcal{G}_{\underline{P}} := \{f - \underline{P}(f) : f \in \mathcal{K}\} \text{ and } \mathcal{A}_{\underline{P}} := \{f + \alpha : f \in \mathcal{G}_{\underline{P}}, \alpha > 0\}, \tag{2.6}$$

then specifying the lower prevision $\underline{P}$ effectively corresponds making the assessment that all gambles in $\mathcal{G}_{\underline{P}}$ are marginally desirable and those in $\mathcal{A}_{\underline{P}}$ are desirable (see Section 1.6.4).

If we look closer at Condition (2.1), then we see that it is equivalent to

$$(\forall g \in \text{ posi } \mathcal{G}_{\underline{P}}) \sup g \geq 0,$$

so the lower prevision $\underline{P}$ avoids sure loss if and only if the set of desirable gambles $\mathcal{A}_{\underline{P}}$ avoids partial loss: posi $\mathcal{A}_{\underline{P}} \cap \mathcal{L}^-(\mathcal{X}) = \emptyset$. A closer scrutiny of Condition (2.2) reveals that $\underline{P}$ is coherent if and only if

$$(\forall g \in \text{posi } \mathcal{G}_{\underline{P}})(\forall f \in \mathcal{G}_{\underline{P}}) \sup[g - f] \geq 0.$$

Although there is an infinite number of coherent sets of desirable gambles associated with the same coherent lower prevision, we can make a one-to-one correspondence between coherent lower previsions and the coherent sets of almost desirable gambles introduced in Section 1.6.4: on the one hand, if $\underline{P}$ is a coherent lower prevision on $\mathcal{K}$, the set

$$\{f : \underline{P}(f) \geq 0\} \tag{2.7}$$

is a coherent set of almost desirable gambles; and, conversely, if $\mathcal{D}_{\sqsupseteq}$ is a coherent set of almost desirable gambles, the lower prevision $\underline{P}$ it induces by means of Equation (1.35) is coherent. It can be checked, moreover, that both procedures commute. We will use this one-to-one correspondence in Chapter 3.

We see that we can express coherent assessments in three different ways: using coherent sets of desirable gambles, coherent lower previsions, and convex compact sets of linear previsions. Of these models, the first is the most general and expressive, and the second and the third are mathematically equivalent. Which of them we should use, will depend on the context, and on how much expressive power we are after. A representation in terms of sets of linear previsions may be more useful if we want to give our model a sensitivity analysis interpretation, while the use of sets of desirable gambles may for instance be more interesting in connection with decision making.

2.2.4 Natural extension

We now turn to making inferences based on lower prevision assessments. Assume that our subject has established supremum acceptable buying prices $\underline{P}(f)$ for all gambles f in some domain $\mathcal{K}$. What do these assessments imply about buying prices for other gambles g in $\mathcal{L}$? In other words, given such a gamble g, what is the supremum buying price that he can infer for g, taking into account his assessments $\underline{P}$, using *only* coherence?

The basic argument is the following. Assume that for a given price μ there are gambles $g_1, \dots, g_n$ in $\mathcal{K}$ and non-negative real numbers $\lambda_1, \dots, \lambda_n$, and $\delta > 0$, such that

$$f(x) - \mu \geq \sum_{i=1}^{n} \lambda_i [g_i(x) - \underline{P}(g_i) + \delta] \text{ for all } x \in \mathcal{X}.$$

All the transactions in the sum on the right-hand side are acceptable to our subject by definition, so we infer from III and II that the left-hand side, which dominates their sum, should be acceptable too. Hence, he should be disposed to pay the price μ for the gamble f, and therefore his supremum acceptable buying price for f should be at least this μ. This argument leads to the following definition:

Definition 2.1 The *natural extension* of $\underline{P}$ is the lower prevision defined on all gambles $f \in \mathcal{L}$ by:

$$\underline{E}(f) := \sup_{\substack{g_i \in \mathcal{K}, \lambda_i \geq 0 \\ i=1,\dots,n, n \in \mathbb{N}}} \inf_{x \in \mathcal{X}} \left[f(x) - \sum_{i=1}^{n} \lambda_i [g_i(x) - \underline{P}(g_i)] \right]. \tag{2.8}$$

The reasoning above tells us that our subject should be disposed to pay a price $\underline{E}(f) - \epsilon$ for the gamble f, and this for all $\epsilon > 0$. Hence, his supremum acceptable buying price should dominate $\underline{E}(f)$. But this value is also sufficient to achieve coherence, as we see next:

Proposition 2.6 [672, Theorem 3.1.2]. *Let $\underline{P}$ be a lower prevision with domain $\mathcal{K}$ that avoids sure loss, and let $\underline{E}$ be its natural extension.*

(i) *$\underline{E}$ is the smallest coherent lower prevision on $\mathcal{L}$ that dominates $\underline{P}$ on $\mathcal{K}$.*

(ii) *$\underline{E}$ agrees with $\underline{P}$ on $\mathcal{K}$ if and only if $\underline{P}$ is coherent, and in that case it is the smallest coherent extension of $\underline{P}$ to $\mathcal{L}$.*

Therefore, $\underline{E}(f)$ is the smallest, or most conservative, or least committal, value we can give to the buying price of f in order to achieve coherence with the assessments in $\underline{P}$. There may be other coherent extensions, which may be interesting in some situations. However, any such less conservative coherent extensions will represent stronger commitments than the ones that can be derived purely from $\underline{P}$ and coherence. This is why it makes good sense for the subject to adopt $\underline{E}$ as his inferred model.

Example 2.6 Our subject is offered yet another gamble h, yielding 2 euros if he draws a green ball, 3 if he draws a red ball, and 4 euros if he draws a black ball. Taking into account his previous assessments, $\underline{P}(f) = 5$ and $\underline{P}(g) = 4$, he should at least pay

$$\underline{E}(h) = \sup_{\lambda_1, \lambda_2 \geq 0} \min\{2 - 5\lambda_1 + 4\lambda_2, 3 - \lambda_2, 4 + 5\lambda_1 - 5\lambda_2\}.$$

This supremum is achieved for $\lambda_1 = 0$ and $\lambda_2 = \frac{1}{5}$, leading to $\underline{E}(h) = \frac{14}{5}$ as the supremum acceptable buying price for h that our subject can derive from his previous assessments and coherence. ♦

If the lower prevision $\underline{P}$ does not avoid sure loss, then Equation (2.8) yields $\underline{E}(f) = +\infty$ for all $f \in \mathcal{L}$. The idea is that if our subject's initial assessments are such that he can end up losing utiles no matter the outcome of the experiment, he may also be forced to lose utiles by offering him a suitable combination of gambles together with any other gamble offered to him. Because of this, the first thing we have to verify is whether the initial assessments avoid sure loss, and only then can we consider their consequences on other gambles.

When $\underline{P}$ avoids sure loss, $\underline{E}$ is the smallest coherent lower prevision on all gambles that dominates $\underline{P}$ on $\mathcal{K}$, in the sense that $\underline{E}$ is coherent and any other coherent lower prevision $\underline{Q}$ on $\mathcal{L}$ such that $\underline{Q}(f) \geq \underline{P}(f)$ for all $f \in \mathcal{K}$ will satisfy $\underline{Q}(f) \geq \underline{E}(f)$ for all f in $\mathcal{L}$. $\underline{E}$ is not in general an extension of $\underline{P}$; it will only be so when $\underline{P}$ is coherent itself. Otherwise, the natural extension will correct the assessments present in $\underline{P}$ into the smallest possible coherent lower prevision. Hence, natural extension can be used to modify the initial assessments into other assessments that satisfy coherence, and it does so in a least-committal way, i.e., it provides the smallest coherent lower prevision with the same property.

Example 2.7 Let us again consider the assessments $\underline{P}(f) = 5$, $\underline{P}(g) = 4$ and $\overline{P}(g) = 6$. These imply the acceptable buying transactions in Example 2.3, which, as we showed, avoid sure loss but are incoherent. If we apply Equation (2.8) to them we obtain that their natural extension in these gambles is $\underline{E}(f) = 5$, $\underline{E}(g) = 4$ and $\overline{E}(g) = 5$. Hence, it is a consequence of coherence that our subject should be disposed to sell the gamble g for any amount of money greater than 5 euros. ♦

The natural extension of the assessments given by a coherent lower prevision $\underline{P}$ can also be calculated in terms of the equivalent representations given in Sections 2.2.2 and 2.2.3.

Consider a coherent lower prevision $\underline{P}$ with domain $\mathcal{K}$, and let $\mathcal{A}_{\underline{P}}$ be the set of desirable gambles associated with the lower prevision $\underline{P}$ by Equation (2.6):

$$\mathcal{A}_{\underline{P}} := \{f - \underline{P}(f) + \alpha : f \in \mathcal{K} \text{ and } \alpha > 0\}.$$

The natural extension $\mathcal{E}(\mathcal{A}_{\underline{P}})$ of $\mathcal{A}_{\underline{P}}$ provides the smallest set of desirable gambles that includes $\mathcal{A}_{\underline{P}}$ and is coherent. It is the closure under positive linear combinations of the set

$\mathcal{A}_{\underline{P}} \cup \mathcal{L}^+$: the smallest convex cone that includes $\mathcal{A}_{\underline{P}}$ and all non-negative gambles. The natural extension of $\underline{P}$ to all gambles is then given by

$$\underline{E}(f) = \sup\{\mu : f - \mu \in \mathcal{E}(\mathcal{A}_{\underline{P}})\}.$$

If we consider the set $\mathcal{M}(\underline{P})$ of linear previsions that dominate $\underline{P}$ on $\mathcal{K}$, then also

$$\underline{E}(f) = \min\{P(f) : P \in \mathcal{M}(\underline{P})\}. \tag{2.9}$$

This last expression also makes sense if we consider the sensitivity analysis interpretation given to coherent lower previsions in Section 2.2.2: there is a linear prevision modelling our subject's information, but his imperfect knowledge of it makes him consider a set of linear previsions $\mathcal{M}(\underline{P})$, whose lower envelope is $\underline{P}$. If he wants to extend $\underline{P}$ to a larger domain, he should consider all the linear previsions in $\mathcal{M}(\underline{P})$ as possible models (he has no additional information allowing to disregard any of them), or equivalently their lower envelope. He then obtains that $\mathcal{M}(\underline{P}) = \mathcal{M}(\underline{E})$.

On the other hand, we can also consider the natural extension of a lower prevision $\underline{P}$ from a domain $\mathcal{K}$ to a bigger domain $\mathcal{K}_1$ (not necessarily equal to $\mathcal{L}$). It can be checked then that the procedure of natural extension is *transitive*, or *commutes*, in the following sense: if $\underline{E}_1$ denotes the natural extension of $\underline{P}$ to $\mathcal{K}_1$ and we later consider the natural extension $\underline{E}_2$ of $\underline{E}_1$ to some bigger domain $\mathcal{K}_2 \supset \mathcal{K}_1$, then $\underline{E}_2$ agrees with the natural extension of $\underline{P}$ from $\mathcal{K}$ to $\mathcal{K}_2$: in both cases we are only considering the behavioural consequences of the assessments on $\mathcal{K}$ and the condition of coherence. This is easiest to see using Equation (2.9): we have that $\mathcal{M}(\underline{E}_1) = \mathcal{M}(\underline{E}_2) = \mathcal{M}(\underline{P})$.

2.3 Conditional lower previsions

Let B be a subset of the sampling space $\mathcal{X}$, and consider a gamble f on $\mathcal{X}$. Walley's theory of coherent lower previsions gives two different interpretations of the *conditional lower prevision* $\underline{P}(f|B)$ of f given B: the *updating* and the *contingent* one. Under the contingent interpretation, $\underline{P}(f|B)$ is our subject's current supremum buying price for the gamble f contingent on B, that is, the supremum value of μ such that the gamble $I_B(f - \mu)$ is desirable for our subject.

In order to relate the subject's current dispositions on a gamble f contingent on B with his current dispositions towards this gamble if he came to know that the outcome of the experiment belongs to B, Walley introduces the so-called *Updating Principle*. We say a gamble f is B-desirable for our subject when he is currently disposed to accept f if he observed that the outcome belongs to B. Then the Updating Principle requires that a gamble is B-desirable if and only if $I_B f$ is desirable.

Under the updating interpretation of conditional lower previsions, $\underline{P}(f|B)$ is the subject's supremum acceptable buying price he would pay for the gamble f now if he came to know that the outcome belonged to the set B, and nothing more. It coincides with the value determined by the contingent interpretation of $\underline{P}(f|B)$ because of the Updating Principle.

It should be stressed at this point that the Updating Principle does not allow us to give a dynamic interpretation to conditional lower previsions: $\underline{P}(f|B)$ is not to be interpreted as the supremum acceptable buying price for the gamble f *after* observing B.[5]

[5] We refer to [722] for a dynamic study of coherent lower previsions and a discussion of the related literature.

Let $\mathcal{B}$ be a partition of our sampling space $\mathcal{X}$, and consider an element B of this partition. This partition could be for instance a class of categories of the set of outcomes. Assume that our subject has given conditional assessments $\underline{P}(f|B)$ for all gambles f on some domain $\mathcal{K}_B$, which we will assume includes all constant gambles. As was the case for (unconditional) lower previsions, we should require that these assessments are consistent with each other.

Definition 2.2 We say that the conditional lower prevision $\underline{P}(\cdot|B)$ is *separately coherent* when the following two conditions are satisfied:

(i) it is coherent as an unconditional prevision, i.e.,

$$\sup_{x\in\mathcal{X}} \sum_{i=1}^{n} \left[f_i(x) - \underline{P}(f_i|B)\right] - m\left[f_0(x) - \underline{P}(f_0|B)\right] \geq 0 \tag{2.10}$$

for all non-negative integers n, m and all gambles $f_0, \ldots ,f_n$ in $\mathcal{K}_B$;

(ii) the indicator function of B belongs to $\mathcal{K}_B$ and $\underline{P}(B|B) = 1$.

The coherence requirement (2.10) can be given a behavioural interpretation in the same way as we did with (unconditional) coherence in Equation (2.2): if it does not hold for some non-negative integers n, m and gambles $f_0, \ldots ,f_n$ in $\mathcal{K}_B$, then it can be checked that either: (i) there is some $\delta > 0$ such that the desirable gamble $\sum_{i=1}^{n}[f_i - \underline{P}(f_i|B)] + \delta$ incurs in a sure loss (if $m = 0$) or (ii) we can raise $\underline{P}(f_0|B)$ in some positive quantity δ, contradicting our interpretation of it as his supremum acceptable buying price (if $m > 0$).

On the other hand, Equation (2.10) already implies that $\underline{P}(B|B)$ should be smaller than, or equal to, 1. That we require it to be equal to one means that our subject should currently be disposed to bet at all odds on the occurrence of the event B if he observed it.

In this way, we can obtain separately coherent conditional lower previsions $\underline{P}(\cdot|B)$ with domains $\mathcal{K}_B$ for all events B in the partition $\mathcal{B}$. It is a consequence of separate coherence that the conditional lower prevision $\underline{P}(f|B)$ only depends on the values that f takes on B, i.e, for every two gambles f and g such that $I_B f = I_B g$, we have $\underline{P}(f|B) = \underline{P}(g|B)$. Taking into account this property, we will assume that the domain of $\underline{P}(\cdot|B)$ is the set of gambles

$$\mathcal{H}_B := \{I_B f : f \in \mathcal{K}_B\}$$

rather than $\mathcal{K}_B$. Since we assumed before that $\mathcal{K}_B$ contains all constant gambles, $\mathcal{H}_B$ contains the indicator function of the set B and the zero gamble. From this we deduce that all the domains $\mathcal{H}_B$, $B \in \mathcal{B}$, can be extended to the common domain

$$\mathcal{H} := \left\{ \sum_{B\in\mathcal{B}} f_B : (\forall B \in \mathcal{B})\, f_B \in \mathcal{H}_B \right\}, \tag{2.11}$$

and we can define on $\mathcal{H}$ a *conditional lower prevision* $\underline{P}(\cdot|\mathcal{B})$ by

$$\underline{P}(f|\mathcal{B}) := \sum_{B\in\mathcal{B}} I_B \underline{P}(f|B),$$

i.e., $\underline{P}(f|\mathcal{B})$ is the gamble on $\mathcal{X}$ that assumes the value $\underline{P}(f|B)$ on all elements of B. This conditional lower prevision is then called *separately coherent* when $\underline{P}(\cdot|B)$ is separately coherent for all $B \in \mathcal{B}$. It provides the updated supremum buying price after learning that the outcome of the experiment belongs to some particular element of $\mathcal{B}$. We will also use the notations

$$G(f|B) := I_B(f - \underline{P}(f|B)) \text{ and } G(f|\mathcal{B}) := \sum_{B \in \mathcal{B}} G(f|B) = f - \underline{P}(f|\mathcal{B}), \quad (2.12)$$

and we shall see that these are *marginally desirable* gambles in the terminology used in Section 1.6.4.

Proposition 2.7 [672, § 6.2.7]. *Consider a conditional lower prevision $\underline{P}(\cdot|\mathcal{B})$ defined on a domain $\mathcal{H}$ given by Equation (2.11) that is moreover a linear space. Then $\underline{P}(\cdot|\mathcal{B})$ is separately coherent if and only if the following conditions hold for all positive real λ and all gambles f, g in $\mathcal{H}$:*

SC1. $\underline{P}(f|\mathcal{B}) \geq \inf(f|\mathcal{B})$, *where the gamble* $\inf(f|\mathcal{B})$ *is given by* $\inf(f|\mathcal{B})(x) := \inf_{y \in B} f(y)$ *for every $x \in B$ and every $B \in \mathcal{B}$;*

SC2. $\underline{P}(f + g|\mathcal{B}) \geq \underline{P}(f|\mathcal{B}) + \underline{P}(g|\mathcal{B})$;

SC3. $\underline{P}(\lambda f|\mathcal{B}) = \lambda \underline{P}(f|\mathcal{B})$.

The first requirement shows that the conditional lower prevision on B should only depend on the behaviour of f on this set; conditions SC2 and SC3 are the counterparts of the requirements C2 and C3 we have imposed on unconditional lower previsions, respectively.

Example 2.8 Consider the gambles f and g whose reward in terms of the colour of the ball drawn is given by:

	green	red	black
f	10	5	0
g	0	5	9

Before, our subject had made the coherent assessments $\underline{P}(f) = 5$ and $\underline{P}(g) = 4$. But now he may also establish his supremum acceptable buying prices for these gambles depending on some further information on the colour of the ball drawn. If for instance he is informed that the ball drawn is not green, he should update his lower prevision for the gamble g, because he is sure that in that case he would get at least a prize of 5 euros out of it. On the other hand, if he keeps the supremum buying prize of 5 euros for f he is implying that he is sure that the ball that has been drawn is red once he comes to know that it is not green.

If, for instance, he considers as possible models the ones in Example 2.5 and updates them using Bayes' Rule, so that given any such model P,

$$P(f|\text{not green}) = \frac{f(\text{red})P(\text{red}) + f(\text{black})P(\text{black})}{P(\text{red}) + P(\text{black})},$$

then the updated supremum buying prices he should give by taking lower envelopes would be $\underline{P}(f|\text{not green}) = 0$ and $\underline{P}(g|\text{not green}) = 5$. ♦

It is often interesting and useful to consider lower previsions that are conditional on the values of some variables. Consider for instance variables $X_1, \dots, X_n$ taking values in respective spaces $\mathcal{X}_1, \dots, \mathcal{X}_n$. Then given a set of indices $I \subseteq \{1, \dots, n\}$, we can also consider the joint random variable $X_I = (X_i)_{i \in I}$ that takes values in the product space $\mathcal{X}_I := \times_{i \in I} \mathcal{X}_i$. If we now let I and O be disjoint subsets of $\{1, \dots, n\}$, we may want to define a conditional lower prevision $\underline{P}_O(\cdot|X_I)$ that represents the behavioural assessments for gambles that depend on the value that X_O takes, provided we know the value taken by X_I. Specifically, the domain of $\underline{P}_O(\cdot|X_I)$ would be those gambles on $\mathcal{X}_{\{1,\dots,n\}}$ whose value depend on the components in $I \cup O$ only.

To see that this idea can be embedded in the previous formulation, consider the partition of $\mathcal{X}_{\{1,\dots,n\}}$ given by $\mathcal{B} := \{\pi_O^{-1}(x) : x \in \mathcal{X}_O\}$, where π_O denotes the projection operator and where as a consequence $\pi_O^{-1}(x)$ is the set of elements from $\mathcal{X}_{\{1,\dots,n\}}$ whose components from O coincide with those of x. Then the functional $\underline{P}_O(\cdot|X_I)$ would be a conditional lower prevision $\underline{P}(\cdot|\mathcal{B})$ with the domain described above.

As a consequence, the theory of conditional lower previsions we are going to outline next can be applied in particular to model the conditional assessments for a number of variables; see [207, 468, 469] for some papers using this approach.

2.3.1 Coherence of a finite number of conditional lower previsions

In practice, it is not uncommon to have lower previsions conditional on different partitions $\mathcal{B}_1, \dots, \mathcal{B}_n$ of $\mathcal{X}$. We can think for instance of different sets of relevant categories, or of information provided in a sequential way. We end up, then, with a finite number of separately coherent conditional lower previsions $\underline{P}_1(\cdot|\mathcal{B}_1), \dots, \underline{P}_n(\cdot|\mathcal{B}_n)$ with respective domains $\mathcal{H}_1, \dots, \mathcal{H}_n$, which are all subsets of $\mathcal{L}$. We assume that these conditional lower previsions are separately coherent, which allows us to define all the $\underline{P}_j(\cdot|B_j)$ on the same domain $\mathcal{H}_j$ for all $B_j \in \mathcal{B}_j$, and this for $j = 1, \dots, m$, as explained in the previous section.

As with unconditional lower previsions, before making any inferences based on these assessments, we have to verify that they are consistent with each other. And again, by 'consistent' we will mean that a combination of acceptable buying prices should neither lead to a sure loss, nor to an increase of the supposedly supremum acceptable buying price for a gamble f.

To see which form coherence takes now, we need to erect the second pillar of Walley's theory of conditional previsions (the first is the Updating Principle): the *Conglomerative Principle*. This rationality principle requires that if a gamble is B-desirable for every set B in a partition $\mathcal{B}$ of $\mathcal{X}$, then f is also desirable. Taking into account the Updating Principle, this means that if $I_B f$ is desirable for every B in $\mathcal{B}$, then f should be desirable.

It follows from this principle that for every gamble f in the domain of $\underline{P}(\cdot|\mathcal{B})$, the gamble $G(f|\mathcal{B})$ given by Equation (2.12) should be marginally desirable. This is the basis for the following definition of coherence. For simplicity, we will assume that the domains $\mathcal{H}_1, \dots, \mathcal{H}_n$ are linear spaces of gambles, which allows us to apply Proposition 2.7. A possible generalization to nonlinear domains can be found in [466]. Given $f_j \in \mathcal{H}_j$, we denote by $S_j(f_j) := \{B \in \mathcal{B}_j : I_B f_j \neq 0\}$ the *$\mathcal{B}_j$-support* of f_j. It is the set of elements of $\mathcal{B}_j$ where f_j

is not identically zero. It follows from the separate coherence of $\underline{P}_j(\cdot|\mathcal{B}_j)$ that $\underline{P}_j(0|\mathcal{B}_j) = 0$ for all $j = 1, \dots, n$, and as a consequence the gamble $\underline{P}_j(f|\mathcal{B}_j)$ (as well as $G_j(f|\mathcal{B}_j)$, for that matter) is identically zero outside $S_j(f_j)$.

Definition 2.3 We say that $\underline{P}_1(\cdot|\mathcal{B}_1), \dots, \underline{P}_n(\cdot|\mathcal{B}_n)$ are (jointly) *coherent* when for all $f_j \in \mathcal{H}_j, j = 1, \dots, n$, all $j_0 \in \{1, \dots, n\}$, all $f_0 \in \mathcal{H}_{j_0}$ and all $B_0 \in \mathcal{B}_{j_0}$:

$$\sup_{x \in B} \left[\sum_{j=1}^{n} G_j(f_j|\mathcal{B}_j) - G_{j_0}(f_0|B_0) \right](x) \geq 0 \text{ for some } B \in \{B_0\} \cup \bigcup_{j=1}^{n} S_j(f_j). \tag{2.13}$$

Assume that condition (2.13) does not hold for some $f_j \in \mathcal{H}_j$, $j = 1, \dots, n$, some $j_0 \in \{1, \dots, n\}$, some $f_0 \in \mathcal{H}_{j_0}$ and some $B_0 \in \mathcal{B}_{j_0}$. For any $B \in \{B_0\} \cup \bigcup_{j=1}^{n} S_j(f_j)$, let $\epsilon_B := \sup_{x \in B}[\sum_{j=1}^{n} G_j(f_j|\mathcal{B}_j) - G_{j_0}(f_0|B_0)](x) < 0$ and $\delta_B := -\frac{\epsilon_B}{n+1}$. Then it holds that

$$\left[\sum_{j=1}^{n} G_j(f_j|\mathcal{B}_j) - G_{j_0}(f_0|B_0) + \sum_{j=1}^{n} \sum_{B_j \in \mathcal{B}_j} I_{B_j}\delta_{B_j} + I_{B_0}\delta_{B_0} \right] \leq 0,$$

so $G_{j_0}(f_0|B_0) - I_{B_0}\delta_{B_0}$ dominates the desirable gamble $\sum_{j=1}^{n}[G_j(f_j|\mathcal{B}_j) + \sum_{B_j \in \mathcal{B}_j} I_{B_j}\delta_{B_j}]$. This means, by II, that $\underline{P}_{j_0}(f_0|B_0) + \delta_{B_0}$ should be an acceptable buying price for f_0, contingent on B_0. This is inconsistent with the behavioural interpretation of $\underline{P}_{j_0}(f_0|B_0)$ as a supremum acceptable buying price (if we accept the Updating Principle).

There are two other consistency conditions that are weaker than joint coherence, but which possess interesting mathematical properties. The first one is called *avoiding partial loss*:

Definition 2.4 Consider separately coherent $\underline{P}_1(\cdot|\mathcal{B}_1), \dots, \underline{P}_n(\cdot|\mathcal{B}_n)$ with linear domains $\mathcal{H}_1, \dots, \mathcal{H}_n$. We say that they *avoid partial loss* when for all $f_j \in \mathcal{H}_j$, $j = 1, \dots, n$, there some event $B \in \bigcup_{j=1}^{n} S_j(f_j)$ such that

$$\sup_{x \in B} \left[\sum_{j=1}^{n} G_j(f_j|\mathcal{B}_j) \right](x) \geq 0. \tag{2.14}$$

The idea behind this notion is that a combination of transactions that are acceptable for our subject should not make him lose utiles. It is related to rationality requirement I: a gamble $f \leq 0$ such that $f(x) < 0$ on some $x \in \mathcal{X}$ should not be desirable, which is the notion called avoiding partial loss in Section 1.2.2.

Definition 2.5 Consider separately coherent $\underline{P}_1(\cdot|\mathcal{B}_1), \dots, \underline{P}_n(\cdot|\mathcal{B}_n)$ with respective linear domains $\mathcal{H}_1, \dots, \mathcal{H}_n$. We say that they are *weakly coherent* when for all $f_j \in \mathcal{H}_j$, $j = 1, \dots, n$, all $j_0 \in \{1, \dots, n\}$, all $f_0 \in \mathcal{H}_{j_0}$ and all $B_0 \in \mathcal{B}_{j_0}$:

$$\sup_{x \in \mathcal{X}} \left[\sum_{j=1}^{n} G_j(f_j|\mathcal{B}_j) - G_{j_0}(f_0|B_0) \right](x) \geq 0.$$

With this condition we require that our subject should not be able to raise his supremum acceptable buying price $\underline{P}_{j_0}(f_0|B_0)$ for a gamble f_0 contingent on B_0 by taking into account other conditional assessments. However, under the behavioural interpretation, a number of weakly coherent conditional lower previsions can still present some forms of inconsistency with each other, as [672, Example 7.3.5] shows. This is why for the stronger coherence requirement we need to further restrict the supremum in the equation above, and impose condition (2.13) instead.

One of the most interesting properties of weakly coherent models is the following:

Theorem 2.1 ([468, Theorem 1]). *$\underline{P}_1(\cdot|\mathcal{B}_1), \ldots, \underline{P}_n(\cdot|\mathcal{B}_n)$ are weakly coherent if and only if there is a lower prevision $\underline{P}$ on $\mathcal{L}$ that is pairwise coherent with each conditional lower prevision $\underline{P}_j(\cdot|\mathcal{B}_j)$.*

This theorem, together with [672, Theorem 8.1.8], shows one of the differences between Definitions 2.3 and 2.5: weak coherence is equivalent to the existence of a joint that is coherent with each of the assessments, considered separately.[6] Coherence, on the other hand, is equivalent to the existence of a joint that is coherent with all the assessments, taken together. It is also interesting that we can relate both properties, by means of the following result:

Theorem 2.2 ([471, Theorem 7]). *Let $\mathcal{B}_1, \ldots, \mathcal{B}_n$ be partitions of $\mathcal{X}$, and consider weakly coherent $\underline{P}_1(\cdot|\mathcal{B}_1), \ldots, \underline{P}_n(\cdot|\mathcal{B}_n)$ with respective linear domains $\mathcal{H}_1, \ldots, \mathcal{H}_n$. Let $\underline{P}$ be a coherent prevision such that $\underline{P}$ and $\underline{P}_j(\cdot|\mathcal{B}_j)$ are pairwise coherent for $j = 1, \ldots, n$. If $\underline{P}(B_j) > 0$ for all $B_j \in \mathcal{B}_j$ and all $j = 1, \ldots, n$, then $\underline{P}_1(\cdot|\mathcal{B}_1), \ldots, \underline{P}_n(\cdot|\mathcal{B}_n)$ are coherent.*

This theorem shows that the problem of conditioning on sets of zero lower probability is a relevant issue within the theory of coherent lower previsions. We should mention here an interesting approach to conditioning on sets of probability zero in a coherent setting: the *zero-layers* of Coletti and Scozzafava [119, § 12], which also appear in earlier work by Krauss [409]. Their approach to conditioning is nevertheless slightly different from Walley's, since they consider conditional previsions as previsions whose domain is a class of conditional events. See [117, 118, 119] for further information on Coletti and Scozzafava's work, and [672, § 6.10] and [679, § 1.4] for further details on Walley's approach to conditioning on events of lower probability zero. We will also give some additional comments at the end of this section, when we discuss the issue of conglomerability.

2.3.2 Natural extension of conditional lower previsions

Assume that our subject has provided a finite number of (separately and jointly) coherent lower previsions $\underline{P}_1(\cdot|\mathcal{B}_1), \ldots, \underline{P}_n(\cdot|\mathcal{B}_n)$ on respective linear subspaces $\mathcal{H}_1, \ldots, \mathcal{H}_n$ of $\mathcal{L}$. Then he may wish to see which are the behavioural implications of these assessments on gambles that are not in the domain. The way to do this is through *natural extension*. Given

[6] The coherence of an unconditional lower prevision $\underline{P}$ and a conditional lower prevision $\underline{P}(\cdot|\mathcal{B})$ will be studied in more detail in Section 2.3.3.

$f \in \mathcal{L}$ and $B_0 \in \mathcal{B}_{j_0}$, $\underline{E}_{j_0}(f|B_0)$ is defined as the supremum value of μ for which there are $f_j \in \mathcal{H}_j$ such that

$$\sup_{x \in B} \left[\sum_{j=1}^{n} G_j(f_j|\mathcal{B}_j) - I_{B_0}(f - \mu) \right](x) < 0 \text{ for all } B \text{ in the set } \{B_0\} \cup \bigcup_{j=1}^{n} S_j(f_j). \tag{2.15}$$

In the particular case where we have only an unconditional lower prevision, i.e., when $n = 1$ and $\mathcal{B}_1 := \{\mathcal{X}\}$, this notion coincides with the unconditional natural extension we discussed in Section 2.2.4. Given a number of conditional lower previsions $\underline{P}_1(\cdot|\mathcal{B}_1), \ldots, \underline{P}_n(\cdot|\mathcal{B}_n)$, we can calculate their natural extensions $\underline{E}_1(\cdot|\mathcal{B}_1), \ldots, \underline{E}_n(\cdot|\mathcal{B}_n)$ to all gambles using Equation (2.15). If the partitions $\mathcal{B}_1, \ldots, \mathcal{B}_n$ are finite, then these natural extensions share some of the properties of the unconditional natural extension:

(i) They coincide with $\underline{P}_1(\cdot|\mathcal{B}_1), \ldots, \underline{P}_n(\cdot|\mathcal{B}_n)$ if and only if these conditional lower previsions are coherent.

(ii) They are the smallest coherent extensions of $\underline{P}_1(\cdot|\mathcal{B}_1), \ldots, \underline{P}_n(\cdot|\mathcal{B}_n)$ to all gambles.

(iii) They are the lower envelope of a family of coherent conditional *linear* previsions, $\{P_1^\gamma(\cdot|\mathcal{B}_1), \ldots, P_n^\gamma(\cdot|\mathcal{B}_n) : \gamma \in \Gamma\}$.

However, there are also interesting situations, such as in parametric inference, where we must deal with infinite spaces and where we end up with partitions that have an infinite number of different elements. In that case, it is easy to see that in order to achieve coherence, $\underline{P}_{j_0}(f|B_0)$ must be at least as large as the supremum μ that satisfies the condition (2.15) for some $f_j \in \mathcal{H}_j$, $j = 1, \ldots, n$. However, in contrast with the case of finite partitions, we cannot guarantee that these values provide coherent extensions of $\underline{P}_1(\cdot|\mathcal{B}_1), \ldots, \underline{P}_n(\cdot|\mathcal{B}_n)$ to all gambles: in general these will only be lower bounds of all the coherent extensions. Indeed, when the partitions are infinite, we can have a number of problems:

(i) There may be no coherent extensions, and as a consequence the natural extensions may not be coherent [672, §§ 6.6.6–6.6.7].

(ii) Even if the smallest coherent extensions exist, they may differ from the natural extensions, which are not coherent [672, § 8.1.3].

(iii) The smallest coherent extensions, and as a consequence also the natural extensions, may not be lower envelopes of coherent linear collections [672, §§ 6.6.9–6.6.10].

The natural extensions are the smallest coherent extensions of the lower previsions $\underline{P}_1(\cdot|\mathcal{B}_1), \ldots, \underline{P}_n(\cdot|\mathcal{B}_n)$ if and only if they are jointly coherent themselves. But we need some additional sufficient conditions to guarantee the coherence of $\underline{E}_1(\cdot|\mathcal{B}_1), \ldots, \underline{E}_n(\cdot|\mathcal{B}_n)$. One of these is that all the partitions $\mathcal{B}_i$ are finite. But even when the partitions are infinite it may happen that we are able to characterize the smallest coherent extensions, but that they differ from the natural extensions. One of the reasons for this 'defective' behaviour of the natural extension in the conditional case lies in the requirement of conglomerability that we will discuss in the next section, and which becomes trivial when the partitions are finite.

One interesting particular case where we can determine the smallest coherent extensions of a number of conditional lower previsions $\underline{P}_1(\cdot|\mathcal{B}_1), \ldots, \underline{P}_n(\cdot|\mathcal{B}_n)$ is when the partitions $\mathcal{B}_1, \ldots, \mathcal{B}_n$ are nested and the domain $\mathcal{H}_j$ of $\underline{P}_j(\cdot|\mathcal{B}_j)$ is given by gambles which are constant on the elements of $\mathcal{B}_{j+1}$, for $j = 1, \ldots, n-1$; this is for instance the case when we want to condition on increasing sequences of random variables, as in the laws of large numbers in [207]. Then the smallest coherent extensions to $\mathcal{L}$ are given by the recursion equations:

$$\begin{cases} \underline{M}_n(\cdot|\mathcal{B}_n) := \underline{E}_n(\cdot|\mathcal{B}_n) \\ \underline{M}_j(\cdot|\mathcal{B}_j) := \underline{E}_j(\underline{M}_{j+1}(\cdot|\mathcal{B}_{j+1})|\mathcal{B}_j), \quad j = 1, 2, \ldots, n-1. \end{cases} \tag{2.16}$$

where $\underline{E}_n(\cdot|\mathcal{B}_n)$ is the unconditional natural extension[7] of $\underline{P}_n(\cdot|\mathcal{B}_n)$ to all gambles and $\underline{E}_j(\cdot|\mathcal{B}_j)$ is the unconditional natural extension of $\underline{P}_j(\cdot|\mathcal{B}_j)$ to the set of all gambles that are constant on the elements of $\mathcal{B}_{j+1}$, for $j = 1, \ldots, n-1$. The conditional lower previsions $\underline{M}_1(\cdot|\mathcal{B}_1), \ldots, \underline{M}_n(\cdot|\mathcal{B}_n)$ are called the *marginal extensions* of $\underline{P}_1(\cdot|\mathcal{B}_1), \ldots, \underline{P}_n(\cdot|\mathcal{B}_n)$. See [672, § 6.7.2] and [467] for more details.

2.3.3 Coherence of an unconditional and a conditional lower prevision

Let us consider in more detail the case of a separately coherent conditional lower prevision $\underline{P}(\cdot|\mathcal{B})$ on some domain $\mathcal{H}$, built as in Equation (2.11), and an unconditional coherent lower prevision $\underline{P}$ on some set of gambles $\mathcal{K}$. Assume moreover that these two sets are linear subspaces of $\mathcal{L}$. This assumption is made for mathematical convenience only, because the results can be extended to nonlinear domains [466]. The unconditional lower prevision can be seen as a conditional lower prevision by simply considering the partition $\{\mathcal{X}\}$; then the gamble $G(f|\mathcal{B})$ in Equation (2.12) becomes $G(f) = f - \underline{P}(f)$.

The lower previsions $\underline{P}$ and $\underline{P}(\cdot|\mathcal{B})$ are (jointly) *coherent*, i.e., they satisfy Equation (2.13), if and only if

JC1. $\sup_{x\in\mathcal{X}}[G(f_1) + G(g_1|\mathcal{B}) - G(f_2)] \geq 0$ and

JC2. $\sup_{x\in\mathcal{X}}[G(f_1) + G(g_1|\mathcal{B}) - G(g_2|B_0)] \geq 0$

for all $f_1, f_2 \in \mathcal{K}$, $g_1, g_2 \in \mathcal{H}$ and $B_0 \in \mathcal{B}$. These conditions can be further simplified under some additional assumptions on the domains (see [672, § 6.5] for details).

Again, it can be checked that if any of these conditions fails, the assessments of our subject produce inconsistencies. Assume first that JC1 does not hold. If $f_2 = 0$, then we have a sum of acceptable transactions that produces a sure loss. If $f_2 \neq 0$, then there is some $\delta > 0$ such that the gamble $G(f_2) - \delta$ dominates the desirable gamble $G(f_1) + G(g_1|\mathcal{B}) + \delta$. This means that our subject can be effectively forced to increase his supremum acceptable buying price for f_2 by δ, an inconsistency.

Similarly, if JC2 does not hold and $g_2 = 0$ we have a sum of acceptable transactions that produces a sure loss; and if $g_2 \neq 0$ there is some $\delta > 0$ such that $G(g_2|B_0) - \delta$ dominates $G(f_1) + G(g_1|\mathcal{B}) + \delta$ and is therefore desirable. Hence, our subject should be willing to pay $\underline{P}(g_2|B_0) + \delta$ for g_2 contingent on B_0, an inconsistency.

It is a consequence of the joint coherence of $\underline{P}$ and $\underline{P}(\cdot|\mathcal{B})$ that, given $f \in \mathcal{H}$ and $B \in \mathcal{B}$, $\underline{P}(G(f|B)) = \underline{P}(I_B(f - \underline{P}(f|B))) = 0$, where $G(f|B)$ is defined in Equation (2.12). When $\underline{P}(B) > 0$, there is a unique value μ such that $\underline{P}(I_B(f - \mu)) = 0$, and therefore this μ must

[7] This means that for each $B_n \in \mathcal{B}_n$, $\underline{E}_n(\cdot|B_n)$ is the unconditional natural extension of the lower prevision $\underline{P}_n(\cdot|B_n)$.

be the conditional lower prevision $\underline{P}(f|B)$. This is called the *generalized Bayes rule* (GBR). This rule has a number of interesting properties:

(i) it is a generalization of Bayes's Rule from classical probability theory;

(ii) if $\underline{P}(B) > 0$ and we define $\underline{P}(f|B)$ via the generalized Bayes rule, then it is the lower envelope of the conditional linear previsions $P(f|B)$ that we can define by applying Bayes's Rule to the elements of $\mathcal{M}(\underline{P})$;

(iii) when the partition $\mathcal{B}$ is finite and $\underline{P}(B) > 0$ for all $B \in \mathcal{B}$, then the GBR uniquely determines the conditional lower prevision $\underline{P}(\cdot|\mathcal{B})$.

Example 2.9 Three horses (a, b and c) take part in a race. Our subject's prior lower probability for each horse being the winner is determined by $\underline{P}(\{a\}) = \frac{1}{10}$, $\underline{P}(\{b\}) = \frac{1}{4}$, $\underline{P}(\{c\}) = \frac{3}{10}$, $\underline{P}(\{a,b\}) = \frac{2}{5}$, $\underline{P}(\{a,c\}) = \frac{3}{5}$ and $\underline{P}(\{b,c\}) = \frac{7}{10}$. Since there are rumours that c is not going to take part in the race due to some injury, he may provide his conditional lower probabilities for that case using the generalized Bayes rule. Taking into account that we are dealing with finite spaces and that the conditioning event has positive lower probability, property (ii) above implies that applying the generalized Bayes rule is equivalent to taking the lower envelope of the linear conditional previsions that we obtain by applying Bayes's Rule to the elements of $\mathcal{M}(\underline{P})$. Thus, we obtain:

$$\underline{P}(\{a\}|\{a,b\}) = \inf\left\{\frac{P(\{a\})}{P(\{a,b\})} : P \in \mathcal{M}(\underline{P})\right\} = \frac{\frac{1}{10}}{\frac{1}{2}} = \frac{1}{5}$$

$$\underline{P}(\{b\}|\{a,b\}) = \inf\left\{\frac{P(\{b\})}{P(\{a,b\})} : P \in \mathcal{M}(\underline{P})\right\} = \frac{\frac{1}{4}}{\frac{11}{20}} = \frac{5}{11}. \qquad \blacklozenge$$

The conditional lower prevision $\underline{P}(f|B)$ is uniquely determined by the generalized Bayes rule if $\underline{P}(B) > 0$. If $\underline{P}(B) = 0$, however, there will typically be more than one value for $\underline{P}(f|B)$ that achieves coherence. The smallest conditional lower prevision $\underline{P}(\cdot|B)$ that is coherent with $\underline{P}$ is the vacuous conditional prevision, given by $\underline{P}(f|B) := \inf_{x \in B} f(x)$, and if we want to have more informative assessments, we may need to impose some additional requirements, or make additional assumptions or assessments. Indeed, the approach to conditioning on sets of probability zero is one of the differences between the approaches to conditioning by Walley (and also de Finetti and Williams) and by Kolmogorov. In Kolmogorov's approach, conditioning is made on a σ-field $\mathcal{A}$, and a conditional prevision $P(f|\mathcal{A})$ is any $\mathcal{A}$-measurable gamble that satisfies $P(gP(f|\mathcal{A})) = P(gf)$ for every $\mathcal{A}$-measurable gamble g. In particular, if we consider an event B of probability zero, Kolmogorov allows the prevision $P(\cdot|B)$ to be completely arbitrary. Walley's coherence condition is more general because it can be applied on previsions conditional on partitions, and is more restrictive when dealing with sets of probability zero than Kolmogorov's. See [672, § 6.10] for more information.

From the discussion of natural extension in Section 2.3.2, we may infer that a number of conditional lower previsions may be coherent and still have some undesirable properties, when the partitions are infinite. Something similar applies to the case of a conditional and an unconditional lower prevision. For instance, a coherent pair $\underline{P}$ and $\underline{P}(\cdot|\mathcal{B})$ is not necessarily the lower envelope of coherent pairs of linear unconditional and conditional previsions, P and $P(\cdot|\mathcal{B})$ (these even may not exist). As a consequence, the sensitivity analysis interpretation we mentioned in Section 2.2.2 for unconditional coherent lower previsions is no longer valid in the conditional case.

On the other hand, there are linear previsions P for which there is no conditional linear prevision $P(\cdot|\mathcal{B})$ such that P and $P(\cdot|\mathcal{B})$ are coherent in the sense of Equation (2.13), i.e., there are linear previsions that cannot be updated in a coherent way to a linear conditional prevision $P(\cdot|\mathcal{B})$, but which can be updated to a conditional lower prevision. See [672, § 6.6.10] and [238] for more information.

Taking this into account, given an unconditional prevision $\underline{P}$ representing our subject's beliefs and a partition $\mathcal{B}$ of $\mathcal{X}$, we may be interested in the conditional lower previsions $\underline{P}(\cdot|\mathcal{B})$ that are coherent with $\underline{P}$, i.e., those for which conditions JC1 and JC2 are satisfied. A necessary and sufficient condition for the existence of such $\underline{P}(\cdot|\mathcal{B})$ is that $\underline{P}$ should be $\mathcal{B}$*-conglomerable*: this is the case when given different sets $B_1, B_2, \ldots$ in $\mathcal{B}$ such that $\underline{P}(B_n) > 0$ for all n and a gamble f such that $\underline{P}(I_{B_n} f) \geq 0$ for all n, it holds that $\underline{P}(I_B f) \geq 0$, where $B := \bigcup_n B_n$. This is related to the notion of $\mathcal{B}$-conglomerability for sets of desirable gambles considered in Equation (1.15), although, as shown in [472], both conditions are not entirely equivalent.

Using this $\mathcal{B}$-conglomerability condition, it can be shown [672, Theorem 6.5.3] that when we consider an unconditional coherent lower prevision $\underline{P}$ and a separately coherent conditional lower prevision $\underline{P}(\cdot|\mathcal{B})$ on $\mathcal{K}$, they are jointly coherent if and only if

GBR. $\underline{P}(G(f|B)) = 0$ for all $f \in \mathcal{K}$ and $B \in \mathcal{B}$

and

CONG. $\underline{P}(G(f|\mathcal{B})) \geq 0$ for all $f \in \mathcal{K}$.

The first of these conditions corresponds to the generalized Bayes rule, while the second is related to $\mathcal{B}$-conglomerability: GBR implies that the gambles $G(f|B)$ are almost desirable for all $B \in \mathcal{B}$, and CONG that their sum $G(f|\mathcal{B})$ is almost desirable as well. This is called *strong* $\mathcal{B}$*-conglomerability* in [472].

$\mathcal{B}$-conglomerability holds trivially when the partition $\mathcal{B}$ is finite, or when $\underline{P}(B) = 0$ for every set B in the partition. It only becomes nontrivial when we consider a partition $\mathcal{B}$ for which there are infinitely many elements B satisfying $\underline{P}(B) > 0$. It makes sense as a rationality axiom once we accept the Updating and Conglomerative Principles: to see this, note that if $\{B_n\}_n$ is a partition of $\mathcal{X}$ with $\underline{P}(B_n) > 0$ and $\underline{P}(I_{B_n} f) \geq 0$, then for every $\delta > 0$ the gamble $I_{B_n}(f + \delta)$ is desirable. The Updating and Conglomerative Principles imply then that $f + \delta$ is desirable, whence $\underline{P}(f + \delta) \geq 0$. Since this holds for all $\delta > 0$, we deduce that $\underline{P}(f) \geq 0$.

More generally, Walley says that a coherent lower prevision $\underline{P}$ is *fully conglomerable* when it is $\mathcal{B}$-conglomerable for every partition $\mathcal{B}$. A fully conglomerable coherent lower prevision can be coherently updated to a conditional lower prevision $\underline{P}(\cdot|\mathcal{B})$ for any partition $\mathcal{B}$ of $\mathcal{X}$. Again, full conglomerability can be accepted as an axiom of rationality provided we accept the Updating and Conglomerative Principles, and also provided that when we define our coherent lower prevision we want to be able to update it for all possible partitions of our set of values.

There is an important connection between full conglomerability and countable additivity [672, § 6.9]: a linear prevision P on $\mathcal{L}$ taking infinitely many values is fully conglomerable if and only for every countable partition $\{B_n\}_n$ of $\mathcal{X}$ it satisfies $\sum_n P(B_n) = 1$. Full conglomerability is one of the points of disagreement between Walley's and de Finetti's work. De Finetti rejects the requirement of countable additivity on probabilities and, taking into account the above relationship, also the property of full conglomerability. One key observation here is that de Finetti does not assume the Conglomerative Principle as a rationality axiom, and full conglomerability may be seen as a consequence of it. In this sense, it is worth remarking that a justification of conglomerability that uses only finite sums of desirable gambles has been recently established in [722].

Another issue is that, even if the coherence of a conditional and an unconditional lower prevision given above takes into account conglomerability, this property is not embedded by the natural extension defined in Section 2.3.2, as shown in [472]. This is the motivation for the recent work in [471].

2.3.4 Updating with the regular extension

Consider an unconditional lower prevision $\underline{P}$ defined on $\mathcal{L}$, and a partition $\mathcal{B}$ of $\mathcal{X}$. If this lower prevision is $\mathcal{B}$-conglomerable, then there is a smallest (separately coherent) conditional lower prevision that is coherent with $\underline{P}$ [672, Theorem 6.8.2]: the *natural extension* $\underline{E}(\cdot|\mathcal{B})$ of $\underline{P}$ to a conditional lower prevision, where for all gambles f and all $B \in \mathcal{B}$:

$$\underline{E}(f|B) := \begin{cases} \max\left\{\mu \in \mathbb{R} : \underline{P}(I_B(f-\mu)) \geq 0\right\} & \text{if } \underline{P}(B) > 0 \\ \inf_{x\in B} f(x) & \text{if } \underline{P}(B) = 0. \end{cases}$$

Actually, when $\underline{P}(B) > 0$, $\underline{E}(f|B)$ is the unique solution μ of the GBR equation $\underline{P}(I_B(f - \mu)) = 0$, and is therefore uniquely coherent.

We would like to conclude this section by discussing a procedure other than natural extension for deriving a conditional lower prevision from an unconditional one: the *regular extension*, which coincides with the natural extension when $\underline{P}(B) > 0$ – as is necessary for coherence – but is less conservative when $\underline{P}(B) = 0 < \overline{P}(B)$. The conditional lower prevision $\underline{R}(\cdot|\mathcal{B})$ defined by regular extension is given, for any $f \in \mathcal{L}$ and any $B \in \mathcal{B}$ by

$$\underline{R}(f|B) := \begin{cases} \max\left\{\mu \in \mathbb{R} : \underline{P}(I_B(f-\mu)) \geq 0\right\} & \text{if } \overline{P}(B) > 0 \\ \inf_{x\in B} f(x) & \text{if } \overline{P}(B) = 0. \end{cases}$$

When $\overline{P}(B) > 0$, also

$$\underline{R}(f|B) = \inf\left\{\frac{P(fB)}{P(B)} : P \geq \underline{P} \text{ and } P(B) > 0\right\} :$$

the regular extension is the lower envelope of the updated linear previsions using Bayes's Rule. Regular extension has been used as an updating rule in a number of papers [204, 215, 268, 377, 414, 673].

Even when $\underline{P}$ is $\mathcal{B}$-conglomerable, the conditional lower prevision defined using regular extension need not be coherent with the unconditional lower prevision it is derived from, because it is based on a stronger consistency axiom for desirable gambles, as discussed in

[672, Appendix J]. However, $\underline{P}$ and $\underline{R}(\cdot|\mathcal{B})$ are guaranteed to be coherent when the partition $\mathcal{B}$ is finite. From this result, we deduce that if we let $\mathcal{B}_1, \dots, \mathcal{B}_n$ be finite partitions of $\mathcal{X}$ and use regular extension to define conditional lower previsions $R_1(\cdot|\mathcal{B}_1), \dots, R_n(\cdot|\mathcal{B}_n)$ from an unconditional $\underline{P}$, then $\underline{P}, R_1(\cdot|\mathcal{B}_1), \dots, R_n(\cdot|\mathcal{B}_n)$ are weakly coherent. Moreover, if we consider any other conditional lower previsions $\underline{P}_1(\cdot|\mathcal{B}_1), \dots, \underline{P}_n(\cdot|\mathcal{B}_n)$ that are weakly coherent with $\underline{P}$, it must hold that $R_j(\cdot|\mathcal{B}_j) \geq \underline{P}_j(\cdot|\mathcal{B}_j)$ for $j = 1, \dots, n$. Hence, the procedure of regular extension provides the greatest, or most informative, updated lower previsions that are weakly coherent with $\underline{P}$. Moreover, the following results can be proved by making a straightforward adaptation of [461, Theorem 6] for partitions.

Theorem 2.3 *Let $\underline{P}$ be a coherent lower prevision on $\mathcal{L}$, and consider finite partitions $\mathcal{B}_j$ of $\mathcal{X}$ for $j = 1, \dots, n$. Assume that $\overline{P}(B_j) > 0$ for all $B_j \in \mathcal{B}_j$, and let us define $R_1(\cdot|\mathcal{B}_1), \dots, \underline{R}_n(\cdot|\mathcal{B}_n)$ using regular extension. Then the lower previsions $\underline{P}$, $\underline{R}_1(\cdot|\mathcal{B}_1), \dots, \underline{R}_n(\cdot|\mathcal{B}_n)$ are coherent.*

The regular extension derived from $\underline{P}$ may not coincide with the natural extension when the conditioning event has zero lower probability and positive upper probability (see [461] for more details). A connection between the natural and regular extensions of conditional lower previsions is established in [679] and [469].

2.4 Further reading

We conclude this chapter by summarising briefly a number of mathematical models that can be related to the behavioural theory of coherent lower previsions. We refer to [672, § 5] for additional discussion. Another related approach, which we will not go into here, is the game-theoretic one: it will be discussed in detail in Chapter 6.

2.4.1 The work of Williams

A first approach to generalizing de Finetti's work, taking into account the presence of imprecision and indecision, was taken by Peter Williams in [701, 702, 703]; see also [659] for a recent review of Williams's work. Williams defines a coherent conditional lower prevision by means of a set of acceptable gambles, in the sense that $\underline{P}(f|B)$ is defined as the supremum value of μ such that $I_B(f - \mu)$ is an acceptable transaction for our subject, similarly to what is discussed in Section 1.3.3 and in [672, Appendix F].

In order to define coherence, he requires the set of acceptable gambles to satisfy conditions that are slightly weaker than the ones introduced in Chapter 1. He obtains a condition that we will call *W-coherence*.

The use of W-coherence for conditional lower previsions has a number of technical advantages over Walley's coherence condition:

(i) W-coherent conditional lower previsions are always lower envelopes of sets of W-coherent conditional linear previsions ([703, Theorem 2], see also [672, § 8.1]), and as a consequence they can be given a sensitivity analysis interpretation;

(ii) we can always construct the smallest W-coherent extension of a W-coherent conditional lower prevision [703, Theorem 1]. Moreover, this result can be generalized towards unbounded gambles [629].

But a number of remarks are in order:

(i) the condition of W-coherence for conditional lower previsions is similar to Equation (2.13), but considering gambles of the type $G_j(f_j|B_j)$ instead of $G_j(f_j|\mathcal{B}_j)$.

(ii) W-coherent conditional lower previsions do not in general satisfy the conglomerability condition from Walley's theory, and as a consequence they may not satisfy a number of notions derived from it, such as avoiding sure loss [672, Example 6.6.9]. It may be argued [510, § 4.2.2], however, that this is not actually a drawback, but simply a logical consequence of the rejection of the conglomerative axiom which is at the core of Walley's theory of conditional lower previsions. It can be checked that if $\underline{P}$ and $\underline{P}(\cdot|\mathcal{B})$ are a coherent unconditional lower prevision and a separately coherent conditional lower prevision, respectively, they are W-coherent if and only if they satisfy GBR [703].

A more detailed account of Williams's work will be given in Chapters 12.4 and 16.2.

2.4.2 The work of Kuznetsov

In a relatively unknown book published in Russian [414], the late Vladimir Kuznetsov established, in parallel to Walley, a theory of interval-valued probabilities and previsions that has many things in common with the behavioural theory we have reviewed here. Starting from axioms that seem to be closely related to Walley's coherence requirements, he deduces many of the properties that can also be found in Walley's book. He also obtains interesting limit laws for coherent lower and upper previsions. The main differences between both theories are:

(i) Kuznetsov does not consider a behavioural interpretation for coherent lower and upper previsions;

(ii) he makes some assumptions about the domains, which are not structure-free;

(iii) his theory is valid for unbounded gambles;

(iv) when conditioning on a set B of lower probability zero and positive upper probability, he suggests using the regular extension introduced in Section 2.3.4.

We refer to [415, 416] for two short papers in English presenting Kuznetsov's work, and to [161, 210, 649] for some work inspired by his approach.

2.4.3 The work of Weichselberger

Kurt Weichselberger and some of his colleagues have also established a theory of interval-valued probabilities [689, 691, 692] that is more directly in line with the classical measure-theoretic approach to probability theory, in the sense that it focuses on probabilities instead of previsions and that it also makes some measurability assumptions that are not present in Walley's. Another difference is that Weichselberger's work aims at being free from interpretation, and does not hinge on a behavioural interpretation of lower and upper probabilities. It is also based on sets of σ-additive probabilities, and not merely finitely additive ones, and as such it relates to the work in [408]. While on finite spaces Weichselberger's notion of R-probability and F-probability are technically closely related to the concepts of avoiding sure loss and coherence of lower and upper probabilities, the treatment of conditional lower and upper probabilities and of independence is quite different [696].[8] See [36, 693] for some recent work in the spirit of this approach.

Acknowledgements

Work supported by project MTM2010-17844 and by the SBO project 060043 of the IWT-Vlaanderen. We would like to thank Matthias Troffaes for his helpful comments.

[8] On statistical aspects of Weichselberger's work see also Page 179.

3

Structural judgements

Enrique Miranda[1] **and Gert de Cooman**[2]

[1]*Department of Statistics and Operations Research, University of Oviedo, Spain*
[2]*SYSTeMS Research Group, Ghent University, Belgium*

3.1 Introduction

In Chapters 1 and 2, we have come across a number of *local* probability assessments: stating that certain gambles are desirable, or providing lower and upper previsions for given gambles. We have learnt how to make inferences based on such local assessments, by deriving the least committal coherent models that are compatible with them – meaning that they include them (for sets of desirable gambles) or dominate them (for lower previsions). The present chapter deals with *structural judgements*: assessments that a subject makes about global properties of her belief model. They include assessments of independence and symmetry, and they are important when dealing with imprecise probability models, where they have to be combined with local assessments to find the least committal models that are compatible with these local assessments and still satisfy the global properties.

As we know from previous chapters, imprecise probability assessments can take many different forms: coherent lower previsions, sets of desirable gambles or sets of linear previsions. In this chapter, we will briefly describe how two specific types of structural judgements – independence and symmetry – can be incorporated into these formulations.

Section 3.2 deals with irrelevance and independence. When working with precise probabilities, independence can be modelled in two different and roughly equivalent ways: we can require that the marginal and the conditional models agree, or we can impose that a joint probability distribution factorises as a product of marginal distributions. These two approaches are no longer (even roughly) equivalent when considering imprecise probability

Introduction to Imprecise Probabilities, First Edition.
Edited by Thomas Augustin, Frank P. A. Coolen, Gert de Cooman and Matthias C. M. Troffaes.

models, and this gives rise to two different ways of dealing with independence for imprecise probabilities, which we will call the epistemic and the formalist one, respectively. Moreover, within each of these two approaches there are several possible generalizations of the classical notion of independence. We will describe a few of them and point to the literature for others.

In Section 3.3, we deal with invariance, and describe two other types of structural judgement: the first – weak invariance – states that our subject's beliefs are invariant under a set of transformations, and the second – strong invariance – captures that a subject believes there is symmetry, associated with a set of transformations. Roughly speaking, we can think of these assessments as symmetry of beliefs and beliefs of symmetry, respectively. We will make no assumptions about the set of transformations, other than that it is closed under composition, and contains the identity map: it constitutes a monoid.

Finally, in Section 3.4 we study in detail a particular type of strongly invariant models: the *exchangeable* ones, whose counterparts in the case of precise probabilities were first studied by de Finetti.

In addition to exploring how to model both types of structural judgements in an imprecise context, we also determine to what extent they are compatible with the notion of natural extension: we study which are the consequences, under coherence, of local and structural assessments.

3.2 Irrelevance and independence

We begin with a discussion of the notions of irrelevance and independence. To fix ideas, recall the definition of independence for two random variables in the precise-probabilistic case. Consider two random variables X_1 and X_2 taking values in the respective sets $\mathcal{X}_1$ and $\mathcal{X}_2$, and let P be the probability that models beliefs about the values that (X_1, X_2) assume jointly. For precise probabilities, there are, roughly speaking, two possible approaches to defining independence for these two variables:[1]

Irrelevance: Given a conditional probability measure *compatible* with P, in the sense that

$$P(X_1 = x_1 | X_2 = x_2) := \frac{P(X_1 = x_1, X_2 = x_2)}{P(X_2 = x_2)} \text{ when } P(X_2 = x_2) > 0,$$

it holds that $P(X_1 = x_1 | X_2 = x_2) = P(X_1 = x_1)$ for all $(x_1, x_2) \in \mathcal{X}_1 \times \mathcal{X}_2$.

Factorization: $P(X_1 = x_1 \text{ and } X_2 = x_2) = P(X_1 = x_1)P(X_2 = x_2)$ for all $(x_1, x_2) \in \mathcal{X}_1 \times \mathcal{X}_2$.

These two notions are sometimes referred to as *casual* and *decompositional* independence (see Section 5.3.3). The second criterion is trivially satisfied when either $P(X_1 = x_1)$ or $P(X_2 = x_2)$ are equal to zero, while there are in that case many possible values of $P(X_1 = x_1 | X_2 = x_2)$ for which the first criterion does not hold. The two criteria are equivalent provided the marginal mass functions are everywhere nonzero, that is,

[1] In this definition we assume that the sets $\mathcal{X}_1$ and $\mathcal{X}_2$ are finite in order to simplify the notation; in the infinite case we could consider density functions instead of probability mass functions. We also use $X_1 = x_1$ to denote the event $X_1^{-1}(\{x_1\})$.

provided there is no conditioning on sets of probability zero. Independence under the first criterion then becomes a symmetrical notion too. If it holds, then so does its symmetrical counterpart:

$$P(X_2 = x_2 | X_1 = x_1) = P(X_2 = x_2) \text{ for all } x_1 \in \mathcal{X}_1 \text{ and } x_2 \in \mathcal{X}_2.$$

The classical approach by Kolmogorov [405], where the conditional probability is derived from the unconditional P, takes factorization as a definition of independence. An alternative approach uses *full conditional measures* as defined in [238] and works with the irrelevance condition to define independence.

When knowledge about the value that X_1 and X_2 assume jointly is represented by a coherent lower prevision $\underline{P}$ on $\mathcal{L}(\mathcal{X}_1 \times \mathcal{X}_2)$, there is no unique way to extend the classical notion of independence. The criteria of factorization and irrelevance are not equivalent – as we have seen above they are, strictly speaking, not even equivalent classically, or at least some additional care must be taken to ensure that they are – and moreover symmetry is no longer immediate, meaning that we must distinguish between irrelevance (an asymmetrical notion) and independence (its symmetric counterpart). Also, all our definitions must be formulated using gambles rather than just events, since lower previsions are in general not completely determined by the values they assume on events, as was explained right before Equation (2.5).

In this section, we discuss some of the more important generalizations of the classical notion of independence proposed in the literature, and the relationships between them. To fix the context, we consider n random variables $X_1, \dots, X_n$ taking values in the respective sets $\mathcal{X}_1, \dots, \mathcal{X}_n$. We only consider modelling independence assumptions for finite sequences of random variables; see [207, 214] for some work involving infinite sequences. For any subset J of $N := \{1, \dots, n\}$, we denote by X_J the J-tuple of random variables that assumes values in the (Cartesian) product space $\mathcal{X}_J := \times_{j \in J} \mathcal{X}_j$. If x_N denotes a generic value for an element of $\mathcal{X}_N$, then we agree to denote by x_J the subtuple of x_N whose indices belong to the subset J of N. We assume throughout that the random variables $X_1, \dots, X_n$ are *logically independent*, meaning that X_N can assume any value in the product space $\mathcal{X}_N$.

Let $\underline{P}_N$ be a coherent lower prevision on $\mathcal{L}(\mathcal{X}_N)$ representing a subject's beliefs about the value that $X_1, \dots, X_n$ assume jointly.[2] Using $\underline{P}_N$, we define for any nonempty $J \subseteq N$ the *marginal* coherent lower prevision $\underline{P}_J$ on $\mathcal{L}(\mathcal{X}_J)$ by

$$\underline{P}_J(f) := \underline{P}_N(\tilde{f}) \text{ for all } f \in \mathcal{L}(\mathcal{X}_J),$$

where $\tilde{f}$ is the cylindrical extension of f from $\mathcal{X}_J$ to $\mathcal{X}_N$, given by $\tilde{f}(x_N) := f(x_J)$ for all $x_N \in \mathcal{X}_N$; see also Equation (1.13). We also call $\underline{P}_J$ the $\mathcal{X}_J$-marginal of $\underline{P}_N$. In what follows we will use the same notation f for a gamble and its cylindrical extension $\tilde{f}$.

We will consider two types of definitions of independence: those based on the irrelevance criterion, which lead to what we call the *epistemic* approach to independence; and those based on the factorization criterion, which constitute the *formalist* approach. We intend to see these criteria expressed in terms of coherent lower previsions, and in some cases in terms of the equivalent representations (sets of desirable gambles, sets of linear previsions) mentioned in Section 2.2.

[2] We are assuming for simplicity that the domain of $\underline{P}_N$ is $\mathcal{L}(\mathcal{X}_N)$, but all the developments in this section can be generalized to the case where the domain of $\underline{P}_N$ is some subset $\mathcal{K}$ of $\mathcal{L}(\mathcal{X}_N)$.

3.2.1 Epistemic irrelevance

We begin by exploring the epistemic approach. The basic notion, from which all other notions in this approach can be derived, is that of *conditional epistemic irrelevance*. Let I, O and C be three mutually disjoint subsets of N, of which O is assumed to be nonempty. We say that a subject judges X_I *to be epistemically irrelevant to* X_O *conditional on* X_C when she assumes that, when knowing the value of X_C, learning in addition which value X_I assumes in $\mathcal{X}_I$ will not affect her beliefs about the value that X_O assumes in $\mathcal{X}_O$.

In order to keep this discussion as simple as possible, however, we will restrict it to unconditional epistemic irrelevance, which is the special case that $C = \emptyset$: a subject judges X_I *to be epistemically irrelevant to* X_O when she assumes that learning which value X_I assumes in $\mathcal{X}_I$ will not affect her beliefs about the value that X_O assumes in $\mathcal{X}_O$.

On the epistemic approach, an assessment of epistemic irrelevance is useful, because it allows specific new models to be inferred from existing ones. To illustrate what this means, assume that our subject has assessed a coherent lower prevision $\underline{P}_N$ on $\mathcal{L}(\mathcal{X}_N)$ representing her beliefs about X_N. Her assessment that X_I is epistemically irrelevant to X_O allows her to infer from the joint model $\underline{P}_N$ – or rather from its $\mathcal{X}_O$-marginal $\underline{P}_O$ – a conditional model $\underline{P}_O(\cdot|X_I)$ on the set of gambles $\mathcal{L}(\mathcal{X}_{O\cup I})$ by[3]

$$\underline{P}_O(f|x_I) := \underline{P}_O(f(\cdot, x_I)) = \underline{P}_N(f(\cdot, x_I)) \text{ for all gambles } f \text{ on } \mathcal{X}_{O\cup I} \text{ and all } x_I \in \mathcal{X}_I. \quad (3.1)$$

Observe that $\underline{P}_O(f|x_I) = \underline{P}_O(f(\cdot, x_I)|x_I)$, so we may as well consider $\underline{P}_O(\cdot|X_I)$ as a conditional lower prevision on $\mathcal{L}(\mathcal{X}_O)$, as its behaviour is completely determined by the values it assumes on (cylindrical extensions of) gambles on $\mathcal{X}_O$. We will also sometimes denote $\underline{P}_O(f|x_I)$ by $\underline{P}_O(f|X_I = x_I)$.

Example 3.1 We consider three urns with green and red balls. The first urn has one red (R) ball and one green (G) ball. The second and the third urns also contain red and green balls, but in an unknown distribution. We randomly draw a ball from the first urn. If it is green, then we randomly draw a ball from the second urn; if it is red, we draw a ball from the third urn. Let X_1 be the colour of the first ball drawn and X_2 the colour of the second ball.

Now, if we assess that the colour of the first ball is epistemically irrelevant to the colour of the second ball, this means effectively that our information about the proportion of green and red balls in the second and third urn should be the same, so:

$$\underline{P}(X_2 = G|X_1 = G) = \underline{P}(X_2 = G|X_1 = R) \text{ and}$$
$$\overline{P}(X_2 = G|X_1 = G) = \overline{P}(X_2 = G|X_1 = R).$$

This would not be the case, for instance, if we knew that in the second urn, we have two green balls, one red ball and two other balls of unknown colour – they may be red or green – and that in the third urn, we have one green ball, two red balls and two other balls of unknown colour: in that case we would have

$$\underline{P}(X_2 = G|X_1 = G) = \frac{2}{5} \text{ and } \underline{P}(X_2 = G|X_1 = R) = \frac{1}{5},$$

and X_1 would not be epistemically irrelevant to X_2. ♦

[3] See the discussion in Section 2.3 for more information about conditional lower previsions with respect to random variables.

Epistemic irrelevance can also be expressed in terms of sets of linear previsions or coherent sets of desirable gambles. For instance, if we have two random variables X_1 and X_2 and we assess that X_1 is epistemically irrelevant to X_2, then this leads to

$$\{P_2(\cdot|x_1) : P_2 \in \mathcal{M}(\underline{P})\} = \mathcal{M}(\underline{P}_2) \text{ for all } x_1 \text{ in } \mathcal{X}_1,$$

where we use the simpler notations P_2 and $\underline{P}_2$ rather than $P_{\{2\}}$ and $\underline{P}_{\{2\}}$, respectively. Recall that, taking into account the comments after Equation (3.1), we can assume that $P_2(\cdot|x_1)$ is defined on $\mathcal{L}(\mathcal{X}_2)$. The above equation means that learning the value of the first random variable should not change our uncertainty model (the *set* of possible precise models) about the second; see also [158, 159, 204] for more discussion. In terms of sets of desirable gambles, epistemic irrelevance means that the set of desirable gambles for X_2 should not change after learning the value of X_1; see [209, 482] for more details.

This brings us to the topic of *irrelevant joints and products*. Since in the context of imprecise probability models, due to the lack of symmetry in the notion of epistemic irrelevance (see Example 3.2 further on), there is more variety than in a precise context, we must start with a more general approach and definition.

Consider a number m of epistemic irrelevance assessments: we start with m (different) pairs of proper disjoint subsets I_k and O_k of N, and the assessment that X_{I_k} is epistemically irrelevant to X_{O_k}, for $k = 1, \dots, m$. If we have a joint coherent lower prevision $\underline{P}_N$ on $\mathcal{L}(\mathcal{X}_N)$, then these assessments lead to the introduction of m conditional lower previsions $\underline{P}_{O_k}(\cdot|X_{I_k})$ on $\mathcal{L}(\mathcal{X}_{O_k} \times \mathcal{X}_{I_k})$ for $k = 1, \dots, m$, through Equation (3.1).

Definition 3.1 (Irrelevant joint). A coherent lower prevision is called an (epistemically) *irrelevant joint* (with respect to these m assessments) if it is coherent with the conditional lower previsions $\underline{P}_{O_k}(\cdot|X_{I_k})$ on $\mathcal{L}(\mathcal{X}_{O_k} \times \mathcal{X}_{I_k})$ for $k = 1, \dots, m$ that it induces through Equation (3.1).

Another possibility is that we start out, besides these m epistemic irrelevance assessments, with m (so-called marginal) coherent lower previsions $\underline{P}_{O_k}$ on $\mathcal{L}(\mathcal{X}_{O_k})$, where for consistency we make the assumption that $\underline{P}_{O_k} = \underline{P}_{O_\ell}$ whenever O_k happens to coincide with O_ℓ. Each such lower prevision $\underline{P}_{O_k}$ then induces a conditional lower prevision $\underline{P}_{O_k}(\cdot|X_{I_k})$ on $\mathcal{L}(\mathcal{X}_{O_k} \times \mathcal{X}_{I_k})$ through the corresponding counterpart of Equation (3.1):

$$\underline{P}_{O_k}(f|x_{I_k}) := \underline{P}_{O_k}(f(\cdot, x_{I_k})) \text{ for all gambles } f \text{ on } \mathcal{X}_{O_k \cup I_k} \text{ and all } x_{I_k} \in \mathcal{X}_{I_k}. \tag{3.2}$$

Definition 3.2 (Irrelevant product). A coherent lower prevision is called an (epistemically) *irrelevant product* (with respect to these m assessments) of the marginals $\underline{P}_{O_k}$ if it is coherent with the conditional lower previsions $\underline{P}_{O_k}(\cdot|X_{I_k})$ on $\mathcal{L}(\mathcal{X}_{O_k} \times \mathcal{X}_{I_k})$ for $k = 1, \dots, m$ that it induces through Equation (3.2), and at the same time has $\underline{P}_{O_k}$ as its $\mathcal{X}_{O_k}$-marginal for $k = 1, \dots, m$.

A collection of marginals typically has either no, or – unless the marginals are precise – more than one (typically an infinity of) irrelevant products. But it may happen that there is a smallest irrelevant product, which will then be called the *irrelevant natural extension* of the marginals.

One interesting and fairly simple particular example is the notion of forward irrelevance studied in [208], which seems particularly useful to study independence in discrete-time random processes using imprecise probabilities. Here we consider random variables $X_1, \dots, X_n$

and an assessment of *forward irrelevance*, meaning that the random variable $X_{1,\dots,k}$ is epistemically irrelevant to X_{k+1}, for $k = 1, \dots, n-1$: we do not change our beliefs about future random variables after observing present and past random variables. Given the marginal coherent lower previsions $\underline{P}_\ell$ on $\mathcal{X}_\ell$ for each random variable X_ℓ, it can be shown that these always have a smallest forward irrelevant product, which is actually the *marginal extension*[4] of the induced conditional lower previsions

$$\underline{P}_{k+1}(f|x_{1,\dots,k}) := \underline{P}_{k+1}(f(\cdot, x_{1,\dots,k})) \text{ for all } f \in \mathcal{L}(\mathcal{X}_{1,\dots,k+1}) \text{ and all } x_{1,\dots,k} \in \mathcal{X}_{1,\dots,k}.$$

We refer to [208] for more details. This type of forward irrelevance assessments and products lies at the basis of the laws of large numbers for coherent lower previsions established in [207] (and, based on that, [162]). This paper arguably provides the most generally valid formulation of the laws of large numbers extant in the literature. An extension of the Central Limit Theorem in a game-theoretic context is discussed in [582].

3.2.2 Epistemic independence

Epistemic irrelevance is an asymmetric notion, meaning that X_1 can be epistemically irrelevant to X_2 while X_2 is not epistemically irrelevant to X_1; see Example 3.2. When a subject judges that X_1 and X_2 are epistemically irrelevant to each other, she assesses the two random variables to be *epistemically independent*. This means that she imposes the following equalities:

$$\underline{P}_2(f|X_1 = x_1) = \underline{P}_2(f(x_1, \cdot)) \text{ and } \underline{P}_1(g|X_2 = x_2) = \underline{P}_1(g(\cdot, x_2))$$

for all gambles f, g on $\mathcal{X}_{\{1,2\}}$ and all $(x_1, x_2) \in \mathcal{X}_1 \times \mathcal{X}_2$. To see that indeed epistemic irrelevance and independence are not equivalent, consider the following example:

Example 3.2 Go back to the situation described in Example 3.1. Assume that in both the second and the third urn we have five balls, which can be either red or green. If we assess that the colour of the first ball drawn is epistemically irrelevant to the colour of the second one, this means that we have the same information in both cases about the colour of the unknown balls. For instance, if we knew nothing about the composition of the urns, we would have

$$\underline{P}(X_2 = G|X_1 = G) = \underline{P}(X_2 = G|X_1 = R) = \underline{P}(X_2 = G) = 0,$$

and similarly for the lower probability of drawing a red ball in the second case.

However, an assumption of epistemic independence between the two random variables means something more: if we assess that the colour of the second ball does not give any information about the colour of the first one, then we must necessarily also believe that the second and the third urns must have exactly the same composition: otherwise, if there were a possible composition for the second and third urns with a different proportion of red and green balls in each, we would have

$$\underline{P}(X_1 = G|X_2 = G) \neq \overline{P}(X_1 = G|X_2 = G),$$

[4] Marginal extension is discussed near the end of Section 2.3.2, see Equation (2.16).

which implies that we cannot have

$$\underline{P}(X_1 = G|X_2 = G) = \overline{P}(X_1 = G|X_2 = G)$$
$$= \underline{P}(X_1 = G|X_2 = R) = \overline{P}(X_1 = G|X_2 = R) = P(X_1 = G) = \frac{1}{2};$$

and therefore the second random variable would not be epistemically irrelevant to the first, contradicting epistemic independence. ♦

More generally, given a finite number of random variables $X_1, \dots, X_n$ taking values in the respective sets $\mathcal{X}_1, \dots, \mathcal{X}_n$, there are a number of ways to define epistemic independence [210].

Many-to-many independence: We say that a subject judges the random variables X_k, $k = 1, \dots, n$ to be *epistemically many-to-many independent* when she assesses that learning the value of any number of these random variables will not affect her beliefs about any of the others. In other words, she judges for any disjoint proper subsets I and O of N that X_I is epistemically irrelevant to X_O.

Starting from a coherent lower prevision $\underline{P}_N$ on $\mathcal{L}(\mathcal{X}_N)$, these irrelevance assessments lead to a family of conditional models

$$\{\underline{P}_O(\cdot|X_I) : I \text{ and } O \text{ disjoint proper subsets of } N\}, \tag{3.3}$$

through the application of Equation (3.1), and in the spirit of Definition 3.1, the joint $\underline{P}_N$ is called *many-to-many independent* if it is coherent with this family of conditional lower previsions. Similarly, in the spirit of Definition 3.2, any many-to-many independent joint whose $\mathcal{X}_k$-marginals coincide with given coherent lower previsions $\underline{P}_k$ on $\mathcal{L}(\mathcal{X}_k)$ for $k = 1, \dots, n$ is called a *many-to-many independent product* of $\underline{P}_1, \dots, \underline{P}_n$.

Many-to-one independence: We say that a subject judges the random variables X_k, $k = 1, \dots, n$ to be *epistemically many-to-one independent* when she assesses that learning the value of any number of these random variables will not affect her beliefs about any *single* other. In other words, she judges for any $o \in N$ and any subset I of $N\setminus\{o\}$ that X_I is epistemically irrelevant to X_o.

Starting from a coherent lower prevision $\underline{P}_N$ on $\mathcal{L}(\mathcal{X}_N)$, these irrelevance assessments lead to a family of conditional models

$$\{P_O(\cdot|X_I) : o \in N \text{ and } o \notin I \subseteq N\},$$

through the application of Equation (3.1), and in the spirit of Definition 3.1, the joint $\underline{P}_N$ is called *many-to-one independent* if it is coherent with this family of conditional lower previsions. Similarly, in the spirit of Definition 3.2, any many-to-one independent joint whose $\mathcal{X}_k$-marginals coincide with given coherent lower previsions $\underline{P}_k$ on $\mathcal{L}(\mathcal{X}_k)$ for $k = 1, \dots, n$ is called a *many-to-one independent product* of $\underline{P}_1, \dots, \underline{P}_n$. Many-to-one and many-to-many independent models have been studied in detail in [210]. It was proved there that for any coherent lower previsions $\underline{P}_1, \dots, \underline{P}_n$ for respective random variables $X_1, \dots, X_n$ assuming values in *finite* spaces $\mathcal{X}_1, \dots, \mathcal{X}_n$, the smallest many-to-many and many-to-one independent products always exist and coincide, which allows us to simply call both their *independent natural extension* $\otimes_{k=1}^{n}\underline{P}_k$. It is given by

$$(\otimes_{k=1}^{n}\underline{P}_k)(f) = \sup_{\substack{h_k\in\mathcal{L}(\mathcal{X}_N),\\ k=1,\ldots,n}} \min_{x_N\in\mathcal{X}_N} \left[f(x_N) - \sum_{k=1}^{n}[h_k(x_N) - \underline{P}_k(h_k(\cdot, x_{N\setminus\{k\}}))] \right]$$

for all gambles f on $\times_{k=1}^{n}\mathcal{X}_k$, and it represents the behavioural consequences of a structural judgement of epistemic independence for given random variables $X_1, \ldots, X_n$. It possesses a number of interesting properties: for instance, it is *associative*, meaning that given any partition N_1 and N_2 of $\{1, \ldots, n\}$, then

$$\otimes_{k=1}^{n}\underline{P}_k = (\otimes_{k\in N_1}\underline{P}_k) \otimes (\otimes_{k\in N_2}\underline{P}_k),$$

and it is *strongly externally additive*: if I and O are any disjoint subsets of N, then $(\otimes_{k=1}^{n}\underline{P}_k)(f+g) = (\otimes_{k=1}^{n}\underline{P}_k)(f) + (\otimes_{k=1}^{n}\underline{P}_k)(g) = (\otimes_{i\in I}\underline{P}_i)(f) + (\otimes_{o\in O}\underline{P}_o)(g)$ for all $f \in \mathcal{L}(\mathcal{X}_I)$ and $g \in \mathcal{L}(\mathcal{X}_O)$. The last equality follows from the *marginalization* property of the independent natural extension: for any proper subset I of N and any gamble f on $\mathcal{X}_I$, $(\otimes_{k\in N}\underline{P}_k)(f) = (\otimes_{i\in I}\underline{P}_i)(f)$. The condition is trivial when I or O are empty.

The notion of independent natural extension has also been studied in the more general context of sets of desirable gambles in [209]. See [656, 672] for earlier discussions of epistemic independence.

3.2.3 Envelopes of independent precise models

When imprecise marginals have an independent product, they usually have an infinity of them, the smallest of which is the independent natural extension. In the literature on independent imprecise probability models, we come across methods for defining other types of independent joints, such as lower envelopes of independent precise models. Underlying these methods is a simple result, which states lower envelopes of (many-to-many and many-to-one) independent joints are still independent (many-to-many and many-to-one, respectively): *taking lower envelopes preserves epistemic irrelevance and independence* [210, § 4.3].

Consider a coherent lower prevision $\underline{P}_N$ on $\mathcal{L}(\mathcal{X}_N)$ representing a subject's assessments about the joint behaviour of the random variables $X_1, \ldots, X_n$. We call $\underline{P}_N$ an *independent envelope* when every extreme point P_N of $\mathcal{M}(\underline{P}_N)$ factorizes as $P_N = P_1 \times P_2 \times \ldots \times P_n$ (is an independent precise probability model), where P_k denotes the $\mathcal{X}_k$-marginal of P_N, for $k = 1, \ldots, n$. The independence property expressed by this requirement is called *independence in the selection* [158], or *type-2 independence* [196].

It is easier to understand the ideas behind this notion if we consider the sensitivity analysis interpretation of lower previsions. Assume that we have n random variables $X_1, \ldots, X_n$, and that all we know about the precise probability model for the k-th random variable is that it belongs to some set of precise models $\mathcal{M}(\underline{P}_k)$, for $k = 1, \ldots, n$. Let $\mathcal{M}(\underline{P}_N)$ be the set of possible precise models for the behaviour of the n random variables, taken together. As we have seen in Proposition 2.5, this set – and the corresponding lower prevision – can be characterized by its extreme points. Then $\underline{P}_N$ is an independent envelope when each of these extreme points satisfies the classical notion of independence, in the sense that it can be written as a product of its marginals.

If $\underline{P}_N$ is an independent envelope, it is also many-to-many independent as a lower envelope of (precise) independent models, and as a consequence this type of independence can be seen as a special case of epistemic (many-to-many and many-to-one) independence. The following example shows that the two notions are not equivalent, however.

Example 3.3 Assume now that in the first urn we have one red ball, one green ball and one ball of unknown colour, and that in the second and third urns we have two green balls, one red ball and two balls of unknown colour. Then the models for the random variables X_1 and X_2 satisfy the equalities for epistemic independence:

$$\underline{P}(X_2 = G) = \underline{P}(X_2 = G|X_1 = R) = \underline{P}(X_2 = G|X_1 = G) = \frac{2}{5}$$

$$\underline{P}(X_2 = R) = \underline{P}(X_2 = R|X_1 = R) = \underline{P}(X_2 = R|X_1 = G) = \frac{1}{5}$$

$$\underline{P}(X_1 = G) = \underline{P}(X_1 = G|X_2 = R) = \underline{P}(X_1 = G|X_2 = G) = \frac{1}{3}$$

$$\underline{P}(X_1 = R) = \underline{P}(X_1 = R|X_2 = R) = \underline{P}(X_1 = R|X_2 = G) = \frac{1}{3}.$$

Let P be the probability distribution associated with the following composition or the urns:

Urn 1	**Urn 2**	**Urn 3**
$\{R, R, G\}$	$\{R, R, R, G, G\}$	$\{R, G, G, G, G\}$

It is easy to check that this is an extreme point of the set of all the probability distributions compatible with the available information (i.e., those associated to the possible compositions of the urns). However, it does not factorize: we have for instance

$$P(X_2 = G, X_1 = R) = P(X_2 = G|X_1 = R) \cdot P(X_1 = R) = \frac{4}{5} \cdot \frac{2}{3} = \frac{8}{15},$$

while

$$P(X_2 = G) \cdot P(X_1 = R) = \frac{2}{3} \cdot \frac{2}{3} = \frac{4}{9}.$$

Hence, the set of possible models is not an independent envelope. ♦

We stress that independence in the selection does not prevent the existence of some dependence relationship between the random variables: we could think for instance of two urns with one ball of the same colour, but such that we do not know whether it is red or green. Then knowing the colour of the ball in the first urn completely determines the colour of the second, so intuitively there is no independence; however, the set of possible models for this situation turns out to be an independent envelope. Indeed, it is a vacuous lower prevision, and any vacuous joint is an independent envelope of vacuous marginals. For a vacuous marginal, the lower probability of any proper subset of the possibility space is zero. Another example, where no conditioning is involved on events with lower probability zero, is given in the next section.

In the definition above, we only require the factorization property to hold for the extreme points of the set $\mathcal{M}(\underline{P}_N)$: these determine the lower prevision $\underline{P}_N$ completely, and it can be checked that the factorization property can never (excpept in the trivial case of linear $\underline{P}_N$) hold for all the elements of a *convex* set $\mathcal{M}(\underline{P}_N)$.

3.2.4 Strong independence

Independence in the selection implies that $\mathcal{M}(\underline{P}_N)$ is included in the convex hull of the product set

$$\{P_1 \times \cdots \times P_n : (\forall k \in N) P_k \in \mathcal{M}(\underline{P}_k)\}, \tag{3.4}$$

but the two sets are not necessarily equal. When we do have the equality, we say that the random variables $X_1, \ldots, X_n$ are *strongly independent* [183, 186, 187]. This notion is also called *type-3 independence* [196]. It represents that if we consider precise models compatible with our beliefs about each of random variables, then we can construct their independent product and obtain a precise model that is compatible with the beliefs about all random variables taken together. So we do not have any information about the random variables that allows us to rule out any of the combinations of the marginal distributions.

In spite of its name, strong independence may not be very informative: for instance, if we make vacuous assessments about each of the random variables, it leads us to consider the set of all previsions on $\mathcal{L}(\mathcal{X}_N)$, and this set satisfies strong independence. What strong independence actually represents, is the absence of information pointing towards some dependence between the random variables.

Trivially, strong independence implies independence in the selection, but the following example shows that the converse is not true:

Example 3.4 Assume we have two urns with one red ball, one green ball and one ball of unknown colour, either red or green, but the same in both cases. The set of possible compositions is given by the following table:

	Urn 1	**Urn 2**
Composition 1	$\{R, R, G\}$	$\{R, R, G\}$
Composition 2	$\{R, G, G\}$	$\{R, G, G\}$

We draw a ball at random from each of the urns, and let X_i be the random variable denoting the colour of the ball drawn in urn i, for $i = 1, 2$. Let $\underline{P}$ be the lower prevision associated with (X_1, X_2), and let $\underline{P}_1$ and $\underline{P}_2$ be its marginals. Then $\underline{P}$ is the lower envelope of the previsions that are determined by the compositions in the table above. Both these previsions factorize, and so $\underline{P}$ is an independent envelope. However, we do not have strong independence: consider the previsions P_1 and P_2 given by $P_1(R) = 1/3$ and $P_2(R) = 2/3$. Then $P_1 \geq \underline{P}_1$ and $P_2 \geq \underline{P}_2$. Their product satisfies

$$(P_1 \times P_2)(X_1 = R, X_2 = G) = \frac{1}{9} < \frac{4}{9} = \underline{P}(X_1 = R, X_2 = G).$$

Hence, $\mathcal{M}(\underline{P})$ is a proper subset of the set $\{P_1 \times P_2 : P_j \in \mathcal{M}(\underline{P}_j), j = 1, 2\}$, and therefore strong independence is not satisfied. ♦

Given a number of marginal coherent lower previsions $\underline{P}_1, \ldots, \underline{P}_n$ with respective domains $\mathcal{L}(\mathcal{X}_1), \ldots, \mathcal{L}(\mathcal{X}_n)$, the joint model they determine by means of strong independence is called their *strong product*. It has been used for instance in [721], and has been studied from the theoretical point of view in [210]. It is in particular a many-to-many independent product in the sense of Section 3.2.2.

Strong independence can also be studied in the context of sets of desirable gambles [209]: if $\mathcal{D}_1, \ldots, \mathcal{D}_n$ are maximal sets of desirable gambles as discussed in Section 1.5, their strong product corresponds to the smallest coherent set of desirable gambles that has $\mathcal{D}_1, \ldots, \mathcal{D}_n$ as marginals and satisfies epistemic independence; more generally, given marginal coherent sets of desirable gambles $\mathcal{D}'_1, \ldots, \mathcal{D}'_n$, their strong product is defined as the intersection of all strong products of all the $\mathcal{D}_1, \ldots, \mathcal{D}_n$, where each $\mathcal{D}_j$ is some maximal superset of $\mathcal{D}'_j$, for $j = 1, \ldots, n$.

3.2.5 The formalist approach to independence

We next turn to the more *formalist* approach, and introduce a number of notions that are related to proper generalizations of the factorization property for precise models. We consider a coherent lower prevision $\underline{P}_N$ on $\mathcal{L}(\mathcal{X}_N)$ representing a subject's assessments about the joint behaviour of the random variables $X_1, \ldots, X_n$, and look at a number of possible factorization properties for this joint model. This gives rise to *productive*, *factorizing*, *strongly factorizing*, *Kuznetsov* and *strongly Kuznetsov* coherent lower previsions.

Definition 3.3 (Productivity). A coherent lower prevision $\underline{P}_N$ on $\mathcal{L}(\mathcal{X}_N)$ is called *productive* if for all disjoint subsets I and O of N, all $g \in \mathcal{L}(\mathcal{X}_O)$ and all non-negative $f \in \mathcal{L}(\mathcal{X}_I)$, $\underline{P}_N(f[g - \underline{P}_N(g)]) \geq 0$.

The intuition behind this definition is that a coherent lower prevision $\underline{P}_N$ is productive if multiplying an almost desirable gamble on the variable X_O (the gamble $g - \underline{P}_N(g)$, which has lower prevision zero) with any non-negative gamble f that depends on a different variable X_I, preserves its almost-desirability, in the sense that the lower prevision of the product will still be non-negative. The condition holds trivially when I or O are empty. This condition underlies the very general laws of large numbers for coherent lower previsions established in [207].

Definition 3.4 (Factorization). A coherent lower prevision $\underline{P}_N$ on $\mathcal{L}(\mathcal{X}_N)$ is called

(i) *factorizing* if for all $o \in N$ and all $I \subseteq N \setminus \{o\}$, all $f_o \in \mathcal{L}(\mathcal{X}_o)$ and all non-negative $f_i \in \mathcal{L}(\mathcal{X}_i)$, $i \in I$, $\underline{P}_N(f_I f_o) = \underline{P}_N(f_I \underline{P}_N(f_o))$, where $f_I := \prod_{i \in I} f_i$;

(ii) *strongly factorizing* if $\underline{P}_N(fg) = \underline{P}_N(f\underline{P}_N(g))$ for all $g \in \mathcal{L}(\mathcal{X}_O)$ and non-negative $f \in \mathcal{L}(\mathcal{X}_I)$, where I and O are any disjoint subsets of N.

These two conditions turn out to be very useful in the context of research on credal networks [204].

Finally, there is the property that the late Russian mathematician Vladimir Kuznetsov first drew attention to [161, 414]. In order to define it, we use $\boxtimes$ to denote the (commutative and associative) *interval product* operator defined by:

$$[a, b] \boxtimes [c, d] := \{xy : x \in [a, b] \text{ and } y \in [c, d]\}$$
$$= [\min\{ac, ad, bc, bd\}, \max\{ac, ad, bc, bd\}] \text{ for all } a \leq b \text{ and } c \leq d \text{ in } \mathbb{R}.$$

Moreover, for any gamble f on $\mathcal{X}_N$, we denote by $\overline{\underline{P}}_N(f)$ the interval $[\underline{P}_N(f), \overline{P}_N(f)]$.

Definition 3.5 (Kuznetsov product). We call a coherent lower prevision $\underline{P}_N$ on $\mathcal{L}(\mathcal{X}_N)$

(i) a *Kuznetsov product*, or simply, *Kuznetsov*, if $\overline{\underline{P}}_N(\prod_{n\in N} f_n) = \boxtimes_{n\in N}\overline{\underline{P}}_N(f_n)$ for all $f_n \in \mathcal{L}(\mathcal{X}_n)$, $n \in N$;

(ii) a *strong Kuznetsov product*, or simply, *strongly Kuznetsov*, if $\overline{\underline{P}}_N(fg) = \overline{\underline{P}}_N(f) \boxtimes \overline{\underline{P}}_N(g)$ for all $g \in \mathcal{L}(\mathcal{X}_O)$ and all $f \in \mathcal{L}(\mathcal{X}_I)$, where I and O are any disjoint subsets of N.

These two properties are related to the sensitivity analysis interpretation of coherent lower previsions: if we consider a product of gambles and for each gamble we have an interval of possible values for its expectation, the Kuznetsov property implies that the interval of possible values for the expectation of the product coincides with the product of the intervals of the expectations of the different gambles. As before, the difference between being a Kuznetsov product and being a strong Kuznetsov product resides in whether the gambles we are multiplying depend on one variable only, or on several variables at once.

The relationship between these conditions is the following [210]:

Proposition 3.1 *Consider a coherent lower prevision $\underline{P}_N$ on $\mathcal{L}(\mathcal{X}_N)$. Then*

$$
\begin{array}{ccccc}
\underline{P}_N \text{ is strongly Kuznetsov} & \Rightarrow & \underline{P}_N \text{ is strongly factorizing} & \Rightarrow & \underline{P}_N \text{ is productive.} \\
\Downarrow & & \Downarrow & & \\
\underline{P}_N \text{ is Kuznetsov} & \Rightarrow & \underline{P}_N \text{ is factorizing.} & &
\end{array}
$$

Given marginal coherent lower previsions $\underline{P}_1, \ldots, \underline{P}_n$, there is usually no unique joint with one of these properties: for instance, both the independent natural extension mentioned in Section 3.2.2 and the strong product from Section 3.2.4 are (strongly) factorizing, and they do not coincide in general. In particular, the independent natural extension is not Kuznetsov nor strongly Kuznetsov in general, while the strong product is always strongly Kuznetsov (and, consequently, also Kuznetsov, (strongly) factorizing, and productive).

3.3 Invariance

A second important class of structural assessments concern considerations of symmetry involving a random variable X with set of possible values $\mathcal{X}$. Mathematically speaking, such symmetry is usually expressed as invariance with respect to a set of transformations of the set of possible values $\mathcal{X}$ for X.

Given two transformations T and S of $\mathcal{X}$, we can define new transformations $ST := S \circ T$ and $TS := T \circ S$ by $(ST)x := S(Tx)$ and $TSx := T(Sx)$ for all x in $\mathcal{X}$. So it is natural in this context to consider a set $\mathcal{T}$ of transformations of $\mathcal{X}$ that is a *monoid*, meaning that it is closed under composition – so $(\forall T, S \in \mathcal{T})(TS \in \mathcal{T})$ – , and moreover contains the *identity* transformation $id_{\mathcal{X}}$, defined by $id_{\mathcal{X}}x := x$ for all x in $\mathcal{X}$. In particular, $\mathcal{T}_{\mathcal{X}}$ denotes the monoid of *all* transformations of $\mathcal{X}$.

Since belief models – be they sets of gambles or (lower) previsions – are expressed using gambles f on $\mathcal{X}$, we need a way to turn a transformation of $\mathcal{X}$ into a transformation of

$\mathcal{L}(\mathcal{X})$. This is done by *lifting*: given any gamble f on $\mathcal{X}$, we denote by $T^t f$ the gamble $f \circ T$, meaning that

$$T^t f(x) := f(Tx) \text{ for all } x \text{ in } \mathcal{X}.$$

And given two transformations T and S on $\mathcal{X}$, then for any gamble f on $\mathcal{X}$ we see that

$$(ST)^t f = f \circ (S \circ T) = (f \circ S) \circ T = (S^t f) \circ T = T^t(S^t f),$$

so $(ST)^t = T^t S^t$, and lifting reverses the order of application of the transformations. Observe that any T^t is a *linear* transformation of the linear space $\mathcal{L}(\mathcal{X})$.

In what follows, we will discuss invariance for lower previsions, and for the mathematically equivalent models of sets of almost desirable gambles and credal sets; see Sections 1.6, 2.2.2 and 2.2.3 for more details about these models. It is also possible to discuss invariance for the more powerful and general model of sets of desirable gambles, but doing this in any detail would take us too far beyond the scope of this book; see for instance recent work on exchangeability [213] for suggestions on how to proceed.

3.3.1 Weak invariance

Consider a monoid $\mathcal{T}$ of transformations of $\mathcal{X}$. First of all, we want to express that a belief model about the value that the random variable X assumes in $\mathcal{X}$, exhibits a symmetry that is characterized by the transformations in $\mathcal{T}$. Thus, the notion of (weak) invariance of belief models that we are about to introduce is in a sense a purely mathematical one: it expresses that these belief models are left unchanged under the transformations in $\mathcal{T}$. Weak invariance represents symmetry of belief(model)s.

Definition 3.6 A coherent set of almost desirable gambles $\mathcal{D}_\sqsupseteq$ is called *weakly $\mathcal{T}$-invariant* if it is $\mathcal{T}$-invariant as a set of gambles, i.e., if $T^t\mathcal{D}_\sqsupseteq \subseteq \mathcal{D}_\sqsupseteq$ for all T in $\mathcal{T}$.

Since lifting turns any transformation T of $\mathcal{X}$ into a linear transformation T^t of the linear space $\mathcal{L}(\mathcal{X})$, we see that our definition of invariance is just a special case of a notion that is quite common in the mathematical literature. If $\mathcal{T}$ is a group (or at least left-cancellable), then the weak invariance condition is actually equivalent to $T^t\mathcal{D}_\sqsupseteq = \mathcal{D}_\sqsupseteq$ (with equality) for all T in $\mathcal{T}$.

Weak invariance is a mathematical notion that states that a subject's behavioural dispositions, as represented by a belief model $\mathcal{D}_\sqsupseteq$, are invariant under certain transformations. If we posit that a subject's dispositions are in some way a reflection of the evidence available to her, we see that weak invariance is a way to model 'symmetry of evidence'.

As we argued in Section 2.2, we can use coherent sets of almost desirable gambles, coherent lower previsions, and credal sets of linear previsions as mathematically equivalent representations of the behavioural dispositions of a subject. To show how to define weak invariance in terms of these alternative models, we first introduce the notion of transformation of a functional.

Definition 3.7 Let T be a transformation of $\mathcal{X}$ and let Λ be a real-valued functional defined on a T-invariant set of gambles $\mathcal{K} \subseteq \mathcal{L}(\mathcal{X})$. Then the *transformation* $T\Lambda$ of Λ is the real-valued functional defined on $\mathcal{K}$ by $T\Lambda := \Lambda \circ T^t$, or equivalently, by $T\Lambda(f) := \Lambda(T^t f) = \Lambda(f \circ T)$ for all gambles f in $\mathcal{K}$.

This allows us to establish the following result:

Theorem 3.1 ([206, Theorem 4]). *Let $\underline{P}$ be a coherent lower prevision on $\mathcal{L}(\mathcal{X})$, $\mathcal{D}_{\sqsupseteq}$ a coherent set of almost desirable gambles, and $\mathcal{M}$ a credal set on $\mathcal{L}(\mathcal{X})$. Assume that these belief models are equivalent. Then the following statements are equivalent:*

(i) *$\mathcal{D}_{\sqsupseteq}$ is weakly $\mathcal{T}$-invariant, in the sense that $T^t\mathcal{D}_{\sqsupseteq} \subseteq \mathcal{D}_{\sqsupseteq}$ for all T in $\mathcal{T}$;*

(ii) *$\underline{P}$ is* weakly $\mathcal{T}$-invariant, *in the sense that $T\underline{P} \geq \underline{P}$ for all T in $\mathcal{T}$, or equivalently $\underline{P}(T^t f) \geq \underline{P}(f)$ for all T in $\mathcal{T}$ and f in $\mathcal{L}(\mathcal{X})$;*

(iii) *$\mathcal{M}$ is* weakly $\mathcal{T}$-invariant, *in the sense that $T\mathcal{M} \subseteq \mathcal{M}$ for all T in $\mathcal{T}$, or equivalently, $TP \in \mathcal{M}$ for all P in $\mathcal{M}$ and all T in $\mathcal{T}$.*

In particular, a coherent prevision P on $\mathcal{L}(\mathcal{X})$ is weakly $\mathcal{T}$-invariant if and only if $TP = P$ for all T in $\mathcal{T}$. So for coherent previsions, we have an equality in the weak invariance condition.

If a coherent lower prevision is weakly $\mathcal{T}$-invariant, it is also weakly $\mathcal{T}'$-invariant for any sub-monoid of transformations $\mathcal{T}' \subseteq \mathcal{T}$. Hence, as we add transformations, the collection of weakly invariant belief models will not increase. The limit case is when we consider the class $\mathcal{T}_{\mathcal{X}}$ of all transformations on $\mathcal{X}$:

Example 3.5 Let $\mathcal{T}_{\mathcal{X}}$ be the monoid of all transformations of $\mathcal{X}$. Then the vacuous lower prevision $\underline{P}$ introduced in Section 2.2.2, and given by $\underline{P}(f) := \inf f$ for all $f \in \mathcal{L}(\mathcal{X})$ is the only weakly $\mathcal{T}_{\mathcal{X}}$-invariant coherent lower prevision. Equivalently, we can see that the vacuous coherent set of almost desirable gambles $\mathcal{L}^+(\mathcal{X})$ – respectively the credal set $\mathbb{P}(\mathcal{X})$ – is the only coherent set of almost desirable gambles – respectively credal set – to be weakly $\mathcal{T}_{\mathcal{X}}$-invariant. ♦

This example shows that a coherent lower prevision that is weakly invariant under a set of transformations may not be the lower envelope of a set of weakly invariant linear previsions, and therefore a structural judgement of weak invariance may not be compatible with a sensitivity analysis interpretation.

Interestingly, the natural extension of a weakly invariant coherent lower prevision is still weakly invariant. Therefore, if our subject wants to see which are the consequences of her assessments taking into account the implications of coherence and weak invariance, it suffices to consider only the first of these two properties when starting out with a weakly invariant assessment.

Theorem 3.2 ([206, Theorem 12]). *Let $\underline{P}$ be a lower prevision on some weakly $\mathcal{T}$-invariant set of gambles $\mathcal{K}$ that avoids sure loss and is weakly $\mathcal{T}$-invariant in the sense that $T\underline{P} \geq \underline{P}$ for all $T \in \mathcal{T}$. Then its natural extension $\underline{E}$ is still weakly $\mathcal{T}$-invariant, and consequently it is the point-wise smallest weakly $\mathcal{T}$-invariant coherent lower prevision on $\mathcal{L}(\mathcal{X})$ that dominates $\underline{P}$ on its domain $\mathcal{K}$.*

3.3.2 Strong invariance

Next, we turn to a stronger form of invariance, which is intended to model that a subject has *evidence of symmetry*: she believes that the (phenomenon underlying the) random variable X is subject to symmetry with respect to the transformations T in $\mathcal{T}$. This means that she has *reason not to distinguish* between a gamble f and its transformation $T^t f$, or in other words, that she is indifferent between them, in the sense discussed in Section 1.4.2. Let us consider the linear space

$$\mathcal{I}_{\mathcal{T}} := \operatorname{span}\{f - T^{t}f : f \in \mathcal{L}(\mathcal{X}) \text{ and } T \in \mathcal{T}\},$$

where $\operatorname{span}\mathcal{A} := \operatorname{posi}(\mathcal{A} \cup -\mathcal{A} \cup \{0\})$ denotes the *linear span* of the set of gambles $\mathcal{A}$: the linear space of all linear combinations of gambles in $\mathcal{A}$, or equivalently, the smallest linear subspace that includes $\mathcal{A}$. This leads to the following definition.

Definition 3.8 A coherent set of almost desirable gambles $\mathcal{D}_{\sqsupseteq}$ is called *strongly $\mathcal{T}$-invariant* if $f - T^{t}f \in \mathcal{D}_{\sqsupseteq}$ for all f in $\mathcal{L}(\mathcal{X})$ and all T in $\mathcal{T}$, or equivalently, if $\mathcal{I}_{\mathcal{T}} \subseteq \mathcal{D}_{\sqsupseteq}$.

The following theorem gives equivalent characterizations of strong invariance in terms of the alternative but equivalent types of belief models.

Theorem 3.3 ([206, Theorem 8]). *Let $\underline{P}$ be a coherent lower prevision on $\mathcal{L}(\mathcal{X})$, $\mathcal{D}_{\sqsupseteq}$ a coherent set of almost desirable gambles, and $\mathcal{M}$ a credal set on $\mathcal{L}(\mathcal{X})$. Assume that these belief models are equivalent. Then the following statements are equivalent:*

(i) $\mathcal{D}_{\sqsupseteq}$ *is strongly $\mathcal{T}$-invariant, in the sense that $\mathcal{I}_{\mathcal{T}} \subseteq \mathcal{D}_{\sqsupseteq}$;*

(ii) $\underline{P}$ *is* strongly $\mathcal{T}$-invariant, *in the sense that $\underline{P}(f - T^{t}f) \geq 0$ and $\underline{P}(T^{t}f - f) \geq 0$, and therefore $\underline{P}(f - T^{t}f) = \underline{P}(T^{t}f - f) = 0$ for all f in $\mathcal{L}(\mathcal{X})$ and T in $\mathcal{T}$;*

(iii) $\mathcal{M}$ *is* strongly $\mathcal{T}$-invariant, *in the sense that $TP = P$ for all P in $\mathcal{M}$ and T in $\mathcal{T}$.*

We deduce from this theorem that for precise models we cannot distinguish between symmetry of evidence (weak invariance) and evidence of symmetry (strong invariance): a linear prevision P is then called $\mathcal{T}$-invariant, and satisfies $TP = P$ for all T in $\mathcal{T}$; it is only for imprecise models that we can distinguish between the two notions.[5]

We also see that a strongly $\mathcal{T}$-invariant coherent lower prevision is always the lower envelope of some credal set of $\mathcal{T}$-invariant linear previsions. As a consequence, unlike weak invariance, strong invariance is guaranteed to be compatible with the sensitivity analysis interpretation of coherent lower previsions.

There are monoids $\mathcal{T}$ for which there are no strongly $\mathcal{T}$-invariant coherent (lower) previsions. Indeed, there are $\mathcal{T}$-invariant coherent previsions if and only if the set of almost desirable gambles $\mathcal{I}_{\mathcal{T}}$ avoids sure loss, which, taking into account Section 1.2.2 and the fact that $\mathcal{I}_{\mathcal{T}}$ is a linear space, is equivalent to the condition

$$\sup g \geq 0 \text{ for all } g \in \mathcal{I}_{\mathcal{T}}.$$

In that case, the smallest coherent set of almost desirable gambles that includes $\mathcal{I}_{\mathcal{T}}$, namely

$$\mathcal{I}'_{\mathcal{T}} := \bigcap_{\epsilon>0}\{f \in \mathcal{L}(\mathcal{X}) : f - \epsilon \geq g \text{ for some } g \in \mathcal{I}_{\mathcal{T}}\}, \tag{3.5}$$

the smallest coherent and strongly $\mathcal{T}$-invariant set of almost desirable gambles, or in other words, the belief model that represents evidence of symmetry involving the monoid $\mathcal{T}$.

[5] Compare also to Footnote 49, Chapter 7.

The corresponding lower prevision, defined by

$$\underline{E}_{\mathcal{T}}(f) = \sup\{\mu \in \mathbb{R} : f - \mu \in \mathcal{I}'_{\mathcal{T}}\} \tag{3.6}$$

is the point-wise smallest (least committal) strongly $\mathcal{T}$-invariant coherent lower prevision on $\mathcal{L}(\mathcal{X})$, and if we combine Equations (3.5) and (3.6), we find that

$$\underline{E}_{\mathcal{T}}(f) = \sup\{\inf[f - g] : g \in \mathcal{I}'_{\mathcal{T}}\} = \sup\{\inf[f - g] : g \in \mathcal{I}_{\mathcal{T}}\}.$$

These results expose some of the fundamental differences between weak and strong invariance: while strong invariance with respect to a larger number of transformations means that we must refine our beliefs – make them more precise –, this is not the case with weak invariance. Another difference is that by Theorem 3.3, if $\underline{P}$ is a coherent lower prevision that is strongly $\mathcal{T}$-invariant and $\underline{Q}$ dominates $\underline{P}$, then $\underline{Q}$ is also strongly $\mathcal{T}$-invariant. Example 3.5 shows that a similar property does not hold for weak invariance.

Example 3.6 One interesting example is that of the shift-invariant distributions on the natural numbers, which are those coherent lower previsions $\underline{P}$ on $\mathcal{L}(\mathbb{N}_0)$ that are invariant with respect to the monoid $\mathcal{T} := \{\theta^n : n \in \mathbb{N}_0\}$, where the map $\theta : \mathbb{N}_0 \to \mathbb{N}_0$ is given by $\theta(n) := n + 1$. The linear previsions that are (weakly) $\mathcal{T}$-invariant are called *Banach limits* in [74] and [672].

It follows from the results in this section that the lower envelope of all Banach limits is the smallest coherent lower prevision that is strongly $\mathcal{T}$-invariant. It is given by

$$\underline{E}_{\mathcal{T}}(f) = \liminf_{n} \inf_{k \geq 0} \sum_{\ell=k}^{k+n-1} f(\ell) \text{ for all gambles } f \text{ on } \mathbb{N}_0;$$

it does not coincide with the smallest weakly $\mathcal{T}$-invariant coherent lower prevision: the vacuous lower prevision on $\mathcal{L}(\mathbb{N}_0)$. See [206, § 8.1] for more information. ♦

3.4 Exchangeability

To give a further illustration of why invariance is important, we now focus on a particular instance of strongly invariant models: we consider a joint model for some random variables and we assume that this joint model is *strongly invariant* under permutations of them, so it represents the belief that the order of the variables does not matter. Such models are called *exchangeable*, and they were first studied in the precise case by Bruno de Finetti [216]. Exchangeability was later extended to the theory of coherent lower previsions by Walley [672, § 9.5]. Here, we follow the more detailed and extensive treatment by De Cooman et al. [213, 214].

By virtue of de Finetti's Representation Theorem [216], an exchangeable model can be seen as a convex mixture of multinomial models. This has given some ground [177, 216, 218] to the claim that aleatory probabilities and IID processes can be eliminated from statistics, and that we can restrict ourselves to considering exchangeable sequences instead.[6]

Consider $n \geq 1$ random variables $X_1, \ldots, X_n$ taking values in the same nonempty and finite set $\mathcal{X}$. A subject's beliefs about the values that these random variables assume jointly

[6] For a critical discussion of this claim, see [672, § 9.5.6].

in $\mathcal{X}_N$ is given by their (joint) distribution, which is a coherent lower prevision $\underline{P}_N$ defined on the set $\mathcal{L}(\mathcal{X}_N)$ of all gambles on $\mathcal{X}_N$.

Let us denote by $\mathcal{P}_N$ the set of all permutations of the set of indices $N := \{1, \dots, n\}$. With any such permutation π we can associate a permutation of $\mathcal{X}_N$, also denoted by π, that maps any sequence of observations $x_N = (x_1, \dots, x_n)$ in $\mathcal{X}_N$ to the permuted sequence $\pi x_N := (x_{\pi(1)}, \dots, x_{\pi(n)})$. Similarly, with any gamble f on $\mathcal{X}_N$, we can consider the permuted gamble $\pi^t f := f \circ \pi$, with in other words $(\pi^t f)(x_N) = f(\pi x_N)$ for all $x_N \in \mathcal{X}_N$, in the manner described in Section 3.3.

A subject judges the random variables $X_1, \dots, X_n$ to be *exchangeable* when she is disposed to exchange any gamble f for the permuted gamble $\pi^t f$ in return for any strictly positive price, meaning that $\underline{P}_N(\pi^t f - f) \geq 0$, for any permutation $\pi \in \mathcal{P}_N$. Taking into account the properties of coherence, this means that:

Definition 3.9 (Exchangeability). The random variables $X_1, \dots, X_n$ in $\mathcal{X}$ with joint lower prevision $\underline{P}_N$ are called *exchangeable* if:

$$\underline{P}_N(g) = \overline{P}_N(g) = 0 \text{ for all } g \in \mathcal{I}_{\mathcal{P}_N} := \operatorname{span}\{f - \pi^t f : f \in \mathcal{L}(\mathcal{X}_N) \text{ and } \pi \in \mathcal{P}_N\}.$$

In this case, we also call the joint coherent lower prevision $\underline{P}_N$ *exchangeable*.

A subject will make an assumption of exchangeability when she has evidence that the processes generating the values of the random variables are (physically) similar [672, § 9.5.2], and consequently the order in which the variables are observed is not important.

When $\underline{P}_N$ is in particular a linear prevision P_N, exchangeability is equivalent to having $P_N(\pi^t f) = P_N(f)$ for all gambles $f \in \mathcal{L}(\mathcal{X}_N)$ and all permutations $\pi \in \mathcal{P}_N$. In terms of the (probability) *mass function* p_N of P_N, defined by $p_N(x_N) := P_N(\{x_N\})$, this is equivalent to having $p_N(x_N) = p_N(\pi x_N)$ for all x_N in $\mathcal{X}_N$ and all $\pi \in \mathcal{P}_N$; in other words, the mass function p_N should be invariant under any permutation of the indices. This is essentially de Finetti's [216] definition for the exchangeability of a prevision. The following proposition, mentioned in [672, § 9.5], establishes an even stronger link between Walley's and de Finetti's notions of exchangeability.

Proposition 3.2 *Any coherent lower prevision on $\mathcal{L}(\mathcal{X})$ that dominates an exchangeable coherent lower prevision, is also exchangeable. Moreover, let $\underline{P}_N$ be the lower envelope of some set of linear previsions $\mathcal{M}$, and let $\mathcal{D}_{\sqsupseteq}$ be the coherent set of almost desirable gambles that is equivalent to $\underline{P}_N$. Then $\underline{P}_N$ is exchangeable if and only if all the linear previsions P_N in $\mathcal{M}$ are exchangeable, and if and only if $\mathcal{I}_{\mathcal{P}_N} \subseteq \mathcal{D}_{\sqsupseteq}$.*

This result is a immediate consequence of Theorem 3.3: exchangeability means strong invariance with respect to the set of all permutations of N, and as such it is equivalent to the invariance of all the dominating models.

3.4.1 Representation theorem for finite sequences

Consider any $x_N \in \mathcal{X}_N$, then the so-called (permutation) *invariant atom*

$$[x_N] := \{\pi x_N : \pi \in \mathcal{P}_N\}$$

is the smallest subset of $\mathcal{X}_N$ that contains x_N and is invariant under all permutations π in $\mathcal{P}_N$. We denote the partition of all permutation invariant atoms of $\mathcal{X}_N$ by $\mathcal{A}^N$. We can characterize the invariant atoms using the *counting maps* $T_x^N : \mathcal{X}_N \to \mathbb{N}_0$ defined for all x in $\mathcal{X}$ in such a way that

$$T_x^N(z_N) = T_x^N(z_1, \dots, x_n) := |\{k \in N : z_k = x\}|$$

is the number of components of the N-tuple z_N that assume the value x. $T_{\mathcal{X}}^n$ denotes the map from $\mathcal{X}_N$ to $\mathbb{N}_0^{\mathcal{X}}$ whose component maps are the T_x^n, $x \in \mathcal{X}$. Observe that $T_{\mathcal{X}}^n$ actually assumes values in the set of *count vectors*

$$\mathcal{N}^n := \left\{ m \in \mathbb{N}_0^{\mathcal{X}} : \sum_{x \in \mathcal{X}} m_x = n \right\}.$$

Since permuting the components of a sequence leaves the counts invariant – meaning that $T_{\mathcal{X}}^n(z_N) = T_{\mathcal{X}}^n(\pi z_N)$ for all $z_N \in \mathcal{X}_N$ and $\pi \in \mathcal{P}_N$ – , we see that for all y_N and z_N in $\mathcal{X}_N$:

$$y_N \in [z_N] \Leftrightarrow T_{\mathcal{X}}^n(y_N) = T_{\mathcal{X}}^n(z_N).$$

The counting map $T_{\mathcal{X}}^n$ can therefore be interpreted as a bijection between the set of invariant atoms $\mathcal{A}^n$ and the set of count vectors $\mathcal{N}^n$, and we can identify any invariant atom $[z_N]$ by the count vector $m = T_{\mathcal{X}}^n(z_N)$ of any (and therefore all) of its elements. We therefore also denote this atom by $[m]$; and clearly $y_N \in [m]$ if and only if $T_{\mathcal{X}}^n(y_N) = m$. The number of elements $\nu(m)$ in any invariant atom $[m]$ is given by the number of different ways in which the components of any z_N in $[m]$ can be permuted:

$$\nu(m) := \binom{n}{m} = \frac{n!}{\prod_{x \in \mathcal{X}} m_x!}.$$

If the random variable $X_N = (X_1, \dots, X_n)$ assumes the value z_N in $\mathcal{X}_N$, then the corresponding count vector assumes the value $T_{\mathcal{X}}^n(z_N)$ in $\mathcal{N}^n$: we can see $T_{\mathcal{X}}^n(X_N)$ as a random variable in $\mathcal{N}^n$. If the available information about the values that X_N assumes in $\mathcal{X}_N$ is given by the coherent exchangeable lower prevision $\underline{P}_N$, then the corresponding uncertainty model for the values that $T_{\mathcal{X}}^n(X_N)$ assumes in $\mathcal{N}^n$ is given by the coherent *induced* lower prevision $\underline{Q}_N$ on $\mathcal{L}(\mathcal{N}^n)$ – the distribution of $T_{\mathcal{X}}^n(X_N)$ – , given by

$$\underline{Q}_N(h) := \underline{P}_N(h \circ T_{\mathcal{X}}^n) = \underline{P}_N\left(\sum_{m \in \mathcal{N}^n} h(m) I_{[m]} \right) \text{ for all gambles } h \text{ on } \mathcal{N}^n.$$

Conversely, any exchangeable coherent lower prevision $\underline{P}_N$ is in fact *completely determined* by the corresponding distribution $\underline{Q}_N$ of the count vectors, also called its *count distribution*. This establishes a relationship between exchangeability and sampling without replacement.

In order to make this clear, we must introduce first the linear prevision $\mathrm{Hyp}^n(\cdot|m)$ in the following way:

$$\mathrm{Hyp}^n(f|m) := \frac{1}{\nu(m)} \sum_{z_N \in [m]} f(z_N) \text{ for any gamble } f \text{ on } \mathcal{X}_N.$$

This is the expectation operator for the uniform distribution on the invariant atom $[m]$, and also the expectation operator associated with a *multiple hyper-geometric distribution* [383, § 39], corresponding to sampling without replacement from an urn with n balls, whose possible types correspond to the elements of $\mathcal{X}$, and whose composition is determined by the count vector m: there are m_x balls of type x for any $x \in \mathcal{X}$.

The following theorem implies that any exchangeable coherent lower prevision on $\mathcal{X}_N$ can be associated with – or equivalently, that any collection of n exchangeable random variables in $\mathcal{X}$ can be seen as the result of – n random draws without replacement from an urn with n balls whose types are characterized by the elements x of $\mathcal{X}$, and whose composition M is unknown – a random variable in $\mathcal{N}^n$ – , but for which the available information about this composition is modelled by a coherent lower prevision on $\mathcal{L}(\mathcal{N}^n)$. It concerns the linear transformation $\text{proj}_{\mathcal{P}_N}$ of the linear space $\mathcal{L}(\mathcal{X}_N)$, defined by letting

$$\text{proj}_{\mathcal{P}_N}(f) := \frac{1}{|\mathcal{P}_N|} \sum_{\pi \in \mathcal{P}_N} \pi^t f \text{ for all gambles } f \text{ on } \mathcal{X}_N.$$

Theorem 3.4 (Finite Representation [214, Theorem 2]). *Let $\underline{P}_N$ be a coherent exchangeable lower prevision on $\mathcal{L}(\mathcal{X}_N)$, and f any gamble on $\mathcal{X}_N$. Then the following statements hold:*

(i) *The gamble $\text{proj}_{\mathcal{P}_N}(f)$ is permutation invariant, so $\pi^t \text{proj}_{\mathcal{P}_N}(f) = \text{proj}_{\mathcal{P}_N}(f)$ for all $\pi \in \mathcal{P}_N$. It is therefore constant on the permutation invariant atoms of $\mathcal{X}_N$, and can also be expressed by*

$$\text{proj}_{\mathcal{P}_N}(f) = \sum_{m \in \mathcal{N}^n} I_{[m]} \text{Hyp}^n(f|m).$$

(ii) $\underline{P}_N(f - \text{proj}_{\mathcal{P}_N}(f)) = \underline{P}_N(\text{proj}_{\mathcal{P}_N}(f) - f) = 0$, *whence* $\underline{P}_N(f) = \underline{P}_N(\text{proj}_{\mathcal{P}_N}(f))$.

(iii) $\underline{P}_N(f) = \underline{Q}_N(\text{Hyp}^n(f))$, *where* $\text{Hyp}^n(f)$ *is the gamble on* $\mathcal{N}^n$ *that assumes the value* $\text{Hyp}^n(f|m)$ *in* $m \in \mathcal{N}^n$.

Consequently a lower prevision on $\mathcal{L}(\mathcal{X}_N)$ is exchangeable if and only if it has the form $\underline{Q}_N \circ \text{Hyp}^n$, where $\underline{Q}_N$ is any coherent lower prevision on $\mathcal{L}(\mathcal{N}^n)$.

The lower prevision $\underline{Q}_N$ that represents the beliefs about the composition M of the urn is unique, and called the *count distribution* for the lower prevision $\underline{P}_N$.

3.4.2 Exchangeable natural extension

What will usually happen in practice, is that a subject makes an assessment that n random variables $X_1, \ldots, X_n$ taking values in a finite set $\mathcal{X}$ are exchangeable, and in addition specifies supremum acceptable buying prices $\underline{P}_N(f)$ for all gambles in some (typically finite, but not necessarily so) set of gambles $\mathcal{K} \subseteq \mathcal{L}(\mathcal{X}_N)$. The question then is: *can we turn these assessments into an exchangeable coherent lower prevision $\underline{E}_{\underline{P}_N,\mathcal{P}_N}$ defined on all of $\mathcal{L}(\mathcal{X}_N)$, that is furthermore as small (least-committal, conservative) as possible?*

It turns out that it is possible to determine the coherent lower prevision $\underline{E}_{\underline{P}_N,\mathcal{P}_N}$, which we will call the *exchangeable natural extension* of $\underline{P}_N$. Let $\underline{Q}_N$ be the lower prevision on the set $\text{Hyp}^n(\mathcal{K}) := \{\text{Hyp}^n(f) : f \in \mathcal{K}\} \subseteq \mathcal{L}(\mathcal{N}^n)$ given by

$$\underline{Q}_N(g) := \sup\{\underline{P}_N(f) : f \in \mathcal{K} \text{ and } \text{Hyp}^n(f) = g\} \text{ for all } g \in \text{Hyp}^n(\mathcal{K}). \tag{3.7}$$

Using the reasoning in Section 3.3.2 and [206, Theorem 16], we obtain the following:

Theorem 3.5 *There are exchangeable coherent lower previsions on $\mathcal{K}$ that dominate $\underline{P}_N$ on $\mathcal{K}$ if and only if the functional $\underline{Q}_N$ defined in Equation (3.7) is a lower prevision on* $\mathrm{Hyp}^n(\mathcal{K})$ *that avoids sure loss. In that case $\underline{E}_{\underline{P}_N,\mathcal{P}_N} = \underline{E}_{\underline{Q}_N} \circ \mathrm{Hyp}^n$: the count distribution for the exchangeable natural extension $\underline{E}_{\underline{P}_N,\mathcal{P}_N}$ of $\underline{P}_N$ is the natural extension $\underline{E}_{\underline{Q}_N}$ of the lower prevision $\underline{Q}_N$.*

Since there are quite efficient algorithms (see [679] and Section 16.2) for calculating the natural extension of a lower prevision based on a finite number of assessments, this theorem not only has intuitive appeal, but it provides us with an elegant and efficient manner to find the exchangeable natural extension, i.e., to combine (finitary) local assessments $\underline{P}_N$ with the structural assessment of exchangeability.

3.4.3 Exchangeable sequences

We next extend the results from Section 3.4.1 from finite to countable exchangeable sequences. Consider a countable sequence $X_1, \dots, X_n, \dots$ of random variables taking values in the same nonempty set $\mathcal{X}$. We can see them as a single random variable $X_{\mathbb{N}}$ assuming values in the set $\mathcal{X}^{\mathbb{N}}$, where $\mathbb{N}$ is the set of the natural numbers (zero not included). In its simplest form, we can model the available information about the value that $X_{\mathbb{N}}$ assumes in $\mathcal{X}^{\mathbb{N}}$ by a collection of coherent lower previsions $\underline{P}_{\{1,\dots,n\}}$ on $\mathcal{L}(\mathcal{X}^n)$ for all $n \in \mathbb{N}$, where each $\underline{P}_{\{1,\dots,n\}}$ models beliefs about the first n variables $X_1, \dots, X_n$.

Clearly, the family of coherent lower previsions $\underline{P}_{\{1,\dots,n\}}$, $n \in \mathbb{N}$ must satisfy the following '*time consistency*' requirement:

$$\underline{P}_{\{1,\dots,n\}}(f) = \underline{P}_{\{1,\dots,n+k\}}(\tilde{f}) \text{ for all } n, k \in \mathbb{N} \text{ and all gambles } f \text{ on } \mathcal{X}^n,$$

where $\tilde{f}$ denotes the cylindrical extension of f to $\mathcal{X}^{n+k}$, meaning that $\underline{P}_{\{1,\dots,n\}}$ is the $\mathcal{X}^n$-marginal of $\underline{P}_{\{1,\dots,n+k\}}$.

The following definition generalizes de Finetti's exchangeability condition:

Definition 3.10 The random sequence $X_{\mathbb{N}}$, or the time-consistent family of distributions $\underline{P}_{\{1,\dots,n\}}$, $n \in \mathbb{N}$ is called *exchangeable* if the distributions $\underline{P}_{\{1,\dots,n\}}$ are exchangeable for all $n \in \mathbb{N}$.

It turns out that exchangeable lower previsions for countable sequences also have a representation result, which generalizes de Finetti's famous Representation Theorem for countable sequences [216]. Consider the $\mathcal{X}$-simplex of all (probability) mass functions θ on $\mathcal{X}$:

$$\Sigma_{\mathcal{X}} := \left\{ \theta \in \mathbb{R}^{\mathcal{X}} : (\forall x \in \mathcal{X})\theta_x \geq 0 \text{ and } \sum_{x \in \mathcal{X}} \theta_x = 1 \right\}.$$

Given $\theta \in \Sigma_{\mathcal{X}}$, we can consider a sequence of n independent and identically distributed variables, each with this mass function θ. The so-called *multinomial* expectation operator for these variables is given by:

$$\begin{aligned}\mathrm{Mn}^n(f|\theta) &:= \sum_{z_N \in \mathcal{X}^n} f(z_N) \prod_{x \in \mathcal{X}} \theta_x^{T_x(z_N)} = \sum_{m \in \mathcal{N}^n} \mathrm{Hyp}^n(f|m)\nu(m) \prod_{x \in \mathcal{X}} \theta_x^{m_x} \\ &= \mathrm{CoMn}^n(\mathrm{Hyp}^n(f)|\theta) \text{ for all gambles } f \text{ on } \mathcal{X}_N,\end{aligned}$$

where the (so-called *multinomial count*) linear prevision $\mathrm{CoMn}^n(\cdot|\theta)$ is defined by

$$\mathrm{CoMn}^n(g|\theta) := \sum_{m \in \mathcal{N}^n} g(m)\nu(m) \prod_{x \in \mathcal{X}} \theta_x^{m_x} = \sum_{m \in \mathcal{N}^n} g(m)B_m(\theta) \text{ for any gamble } g \text{ on } \mathcal{N}^n.$$

The corresponding probability mass for any count vector m is given by

$$\mathrm{CoMn}^n(\{m\}|\theta) = \nu(m) \prod_{x \in \mathcal{X}} \theta_x^{m_x} =: B_m(\theta),$$

and the polynomial function B_m on the $\mathcal{X}$-simplex $\Sigma_{\mathcal{X}}$ is called a (multivariate) *Bernstein basis polynomial* of degree n [71, 440, 518]. These basis polynomials B_m, $n \in \mathcal{N}^n$ constitute a basis for the linear space $\mathcal{V}^n(\Sigma_{\mathcal{X}})$ of all (multivariate) polynomials on $\Sigma_{\mathcal{X}}$ of degree up to n:

$$\mathcal{V}^n(\Sigma_{\mathcal{X}}) = \{\mathrm{CoMn}^n(g) : g \in \mathcal{L}(\mathcal{N}^n)\} = \left\{ \sum_{m \in \mathcal{N}^n} g(m)B_m : g \in \mathcal{L}(\mathcal{N}^n) \right\},$$

where $\mathrm{CoMn}^n(g)$ is the polynomial on $\Sigma_{\mathcal{X}}$ that assumes the value $\mathrm{CoMn}^m(g|\theta)$ in $\theta \in \Sigma_{\mathcal{X}}$. The linear space of all polynomials on $\Sigma_{\mathcal{X}}$ – a subspace of $\mathcal{L}(\Sigma_{\mathcal{X}})$ – is given by:

$$\mathcal{V}(\Sigma_{\mathcal{X}}) := \{\mathrm{CoMn}^n(g) : g \in \mathcal{L}(\mathcal{N}^n), n \in \mathbb{N}\} = \{\mathrm{Mn}^n(f) : f \in \mathcal{L}(\mathcal{X}^n), n \in \mathbb{N}\}.$$

Theorem 3.6 (Representation Theorem [214, Theorem 5]). *Given a time-consistent family of exchangeable coherent lower previsions $\underline{P}_{\{1,\dots,n\}}$ on $\mathcal{L}(\mathcal{X}^n)$ with respective count distributions $\underline{Q}_{\{1,\dots,n\}}$, $n \in \mathbb{N}$, there is a unique coherent lower prevision $\underline{S}$ on the linear space $\mathcal{V}(\Sigma_{\mathcal{X}})$, called its* frequency representation, *such that for all $n \in \mathbb{N}$, $f \in \mathcal{L}(\mathcal{X}^n)$ and $g \in \mathcal{L}(\mathcal{N}^n)$:*

$$\underline{P}_{\{1,\dots,n\}}(f) = \underline{S}(\mathrm{Mn}^n(f)) \quad \textit{and} \quad \underline{Q}_{\{1,\dots,n\}}(g) = \underline{S}(\mathrm{CoMn}^n(g)).$$

Conversely, for any coherent lower prevision $\underline{S}$ on $\mathcal{V}(\Sigma_{\mathcal{X}})$, the lower previsions $\underline{P}_{\{1,\dots,n\}} := \underline{S} \circ \mathrm{Mn}^n$ constitute a time-consistent family of exchangeable coherent lower previsions.

Hence, the belief model for any countable exchangeable sequence of random variables in $\mathcal{X}$ can be completely and uniquely characterized by a coherent lower prevision on the linear space of all polynomial gambles on $\Sigma_{\mathcal{X}}$. In the particular case of a time consistent family of exchangeable *linear* previsions $P_{\{1,\dots,n\}}$ on $\mathcal{L}(\mathcal{X}^n)$, $n \in \mathbb{N}$, the frequency representation $\underline{S}$ will be a linear prevision S on $\mathcal{V}(\Sigma_{\mathcal{X}})$, fully characterized by its values $S(B_m)$ on the Bernstein basis polynomials B_m, $m \in \mathcal{N}^n$, $n \in \mathbb{N}$. This is the essence of de Finetti's Representation Theorem [216].

3.5 Further reading

A first detailed study of structural judgements in the context of imprecise probabilities was given by Walley [672, §§3 and 9]. In addition to some of the judgements summarised in this chapter, such as epistemic independence and exchangeability, Walley also discusses other types of structural judgements such as conditional independence [672, § 9.2.6] and robust Bernoulli models [672, § 9.6].

Below we discuss some additional references more specifically related to the structural judgements considered in this chapter: independence, invariance and exchangeability.

3.5.1 Independence

Besides the notions of independence for imprecise probabilities we have introduced in this chapter, there are other related concepts that may be of interest; see [158, 196] for an interesting overview. Of particular interest is the notion of *conditional independence*, only briefly touched upon in this chapter. This has been studied for instance in [179, 196, 210, 618]. There are results (see [210, §4.4]) that indicate that conditional independence may be modelled by means of a collection of epistemic independence assumptions, allowing us to essentially reduce this notion to the one discussed in Section 3.2.

A more detailed study on the relationships between the epistemic and the formalist approaches to independence in terms of coherent lower previsions can be found in [210]. For the formalist approach based on the generalization of the factorization property to the imprecise case, other interesting possibilities are the type-1 or type-2 products introduced in [672, § 9].

Finally, there are also discussions of independence for some of the special cases of coherent lower previsions to be introduced in Chapter 4: see [55, 157] for a survey of independence concepts within evidence theory, and [121, 200] for the particular case of possibility measures. Further information on this matter will be provided in Section 5.3.3.

3.5.2 Invariance

Our notion of a weakly invariant belief model corresponds to the notion of a 'reasonable (or invariant) class of priors' in [512], rather than to a 'class of reasonable (or invariant) priors', the latter being what our notion of strong invariance corresponds to. On the other hand, Walley [672, Def. 3.5.1] defines a $\mathcal{T}$-invariant lower prevision $\underline{P}$ as one for which $\underline{P}(T^t f) = \underline{P}(f)$ for all $T \in \mathcal{T}$ and all gambles f, so he requires equality rather than inequality, as we do here.

Most of the discussion in Section 3.3 can be found in a more detailed form in [206]. The interested reader can also find in this reference a number of additional results, such as a detailed account of the existence of strongly invariant coherent lower previsions or a connection with Choquet integration.

Finally, an interesting generalization of the notions of weak and strong invariance to a set of Markov operators can be found in [596].

3.5.3 Exchangeability

The first detailed study of exchangeability was made by de Finetti [216] (with the terminology of 'equivalent' events). An overview of de Finetti's work can be found in [218, §11.4] and [115]. Other important work on exchangeability was done by, amongst many others,

[234, 352, 360], and, in the context of the behavioural theory of imprecise probabilities by [672]. We refer to [387, 388] for modern, measure-theoretic discussions of exchangeability. Most of the results we have summarised in this section can be found in [214] (for lower previsions) and [213] (for sets of desirable gambles). In [213] the reader can also find out how to update belief models while maintaining exchangeability.

Acknowledgements

Work supported by project MTM2010-17844 and by the SBO project 060043 of the IWT-Vlaanderen.

4

Special cases

Sébastien Destercke[1] and Didier Dubois[2]

[1]*National Center for Scientific Research (CNRS), Heuristic and Diagnostic Methods for Complex Systems Laboratory (HeuDiaSyC), University of Technology, Compiègne, France*

[2]*National Center for Scientific Research (CNRS), Toulouse Research Institute on Computer Science (IRIT), Paul-Sabatier University, Toulouse, France*

4.1 Introduction

As argued in Chapters 1 and 2, lower previsions and sets of desirable gambles are very general models of uncertainty that have solid foundations and a clear behavioural interpretation. However, this generality goes along with a high computational complexity and a difficulty to easily explain such representations to users that are not experts in imprecise probability theories.

Therefore, in practical applications, simplified representations can greatly improve the applicability of imprecise probability theories. There are three main reasons for using simplified representations:

- to facilitate the elicitation or information collection process;
- to improve the computational tractability of mathematical models;
- to improve the interpretability of results when answering some particular question of interest.

The main objection to the use of such representations, or more precisely for restricting oneself to them, is that they may not be general enough to *exactly* model the available information. Moreover, even if some pieces of information can be exactly modelled by such simplified representations, further processing may result in information no longer

Introduction to Imprecise Probabilities, First Edition.
Edited by Thomas Augustin, Frank P. A. Coolen, Gert de Cooman and Matthias C. M. Troffaes.

exactly representable in a simple form (see Walley [674], for example). Indeed, if one works with processing tools specific to coherent lower previsions (and therefore closed for such representations), one can hardly expect more specific properties to be preserved (this remains true for other theories, see Chapter 5). Nevertheless, summarising complex outputs in the form of simplified representations may make these outputs easier to interpret.

In this chapter, we will review several simplified representations that are special cases of coherent lower previsions (actually, most of them are coherent lower probabilities). To do so, we will adopt the languages of the previous chapters. For each representation, we emphasize its interests and limitations.

4.2 Capacities and n-monotonicity

Capacities are set functions defined over the events of a possibility space $\mathcal{X}$ that can be used to represent uncertainty about an experiment.

Definition 4.1 Given a finite possibility space[1] $\mathcal{X}$, a *capacity* on $\mathcal{X}$ is a function g, defined on $\wp(\mathcal{X})$, such that:

- $g(\emptyset) = 0$, $g(\mathcal{X}) = 1$, and
- for all $A, B \subseteq \mathcal{X}$, $A \subseteq B$ implies $g(A) \leq g(B)$ (monotonicity property).

The monotonicity property stresses the fact that if $x \in A$ implies $x \in B$ then one cannot have more confidence in A than in B. Capacities were first introduced by Choquet [113], and are also known as fuzzy measures [486].

A capacity is said to be *super-additive* if the property

$$A \cap B = \emptyset \Rightarrow g(A \cup B) \geq g(A) + g(B) \tag{4.1}$$

holds for all events $A, B \subseteq \mathcal{X}$. The dual notion, called *sub-additivity*, is obtained by reversing the inequality. In particular, coherent lower and upper probabilities are super-additive and sub-additive capacities, respectively. Among capacities, those satisfying n-monotonicity are particularly interesting.

Definition 4.2 Let $n \in \mathbb{N}_0$, $n \geq 2$. A capacity g is said to be *n-monotone* whenever for any collection $\mathcal{A}_n \subseteq \wp(\mathcal{X})$ of n events, it holds that

$$g\left(\bigcup_{A \in \mathcal{A}_n} A\right) \geq \sum_{\emptyset \neq \mathcal{A}' \subseteq \mathcal{A}_n} (-1)^{|\mathcal{A}'|+1} g\left(\bigcap_{A \in \mathcal{A}'} A\right).$$

A capacity g is said to be ∞-*monotone*, *totally monotone* or *completely monotone*, whenever it is n-monotone for every $n \in \mathbb{N}_0$, $n \geq 2$.[2]

If a capacity is n-monotone, then it is m-monotone for all $m \in \mathbb{N}_0$, $2 \leq m \leq n$. Given any capacity g defined on a finite possibility space $\mathcal{X}$, we can define its Möbius inverse:

[1] (Pre-)Capacities on infinite spaces are discussed in the context of game-theoretic probability, see Page 131.

[2] The conjugate capacity $\bar{g}$ with $\bar{g}(A) := 1 - g(A^c)$ for all $A \subseteq \Omega$ is then *n-alternating*.

Definition 4.3 The *Möbius inverse* $m_g : \wp(\mathcal{X}) \rightarrow \mathbb{R}$ of a capacity g is defined, for every event $E \subseteq \mathcal{X}$, as

$$m_g(E) := \sum_{A \subseteq E} (-1)^{|E \setminus A|} g(A). \tag{4.2}$$

Conversely, from a Möbius inverse m_g, one can retrieve the value of $g(A)$ on any event A by computing

$$g(A) = \sum_{E \subseteq A} m_g(E). \tag{4.3}$$

Chateauneuf and Jaffray [110] have shown that the Möbius inverse m_g and the n-monotonicity of the capacity g are related in the following way: if a capacity g is n-monotone, then its Möbius inverse m_g is positive for every subset $A \subseteq \mathcal{X}$ such that $|A| \leq n$. Also note that a Möbius inverse m_g is always such that $\sum_{E \subseteq \mathcal{X}} m_g(E) = 1$. Two specific kinds of capacities are of particular importance for practical purposes: 2-monotone and ∞-monotone capacities.

4.3 2-monotone capacities

A 2-monotone capacity (a.k.a. convex capacity or supermodular function), that is a capacity that satisfies for every pair of events $A, B \subseteq \mathcal{X}$ the inequality

$$g(A \cup B) + g(A \cap B) \geq g(A) + g(B),$$

is a coherent lower probability $\underline{P}$. This means that it induces a nonempty credal set $\mathcal{M}(\underline{P})$, of which it is the lower envelope on events. In contrast, satisfying only super-additivity (4.1) does not guarantee to have $\mathcal{M}(\underline{P}) \neq \emptyset$ (for a counter-example, see [502]). Due to their practical interest, 2-monotone capacities have received particular attention in the literature [110, 672, 462, 86, 114]. When $\mathcal{X}$ is finite, 2-monotone capacities have the following interesting properties (all detailed in [110]):

- The natural extension $\underline{E}(f)$ of a 2-monotone lower probability $\underline{P}$ for any bounded gamble f is given by the Choquet integral (see Section 16.2.3 on Page 332 for details on the Choquet integral).
- Let $\mathcal{X} = \{x_1, \dots, x_k\}$, and let Σ denote the set of all permutations of $\{1, \dots, k\}$. For any permutation $\sigma \in \Sigma$, we can define a probability distribution P_σ as follows: for any $i \in \{1, \dots, k\}$,

  $$P_\sigma(\{x_{\sigma(i)}\}) := \underline{P}(A_{\sigma(i)}) - \underline{P}(A_{\sigma(i-1)}),$$

  with $A_0 = \emptyset$ and $A_{\sigma(i)} = \{x_{\sigma(1)}, \dots, x_{\sigma(i)}\}$. Then, the set of extreme points of the convex set $\mathcal{M}(\underline{P})$ is given by $\text{ext}(\mathcal{M}(\underline{P})) = \{P_\sigma : \sigma \in \Sigma\}$.

The above properties allow performing various computational tasks (coherence checking, statistical testing[3], ...) in an easier way than with generic coherent lower previsions. However, 2-monotone lower probabilities still require, in the general case, to store $2^{|\mathcal{X}|}$ values. Representations used in practice will, however, be of reduced complexity.

[3] See the so-called Huber-Strassen theory on generalized Neyman-Pearson testing outlined in Section 7.5.2.

4.4 Probability intervals on singletons

Probability intervals on singletons [194, 697] are popular representations that play an important role in graphical models (see Chapter 9) and in classification procedures (see Chapter 10). They are defined as lower and upper probabilities specified on singletons of a finite possibility space $\mathcal{X}$. That is, they correspond to assessments $\underline{P}(\{x\})$ and $\overline{P}(\{x\}) = 1 - \underline{P}(\{x\}^c)$ for every $x \in \mathcal{X}$.

De Campos *et al.* [194] have shown that probability intervals $\underline{P}$ avoid sure loss whenever

$$\sum_{x \in \mathcal{X}} \underline{P}(\{x\}) \leq 1 \leq \sum_{x \in \mathcal{X}} \overline{P}(\{x\}) \tag{4.4}$$

and that $\underline{P}$ is coherent when, for each $x \in \mathcal{X}$, the two following inequalities

$$\overline{P}(\{x\}) + \sum_{y \in \mathcal{X} \setminus \{x\}} \underline{P}(\{y\}) \leq 1 \quad \text{and} \quad \underline{P}(\{x\}) + \sum_{y \in \mathcal{X} \setminus \{x\}} \overline{P}(\{y\}) \geq 1 \tag{4.5}$$

hold. Compared to generic algorithms of Chapter 16, these criteria are easy to check. Provided $\underline{P}$ avoids sure loss (satisfies Equation (4.4)), De Campos *et al.* [194] also describe efficient methods to compute probability intervals that are coherent. From now on, we will only consider such coherent probability intervals.

The natural extension $\underline{E}$ of probability intervals $\underline{P}$ to any event $A \subseteq \mathcal{X}$ can easily be computed using the following formulas:

$$\underline{E}(A) = \max \left\{ \sum_{x \in A} \underline{P}(\{x\}), 1 - \sum_{x \in A^c} \overline{P}(\{x\}) \right\}, \tag{4.6a}$$

$$\overline{E}(A) = 1 - \underline{E}(A^c) = \min \left\{ \sum_{x \in A} \overline{P}(\{x\}), 1 - \sum_{x \in A^c} \underline{P}(\{x\}) \right\}. \tag{4.6b}$$

It can also be shown that probability intervals are 2-monotone, allowing one to use corresponding computational tools (De Campos *et al.* [194] also propose various algorithms optimized for probability intervals).

From any lower prevision $\underline{P}$, one can easily get outer-approximating probability intervals by computing the natural extension of $\underline{P}$ over events $\{x\}$ and $\{x\}^c$. Note that only $2|\mathcal{X}|$ values need to be stored to describe probability intervals, instead of the $2^{|\mathcal{X}|}$ needed values for generic 2-monotone capacities. As usual, this complexity reduction goes along with a limited expressive power (i.e., not all 2-monotone capacities can be expressed by probability intervals).

4.5 ∞-monotone capacities

∞-monotone capacities also play an important role as special cases of coherent lower probabilities. Indeed, given the relation between n-monotonicity and the Möbius inverse, any ∞-monotone lower probability $\underline{P}$ on a finite space $\mathcal{X}$ has a non-negative Möbius inverse $m_{\underline{P}} : \wp(\mathcal{X}) \to [0, 1]$. This is a characteristic property, as any mapping $m : \wp(\mathcal{X}) \setminus \emptyset \to [0, 1]$ such that $\sum_{A \subseteq \mathcal{X}} m(A) = 1$ will induce a ∞-monotone measure by using Equation (4.3).

4.5.1 Constructing ∞-monotone capacities

The non-negativeness of m means that it can be seen as a probability mass function defined over $\wp(\mathcal{X})$. In infinite spaces $\mathcal{X}$, random sets [222, 464] also induce ∞-monotone lower probabilities. Indeed, functions m often result from a situation where the available (statistical) pieces of information only partially determine the quantity of interest. This is typically the case when only a compatibility relation (instead of a mapping) between a probability space and the possibility space $\mathcal{X}$ of interest is available. Suppose there is a multimapping $\Gamma : \mathcal{Y} \to \wp(\mathcal{X})$ that defines for each value $y \in \mathcal{Y}$ of the quantity Y the set $\Gamma(y)$ of possible values of the ill-known quantity x in $\mathcal{X}$. If the subject knows $Y = y$, she only knows that $x \in \Gamma(y)$ and nothing else. From the knowledge of a probability function on $\mathcal{Y}$, only a mass assignment on $\mathcal{X}$ is derived, namely: $\forall E \subseteq \mathcal{X}, m(E) = P(\{y \in \mathcal{Y} : \Gamma(y) = E\})$ if $\exists y \in \mathcal{Y}, E = \Gamma(y)$, and 0 otherwise. The probability space $\mathcal{Y}$ can be considered as a sample space like in the framework of frequentist probabilities, but it is then assumed that observations are imprecise.[4]

Example 4.1 Consider an opinion poll pertaining to a French presidential election. The set of candidates is $\mathcal{X} = \{a, b, c, d, e\}$, going from left-wing ($\{a, b\}$) to right-wing ($\{d, e\}$). There is a population $\mathcal{Y} = \{y_1, \ldots, y_n\}$ of n individuals that supply their preferences. But since the opinion poll takes place well before the election, individuals may not have made a final choice, even if they do have an opinion. The opinion of individual y_i is modelled by the subset $\Gamma(y_i) \subseteq \mathcal{X}$. For instance, a left-wing vote is modelled by $\Gamma(y_i) = \{a, b\}$; for an individual having no opinion, $\Gamma(y_i) = \mathcal{X}$, etc. In this framework, if individual responses of this form are collected, $m(E)$ is the proportion of opinions of the form $\Gamma(y_i) = E$. ◆

Another method for constructing Γ can be devised when the frame $\mathcal{X}$ is multidimensional, i.e., $\mathcal{X} = \mathcal{X}_1 \times \mathcal{X}_2 \times \cdots \times \mathcal{X}_k$, and a probability distribution P is available on part of the frame, like $\mathcal{X}_1 \times \mathcal{X}_2 \times \cdots \times \mathcal{X}_i$, $i < k$, and there is a set of constraints relating the various variables $X_1, X_2, \ldots, X_k$, thus forming a relation R on $\mathcal{X}$. R represents all admissible tuples in $\mathcal{X}$. Let $\mathcal{Y} = \mathcal{X}_1 \times \mathcal{X}_2 \times \cdots \times \mathcal{X}_i$. Then if $y = (x_1, x_2, \ldots, x_i)$ and if $[y]$ denotes its cylindrical extension to $\mathcal{X}$, $\Gamma(y) = R \cap [y]$. Example 4.2 in the next section is of this kind.

4.5.2 Simple support functions

A particular instance of ∞-monotone capacities that plays an important role in other interpretations (see Chapter 5) are simple support functions.

Definition 4.4 A *simple support function* is a mass function where a mass $\alpha = m(A)$ is given to a set A and $1 - \alpha = m(\mathcal{X})$.

Such functions are often used to model the reliability of some source of information, as in the next example.

Example 4.2 Consider an unreliable watch. The failure probability ϵ is known. The set $\mathcal{Y}$ describes the possible states of the watch $\mathcal{Y} = \{KO, OK\}$. The subject cares for the time it is. So, $\mathcal{X}$ is the set of possible time-points (discretised according to the watch precision).

---

[4] See also Section 7.8.

Suppose the watch indicates time x. Then the multimapping Γ is such that $\Gamma(OK) = \{x\}$ (if the watch is in order, it provides the right time), and $\Gamma(KO) = \mathcal{X}$ (if the watch does not work properly, the time it is is unknown). The induced mass assignment on $\mathcal{X}$ is thus $m(\{x\}) = 1 - \epsilon$ and $m(\mathcal{X}) = \epsilon$, which is the probability of not knowing the time it is. ♦

4.5.3 Further elements

Apart from offering practical ways to build coherent lower probabilities from imprecise observations, the fact that m (or Γ) can be interpreted as a probability mass function (or a random variable) over $\wp(\mathcal{X})$ means that usual sampling methods such as Monte-Carlo methods can be used to simulate them. A good review of such methods is given by Wilson[5] [704]. In finite spaces $\mathcal{X}$, ∞-monotone lower probabilities are mathematically equivalent to so-called belief functions [576, 222] (see Chapter 5 for further discussion on belief function interpretations). This means that practical results concerning such models can be transposed to ∞-monotone capacities. For example, the fast Möbius transform [391] can be used to quickly compute $m_{\underline{P}}$, or approximation methods [225] can be used to reduce the representation complexity. Indeed, such models still generally require $2^{|\mathcal{X}|}$ values to be entirely specified, however in practice many sets $A \subseteq \mathcal{X}$ will be such that $m(A) = 0$ (e.g., k-additive belief functions [319], that will not be discussed further in this chapter, give positive masses only to sets whose cardinality is at most k).

In infinite settings, the definition of ∞-monotone lower probabilities poses tricky mathematical problems (initially discussed by Shafer [577] in the setting of belief functions and Matheron [452] in the setting of random sets). Nevertheless, it is possible to define a belief function on the reals, based on a continuous mass density bearing on closed intervals [617] (see Smets [600] for more details and Alvarez [14] for simulation techniques).

Except in special situations (see [697, p. 40] and [194, § 6]), there are no specific relations between ∞-monotone capacities and probability intervals studied in Section 4.4. However, several authors have proposed mappings from probability intervals to ∞-monotone capacities (see Denoeux [226], Hall and Lawry [332] and Quaeghebeur [520]).

4.6 Possibility distributions, p-boxes, clouds and related models

In this section, we study practical representations linked by the fact that they correspond to assessments over some collections of nested sets. Like previous models, they have a limited expressiveness but possess interesting properties making them handy.

4.6.1 Possibility distributions

A possibility distribution [248, 253, 246] is a function $\pi : \mathcal{X} \to [0, 1]$ with $\pi(x) = 1$ for at least one $x \in \mathcal{X}$. From this distribution can be defined a possibility measure $\overline{P}$ such that

$$\overline{P}(A) = \sup_{x \in A} \pi(x). \tag{4.7}$$

[5] The review concentrates on a different interpretation of m, but many tools can be used within a lower prevision approach.

This measure is supremum-preserving [202], in the sense that for any $\mathcal{A} \subseteq \wp(\mathcal{X})$, we have

$$\overline{P}\left(\bigcup_{A\in\mathcal{A}} A\right) = \sup_{A\in\mathcal{A}} \overline{P}(A).$$

$\overline{P}$ is a coherent upper probability, and the dual coherent lower probability (called necessity measure) is defined as $\underline{P}(A) = 1 - \overline{P}(A^c) = 1 - \sup_{x\in A^c} \pi(x)$. When $\mathcal{X}$ is finite, the supremum in Equation (4.7) can be replaced by a maximum.

This supremum-preserving property makes possibility measures very easy to use (i.e., evaluations of $\overline{P}(A)$ or $\underline{P}(A)$ are straightforward). On the other hand, this property induces that the inequality $\underline{P}(A) \leq \overline{P}(A)$ holds in a strong form, as $\underline{P}(A) > 0$ implies $\overline{P}(A) = 1$. In particular, we cannot have $\underline{P}(A) = \overline{P}(A)$ for values other than zero and one, hence possibility measures are unable to model linear previsions except in the most trivial cases, this in contrast with previous simple representations. However, possibility distributions and measures have strong connections with the probabilistic setting, as we will see next.

From a possibility distribution π and for any value $\alpha \in [0, 1]$, the strong and regular α-cuts are subsets respectively defined as $A_{\overline{\alpha}} = \{x \in \mathcal{X} : \pi(x) > \alpha\}$ and $A_\alpha = \{x \in \mathcal{X} : \pi(x) \geq \alpha\}$. These α-cuts are nested, since if $\alpha > \beta$, then $A_\alpha \subseteq A_\beta$. The linear previsions included in the corresponding credal sets $\mathcal{M}(\underline{P})$ can be characterized in a particularly interesting way [243], that links them to the concept of strong α-cut (see [155] for extensions on general spaces):

Proposition 4.1 *Given a possibility distribution π and the corresponding credal set $\mathcal{M}(\underline{P})$, we have for all α in $(0, 1]$, $P \in \mathcal{M}(\underline{P})$ if and only if*

$$1 - \alpha \leq P(A_{\overline{\alpha}})$$

Conversely, given any indexed family of nested intervals $A_\alpha, \alpha \in [0, 1]$, such that $A_\alpha \subseteq A_\beta$ whenever $\alpha \geq \beta$, the credal set $\mathcal{M} = \{P : P(A_\alpha) \geq 1 - \alpha\}$ defines a possibility measure, with distribution $\pi(x) = \inf_{x\notin A_\alpha} \alpha$ [243].

One useful consequence of this result is that any lower prevision $\underline{P}$ can be outer approximated by a possibility distribution, simply specifying a set of nested sets $A_1 \subset \ldots \subset A_n$ and considering the natural extension $\underline{E}(A_i)$ for each A_i. This also means that any probabilistic inequality of the form $\{P([x^* - \alpha a, x^* + \alpha a]) \geq f(\alpha) : \alpha \geq 0\}$, where x^* is some landmark value, is captured by a possibility distribution. For instance, in the Chebychev inequality, x^* is the mean, a is the variance, $f(\alpha) = \frac{a^2}{\alpha^2}$ and $\pi(x) = \min(1, 1 - \frac{a^2}{(x-x^*)^2})$. On finite spaces, the set $\{\pi(x) : x \in \mathcal{X}\}$ is of the form $\alpha_0 = 0 < \alpha_1 < \ldots < \alpha_M = 1$, and there are M distinct α-cuts in this case.[6] In practice, this means that the coherent lower probability induced by a possibility distribution can be expressed (in the finite case) by M lower bounds on the probability of nested events $A_{\overline{\alpha_i}}, i = 0, \ldots, M - 1$.

Finally, we note that lower probabilities induced by possibility distributions are special cases of ∞-monotone lower probabilities. The Möbius inverse m of the lower probability

[6] Note that this is true as long as π only assumes a finite number of values, even on non-finite spaces.

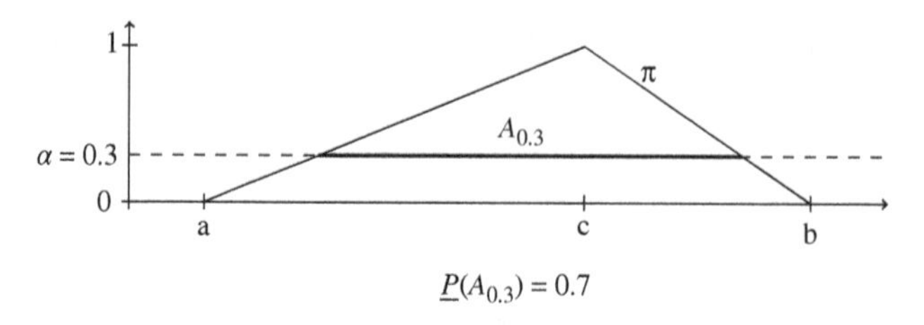

Figure 4.1 Illustration of a triangular fuzzy interval with α-cut of level 0.3.

$\underline{P}$ induced by π can be easily computed: assuming that elements of $\mathcal{X}$ are ranked such that $\pi(x_1) \geq \ldots \geq \pi(x_{|\mathcal{X}|})$, m is such that

$$m(E_i) = \pi(x_i) - \pi(x_{i+1}), i = 1, \ldots, |\mathcal{X}|, \tag{4.8}$$

where $E_i = \{x_1, \ldots, x_i\}$ and letting $\pi(x_{|\mathcal{X}|+1}) = 0$. Note that events E_i are nested. Shafer [576] showed that there is a one-to-one correspondence between possibility measures and upper probabilities $\overline{P}$ whose Möbius inverse is positive on a collection of nested sets only.

4.6.2 Fuzzy intervals

When working with possibility distributions on the real line, fuzzy intervals constitute by far the most usual representation.

Definition 4.5 A *fuzzy interval* $\pi : \mathbb{R} \to [0, 1]$ is a possibility distribution such that for all x and $y \in \mathbb{R}$,

$$\pi(z) \geq \min\{\pi(x), \pi(y)\} \text{ for all } z \in [x, y].$$

By definition a fuzzy interval is normalized in the sense that there is an $x \in \mathbb{R}$ such that $\pi(x) = 1$.

The α-cut A_α of a fuzzy interval π is an interval of the real-line. The piece of information A_α can then be processed by using classical interval analysis or optimization. Figure 4.1 pictures a triangular fuzzy interval together with an α-cut A_α. Fuzzy intervals have been proposed as a natural representation in practical situations where only limited (probabilistic) information is available:

- when the expert gives a set of confidence intervals around a 'best guess' value together with confidence levels (the result is then a possibility distribution with a finite number of values) [553];
- when only the support $[a, b]$ and modal value c of a distribution is known, it can be shown that the credal set induced by the triangular fuzzy set of Figure 4.1 includes all distributions having this support and this mode [50][7];

[7] Note that it also includes multi-modal distributions, and that the full language of lower previsions is needed to exactly represent this information.

- when only a handful of (not necessarily nested) sensor measurements are available [457];
- when considering probabilistic inequalities providing sets of confidence intervals around a central value (for instance the Chebychev inequality).

The features of fuzzy intervals make them a handy tool for various applications, including uncertainty propagation for risk analysis [498], uncertainty analysis in scheduling [242], signal filtering [439].

4.6.3 Clouds

As mentioned earlier, possibility distributions are useful but poorly expressive representations, as they cannot capture linear previsions. Clouds [492] are representations that extend possibility distributions while still remaining simple.

Definition 4.6 A *cloud* $[\delta, \pi]$ on $\mathcal{X}$ is a pair of mappings $\delta : \mathcal{X} \to [0, 1]$, $\pi : \mathcal{X} \to [0, 1]$ such that δ is point-wise less than π (i.e., $\delta \leq \pi$). Moreover, $\pi(x) = 1$ for at least one element x in $\mathcal{X}$, and $\delta(y) = 0$ for at least one element y in $\mathcal{X}$. δ and π are called the lower and upper distributions of the cloud $[\delta, \pi]$, respectively.

From a cloud $[\delta, \pi]$, Neumaier [492] defines the following probabilistic constraints for every $\alpha \in [0, 1]$:

$$P(B_\alpha := \{x \in \mathcal{X} : \delta(x) \geq \alpha\}) \leq 1 - \alpha \leq P(A_{\overline{\alpha}} := \{x \in \mathcal{X} : \pi(x) > \alpha\}). \tag{4.9}$$

These constraints are equivalent to the lower prevision such that $\underline{P}(A_{\overline{\alpha}}) = 1 - \alpha$ and $\underline{P}(B_\alpha^c) = \alpha$ for $\alpha \in [0, 1]$. If $\mathcal{X}$ is continuous, then $\mathcal{M}(\underline{P}) \neq \emptyset$ as soon as $[\delta, \pi]$ satisfies Definition 4.6. On finite spaces, $[\delta, \pi]$ must also satisfy the following condition [232]: $\forall A \subseteq \mathcal{X}$, $\max_{x \in A} \pi(x) \geq \min_{y \notin A} \delta(y)$. Clouds generalize possibility distributions in the sense that if $\delta = 0$, then $\underline{P}$ is the lower probability induced by the possibility distribution π. We also have that $\mathcal{M}(\underline{P}) = \mathcal{M}(\underline{P}_\pi) \cap \mathcal{M}(\underline{P}_{1-\delta})$, with $\underline{P}_\pi$ and $\underline{P}_{1-\delta}$ the lower probabilities induced by the possibility distributions π and $1 - \delta$, respectively.

In general, $\underline{P}$ is not 2-monotone [232], hence harming the practical interest of clouds. However, comonotonic clouds are specific clouds for which $\underline{P}$ is ∞-monotone.

Definition 4.7 A *comonotonic cloud* $[\delta, \pi]$ is defined as a cloud such that δ and π are comonotone, i.e., for any $x, y \in \mathcal{X}$, $\pi(x) > \pi(y) \Rightarrow \delta(x) \geq \delta(y)$

In the finite case, comonotonicity means that there exists a common permutation σ of $\mathcal{X} = \{x_1, \ldots, x_{|\mathcal{X}|}\}$ such that

$$\pi(x_{\sigma(1)}) \geq \pi(x_{\sigma(2)}) \geq \cdots \geq \pi(x_{\sigma(|\mathcal{X}|)})$$

and

$$\delta(x_{\sigma(1)}) \geq \delta(x_{\sigma(2)}) \geq \cdots \geq \delta(x_{\sigma(|\mathcal{X}|)})$$

The simplest comonotonic cloud, that is the cloud summarised by a constraint $\alpha \leq \underline{P}(A) \leq 1 - \beta$ with $\alpha \leq 1 - \beta$, provides an interesting example as it directly extends

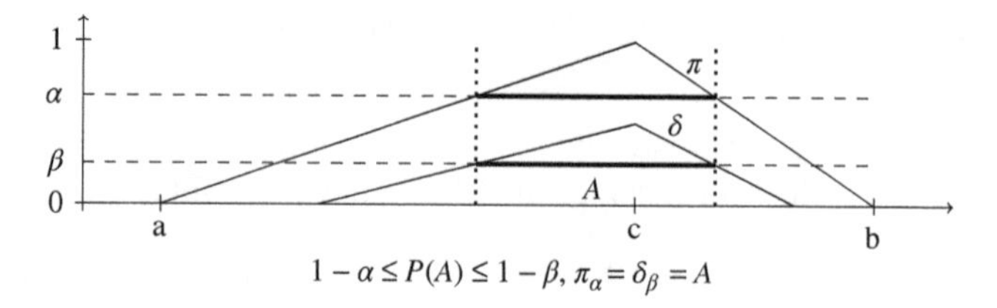

Figure 4.2 Illustration of a comonotonic cloud and of the bounds over event A.

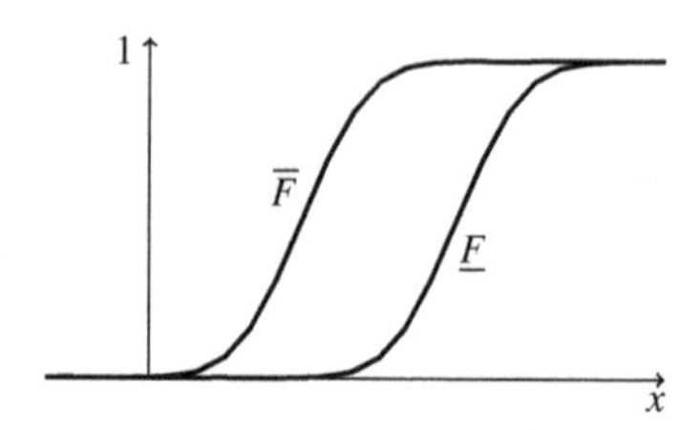

Figure 4.3 Illustration of p-box.

simple support functions of Section 4.5.2. Indeed, its Möbius inverse m is such that $m(A) = \alpha, m(A^c) = \beta, m(\mathcal{X}) = 1 - \alpha - \beta$, and simple support functions are retrieved when $\alpha = 0$ or $\beta = 0$.

A cloud is comonotonic if and only if the sets $\{A_\alpha, B_\alpha : \alpha \in [0, 1]\}$ form a nested sequence (i.e., they are completely ordered w.r.t. inclusion). In the light of constraints in Equation (4.9), this means that comonotonic clouds correspond to collections of nested sets to which are associated upper and lower probabilistic bounds (thus adding an upper bound to the lower bound of possibility distributions). Figure 4.2 illustrates the notion of comonotonic cloud on the real line.

As for possibility distributions, a lower probability induced by a cloud can be described (in the finite case) by M constraints corresponding to the M distinct values assumed by the cloud distributions. Conversely, any lower prevision $\underline{P}$ can be outer-approximated by a comonotonic cloud by simply specifying a set of nested sets $A_1 \subset \ldots \subset A_n$ and considering the lower and upper natural extensions $\underline{E}(A_i)$ and $\overline{E}(A_i)$ for each A_i. Recently, clouds have been proposed as practical representations to deal with robust design problems [306] and signal filtering [233].

4.6.4 p-boxes

A p-box [277] (short for probability box) was originally defined as a pair $[\underline{F}, \overline{F}]$ of cumulative distributions, an upper one $\overline{F} : \mathbb{R} \to [0, 1]$ and a lower one $\underline{F} : \mathbb{R} \to [0, 1]$ such that both $\overline{F}, \underline{F}$ are nondecreasing and $\exists r$ such that $\overline{F}(r) = \underline{F}(r) = 1$; see Fig. 4.3. This model corresponds to specifying lower and upper probabilities on events $(-\infty, x]$ such that $\underline{P}([-\infty, x]) = \underline{F}(x)$ and $\overline{P}([-\infty, x]) = 1 - \underline{P}((x, \infty]) = \overline{F}(x)$. The resulting lower probability $\underline{P}$ is coherent and ∞-monotone.

The concept of p-box has been extended to preordered spaces [231, 633]. This allows them to be defined on nested confidence regions, or to consider orders on $\mathbb{R}$ that differ from the natural order of numbers. In finite spaces $\mathcal{X}$, it has been shown that such p-boxes generalize possibility distributions and are actually equivalent to comonotonic clouds [232].

P-boxes have been proposed as a practical representation in situations where only limited (probabilistic) information is available [277, 50]:

- when the mean value μ and the support I of a random variable X are known. However, as emphasized in [50], it results in a fairly imprecise p-box. Actually, the language of lower previsions is needed to represent exactly this kind of information;
- when only a small number of samples is available. In this case, the use of Kolmogorov-Smirnov confidence limits [271] allow to build bounds over events of the type $[-\infty, x]$;
- when experts provide imprecise assessments about percentiles (already in Walley [672, § 4.6]).

To outer approximate a given lower prevision $\underline{P}$ by a p-box, it is sufficient to define a pre-order $\preceq$ on the possibility space $\mathcal{X}$ and to consider its lower and upper natural extension over events $[0_{\mathcal{X}}, x] := \{y \in \mathcal{X} : 0_{\mathcal{X}} \preceq y \preceq x\}$, where $0_{\mathcal{X}}$ denotes a smallest element of $\mathcal{X}$ with respect to $\preceq$.

P-boxes are most often used as initial or final representations in reliability and safety studies, as well as in (industrial) risk analysis (see, for example, Chapter 13.1). The reason for that is that cumulative distributions answer the question: 'Does quantity X exceeds a given (safety) threshold x or not?'

4.7 Neighbourhood models

Starting from an initial probability measure P_0, neighbourhood models consist in weakening this initial information by means of a parameter $\epsilon \in [0, 1]$ that we will call here *discounting parameter*. The three main neighbourhood models are the so-called *pari-mutuel model*(named after horse races, for it was used to determine betting rates), the *linear-vacuous* model and the *odds-ratio* model.

4.7.1 Pari-mutuel

The pari-mutuel model is described by the following constraint on the upper probability of every event A: $\overline{P}_{pm}(A) = \min\{(1 + \epsilon)P_0(A), 1\}$ and, by duality (conjugacy), $\underline{P}_{pm}(A) = \max\{(1 + \epsilon)P_0(A) - \epsilon, 0\}$. It can be shown (see [672, § 3.2.5]) that this lower probability is 2-monotone, hence previous results concerning this particular representation can be applied to it.

The *pari-mutuel model* inherits its name from the betting world. Consider a gambler betting on an event A and considering $P_0(A)$ as a fair price for a bet that returns 1 if A occurs. In this case, the gamblers gain is $I_A - P_0(A)$, while the house (a bookmaker, an insurance, …) gain is $P_0(A) - I_A$. In real-world situations, the house asks for a positive gain expectation. A way to insure such a positive expectation is to increase the price of the bet by $1 + \epsilon$, transforming house gain into $(1 + \epsilon)P_0(A) - I_A$. The coefficient ϵ is then interpreted as some kind of fee or commission coming from the house. The use of this pari-mutuel model and of the associated upper probability in insurance risk measurements is discussed by Pelessoni *et al.* [511].

4.7.2 Odds-ratio

The odds-ratio model is described by the following inequalities for each pair of events A, B:

$$\frac{P(A)}{P(B)} \leq (1-\epsilon)\frac{P_0(A)}{P_0(B)}.$$

In the finite case, this model corresponds to a finite set of quantitative comparisons between probabilities of events. Except for degenerate cases, it cannot be represented by means of lower probabilities alone (lower previsions are needed).

We will denote by $\underline{P}_{or}$ the lower prevision generated by this neighbourhood model. The odds-ratio neighbourhood model and the more general family of models called density ratio class in the robust Bayesian literature [64, § 4] are particularly interesting for statistical inference purposes (see Chapter 7). Indeed, the posterior generated from the combination of an odds-ratio prior with a likelihood function is also an odds-ratio model, which is not the case for other neighbourhood models.

4.7.3 Linear-vacuous

The linear-vacuous model, also called ϵ-contamination model in the robust Bayesian literature, corresponds to the convex mixture of the linear prevision P_0 (weighted by $(1-\epsilon)$) with the vacuous lower prevision $\inf_{\mathcal{X}}(f) := \inf_{x\in\mathcal{X}} f(x)$ (which is equivalent to the vacuous necessity measure, $N(A) = 0$ for all $A \subset \mathcal{X}$, or to the vacuous mass assignment $m(\mathcal{X}) = 1$). It can be described by the following constraint on the lower probability of every event A: $\underline{P}_{lv}(A) = (1-\epsilon)P_0(A)$. The natural extension $\underline{E}(f)$ of a linear-vacuous model to a gamble $f \in \mathcal{L}(\mathcal{X})$ is given by

$$\underline{E}(f) = (1-\epsilon)E_{P_0}(f) + \epsilon \inf_{\mathcal{X}}(f), \tag{4.10}$$

with E_{P_0} the expectation of f given P_0. Since both linear and vacuous lower previsions are ∞-monotone measures, so is the linear-vacuous model. In this model, $1-\epsilon$ can be seen as the probability that the information P_0 is reliable, hence ϵ is the probability that the source is unreliable, i.e., the probability that we know nothing about a particular variable value (explaining why we combine the initial assessment with the vacuous lower prevision). A generalized version of this scheme is Shafer's discounting technique for belief functions [576]. Example 4.2 is of this kind. Thanks to its simplicity, the linear-vacuous model has been used in many applications [716, 58].

This linear-vacuous model can be extended to the more general case where an information source provides an initial assessment in the form of a lower prevision $\underline{P}$ (or any other model interpretable as such). A lower-vacuous mixture $\underline{P}_\epsilon$ can then be defined, for any gamble f, as $\underline{P}_\epsilon(f) = (1-\epsilon)\underline{E}(f) + \epsilon \inf_{\mathcal{X}}(f)$, where $\underline{E}$ is the natural extension of $\underline{P}$. For the particular case of events, this gives $\underline{P}_\epsilon(A) = (1-\epsilon)\underline{P}(A)$. When $\underline{P}$ is ∞-monotone or when $\overline{P}$ is a possibility measure, this lower-vacuous mixture then coincides with the classical discounting operation of the corresponding theory. Also, simple support functions correspond to the case where $\underline{P}$ is itself vacuous w.r.t. some event E.

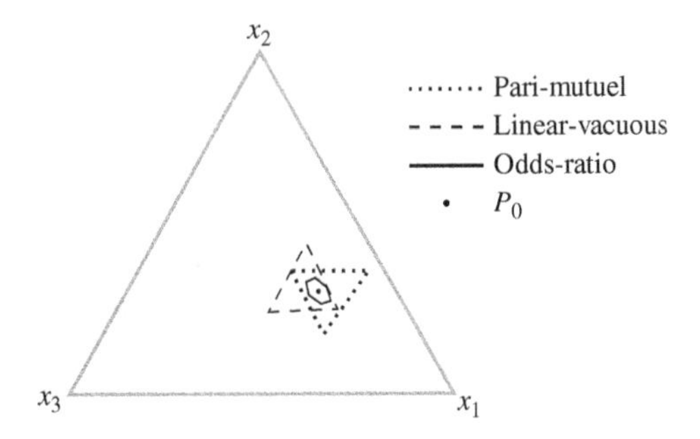

Figure 4.4 Neighbourhood models of Example 4.3.

4.7.4 Relations between neighbourhood models

For given ϵ and P_0, the credal set induced by the odds-ratio model is included in the credal sets induced by the linear-vacuous and the pari-mutuel model, while the two latter generally just overlap, that is, $\max\{\underline{P}_{lv}, \underline{P}_{pm}\} \leq \underline{P}_{or}$.

Example 4.3 (From [672, § 4.6]). Let us consider a probability P_0 on a 3 element space $\mathcal{X} = \{x_1, x_2, x_3\}$ with $P_0(\{x_1\}) = 0.5$, $P_0(\{x_2\}) = 0.3$, $P_0(\{x_3\}) = 0.2$, with a reliability $1 - \epsilon = 0.8$. The different credal sets induced by each neighbourhood model are illustrated in the simplex of Figure 4.4 (each point of the simplex is a probability distribution). Note that the odds-ratio neighbourhood is the only model whose edges are not parallel to one of the simplex edges, showing that it cannot be exactly represented by lower probabilities alone. ♦

4.8 Summary

In this section, we have reviewed the main practical representations that have emerged as instrumental tools for representing imprecise probabilities and computing with them.

However, while they may be helpful in many situations, such as uncertainty propagation in risk analysis or uncertainty assessment procedures, there are still a number of situations where such simplified representations will not be sufficient. For instance, practical representations presented here cannot handle the case where the information consists of quantitative comparative assessments of probabilities [530], nor can lower and upper probabilities. Other typical situations where the use of generic lower previsions may be needed is in the extension of classical statistical notions such as:

- statistical inference (see Chapter 7);
- combination of marginal assessments through independence concepts (see Chapter 3);
- combination of conditional assessments (see Chapter 2);
- extension of notions expressed by the means of expectation operators, such as characteristic or generating functions.

Indeed, for the above problems, even if one starts from simple representations, the representation resulting from information processing will usually not have the same properties.

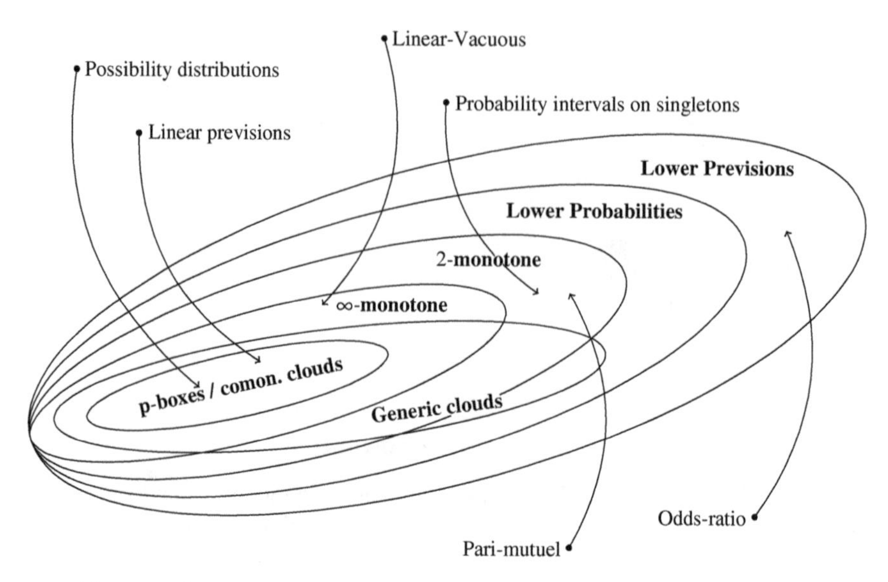

Figure 4.5 Summary of relations between practical representations.

Figure 4.5 summarises the relations between the different representations detailed in this chapter when $\mathcal{X}$ is finite. Note that for clarity purposes, some relations are not present, for instance probability intervals on singletons include linear-vacuous mixtures, which themselves include linear and vacuous previsions as special cases. Similarly simple support functions, that are special cases of possibility distributions, are not in the picture. Note that such relations may not hold in infinite spaces, as in this case the different models may not coincide in some specific situations.

Example 4.4 When $\mathcal{X}$ is infinite, finitely maxitive upper probabilities may fail to possess an underlying possibility distribution, even in the Boolean case. For instance suppose $\mathcal{X}$ is the set of natural integers, and $\overline{P}(A) = 1$ if A is infinite, 0 if finite, then $\overline{P}(\{x\}) = 0, \forall x \in \mathcal{X}$. Similarly, extending Proposition 4.1 is not trivial [155]. Relations between models of Figure 4.5 for infinite spaces largely remain to be explored, but we can mention some results: Miranda *et al.* [463] explore the relation between consonant random sets and possibility measures in the general case, while other works [635] explore the links between p-boxes defined on totally pre-ordered spaces and possibility distributions. ♦

5

Other uncertainty theories based on capacities

Sébastien Destercke[1] and Didier Dubois[2]

[1]*National Center for Scientific Research (CNRS), Heuristic and Diagnostic Methods for Complex Systems Laboratory (HeuDiaSyC), University of Technology, Compiègne, France*

[2]*National Center for Scientific Research (CNRS), Toulouse Research Institute on Computer Science (IRIT), Paul-Sabatier University, Toulouse, France*

Some of the simple mathematical models reviewed in Chapter 4 first emerged as basic building blocks of other uncertainty theories, and not as special instances of coherent lower previsions. In the following, the term *uncertainty theories* refers to several mathematical models and approaches devoted to the numerical or ordinal representation of beliefs induced by uncertainty due to the variability of phenomena, or the presence of incomplete information, or both. As these theories emerged from different areas and motivations, they led to different requirements, settings or axioms and to different tools for information processing. More precisely, we can enumerate several lines of thought motivating the development of uncertainty theories:

- The idea that in Bayesian approaches to statistics, prior probabilities on model parameters are hard to precisely specify, partly because there is no truly satisfying 'noninformative' prior, and partly because probabilities elicited from experts are most of the time imprecise. This is the robust Bayesian approach [64].
- A questioning of the classical fair betting behaviour framework of subjective probability [218] leading to upper and lower previsions for gambles (see Chapter 2).

Introduction to Imprecise Probabilities, First Edition.
Edited by Thomas Augustin, Frank P. A. Coolen, Gert de Cooman and Matthias C. M. Troffaes.

- Systematic deviations of human behaviour from Savage's [555] axioms for rational decision-making (especially the sure-thing principle) in the presence of ambiguity about probabilities (like in Ellsberg's urn experiment [264]). This is the non-additive expected utility trend, where subjective probability is replaced by more general set functions [109] (see Chapter 8).
- The claim that incomplete information is better captured by sets of more or less plausible states of nature. It is a subjective, often ordinal, view of plausibility advocated by Shackle [575], where impossibility is equated to the absence of surprise. This is also the starting point of the modal view of uncertainty first put forward by Hamblin [337] and Lewis [432], that is closely related to possibility theory [249].
- The handling of linguistic information understood as defining flexible constraints on values of parameters of interest (this is Zadeh's view of possibility theory [714]).
- The representation and fusion of uncertain evidence such as unreliable testimonies, a problem that goes back to the origins of probability theory. This is a topic of renewed interest in Artificial Intelligence and Robotics. This is the theory of belief functions by Shafer [576].
- The issue of plausible reasoning in inference tasks under incomplete statistical or commonsense information, where the choice of the proper reference class is the key problem. This concern goes from Kyburg's evidential probabilities to non-monotonic reasoning in Artificial Intelligence [422].

Apart from the approaches relying on convex sets of probability measures and lower expectations, presented in Chapters 1 and 2, the two main uncertainty representations in the literature that tolerate imprecision are possibility distributions and random disjunctive sets. This chapter will devote special attention to the theories that have emerged from them. These formalisms follow different calculi and use different basic building blocks, hence different languages. Even if these building blocks can be ordered in terms of generality and mathematically unified under the umbrella of lower previsions, this difference in language creates discrepancies when processing information that become like impediments preventing a complete unification. Some notions, easily expressible in special cases of lower previsions using the language of the specialized theory become very difficult to express in the more general setting that does not use this language. For instance the setting of possibility theory enables many information fusion operations to be defined, that have no clear counterparts in more general theories. Likewise, Dempster's rule of combination in evidence theory is expressed in terms of set functions that are not used in the theory of lower previsions, even if these set-functions (Möbius transforms, commonality) can be mathematically defined for lower previsions. In some cases, bridges can be built and equivalent notions identified between these frameworks, but sometimes only to a limited extent, as each theory uses different primitive notions. There is therefore a need for both unifying concepts when possible and clarifying the underlying assumptions made by each theory, so as to lay bare situations to which they apply, and better situate them with respect to the lower prevision approach put forward in this book.

The route taken in this chapter is different from the one adopted so far. Indeed, the framework of desirability assumes an additive linear bipolar 'currency' scale (the real line), and derives upper and lower probabilities and previsions from rationality requirements.

By linear bipolar value scale, we mean a scale where the zero plays a key-role for separating the good from the bad [254], and the positive and negative parts of the currency scale are ratio scales.

However, events can exist without the need to express them as numerical functions, and therefore one might want to develop uncertainty theories without such numerical underpinnings – this is particularly true if one starts from symbolic logic. Moreover, there are situations where coherence may not be essential: for instance when one receives new information that contradicts old information, it may be desirable to let go of some of the initial information, and perform a revision step. The result of revision protects the input information but is no longer consistent with the prior one. At least it achieves minimal change of the prior information so as to accept the input one. A special case of this situation is theory revision of Gärdenfors and colleagues in the setting of classical logic [11], a setting subsumed by lower previsions. Another example will be Dempster's rule of conditioning that operates a revision but does not preserve coherence.

In the first part of this chapter, we start from epistemic logic and derive the need for capturing imprecision in information representations. We then bridge the gap between uncertainty theories and epistemic logic showing that imprecise probabilities subsume modalities of possibility and necessity as much as probability. In fact it is the joint probabilistic and set-theoretic (and modal) nature of imprecise probabilities that makes this approach to uncertainty so rich. In the second part of this chapter, we present possibility and evidence theories, their origins, assumptions and semantics, discuss the connections between them and the general framework of imprecise probability. Finally, we point out the remaining discrepancies between the different theories regarding various basic notions, such as conditioning, independence or information fusion and the existing bridges between them.

5.1 Imprecise probability = modal logic + probability

Consider a variable X assuming values in a nonempty state space $\mathcal{X}$. The usual numerical representation of uncertainty consists of attaching to each statement of the form $x \in A \subset \mathcal{X}$ a degree of confidence $g(A)$ to each subset of $\mathcal{X}$. The confidence function thus defined is chosen to be a capacity (see Definition 4.1).

The prototypical example of a confidence function is a probability measure. While the frequentist view of probability looks natural for repeatable events, the notion of subjective probability is more problematic for singular events in a situation of incomplete information [255]. A piece of information concerning X is said to be *incomplete* in a given context if it is not sufficient to answer relevant questions in this context.

5.1.1 Boolean possibility theory and modal logic

The typical form of a piece of incomplete information is '$x \in E$' where E is a subset of $\mathcal{X}$ containing more than one element. An important remark is that elements in E are mutually exclusive, since X takes on a single value. Hence, a piece of incomplete information takes the form of a disjunction of mutually exclusive values. To emphasize this use of sets, we speak of *disjunctive* sets. Just like a subjective probability, a disjunctive set E represents the epistemic state of an agent about the ill-known value of X, and is attached to this agent.

This type of representation of incomplete information can be found in two areas: classical logic and interval analysis. The two settings differ by the type of variable used to describe the state space $\mathcal{X}$: Boolean for logic, numerical for intervals. For instance, to say that *Peter is between 20 and 25 years old*, i.e., $x = age(Peter) \in E = \{20, 21, 22, 23, 24, 25\}$, is to suppose $x = 20$ or $x = 21$ or $x = 22$ or $x = 23$ or $x = 24$ or $x = 25$. On the real line, incomplete information is often of the form $a \leq X \leq b$, where $E = [a, b]$. In logic, information comes in the form of a belief base K containing several propositions ϕ_i expressed in some Boolean language. The corresponding epistemic state is the disjunctive set of models of K.

There exist confidence measures that are more natural than probability measures to handle this situation. From a consistent epistemic state $x \in E \neq \emptyset$, we can define two binary confidence functions, namely a *possibility measure* Π such that $\Pi(A) = 1$ if $E \cap A \neq \emptyset$ and 0 otherwise, and a *necessity measure* N such that $N(A) = 1$ if $E \subseteq A$ and 0 otherwise. It is clear that $\Pi(A) = 1$ if and only if proposition '$x \in A$' is not inconsistent with information item '$x \in E$', and that $N(A) = 1$ if and only if proposition '$x \in A$' is entailed by information item $x \in E$.

Basically, $\Pi(A) = 0$ means that A is impossible if '$x \in E$' is true. $N(A) = 1$ expresses that A is certain if '$x \in E$' is true. Moreover to say that A is impossible ($A \cap E = \emptyset$) is to say that the opposite event A^c is certain. So, functions N and Π are related by the conjugateness property:

$$N(A) = 1 - \Pi(A^c). \tag{5.1}$$

This is the Boolean version of possibility theory [249]. Possibilistic uncertainty is clearly at work in classical logic. If a set of formulas K in propositional logic has $E \subseteq \mathcal{X}$ as a set of models (in this case, $\mathcal{X}$ is the set of interpretations of the language), and if proposition p is the syntactic form of proposition $x \in A$, then $N(A) = 1$ if and only if K implies p, and $\Pi(A) = 0$ if and only if $K \cup \{p\}$ is logically inconsistent.

However, in propositional logic, it cannot be expressed that $N(A) = 0$ nor $\Pi(A) = 1$. To do so, a modal logic is needed (Chellas [111]), that prefixes propositions with modalities such as possible ($\Diamond$) and necessary ($\Box$): in a modal belief base K^{mod}, $\Diamond p \in K^{mod}$ encodes $\Pi(A) = 1$, and $\Box p \in K^{mod}$ encodes $N(A) = 1$ (the latter is encoded by $p \in K$ in classical logic). The well-known conjugateness relation in modal logic reads: $\Diamond p = \neg \Box \neg p$.

Within this setting, possibility and necessity measures naturally satisfy the following equalities, respectively:

$$\Pi(A \cup B) = \max\{\Pi(A), \Pi(B)\} \tag{5.2}$$

$$N(A \cap B) = \min\{N(A), N(B)\} \tag{5.3}$$

besides $N(\mathcal{X}) = \Pi(\mathcal{X}) = 1$ and $\Pi(\emptyset) = N(\emptyset) = 0$. Possibility measures are said to be *maxitive*, while necessity measures are said to be *minitive*. These properties are taken as postulates in the finite setting, even when possibility and necessity values lie in the unit interval $[0, 1]$ and are no longer Boolean. They also do not rely on a numeric setting, as taking the minimum and maximum of N and Π also makes sense when they range on a lattice [198]. In the modal logic setting, these axioms respectively read $\Diamond(p \vee q) = \Diamond p \vee \Diamond q$ and $\Box(p \wedge q) = \Box p \wedge \Box q$ and are well-known to hold in regular modal logics. A simple modal logic for Boolean possibility theory can be devised from a propositional language. It corresponds to a fragment of the logic KD. Atoms of this logic are of the form $\Box p$, where p is any propositional formula. Well-formed formulas in this logic are obtained by applying standard conjunction and negation to these atoms. Models of such modal formulas are epistemic states: for instance E is a model of $\Box p$ means that $E \subseteq A$, where A is the set of models of p [43].

Recall (see Section 4.6.1) that a possibility measure Π (and its dual necessity measure N) is entirely determined by the knowledge of a so-called possibility distribution $\pi : \mathcal{X} \to [0, 1]$ such that $\pi(x) := \Pi(\{x\})$. In the following, we extend this modal framework for possibility and certainty to more general numerical settings.

5.1.2 A unifying framework for capacity based uncertainty theories

All numerical representations of incompleteness-tolerant uncertainty have the common feature that the uncertainty of each event $A \subseteq \mathcal{X}$ is characterized by two, respectively upper and lower, evaluations. We shall respectively call them (adopting a subjectively biased language) degrees of *epistemic possibility* and *certainty* (in order to differentiate them from coherent lower and upper probabilities), that will be denoted Ep and Cer. Epistemic possibility refers to a lack of surprise and generalizes logical consistency (hence the possibility modality and set function Π), while certainty generalizes logical entailment (hence the necessity modality and set function N). These two degrees define monotonic confidence functions (capacities) from the frame $\mathcal{X}$ to $[0, 1]$ such that

$$\forall A \subseteq \mathcal{X}, \mathrm{Cer}(A) \leq \mathrm{Ep}(A). \tag{5.4}$$

They are supposed to be conjugate to each other, namely:

$$\forall A \subseteq \mathcal{X}, \mathrm{Cer}(A) = 1 - \mathrm{Ep}(A^c). \tag{5.5}$$

The first condition, Equation (5.4), postulates that an event must be epistemically possible prior to being certain, and the second condition, Equation (5.5), says that an event is all the more certain as its opposite is less epistemically possible. This framework has the merit of unambiguously encoding three extreme epistemic attitudes pertaining to event A:

- The case when A is certainly true: $\mathrm{Cer}(A) = 1$ (hence $\mathrm{Ep}(A) = 1, \mathrm{Ep}(A^c) = 0, \mathrm{Cer}(A^c) = 0$).
- The case when A is certainly false: $\mathrm{Ep}(A) = 0$ (hence $\mathrm{Cer}(A) = 0$).
- The case when the agent does not know if A is true or false: $\mathrm{Cer}(A) = 0$ and $\mathrm{Ep}(A) = 1$ (then $\mathrm{Ep}(A^c) = 1$; $\mathrm{Cer}(A^c) = 0$).

It is clear that probability measures, as well as Boolean possibility and necessity measures are special cases of this setting. In the case of probability measures, $\mathrm{Cer}(A) = \mathrm{Ep}(A) = P(A)$. Similarly, lower and upper probabilities on event A are important general examples of epistemic certainty ($\mathrm{Cer}(A) = \underline{P}(A)$) and of epistemic possibility ($\mathrm{Ep}(A) = \overline{P}(A)$), respectively. It is clear that super-additivity implies $g(A) + g(A^c) \leq 1$, so that a super-additive capacity is a certainty function Cer, while a sub-additive capacity is an epistemic possibility function.

5.2 From imprecise probabilities to belief functions and possibility theory

All set functions proposed for the representation of uncertainty obey the above conventions for epistemic certainty and possibility functions, hence extending the framework detailed in

Section 5.1.1. As seen in Chapter 4, such set functions can be interpreted (without considering conditioning) as special cases of coherent upper and lower probabilities. However, they also stand as proper extensions of Boolean possibility theory, as discussed in this section.

A first idea to extend the framework of Section 5.1.1 is to consider a probability distribution over the set of epistemic states. A second idea is to turn epistemic sets into fuzzy sets, allowing for degrees of possibility.

5.2.1 Random disjunctive sets

The approach adopted in the theory of evidence (Shafer [576]) is somewhat opposite to the one of the lower prevision school. Instead of augmenting the probabilistic approach with higher order uncertainty due to incompleteness, described by sets of probabilities or equivalent models, the idea is to add higher order probabilistic information to the disjunctive set approach to incompleteness. So, instead of a single representation of the form $x \in E$, a (generally) discrete probability distribution is defined over the various possible assertions of the form $x \in E$ (assuming a finite frame $\mathcal{X}$).

5.2.1.1 Mass functions

Let m be a probability distribution over the power set $\wp(\mathcal{X})$. The function m is called *mass assignment*, $m(E)$ is the belief mass allocated to the set E, and any subset E of $\mathcal{X}$ such that $m(E) > 0$ is called a *focal set*. Let $\mathcal{F} \subseteq \wp(\mathcal{X})$ be the collection of focal sets of a given m. Usually, no positive mass is assigned to the empty set ($m(\emptyset) = 0$ is assumed). However, the Transferable Belief Model, after Smets [601], does not make this assumption. In this case $m(\emptyset)$ represents the degree of internal conflict of the mass assignment. This is an essential difference with the theory of lower previsions, since the notion of coherence is central in the latter, while allowing for $m(\emptyset) > 0$ precisely means that some degree of incoherence can be accepted in the Transferable Belief Model. Indeed, the condition $m(\emptyset) = 0$ is a form of normalization. As m is a probability distribution, the condition $\sum_{E \subseteq \mathcal{X}} m(E) = 1$ must hold anyway. Also note that the mass assignment m is here considered as the basic building block of the theory, while in Section 4.5 it is just the Möbius transform of a lower probability, whose positivity is a useful by-product of the ∞-monotone property.

In this hybrid representation of uncertainty, it is important to understand the meaning of the mass function, and it is essential not to confuse $m(E)$ with the probability of occurrence of event E.

- Shafer [576] says that $m(E)$ is the belief mass assigned to E only and to none of its subsets.
- One may also see $m(E)$ as the amount of probability to be shared between elements of E without being assigned yet, by lack of knowledge. This is in agreement with a frequentist view of the mass function and the idea of incomplete observation.[1]
- An explanation in the subjective line consists in saying that $m(E)$ is the probability that the subject only knows that $x \in E$. So, there is an epistemic modality implicitly present in $m(E)$, but absent from $P(A)$ (see [43] for the connection between modal logic and mass assignments).

[1] See also the construction in the context of data imprecision, most notably Sections 7.8.2 and 7.8.3.

These views explain why the function m is not required to be inclusion-monotonic. It is allowed to have $m(E) > m(F) > 0$ even if $E \subset F$, when the subject is sure enough that what is known is of the form $x \in E$. In particular, $m(\mathcal{X})$ is the probability that the agent is completely ignorant about the value of a variable X. We have already mentioned in Section 4.5 a number of situations from which mass functions can arise, the most well-known being a multivalued mapping between a probability space and a space $\mathcal{X}$, as proposed by Dempster [222].

Note that m can be formally associated to a random set such that $m(E)$ is the probability of observing the (random) disjunctive set E. However, in Dempster's, Shafer's and Smets's views, this is a random disjunctive set, in contrast with other views of random sets [390, 452], where sets are conjunctions of their elements, i.e., they represent real objects (like a region in a land). This difference is crucial when considering derived set-functions and operations such as conditioning, which may differ in both situations.

5.2.1.2 Belief and plausibility functions

A mass assignment m induces two set-functions, respectively a belief function Bel and a plausibility function Pl, defined by:

$$\mathrm{Bel}(A) = \sum_{E \subseteq A, E \neq \emptyset} m(E); \tag{5.6}$$

$$\mathrm{Pl}(A) = \sum_{E \cap A \neq \emptyset} m(E). \tag{5.7}$$

When $m(\emptyset) = 0$, it is clear that $\mathrm{Bel}(\mathcal{X}) = \mathrm{Pl}(\mathcal{X}) = 1$, $\mathrm{Pl}(\emptyset) = \mathrm{Bel}(\emptyset) = 0$, and $\mathrm{Bel}(A) = 1 - \mathrm{Pl}(A^c)$ so that these functions are another example of certainty (Cer = Bel) and epistemic possibility (Ep = Pl).

At the interpretive level, Bel and Pl can be viewed as direct extensions of Boolean necessity and possibility functions respectively. They are recovered when $m(E) = 1$ for some focal subset E. In fact, $\mathrm{Bel}(A)$ is an expected necessity degree (in the sense of m), or a probability of provability [506] (the probability that m logically implies A). Likewise, $\mathrm{Pl}(A)$ is an expected possibility degree (in the sense of m), or a probability of consistency (the probability that m is logically consistent with A). In terms of modal operations $\Box, \Diamond$ introduced earlier, it is clear that $\mathrm{Bel}(A) = P(\Box A)$ and $\mathrm{Pl}(A) = P(\Diamond A)$.

5.2.1.3 Belief functions and lower probabilities

If $m(\emptyset) = 0$, belief functions coincide with ∞-monotone capacities. However, if $m(\emptyset) > 0$ then $\mathcal{M}(\mathrm{Bel}) = \emptyset$ and the connection with imprecise probability theory is lost.

It is also important to note that, when a mass assignment m is generated by a random variable Y imprecisely observed through a multivalued mapping $\Gamma : \mathcal{Y} \to \wp(\mathcal{X})$, the credal set $\mathcal{M}(\mathrm{Bel})$ is usually less informative than Γ [464]. In that case, the belief function is viewed as a disjunctive set of random variables on $\mathcal{X}$, each being defined by a selection function $\phi : \mathcal{Y} \to \mathcal{X}$ of Γ. So, for each element $y \in \mathcal{X}$, the probability mass $P(\{y\})$ should be entirely transferred to one element $x = \phi(y) \in \Gamma(y)$, and the resulting set of probabilities $\mathcal{M}(\Gamma) = \{P_\phi : \phi \in \Gamma\}$ is not generally convex. In contrast, when building the credal set $\mathcal{M}(\mathrm{Bel})$, masses on focal subsets can be redistributed arbitrarily among the elements of the focal sets, and $\mathcal{M}(\Gamma) \subset \mathcal{M}(\mathrm{Bel})$.

Example 5.1 Consider a set of ages discretized in periods of 5 years ($\mathcal{Y} = \{y_1 = [16, 20], y_2 = [21, 25], \ldots\}$). We are interested in the knowledge on the precise age X of Pierre ($\mathcal{X} = \{15, \ldots, 100\}$). All we know is that *Pierre is between 21 and 25 years old*. Y is reduced to the Dirac measure $P(\{y_2\}) = 1$, with $\Gamma(y_2) = A = \{21, \ldots, 25\}$. The set $\mathcal{M}(\Gamma)$ corresponds to the five Dirac probabilities focusing on singletons in A. The corresponding credal set contains all probability measures with support within A. ♦

Since the convex hull of $\mathcal{M}(\Gamma)$ equals $\mathcal{M}(\text{Bel})$, they have the same behavioural consequences in the theory of lower previsions, i.e., expectations of any function $f : \mathbb{R} \to \mathcal{X}$ have the same lower and upper bounds for $\mathcal{M}(\Gamma)$ and $\mathcal{M}(\text{Bel})$. However, this equality may not hold for other (nonlinear) functionals like variance. In the above example, the set of possible variances of probability distributions in $\mathcal{M}(\Gamma)$ is reduced to the singleton $\{0\}$, while it is not so for distributions in $\mathcal{M}(\text{Bel})$ (for instance, the uniform distribution, which has a non-null variance, is in $\mathcal{M}(\text{Bel})$). See [154] for a discussion about variance in Dempster-Shafer theory.

5.2.2 Numerical possibility theory

The roots of possibility theory can be traced back to the degree of potential surprise introduced by the economist Shackle [575], and its revival is mainly due to Zadeh [714] (starting from the representation of linguistic information). Other scholars dealing with various aspects of possibility theory are Lewis [432], Dubois and Prade [248, 253, 246], and De Cooman [198, 199, 200]. In this section, we focus on the numerical representation of possibility and necessity.

5.2.2.1 Possibility distributions

The basic idea of possibility theory is to refine epistemic states $X \in E$ by introducing degrees of epistemic possibility of elements. The disjunctive set E is replaced by a fuzzy set, namely by a mapping $\pi : \mathcal{X} \to L$, where L is a possibility scale. L must be at least a complete lattice bounded from above and from below by 1 and 0 respectively; more often it is a totally ordered set, for instance the unit interval. It is generally assumed that $\pi(x) = 1$ for some x. This is not always required, meaning that the theory permits the presence of some inconsistency degree, as for the case of belief functions. More generally, following Shackle's ideas, $1 - \pi(x)$ is the amount of surprise that $X = x$, equating impossibility to full surprise. A possibility distribution is a purely epistemic construction representing in a graded way the knowledge of an agent.

This kind of representation has been found natural in the qualitative case where the possibility scale is not numerical (only an ordering of more or less possible values of X is useful in this case); this trend was initiated by Lewis [432] and pursued in the literature of non-monotonic reasoning and belief revision [240].

Zadeh [714] considers that natural language propositions modelled by fuzzy sets generate possibility distributions on domains of interest. For instance, the statement *John is tall* involves the gradual predicate *tall* delimiting an elastic constraint on the set of human heights $\mathcal{X}$, modelled by the membership function $\mu_{tall} : \mathcal{X} \to [0, 1]$. Then the piece of information *John is tall* defines the possibility distribution $\pi = \mu_{tall}$. Beyond these original motivations, possibility distributions seem to be ubiquitous in uncertainty theories, in connection with probability theory as seen in Chapter 4 and later on.

5.2.2.2 Possibility and necessity measures

From the knowledge of π ranging on the unit interval, we can define two confidence functions, namely a *possibility measure* Π, already encountered in the previous Chapter 4 (see Section 4.6.1) such that

$$\Pi(A) = \sup_{x \in A} \pi(x). \tag{5.8}$$

and a *necessity measure* N such that

$$N(A) = \inf_{x \notin A} 1 - \pi(x). \tag{5.9}$$

The possibility distribution π is here the primitive concept from which are induced the two measures Π and N. Again, $\Pi(A)$ measures the degree of consistency of proposition '$x \in A$' with information item π, and $N(A)$ is a degree of logical entailment of proposition '$x \in A$' by information item π, thus extending the Boolean version of possibility theory [249]. The same conventions for extreme values of Π and N are valid: in fact, they are yet another example of certainty (Cer $= N$) and epistemic possibility (Ep $= \Pi$).

Outside their interpretation as special cases of ∞-monotone capacities, another interpretation of numerical possibility distributions is the likelihood function in non-Bayesian statistics (Smets [598], Dubois *et al.* [244]). In the framework of an estimation problem, one is interested in determining the value of some parameter $\theta \in \Theta$ that defines a probability distribution $P(\cdot|\theta)$ over $\mathcal{X}$. Suppose that we observed the event A. Then $L(\theta) := P(A|\theta)$, as a function of $\theta \in \Theta$, is not a probability distribution, but a likelihood function[2]: a value θ is considered as being all the more plausible as $P(A|\theta)$ is higher, and the hypothesis θ will be rejected if $P(A|\theta) = 0$ (or is below some threshold).

Often, this function is renormalized so that its maximum is 1. We are allowed to let $\pi(\theta) := P(A|\theta)$ (thanks to this renormalization) and to interpret this likelihood function in terms of possibility degrees. In particular, it can be checked that for all $B \subseteq \Theta$, bounds for the value of $P(A|B)$ can be computed as:

$$\inf_{\theta \in B} P(A|\theta) \le P(A|B) \le \sup_{\theta \in B} P(A|\theta)$$

The reason for these inequalities is that $P(A|B)$ is a weighted arithmetic mean of the $P(A|\theta)$ where the weights are proportional to (unknown) prior probabilities attached to the parameter values $\theta \in B$. It shows that adopting the maxitivity axiom (Equation (5.2)) corresponds to an optimistic computation of $P(A|B) = \Pi(B)$. It is easy to check that letting $P(A|B) := \sup_{\theta \in B} P(A|\theta)$ is the only way of building a confidence function over Θ from $P(A|\theta), \theta \in \Theta$.[3] Indeed, the monotonicity w.r.t. inclusion of the likelihood function $L(\cdot) = P(A|\cdot)$ forces $P(A|B) \ge \sup_{\theta \in B} P(A|\theta)$ to hold [120]. Also note that the maximum likelihood principle originally due to Fisher, which consists in choosing the value $\theta = \theta^*$ that maximizes $P(A|\theta)$, is in total agreement with possibility theory.

[2] See also 7.1.4

[3] See also the profile-likelihood approach in Section 7.7

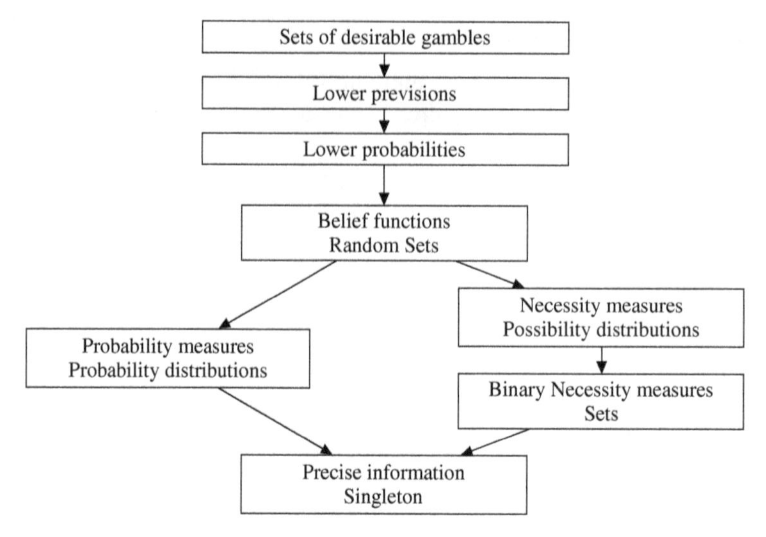

Figure 5.1 Summary of uncertainty theories and basic representations.

5.2.3 Overall picture

We can now draw a picture (Figure 5.1) of the different imprecision tolerant basic uncertainty representations and their formal relationships. It docs not reflect the variety of existing viewpoints but it helps to situate the different theories in a common framework.

Once we have Figure 5.1 in mind, notions such as set intersection and unions then needs to be extended to possibility theory, random sets, lower previsions, etc. Similarly, probabilistic notions such as conditioning or independence modelling needs to be properly and meaningfully extended to other frameworks. Since the new frameworks are usually richer than basic ones, such extensions are usually not unique, whence the apparition of discrepancies that need to be understood and reconciled across the different theories.

5.3 Discrepancies between uncertainty theories

The generality of desirable gambles and lower previsions may suggest that the apparent diversity of uncertainty theories is delusive, and that a single formalism is at work, the choice of a representation framework depending on practical considerations. However this is not the end of the story.

On the one hand, each uncertainty theory has its own history and was motivated by different problems, so a proper unification also needs to figure out whether these various stories fit together. Already in standard probability theory, there is more than one story leading to a probability measure: noticeably, the frequentist and the subjectivist stance [290]. Each one leads to a different definition of probability (for instance, sigma-additivity, central in Kolmogorov approach, is not part of the subjectivist approach after De Finetti). This distinction between frequentist and subjectivist views is still present across all uncertainty theories.

On the other hand, each uncertainty theory can be seen as a language for describing and solving problems. These languages rely on primitive notions that are used in all subsequent definitions, and these are not equally expressive. To summarize, one may say that the basic building block in possibility theory is the possibility distribution; in evidence theory it is the Möbius mass function; in Walley theory the basic building block is the coherent lower prevision, to which a corresponding credal set can be associated. In each theory, everything is defined by means of the primitive building block. So there will be some discrepancies on important concepts, such as conditioning or fusion operations. This section provides some examples of such discrepancies.

5.3.1 Objectivist vs. Subjectivist standpoints

A very important distinction must be drawn between singular and generic information. It underlies the opposition between frequentist and subjectivist sides of uncertainty theories, but is in some sense more fundamental. *Generic* information refers to a collection of cases or a population of situations. It could be a physical law, a statistical model built from a sample of observations, or a piece of commonsense knowledge like 'birds fly'. A probability distribution can be used to represent the behaviour of a population, and has then a frequentist flavour. One can then speak of the true (or the best) probability distribution underlying this population. A credal set may represent incomplete knowledge of this best model and it then underlies a currently unreachable probability measure induced by this population. However, the population need not even be well-defined for the concept of generic information to make sense.

In contrast, *singular* information refers to a particular situation, the occurrence of a single event: for instance, an observation (a patient has fever at a given time point), or a testimony (the crazy driver's car was blue). A probability degree attached to a piece of singular information is a degree of belief, and is not necessarily related to frequentist information. The idea of probability degrees justified by betting odds precisely aims at attaching probabilities to unique events, bypassing the need to resort to frequentist notions. This is the spirit of Walley's lower previsions. However an agent often has some experience about the world, and possesses some generic knowledge, be it imprecise. It has been argued that one can use generic knowledge on the frequency of an event to derive a degree of belief in the singular occurrence of this event in a specific situation (Hacking [326]). This reasoning looks valid for imprecise probabilities as well.

Evidence theory after Shafer clearly stands on the subjectivist side, degrees of belief being derived from the perceived reliability of testimonies about singular events. Like in Walley's approach, a belief function is a full-fledged representation of belief, and Smets has insisted that referring to an ill-known probability degree $P(A)$ lower bounded by $\mathrm{Bel}(A)$ made no practical sense in the approach. On the contrary Dempster's view is fully compatible with a frequentist interpretation of belief functions. Indeed, if the underlying probability space $\mathcal{Y}$ reflects statistical information, the corresponding belief function is a lower probability having a frequentist flavour. So the random set underlying a belief function may have frequentist or subjectivist interpretations.

Similar comments make sense for possibility distributions as well. It has been shown that while they can be built from subjective linguistic information as suggested by Zadeh, they are present in frequentist statistics, in the form of nested prediction, dispersion or confidence intervals or likelihood functions. But a possibility measure can serve as a special type of upper prevision in the subjectivist view as well [202].

5.3.2 Discrepancies in conditioning

Most of the time, a probability distribution refers to a population. This is a form of generic information, typically frequentist. This information can be used for inferring beliefs about a particular situation for which we have incomplete but clear-cut observations. This is called prediction. In this framework, $P(A|C)$ is the (frequentist) probability of event A in context C and it can quantify the belief of the agent in A when it has been observed that C is true in the current situation and nothing else (assuming that the situation is typical of environment C). The degree of belief of the agent in context C has changed from $P(A)$ to $P(A|C)$. Conditioning here is used for updating the beliefs of the agent about the current situation by exploiting (but not modifying) generic information. One only applies the available generic knowledge to a reference class C, what is called *focusing* by Dubois, Moral, and Prade [245]. For instance, the probability measure P represents medical knowledge (often compiled under the form of a Bayesian network). The singular information item C represents the results of tests for a patient. $P(A|C)$ is the probability of having disease A for patients for whom C has been observed; this value also estimates the singular probability (belief) that this patient has this disease.

When probability P is subjective, it has a singular nature (when betting on the occurrence of a nonrepeatable event). In this case conditioning can be interpreted as the revision of the singular probability function *by a piece of information of the same nature*. In this case, information C is interpreted as $P(C) = 1$, which represents a constraint that has to be taken into account when revising P. For instance [255], in a criminal affair where the investigator suspects Peter, Paul and Mary with probabilistic confidence degrees $\frac{1}{4}$, $\frac{1}{4}$ and $\frac{1}{2}$ respectively, and then learns that Peter has an alibi, i.e. $P(\{\text{Mary, Paul}\}) = 1$. We then have to revise these singular probabilities. The use of conditional probability for handling this revision of the probabilities is often proposed (and justified by the minimal change principle, already mentioned above), which yields probabilities $\frac{1}{3}$ and $\frac{2}{3}$ for Paul and Mary respectively. However, the problem of revising P is different from the one of updating singular beliefs on the basis of generic information.

In the case where the generic knowledge of the agent is represented by imprecise probabilities, Bayesian prediction is generalized by performing a sensitivity analysis on the conditional probability. Let $\mathcal{M}$ be a generic credal set on $\mathcal{X}$. For each proposition A a lower bound $\underline{P}(A)$ and an upper bound $\overline{P}(A)$ of the probability degree of A are known. In the presence of singular observations summarized under the form of a context C, the belief of an agent in a proposition A could be represented by the interval $[\underline{P}(A|C), \overline{P}(A|C)]$ defined by

$$\underline{P}(A|C) = \inf\{x : P(A \cap C) = x \cdot P(C), P \in \mathcal{M}\}$$

$$\overline{P}(A|C) = \sup\{x : P(A \cap C) = x \cdot P(C), P \in \mathcal{M}\}.$$

which is the conditioning rule obtained by means of the natural extension of Section 2.3. It may happen that the interval $[\underline{P}(A|C), \overline{P}(A|C)]$ is larger than $[\underline{P}(A), \overline{P}(A)]$, which corresponds to a loss of information (a dilation in the sense of Seidenfeld and Wasserman [573]) in specific contexts. This property reflects the idea that the more singular information is available about a situation, the less informative is the application of generic knowledge to it (since the number of relevant statistical data, i.e. where C is true, may become very small). We see that this form of conditioning does not correspond at all to the idea of enriching generic information, it is only a matter of querying it. Indeed the input information pertains to a singular situation, and focuses generic knowledge (the credal set) on a specific subset of possible worlds, on which the credal set may be totally uninformative (typically if the lower probability of C is zero, that is when nothing supports C in the credal set, we get $[\underline{P}(A|C), \overline{P}(A|C)] = [0, 1]$).

If belief and plausibility functions are used as generic knowledge, $m(E)$ represents, for instance, the proportion of imprecise results, of the form $x \in E$, in a statistical test on a random variable X. In this framework, prediction in context C consists in evaluating the weight function $m(\cdot|C)$ induced by the mass function m on the subset C of states, taken as the new frame. Three cases should be considered [195]:

- $E \subseteq C$: In this case, $m(E)$ remains assigned to E.
- $E \cap C = \emptyset$: In this case, $m(E)$ no longer matters and is eliminated.
- $E \cap C \neq \emptyset$ and $\overline{E} \cap C \neq \emptyset$: In this case, some fraction $\alpha_E \cdot m(E)$ of $m(E)$ remains assigned to $E \cap C$ and the rest, i.e. $(1 - \alpha_E) \cdot m(E)$, is allocated to $\overline{E} \cap C$. But this sharing process is unknown.

The third case corresponds to incomplete observations E that neither confirm nor disconfirm C. The scalar value α_E is the proportion of the population for which C is true among those observations for which all that is known is that E holds. We have not enough information in order to know if, for each of these observations, C is true or not, since only E is known and it overlaps both C and its complement. Suppose that the values $\{\alpha_E, E \subseteq S\}$ were known. Note that $\alpha_E = 1$ and $\alpha_E = 0$ in the first and second cases, respectively. Then, we can build a mass function $m_\alpha^C(\cdot)$. Note that a renormalization of this mass function is necessary, in general, as soon as $\mathrm{Pl}(C) < 1$ (letting $m_\alpha(\cdot|C) = \frac{m_\alpha^C(\cdot)}{Pl(C)}$ with $m_\alpha^C(A) = \sum_{E:A\cap C=E} \alpha_E m(E)$). If one denotes by $\mathrm{Bel}_\alpha(A|C)$ and $\mathrm{Pl}_\alpha(A|C)$ the belief and plausibility functions obtained by focusing on C, based on the allocation vector α, the conditional belief and plausibility degrees on C are defined by [195]:

$$\mathrm{Bel}(A|C) = \inf_\alpha Bel_\alpha(A|C), \tag{5.10}$$

and

$$\mathrm{Pl}(A|C) = \sup_\alpha Pl_\alpha(A|C). \tag{5.11}$$

One still obtains belief and plausibility functions (Jaffray [377]), and necessity and possibility measures if we start with such measures (Dubois and Prade [252]). The following results show that what is obtained is a generalization of Bayesian inference [268]: if $\mathrm{Bel}(C) > 0$, then

$$\mathrm{Bel}(A|C) = \inf\{x : P(A \cap C) = x \cdot P(C), P \geq \mathrm{Bel}\} = \frac{Bel(A \cap C)}{\mathrm{Bel}(A \cap C) + \mathrm{Pl}(\overline{A} \cap C)},$$

$$\mathrm{Pl}(A|C) = \sup\{x : P(A \cap C) = x \cdot P(C), P \geq \mathrm{Bel}\} = \frac{Pl(A \cap C)}{\mathrm{Pl}(A \cap C) + \mathrm{Bel}(\overline{A} \cap C)}.$$

It corresponds again to Walley's updating rule for upper and lower probabilities in Section 2.3 based on the natural extension. It is easy to see that $Pl(A|C) = 1 - Bel(\overline{A}|C)$, and that these formulas generalize probabilistic conditioning. This closed form already appears in the first report by Walley [669], and in [195, 268]. It is present with a different notation in the seminal paper of Dempster [222]. Jaffray [377] expresses this conditioning in terms of mass functions, laying bare the Möbius transform of $\mathrm{Bel}(\cdot|C)$ in terms of m.

Note that if $\mathrm{Bel}(C) = 0$ and $\mathrm{Pl}(s) > 0, \forall s \in C$ (no value in C is impossible but C has no certainty) then all the focal sets of m either overlap (possibly contain) C or are disjoint from C, none being contained in it. In this case, from Equations (5.10) and (5.11), $Bel(A|C) = 0$ and $\mathrm{Pl}(A|C) = 1$ for all nontrivial (i.e. $A \neq C$ and $A \neq \emptyset$) events $A \subset C$: one cannot infer anything in context C. However, if $\mathrm{Pl}(A) = 0$ for some $A \subset C$, then $\mathrm{Pl}(A|C) = 0$ as well. So this kind of conditioning, usually expressed using credal sets, can be couched in terms of mass functions as well.

In contrast, the usual type of conditioning in evidence theory, called 'Dempster rule of conditioning', proposed by Shafer [576] and Smets [601], systematically assumes $\alpha_E = 1$ as soon as $E \cap C \neq \emptyset$ in the above construction. It supposes a transfer of the full mass of each focal set E to $E \cap C \neq \emptyset$ (followed by a renormalization). This means that we interpret the new information C as allowing us to restrict to probability functions such that $P(C)$ is maximal: situations where C is false are considered as impossible. If one denotes by $\mathrm{Pl}(A||C)$ the plausibility function based on this assumption, we have:

$$\mathrm{Pl}(A||C) = \frac{\mathrm{Pl}(A \cap C)}{\mathrm{Pl}(C)}.$$

This constitutes another generalization of probabilistic conditioning. The conditional belief is then obtained by duality $\mathrm{Bel}(A||C) = 1 - \mathrm{Pl}(\overline{A}||C)$. With this conditioning, the size of focal sets diminishes, thus information becomes more precise, and when $\mathrm{Pl}(C) = 1$, the intervals $[\mathrm{Bel}, \mathrm{Pl}]$ become tighter, in the sense that $[\mathrm{Bel}(A||C), \mathrm{Pl}(A||C)] \subseteq [\mathrm{Bel}(A), \mathrm{Pl}(A)]$ for all $A \subseteq C$ (they are also always tighter than those obtained by focusing [420]).

Dempster rule of conditioning thus corresponds to a process where information is enriched, or more generally significantly altered, which contrasts with focusing. In particular, when $\mathrm{Pl}(C) < 1$ this kind of conditioning violates coherence in the sense of Walley, contrary to the previous one [671].

However, if the credal set represents singular uncertain evidence, this conditioning performs a revision of the underlying credal set. The incurred lack of coherence then is not surprising. When revising our beliefs, if the new information is at odds with previous beliefs, it can be expected that the new beliefs be conflicting with the previous one, not that they be coherent with it. On the contrary, the focusing conditioning operation for prediction of singular events out of generic knowledge preserves coherence because the prediction is a consequence of the credal set, and the latter is never altered, just exploited in a subset of $\mathcal{X}$.

In the more general framework of imprecise probabilities, a blind application of the revision process by a piece of information C consists in adding the supplementary constraint $P(C) = 1$ to the family $\mathcal{M}$, i.e.

$$\underline{P}(A||C) = \inf\{P(A|C), P(C) = 1, P \in \mathcal{M}\};$$

$$\overline{P}(A||C) = \sup\{P(A|C), P(C) = 1, P \in \mathcal{M}\}.$$

If $\overline{P}(C) = 1$ it coincides with Dempster rule of conditioning (and then, coherence is preserved). But clearly, it may happen that the set $\{P \in \mathcal{M}, P(C) = 1\}$ is empty (it is always the case in the classical Bayesian framework since $\mathcal{M}$ is a singleton). Dempster rule of conditioning then corresponds to applying the maximal likelihood principle (Gilboa and

Schmeidler [311]) and we replace the condition $P(C) = 1$ by $P(C) = \overline{P}(C)$ in the above equation. In this way, we generalize Dempster rule (which is recovered if $\overline{P}$ is a plausibility function) to 2-monotone lower probabilities (unless $\overline{P}(C) = 0$).

This type of conditioning has nothing to do with focusing. Indeed, Shafer and Smets view the Möbius mass function m only as singular information, and no generic knowledge is taken into account in the picture [601]. In the crime example (originally due to Smets), suppose that the organizer of the crime tossed a coin for deciding whether a man or a woman is recruited to be the killer. This piece of uncertain singular information is represented by the mass function $m(\{\text{Peter, Paul}\}) = \frac{1}{2}$ (there is no information available about Peter alone and Paul alone), and $m(\{\text{Mary}\}) = \frac{1}{2}$. Now, if we learn that Peter has an alibi, the focal set {Peter, Paul} reduces to {Paul} and we deduce after revision, that $P(\{\text{Mary}\}) = P(\{\text{Paul}\}) = \frac{1}{2}$. We have seen that the Bayesian approach shares the mass $m(\{\text{Peter, Paul}\})$ equally between Peter and Paul, so that Bayesian conditioning then yields $P(\{\text{Mary}\}) = 2 \cdot P(\{\text{Paul}\}) = \frac{2}{3}$. This result may sound debatable when dealing with uncertain singular pieces of information (let alone at a court of law).

As seen from the above discussion, discrepancies between belief functions and lower previsions concerning conditioning can be solved to a large extent. Shafer and Walley use different conditioning rules in their theories because they deal with different kinds of problems. Once it is acknowledged that Walley's conditioning is tailored for prediction, and Dempster rule of conditioning is tailored for revision of singular uncertain information, some kind of reconciliation can take place. In the case of possibility measures, both forms of conditioning make sense and preserve consonance [252, 676].

5.3.3 Discrepancies in notions of independence

Independence plays an important role in uncertainty theories, both for practical and theoretical reasons[4]. When dealing with precise probabilities, independence can be modelled in two conceptually different but formally equivalent ways. Consider two variables X, Y assuming their values in possibility spaces $\mathcal{X}, \mathcal{Y}$. Let P_X, P_Y be the marginal probability measures on $\mathcal{X}, \mathcal{Y}$, and P be the joint probability measure modelling our uncertainty about the value that (X, Y) assume jointly. According to the first point of view, Y is independent of X if learning the value of Y does not change the uncertainty about X, i.e., if

$$P_X(A|B) = P_X(A),\ \forall A \subseteq \mathcal{X}, B \subseteq \mathcal{Y} \quad \text{[marginalization]}. \tag{5.12}$$

We call this definition *causal* independence, since what is expressed here is the absence of (direct or indirect) causal influence of Y on X (this view is called *epistemic* in Chapter 3). Note that Equation (5.12) looks asymmetric. Only the axioms of probability theory make it symmetric when P is non-null. The second point of view, called here *decompositional* independence (and called *formalist* in Chapter 3), requires the joint model to be computable from the marginals alone. This condition is expressed by

$$P(A, B) = P_X(A) \cdot P_Y(B),\ \forall A \subseteq \mathcal{X}, B \subseteq \mathcal{Y} \quad \text{[factorization]}, \tag{5.13}$$

which is symmetric by definition. Equations (5.12) and (5.13) are equivalent when P is assumed to be non-null everywhere, and correspond to stochastic independence. The two

[4] See also the discussion in Section 3.2.

approaches are no longer equivalent in the other uncertainty theories, and they give rise to different independence notions and joint models. Also, the meaning of Equation (5.12) depends on the choice of a conditional capacity $g(A|B)$, hence the corresponding causal independence notion depends on the considered conditioning operation. Since probabilistic conditioning has multiple generalizations, Equation (5.12) gives rise to different notions.

The extension of Equation (5.12) to evidence theory has been investigated by Ben Yaghlane *et al.* [54]. Let $m(\cdot||B)$ denote the mass distribution obtained by applying Dempster's conditioning to m, and m_X denote the marginal mass on $\mathcal{X}$ obtained for any $A \subseteq \mathcal{X}$ by $m_X(A) = \sum_{B \subseteq \mathcal{Y}} m(A \times B)$. In this setting, Y is said to be causally independent from X if the following condition

$$m_X(A||B) = m_X(A), \ \forall A \subseteq \mathcal{X} \tag{5.14}$$

holds for every $B \subseteq \mathcal{Y}$, where $m_X(\cdot||B)$ is the mass obtained by conditioning m and marginalizing it on $\mathcal{X}$. This condition turns out to be symmetric and to be equivalent to Shafer's definition of *cognitive independence* that requires the joint plausibility measure of the Cartesian product of events be the product of marginal plausibilities.

The extension of Equation (5.12) to possibility theory has been studied by De Campos and Huete [193]. In this case, one starts from a joint possibility distribution π defined over $\mathcal{X} \times \mathcal{Y}$. Causal independence of Y from X comes down to stating that for every $x \in \mathcal{X}$, $y \in \mathcal{Y}$, we have

$$\pi(x||y) = \pi_X(x) \tag{5.15}$$

where $\pi_X(x) = \max_{y \in \mathcal{Y}} \pi(x, y)$ is the marginal possibility distribution on X and conditional possibility is defined by

$$\pi(x||y) = \frac{\pi(x, y)}{\pi_Y(y)}. \tag{5.16}$$

Again, this independence notion turns out to be symmetric, and corresponds to requiring that the joint possibility distribution be the product of the marginal distributions. Other extensions using other T-norm operators instead of the product have been studied [83, 192], however, such extensions, e.g. using minimum instead of product, can suffer from continuity problems in the numerical setting [200]. The problem of characterizing the most conservative joint model that has some given marginals and satisfies Conditions (5.14) and (5.16) is still open. Likewise, the causal independence notion associated with the prediction conditioning [268] based on *focusing* [245] has received little attention in the frameworks of possibility theory and evidence theory.

This is in contrast with the theory of lower previsions, where it corresponds to the well-studied notions of epistemic irrelevance and independence presented in Chapter 3, but where extensions of Conditions (5.14) and (5.16) to more general frameworks than belief functions (i.e. coherent lower probabilities and previsions) have not been investigated. This may be explained by the fact that neither n-monotonicity nor supremum-preserving properties of marginals are preserved in the most conservative joint model obtained with such an assumption. For example, Miranda and De Cooman [465] have attempted to embed epistemic independence in possibility theory, reaching the conclusions that the limited expressiveness of possibility distributions may not be sufficient to model epistemic independence. Extensions of Equation (5.13) are usually more straightforward and mostly consist in applying product rules to the basic building block of each theory. In evidence theory, the extension applies the product rule to the marginal mass distributions m_X, m_Y. This comes down to building the joint model m over $\mathcal{X} \times \mathcal{Y}$ such that

$$m(A \times B) = m_X(A)m_Y(B). \tag{5.17}$$

Shafer called this kind of independence *evidential*, but it is often called random set independence. It expresses independence between sources informing about X and Y, respectively. In possibility theory, the obtained joint model [193] from π_X, π_Y using the product rule is simply

$$\pi(x, y) = \pi_X(x)\pi_Y(y). \tag{5.18}$$

and is equivalent to Equation (5.15) when possibility degrees are positive. The decompositional independence in the framework of lower previsions, called *strong independence*, is detailed in Chapter 3. There exist some formal links between the three notions, since the credal set induced by the possibility distribution (5.18) is included in the credal set induced by *strong independence*, while the latter credal set is included in the one induced by the mass distribution given by Equation (5.17) (see Fetz [279], for example). Finally, it is worth mentioning that the so-called noninteractive [714] joint possibility distribution $\pi_\wedge$ that has the following form

$$\pi_\wedge(x, y) = \min(\pi_X(x), \pi_Y(y)) \quad \forall (x, y) \in \mathcal{X} \times \mathcal{Y}, \tag{5.19}$$

does not have strong connections with the previously mentioned joint models [624], except that it is less informative than Equation (5.18). This is due to the fact that it assumes a complete dependence between the sources supplying the possibility distributions (via the choice of the same α-cuts for both sources) instead of independence between them[5], and no knowledge on the dependence between X and Y [49].

Discrepancies between independence notions across the different theories appear more severe than for conditioning. However, works attempting to fill in the gaps between these notions are more recent than for conditioning [157]. As mentioned previously, what form would causal independence take using Dempster conditioning for lower previsions is still unknown. Studies linking the focusing rule with a causal independence notion in possibility and evidence theory should be carried out. Discrepancies among decompositional independence notions are less severe, and the main interest of such a notion is computational.

5.3.4 Discrepancies in fusion operations

Fusion is the operation consisting in merging uncertain information from several sources. This notion emerged first in the theory of belief functions, where the rule of combination originally due to Dempster plays a major role. In probability theory, this question is seldom addressed, due to the fact that it has been traditionally envisaged (since the XIXth century) from a statistical and frequentist point of view, despite the fact that the problem of merging testimonies (i.e., nonstatistical evidence) motivated early research in probability back in the XVIIth and XVIIIth centuries [515].

In subjective probability, one is often concerned with a single agent, however the problem of modelling expert opinions in reliability has prompted probabilists to address the

[5] In which case random set independence is retrieved.

issue of fusion of subjective probabilities supplied by experts [123]. Walley [670] himself devoted a long report to this problem.

Considering that imprecise probabilities generalize both probabilities and sets, the two building blocks of uncertain information fusion are the fusion of probabilities and the fusion of sets. There are basically two approaches to merging probabilities:

- The non-Bayesian approach [123] supposes that n sources (experts) supply n subjective probability distributions P_i pertaining to the location of a variable x. Reliability weights $r_i \in [0, 1]$ such that $\sum_{i=1}^{n} r_i = 1$ are attached to each source. Then the merging rule prescribes to consider the convex mixture $\sum_{i=1}^{n} r_i P_i$ as the result of the fusion. This form of fusion is basically the only possible way of retrieving a probability measure as the result [458]. It presupposes each expert is a kind of random source and the fusion can be seen as the direct extension of the statistical view relying on identical, yet independent processes.
- On the contrary the Bayesian view considers a holistic approach to the fusion, whereby there is a joint probability that each source produces the output x_i when the real value of the quantity under concern is x. The fusion tries to compute the probability distribution $P(x|x_1, x_2, \dots x_n)$, when the observations $x_1, \dots x_n$ are obtained and assuming a known likelihood and prior. This is a very demanding approach in terms of data, and the requirement of a prior is questionable as the reason for undertaking a fusion process is generally a lack of prior knowledge.

In contrast, if information comes in the form of epistemic states modelled by sets, the fusion can be envisaged from a logical perspective [250]. Suppose one source provides the information items $x \in A$ and $x \in B$. There are three fusion modes:

- A first option is to perform the conjunction of the two sets. It presupposes that both sources can be trusted, and the two pieces of information are consistent.
- A second option is to perform the disjunction of the two sets. It is enough to assume that one source is reliable without knowing which one of the two. It then makes sense even if the pieces of information are inconsistent.
- A third option is to count the number of times each value of x is considered possible by the sources. This is in line with the accumulation method of statistics, assuming sources are independent and identical.

In case of $n > 2$ sources, the two first options would often respectively yield inconsistent or uninformative results, and the most natural approach is then to find maximal consistent subsets of sources, perform conjunction among the information items from each source in each group and perform a disjunction of the results, an idea going back to Rescher and Manor [531] in the logical setting. Alternatively one may assume the knowledge of a number of reliable sources, and perform a *k-out-of-n* kind of merging [251].

The above set-based modes of fusion directly extend to possibility and evidence theories. For instance the conjunction mode exploits fuzzy set intersections to combine possibility theories and random set intersections for combining belief functions. Here, we can observe some discrepancies between the two approaches:

- There are many conjunctive fusion operations for merging possibility distributions, usually using T-norms operators [397]. In particular, if dependencies between sources are unknown the minimum idempotent operator applies, and if independence is assumed one can perform the pointwise product of possibility distributions. One may even perform a fusion mode assuming that a prescribed number of sources may lie (Mundici [485]). In general, merging possibility distributions under the conjunctive mode yields a possibility distribution, but the result may be subnormalized ($\pi(x) < 1, \forall x$). A renormalization step is often considered, assuming the remaining values of positive possibility, however implausible, contain the true one.

- There is essentially one conjunctive merging rule in evidence theory (Dempster's rule of combination) and it presupposes independence between sources. Usually the resulting belief function is subnormal in the sense that $m(\emptyset) > 0$, which can be renormalized (however Smets recommends not to do it so as to keep track of the conflict). This combination rule is the cornerstone of Shafer's theory of evidence, and the possibility of decomposing any belief functions into simple support functions is studied in Shafer's book.[6] This situation contrasts with the case of possibility theory where many operations are available. The possibilistic product rule is in some agreement with the unnormalized Dempster rule of combination because the obtained contour functions are proportional to each other. But merging consonant belief functions by the conjunctive rule does not preserve consonance, and the normalization factors in both theories differ. Worse, there is no clear idempotent merging rule in belief function theory; in particular it is generally impossible to find a joint mass function whose marginals are the source information items and whose contour function would be the minimum of the two contour functions [230]. However Denoeux [227] has enlarged the set of combination rules for belief functions, including an idempotent one, by exploiting Smets [599] decomposition of belief functions via Dempster's combination rule.

Disjunctive fusion modes in possibility theory and evidence theory have similar discrepancies. Random set disjunction corresponds to computing the product of belief functions (or necessity measures) [247]. Since the latter applies to probability measures, the enlarged setting enables the product of probability measures to make sense as a disjunctive combination thereof, yielding random sets having at most two-element realizations, a benefit of considering a unified view of uncertainty theories.

Moreover, the convex combination of belief or plausibility functions and necessity or possibility measures makes sense, but does not preserve consonance. However it comes down to extending the counting mode of fusion to possibility distributions, when restricting to contour functions.

Turning to the general setting of credal sets, Walley [670] extensively discusses the fusion problem in the context of upper and lower probabilities. He proposes a number of properties a rational fusion method should satisfy, which can be adapted to possibility theory and belief functions. Given two families $\mathcal{M}_1$ and $\mathcal{M}_2$ of probabilities provided by two reliable sources, it is natural to consider that the result of a fusion is the intersection $\mathcal{M} = \mathcal{M}_1 \cap \mathcal{M}_2$, when nonempty. The most natural way of merging n probability functions is to consider the credal set of which they are the vertices. The latter is an idempotent rule adopting a disjunctive mode of combination. This rule has the merit of satisfying all criteria

[6] This turns out to be possible only for a subclass of so-called support belief functions. This difficulty is overcome by Smets [599] by enlarging the notion of simple support belief function.

requested by Walley. It is clear that the rule consisting of considering the convex hull of $\mathcal{M}_1 \cup \mathcal{M}_2$ (called the unanimity rule by Walley) extends this rule to sets of probabilities. Walley also considers the hybrid fusion rule based on maximal consistent subsets of sources, as well as the credal set extension of the weighted average probability combination rule.

The fusion of imprecise probabilities and the fusion of belief functions are not really in agreement. It sounds hard to justify the Dempster rule of combination in terms of imprecise probabilities and, in the same way, it is not easy to express the mass function induced by $\mathcal{M}_1 \cap \mathcal{M}_2$ in terms of the mass functions induced by $\mathcal{M}_1$ and $\mathcal{M}_2$. Walley rejects Dempster rule of combination on the basis of its failing to satisfy criteria he considers important [671]. One may apply the idempotent fusion of imprecise probabilities to belief functions, by performing the intersection of sets $\mathcal{M}_i = \{P : P(A) \geq \text{Bel}_i(A), \forall A \subseteq S\}$ for $i = 1, 2$. But the lower bounds of the induced probabilities are not belief functions in general. Chateauneuf [108] explores these issues in some detail.

The setting of uncertainty theories explicitly handling incomplete information seems to be more flexible than pure probability theory for the purpose of merging uncertain information. However many questions remain open on this topic, regarding the most appropriate rules of combination of credal sets and their relationships with similar combination modes in more specialized settings.

5.4 Further reading

We have tried in this chapter to present the main uncertainty theories explicitly coming to grips with imprecision, together with their original motivations. In contrast with the theory of lower previsions, the other uncertainty theories rely on the use of special cases of capacities. We have provided a brief account of some discrepancies between the different theories on central concepts such as conditioning, independence and combination rules, pointing out the existing bridges between these discrepancies. We could have done the same with methods of information comparison. On this issue, the reader can consult [237, 238, 220].

We have not discussed other emerging approaches such as info-gap theory [52] or the game-theoretic approach of Shafer and Vovk, described in Chapter 6, for their links with the theories introduced in this chapter largely remain to be studied in more detail.

In this chapter, we have also focused on quantitative representations of uncertainty, leaving aside qualitative representations. However, a purely ordinal framework looks natural when dealing with subjective uncertainty assessments. Indeed, it is easier for an agent to assert that one proposition is more credible than another, rather than providing a numerical belief degree. Ordinal or qualitative approaches represent uncertainty by means of a confidence relation between propositions (or events) obeying some natural properties like consistency with logical deduction. An uncertainty relation formalizes a pairwise comparison of events in terms of relative likelihood. The idea of representing uncertainty by relations over sets of propositions (or events) dates back to De Finetti [216], Koopman [406] and Ramsey [527]. These authors tried to find ordinal counterparts to subjective probabilities. Later on, philosophers such as Lewis [433] have considered other kinds of uncertainty relations in the framework of modal logic, including what eventually turned out to be comparative counterparts to possibility measures [239]. The survey of bridges between (imprecise) probability theory and modal logic [677] is also interesting to consult. Starting from this basic ordinal framework, one may start to redefine the notions of conditioning, independence, fusion, etc. in order to have a full-fledged qualitative uncertainty theory.

Qualitative approaches may also rely on a finite scale of confidence levels. Possibility theory is particularly well adapted to such a qualitative framework, as the minimum and maximum operators are well defined on any lattice (see De Cooman [198, 199, 200] for lattice-valued possibility measures, and [60] and [241] for discussions about qualitative conditioning and qualitative independence, respectively). A recent paper on conditional comparative uncertainty structures is [122].

Conversely, one can search for conditions under which a relation $\leq_\kappa$ on events is compatible with some numerical representations. When only one confidence function $g : \wp(\mathcal{X}) \to [0, 1]$ is considered, then it is natural to require $\leq_\kappa$ to be complete and transitive. A confidence function g is then said to represent the relation $\leq_\kappa$ as soon as

$$A \leq_\kappa B \text{ if and only if } g(A) \leq_\kappa g(B).$$

Conditions under which a relation $\leq_\kappa$ is representable by a probability measure, a possibility measure, a belief function and a lower probability were respectively studied by Koopman [406], Dubois [239], Wong *et al.* [709] and Capotorti *et al.* [101]. The reader may benefit from consulting books like the one by Fine [290] and the survey paper by Fishburn [292], for an account of early results on ordinal versions of probability.

On the other hand, one may relax the completeness of $\leq_\kappa$ and search under which conditions a partial preorder is representable by pairs of dual confidence functions. There are however less results concerning such representations. Section 1.4 relates partial preferences with desirable gambles, however gambles are naturally numerical, and can hardly be characterized from the qualitative standpoint. Recently, Nehring [491] has explored under which conditions a partial preorder $\leq_\kappa$ where $A \leq_\kappa B$ is interpreted as $P(A) \leq_\kappa P(B)$ is representable by a credal set.

6

Game-theoretic probability

Vladimir Vovk[1] and Glenn Shafer[2]
[1]*Department of Computer Science, Royal Holloway, University of London, UK*
[2]*Department of Accounting and Information Systems, Rutgers Business School – Newark and New Brunswick, USA*

6.1 Introduction

The standard theory of probability is based on Kolmogorov's [405] measure-theoretic axioms. A less known alternative is the game-theoretic approach to probability, which is perhaps as old as the measure-theoretic approach (see the end of Section 6.9). There has been a revival of interest in game-theoretic probability lately, and the purpose of this chapter is to give a brief introduction into its current state.

Measure-theoretic probability is, by its nature, precise: according to Kolmogorov's axioms, the probability of every event is a number (not, say, an interval). The usual approach to imprecise probability is also axiomatic: Kolmogorov's axioms are weakened to allow interval probabilities and expectations (cf. the first two sections of Chapter 2). Game-theoretic probability can be precise or imprecise; however, precise game-theoretic probability tends to be equivalent to measure-theoretic probability (we will see an example of this at the end of Section 6.3), and so is not particularly interesting from the point of view of this chapter.

An important difference of game-theoretic probability from measure-theoretic probability, and from the standard approaches to imprecise probability, is its constructive character: for example, whereas limit theorems of measure-theoretic probability theory (and their generalizations in the theory of imprecise probability [162, 207]) assert that various interesting events have probability (or lower probability) one (or close to one), limit theorems of game-theoretic probability often exhibit explicit strategies for one of the players in a game of prediction. We will see examples in this chapter: see Sections 6.6 and 6.7.

Introduction to Imprecise Probabilities, First Edition.
Edited by Thomas Augustin, Frank P. A. Coolen, Gert de Cooman and Matthias C. M. Troffaes.

The formal notion of probability is much less important in game-theoretic probability theory than the formal notion of probability in measure-theoretic probability theory (or even the notions of upper probability and lower probability in the theory of imprecise probability); rather than being the (or at least a) fundamental notion of the theory, it is defined in terms of other more fundamental notions (such as perfect-information games and supermartingales). In fact, the game-theoretic approach to probability had been developing long before game-theoretic probability was formally defined. However, once we have the notion of game-theoretic probability, it becomes natural to study which of the usual axioms it satisfies; article [674] contains a useful review of such axioms. This will be the main focus of this chapter. In particular, we will see that game-theoretic upper probability is indeed a coherent upper probability in the sense of imprecise probabilities (Section 6.3) and, under an additional axiom, is an outer measure (Section 6.4) and a Choquet precapacity (not necessarily two-alternating, as shown in Section 6.8).

To set the stage, we start the chapter by stating a simple game-theoretic strong law of large numbers (Section 6.2). This motivates the introduction of a general forecasting protocol and the definition of game-theoretic expectation and probability in Section 6.3. In Section 6.5 we investigate the robustness of our definitions with respect to various modifications. In Section 6.6 we use results of Section 6.5 to prove the strong law of large numbers of Section 6.2, exhibiting an explicit strategy for demonstrating it. In the same section we list some other limit theorems of game-theoretic probability giving relevant references. The next section, Section 6.7, is devoted to the game-theoretic approach to a different kind of classical results of probability, zero-one laws, where we study in detail Lévy's zero-one law and give references to the relevant literature about other game-theoretic zero-one laws. Section 6.9 gives references for further reading.

We tried to keep the mathematical level of this chapter very basic. In particular, the reader is not supposed to know any measure-theoretic probability theory, except for the paragraphs (explicitly flagged) where game-theoretic probability is compared with measure-theoretic probability.

One recent publication that discusses connections of imprecise probability and game-theoretic probability is [203]. Whereas [203] concentrates on the case of a finite horizon, in this chapter we will deal almost exclusively with the infinite horizon.

6.2 A law of large numbers

One of the simplest prediction protocols is given as Protocol 6.1. As all prediction protocols in this chapter, it is a perfect-information protocol. There are three players: Reality, who chooses "random numbers" $y_1, y_2, \dots$, Forecaster, who chooses forecasts p_t for y_t, $t = 1, 2, \dots$, and Sceptic, who gambles against Forecaster. Issuing a forecast p_t is interpreted as offering for sale tickets that pay $y_t - p_t$ after the outcome y_t becomes known; s_t is the number of tickets (possibly negative) that Sceptic chooses to buy. Sceptic's accumulated capital at the end of trial t is $\mathcal{K}_t$; for now, we assume that he starts with initial capital $\mathcal{K}_0 = 1$. The protocol has one parameter: a constant $C \geq 1$; Reality is constrained to $y_t \in [-C, C]$ (this constraint will allow us to state a simple strong law of large numbers and prove it easily; we will give references to strong laws of large numbers without this assumption in Section 6.6).

This is our simple game-theoretic strong law of large numbers:

Protocol 6.1 Predicting numbers in $[-C, C]$

$\mathcal{K}_0 := 1$;
for $t = 1, 2, \dots$ **do**
Forecaster announces $p_t \in \mathbb{R}$;
Sceptic announces $s_t \in \mathbb{R}$;
Reality announces $y_t \in [-C, C]$;
$\mathcal{K}_t := \mathcal{K}_{t-1} + s_t(y_t - p_t)$;
end for

Theorem 6.1 *Sceptic has a strategy in Protocol 6.1 such that his capital $\mathcal{K}_t$ is always non-negative and either* $\lim_{t\to\infty}\mathcal{K}_t = \infty$ *or*

$$\lim_{T\to\infty} \frac{1}{T} \sum_{t=1}^{T} (y_t - p_t) = 0. \tag{6.1}$$

We will give a constructive proof of this result in Section 6.6; namely we will exhibit a measurable strategy for Sceptic satisfying the required properties.

In this paragraph we will assume that the reader is familiar with the basics of the theory of martingales in measure-theoretic probability. The measure-theoretic counterpart of Theorem 6.1is the following law of large numbers for a probability space equipped with a filtration $(\mathcal{F}_t)_{t=0}^{\infty}$. Suppose y_t is an $\mathcal{F}_t$-measurable random variable and p_t is (any version of) the conditional expectation of y_t given $\mathcal{F}_{t-1}$, $t \in \mathbb{N} := \{1, 2, \dots\}$. Then (6.1) holds with probability one. This measure-theoretic result is a simple corollary of Theorem 6.1. Suppose Reality and Forecaster generate their moves stochastically: y_t and p_t are the realized values of the corresponding random variables. It is clear that in this case $\mathcal{K}_t$ is a non-negative martingale, and so is bounded with probability one. This implies that (6.1) holds with probability one.

The measure-theoretic counterpart of Theorem 6.1 describes the situation where Reality chooses her moves y_t from a given probability measure P and Forecaster chooses his moves p_t as the conditional expectations of y_t given the past (Reality and Forecaster are colluding, to use the expression from [181, § 5.1]). These are some intuitive advantages of Theorem 6.1:

- Theorem 6.1 does not require Forecaster to follow a well-defined strategy when choosing the forecasts, and he can change his plans as time progresses.
- Theorem 6.1 does not require Forecaster to give a full probability forecast for the next observation y_t. Forecaster quotes only one number, p_t, which is intuitively the expected value of y_t given the past, and does not quote, for example, the variance of y_t or the probability of $y_t \geq 0$; the latter simply do not exist in the game-theoretic framework. This is in the spirit of imprecise probabilities (see Chapter 1), which also do not require full probability specifications.
- Theorem 6.1 is more constructive: it asserts the existence of a strategy for Sceptic with a desirable property, and such a strategy is constructed explicitly in the proof (see Section 6.6).

Theorem 6.1 is a mathematical result that can have more than one possible interpretation in the real world (if we ignore the fact that the statement (6.1) involves the limit as $T \to \infty$ and so is not observable; in Section 6.6 we will refer to similar results for finitely many observations). Suppose that we regard it Forecaster's failure if Sceptic multiplies the capital that he risks manyfold (this is called the "fundamental interpretative assumption" in [582] and "Ville's principle" in [579] and [661]). If (6.1) fails to happen, this means that Forecaster failed. On the other hand, if we believe that Forecaster will not fail, we can predict that (6.1) will hold.

The modification of Protocol 6.1 where Reality's moves are constrained to the 2-element set $\{0, 1\}$ is still non-trivial.

$y_t = 1$ may mean that some event happens at trial t and $y_t = 0$ that it fails to happen. Then p_t may be interpreted as Forecaster's probability that this event will happen. This probability can be called subjective if p_t is output by an individual, objective if p_t is computed from a well-tested theory, and intersubjective if p_t is output by a prediction market, such as Iowa Electronic Markets.

Notice that in the latter case we do not really know which term of the disjunction in the statement of Theorem 6.1 we should believe if Sceptic (speculator) plays one of the strategies whose existence is asserted by Theorem 6.1 (e.g., the strategy constructed in Section 6.6). We know that either Sceptic becomes infinitely rich or the property (6.1), which is sometimes called unbiasedness in the large [487, 178], holds. There is, however, no obvious convincing reason for either the market being unbiased in the large or a participant being able to become infinitely rich without risking more than one monetary unit.

Protocol 6.1 with the restriction $y_t \in \{0, 1\}$ will be called the *binary prequential protocol*. The name "prequential" was coined by A. P. Dawid as a portmanteau word that "combines *pro*bability forecasting with *sequential pre*diction" ([176, p. 279]; in the binary case we can regard p_t as a probability forecast, namely the predicted probability of $y_t = 1$).

For future use, we will define two further special cases of Protocol 6.1. First, we can let Forecaster play the fixed strategy $p_t \equiv 0$; since he then becomes superfluous, we can remove him from the protocol, arriving at Protocol 6.2.

Protocol 6.2 Predicting zero-mean numbers in $[-C, C]$

$\mathcal{K}_0 := 1$;
for $t = 1, 2, \ldots$ **do**
  Sceptic announces $s_t \in \mathbb{R}$;
  Reality announces $y_t \in [-C, C]$;
  $\mathcal{K}_t := \mathcal{K}_{t-1} + s_t y_t$;
end for

Protocol 6.2 is important for us since it suffices to prove Theorem 6.1 for it. Indeed, we can assume, without loss of generality, that Forecaster's moves in Protocol 6.1 are $p_t \in [-C, C]$ (since y_t are known to belong to $[-C, C]$, Sceptic can become arbitrarily rich if $p_t \notin [-C, C]$). Let us say that a strategy for Sceptic is *winning* if it satisfies the condition in Theorem 6.1 (with $p_t := 0$ in the case of Protocol 6.2). If we have a winning strategy for Sceptic in Protocol 6.2 with $2C$ in place of C, we can apply the same strategy in Protocol 6.1 using $y_t - p_t$ in place of y_t. It will also be a winning strategy for Sceptic.

We can specialize Protocol 6.2 further, restricting Reality's moves to $y_t \in \{-1, 1\}$. This new protocol will be called the *coin-tossing protocol* (it is a simple linear transformation of the prequential protocol in which Forecaster is required to set $p_t := 1/2$).

6.3 A general forecasting protocol

In this section we introduce a general discrete-time forecasting framework. In game-theoretic probability one often starts from local previsions on a local possibility space (although in this chapter we prefer to start from sets of available gambles on the local possibility space; cf. Chapters 1 and 2), in terms of which global previsions (which we call upper and lower expectations) and probabilities can be defined. As we have already said, the last step is optional; there are many results, such as Theorem 6.1, that can be stated without explicitly using global previsions.

Let $\mathcal{X}$ be a fixed set, which we will call the (local) *possibility space*, and let $\mathcal{L}(\mathcal{X})$ stand for the set of all gambles on $\mathcal{X}$, i.e., bounded functions $f : \mathcal{X} \to \mathbb{R}$; we will sometimes refer to the elements of $\mathcal{L}(\mathcal{X})$ as *local gambles*. We will be interested in sets $\mathcal{A} \subseteq \mathcal{L}(\mathcal{X})$, which will play the role of sets of available gambles (for Sceptic). We impose the following axioms on sets of available gambles (very similar to the axioms (1.1), (1.2), (1.4), and (1.5) for sets of desirable gambles in Chapter 1):

AG1. If $f \in \mathcal{A}$ and $\lambda > 0$, then $\lambda f \in \mathcal{A}$.

AG2. If $f, g \in \mathcal{A}$, then $f + g \in \mathcal{A}$.

AG3. If $f \in \mathcal{L}(\mathcal{X})$ is such that $\sup f \leq 0$, then $f \in \mathcal{A}$.

AG4. If $f \in \mathcal{L}(\mathcal{X})$ is such that $\inf f > 0$, then $f \notin \mathcal{A}$.

Let us now discuss in more detail the connection between sets of available gambles and coherent sets of desirable gambles, defined in Chapter 1. Essentially, sets of available gambles $\mathcal{A}$ are sets of the form $-\mathcal{D}$, where $\mathcal{D}$ is a coherent set of desirable games (see Section 1.2.3). We can imagine that Forecaster allows Sceptic to pay him f for any $f \in \mathcal{D}$; such a payment increases Sceptic's capital $\mathcal{K}$ by $-f$. AG1 and AG2 correspond to (1.1) and (1.2), respectively. AG3 corresponds to $\inf f \geq 0 \Rightarrow f \in \mathcal{D}$, which is identical to (1.4) except that we also include 0 in $\mathcal{D}$ ($0 \in \mathcal{D}$ makes sense in our current context as we can't deny Sceptic's right not to enter any deals). And AG4 corresponds to $\sup f < 0 \Rightarrow f \notin \mathcal{D}$, which is weaker than (1.5).

Let $\mathbb{A}$ be the family of all sets $\mathcal{A} \subseteq \mathcal{L}(\mathcal{X})$ satisfying AG1–AG4. The most general protocol that we consider in this chapter is given as Protocol 6.3.

Protocol 6.3 General forecasting protocol

Sceptic announces $\mathcal{K}_0 \in \mathbb{R}$;
for $t = 1, 2, \ldots$ **do**
  Forecaster announces $\mathcal{A}_t \in \mathbb{A}$;
  Sceptic announces $f_t \in \mathcal{A}_t$;
  Reality announces $y_t \in \mathcal{X}$;
  $\mathcal{K}_t := \mathcal{K}_{t-1} + f_t(y_t)$;
end for

We call the set $\Omega := (\mathbb{A} \times \mathcal{X})^\infty$ of all infinite sequences of Forecaster's and Reality's moves the *sample space* (it serves as our global possibility space). The elements of the set $\bigcup_{t=0}^{\infty} (\mathbb{A} \times \mathcal{X})^t$ of all finite sequences of Forecaster's and Reality's moves are called *clearing situations*, and the elements of the set $\bigcup_{t=0}^{\infty}((\mathbb{A} \times \mathcal{X})^t \times \mathbb{A})$ are called *betting situations*. If x is a betting situation and $y \in \mathcal{X}$, xy (i.e., x extended by adding y) is a clearing situation, and if x is a clearing situation and $\mathcal{A} \in \mathbb{A}$, $x\mathcal{A}$ is a betting situation. The term *situation* will be applied to both betting situations and clearing situations. For each situation x we let $\Gamma(x) \subseteq \Omega$ stand for the set of all infinite extensions in Ω of x (i.e., $\Gamma(x)$ is the set of all $\omega \in \Omega$ such that x is a prefix of ω). Let $\square$ be the empty situation (which is a clearing situation).

The *level* of a situation x is the number of predictions in x. In other words, t is the level of betting situations of the form $\mathcal{A}_1 y_1 \dots \mathcal{A}_{t-1} y_{t-1} \mathcal{A}_t$ and of clearing situations of the form $\mathcal{A}_1 y_1 \dots \mathcal{A}_t y_t$. The level of $\square$ is 0. If $\omega \in \Omega$ and $t \in \mathbb{N}_0 := \{0, 1, \dots\}$, ω^t is defined to be the unique clearing situation of level t that is a prefix of ω.

If we fix a strategy Σ for Sceptic, Sceptic's capital $\mathcal{K}_t$ becomes a function of the current clearing situation x of level t. We write $\mathcal{K}^\Sigma(x)$ for $\mathcal{K}_t$ resulting from Sceptic following Σ and from Forecaster and Reality playing x. The function $\mathcal{K}^\Sigma$, defined on the set of all clearing situations and taking values in $\mathbb{R}$, will be called the *capital process* of Σ; we will omit Σ when it is clear from context. A function $\mathcal{S}$ is called a (game-theoretic) *supermartingale* if it is the capital process, $\mathcal{S} = \mathcal{K}^\Sigma$, of some strategy Σ for Sceptic. In other words, a real-valued function on the clearing situations is called a supermartingale if $\mathcal{S}(x\mathcal{A}\cdot) - \mathcal{S}(x) \in \mathcal{A}$ for all clearing situations x and all $\mathcal{A} \in \mathbb{A}$. A supermartingale $\mathcal{S}$ is a *test supermartingale* if it is non-negative and $\mathcal{S}(\square) = 1$. Sometimes we will extend the domain of supermartingales $\mathcal{S}$ to include betting situations: if x is a betting situation, $\mathcal{S}(x)$ is interpreted as $\mathcal{S}(x^-)$, where x^- is x with the last prediction removed. We will often write $\mathcal{S}_t(\omega)$ to mean $\mathcal{S}(\omega^t)$ (to make our notation closer to the one used in measure-theoretic probability theory and to facilitate omitting ω without loss of clarity).

Remark 6.1 We could define a *martingale* as a supermartingale $\mathcal{S}$ such that $-\mathcal{S}$ is also a supermartingale. However, this class of martingales would be too narrow for our purposes and we never use it in this book.

For each bounded function $\xi : \Omega \to \mathbb{R}$ and each (betting or clearing) situation x, we define the (conditional) *upper expectation* of ξ given x by

$$\overline{\mathrm{E}}(\xi|x) := \inf \left\{ \mathcal{S}(x) : (\forall \omega \in \Gamma(x)) \liminf_{t\to\infty} \mathcal{S}_t(\omega) \geq \xi(\omega) \right\}, \tag{6.2}$$

where $\mathcal{S}$ ranges over the supermartingales, and we define the (conditional) *lower expectation* of ξ given x by

$$\underline{\mathrm{E}}(\xi|x) := -\overline{\mathrm{E}}(-\xi|x).$$

(In the case of a finite horizon and a fixed strategy for Forecaster, the interpretation of $\overline{\mathrm{E}}(\xi|x)$ in terms of natural extension is given in [203].)

If E is any subset of Ω, its *upper* and *lower probability* given a situation x are defined by

$$\overline{\mathrm{P}}(E|x) := \overline{\mathrm{E}}(I_E|x) \quad \text{and} \quad \underline{\mathrm{P}}(E|x) := \underline{\mathrm{E}}(I_E|x), \tag{6.3}$$

respectively. In what follows we sometimes refer to bounded functions $\xi : \Omega \to \mathbb{R}$ as *global gambles* and to sets $E \subseteq \Omega$ as *events*.

It is straightforward to check that, for each situation x, $\overline{\mathrm{E}}(\xi|x)$ satisfies the following properties of an upper prevision (cf. the 'dual' conditions C1–C3 for being a coherent lower prevision given in Chapter 2 and corresponding to the case $x = \square$):

DC1. For any $\xi \in \mathcal{L}(\Omega)$, any situation x, and any $\lambda > 0$, $\overline{\mathrm{E}}(\lambda\xi|x) = \lambda\overline{\mathrm{E}}(\xi|x)$.

DC2. For any $\xi_1, \xi_2 \in \mathcal{L}(\Omega)$ and any situation x, $\overline{\mathrm{E}}(\xi_1 + \xi_2|x) \leq \overline{\mathrm{E}}(\xi_1|x) + \overline{\mathrm{E}}(\xi_2|x)$.

DC3. For any $\xi \in \mathcal{L}(\Omega)$ and any situation x, $\overline{\mathrm{E}}(\xi|x) \leq \sup_{\omega\in\Gamma(x)}\xi(\omega)$.

DC1 follows from $\lambda\mathcal{S}$ being a supermartingale for any $\lambda > 0$ and any supermartingale $\mathcal{S}$, and DC2 follows from $\mathcal{S}^{(1)} + \mathcal{S}^{(2)}$ being a supermartingale for any supermartingales $\mathcal{S}^{(1)}$ and $\mathcal{S}^{(2)}$. However, the fact that $\overline{\mathrm{E}}(\cdot|x)$ satisfies DC1–DC3 does not mean that it is a coherent upper prevision. The catch is that we still need to show that $\overline{\mathrm{E}}(\cdot|x)$ takes values in $\mathbb{R}$ (and not, e.g., the value $-\infty$). In the next lemma we will prove the following property.

DC4. For any $\xi \in \mathcal{L}(\Omega)$ and any situation x, $\overline{\mathrm{E}}(\xi|x) \geq \inf_{\omega\in\Gamma(x)}\xi(\omega)$.

By DC3 and DC4,

$$\overline{\mathrm{E}}(\xi|x) \in \left[\inf_{\omega\in\Gamma(x)} \xi(\omega), \sup_{\omega\in\Gamma(x)} \xi(\omega)\right],$$

and it is clear that this inclusion also holds for $\underline{\mathrm{E}}(\xi|x)$ in place of $\overline{\mathrm{E}}(\xi|x)$.

Lemma 6.1 *For all situations x and all global gambles $\xi : \Omega \to \mathbb{R}$, $\overline{\mathrm{E}}(\xi|x) \geq \inf_{\omega\in\Gamma(x)}\xi(\omega)$.*

Proof. Set $c := \inf_{\omega\in\Gamma(x)}\xi(\omega)$ and suppose $\overline{\mathrm{E}}(\xi|x) < c$. Then there exist $\epsilon > 0$ and a supermartingale $\mathcal{S}$ such that $\mathcal{S}(x) < c - \epsilon$ and, on $\Gamma(x)$, $\liminf_t \mathcal{S}_t \geq \xi$. Let us show that this is impossible. By AG4, Reality can choose the outcomes after the situation x so that

$$\mathcal{S}(x) = \mathcal{S}(\omega^k) \geq \mathcal{S}(\omega^{k+1}) - \frac{1}{2}\epsilon \geq \mathcal{S}(\omega^{k+2}) - \frac{3}{4}\epsilon \geq \mathcal{S}(\omega^{k+3}) - \frac{7}{8}\epsilon \geq \cdots,$$

where k is the level of x and $\omega \in \Gamma(x)$ is the realized path; this path ω will satisfy $\liminf_t \mathcal{S}_t(\omega) \leq \limsup_t \mathcal{S}_t(\omega) \leq \mathcal{S}_k(\omega) + \epsilon < c \leq \xi(\omega)$. □

Lemma 6.1 justifies the adjectives 'upper' and 'lower' that we use in front of 'expectation' and 'probability':

Corollary 6.1 *For all situations x and all global gambles $\xi : \Omega \to \mathbb{R}$, $\underline{\mathrm{E}}(\xi|x) \leq \overline{\mathrm{E}}(\xi|x)$. In particular, $\underline{\mathrm{P}}(E|x) \leq \overline{\mathrm{P}}(E|x)$ for all events $E \subseteq \Omega$.*

Proof. Suppose $\underline{\mathrm{E}}(\xi|x) > \overline{\mathrm{E}}(\xi|x)$, i.e., $\overline{\mathrm{E}}(\xi|x) + \overline{\mathrm{E}}(-\xi|x) < 0$. By DC2, this implies $\overline{\mathrm{E}}(0|x) < 0$, which contradicts DC4. □

Important special cases are where $x = \square$ (unconditional upper and lower expectation and probability). We set $\overline{\mathrm{E}}(\xi) := \overline{\mathrm{E}}(\xi|\square)$, $\underline{\mathrm{E}}(\xi) := \underline{\mathrm{E}}(\xi|\square)$, $\overline{\mathrm{P}}(E) := \overline{\mathrm{P}}(E|\square)$, and $\underline{P}(E) := \underline{\mathrm{P}}(E|\square)$. We say that an event E is *almost certain*, or happens *almost surely* (a.s.), if $\underline{P}(E) = 1$; in this case we will also say that E, considered as a property of $\omega \in \Omega$, holds for *almost all* ω. An event E is *almost impossible*, or *null*, if $\overline{\mathrm{P}}(E) = 0$.

The lower probability of an event E is often defined as $1 - \overline{\mathrm{P}}(E^c)$. It is not difficult to check that this definition is equivalent to the one given above.

The definition of upper expectation and probability is fairly robust. As will be shown in Section 6.5, the definition of upper expectation will not change if we replace the lim inf in (6.2) by lim sup. This definition is also equivalent to

$$\overline{\mathrm{E}}(\xi|x) := \inf\left\{ \mathcal{S}(x) : (\forall \omega \in \Gamma(x)) \lim_{t\to\infty} \mathcal{S}_t(\omega) \geq \xi(\omega) \right\}$$

where $\mathcal{S}$ ranges over the class of all bounded supermartingales for which $\lim_{t\to\infty} \mathcal{S}_t(\omega)$ exists for all $\omega \in \Omega$. The definition of upper probability, which can be written in the form

$$\overline{\mathrm{P}}(E|x) := \inf\left\{ \mathcal{S}(x) : (\forall \omega \in E \cap \Gamma(x)) \liminf_{t\to\infty} \mathcal{S}_t(\omega) \geq 1 \right\}, \tag{6.4}$$

where $\mathcal{S}$ ranges over the non-negative supermartingales, is even more robust: we can replace the $\liminf_{t\to\infty}$ in (6.4) by $\sup_t$.

Let us now discuss the connection between Protocols 6.1 and 6.3. First we modify Protocol 6.1 by allowing Sceptic to announce any $\mathcal{K}_0 \in \mathbb{R}$, not necessarily $\mathcal{K}_0 := 1$. Even after this extension Protocol 6.1 is still a special case of Protocol 6.3. The set of available gambles $\mathcal{A} = \mathcal{A}_p$ corresponding to a prediction $p \in [-C, C]$ is defined as the smallest set satisfying AG1–AG4 and containing the functions $y \mapsto y - p$ and $y \mapsto p - y$ (choosing $p \notin [-C, C]$ by Forecaster would violate AG4 straight away). Essentially, $\mathcal{A}$ is the natural extension (on the part of Sceptic) of the assessment that the prevision of the identity map is p; cf. (1.9). It is easy to see that $\mathcal{A}$ consists of all functions f such that $(\exists s)(\forall y) f(y) \leq s(y - p)$; it clearly satisfies AG1–AG4. Each supermartingale in this specialization of Protocol 6.3 is a supermartingale in the modified Protocol 6.1 and vice versa. Therefore, all concepts (such as upper and lower expectation and probability) defined in terms of supermartingales coincide for the two protocols, and all results that can be stated in terms of supermartingales (such as Theorem 6.1) hold or fail to hold in the two protocols simultaneously.

For a general discussion of game-theoretic probability in the rest of this section we will use the special cases of Protocol 6.3 given by Protocol 6.1 and its three special cases discussed in Section 6.2 (with $\mathcal{K}_0$ chosen by Sceptic). In the case of Protocol 6.1, game-theoretic probability is imprecise probability for at least two reasons. First, nobody forecasts Forecaster's moves; for example, for any set $\emptyset \subset E \subset \mathbb{R}$ ($\subset$ standing for proper inclusion) and any $t \in \mathbb{N}$, we have $0 = \underline{\mathrm{P}}(p_t \in E) < \overline{\mathrm{P}}(p_t \in E) = 1$. And second, even Forecaster's forecasts are incomplete: the $\mathcal{A}_t$ corresponding to p_t fall short of being maximally committal (as defined in Section 1.5). Imprecision diminishes when we move to the prequential protocol (where $\mathcal{A}_t$ become maximally committal) and Protocol 6.2 (where the uncertainty of Forecaster's moves disappears). Eventually imprecision disappears altogether in the coin tossing protocol. In the latter, we redefine Ω by omitting Forecaster's moves, which are no longer informative, and then for each Borel set $E \subseteq \Omega = \{-1, 1\}^\infty$ we have $\underline{P}(E) = \overline{\mathrm{P}}(E) = U^\infty(E)$, where U is the uniform distribution on the set $\{-1, 1\}$. (This is spelled out in detail in [582, § 8.2].) In fact, precise game-theoretic probability is a standard pedagogical device and computational tool in mathematical finance: see, e.g., the discussion of binomial trees in [582, § 1.5].

6.4 The axiom of continuity

The main purpose of this section (most of which can be skipped alongside Section 6.8 without interrupting the flow of ideas) is to discuss the following additional axiom (the 'axiom of continuity') for sets of available gambles:

AG5. For any uniformly bounded and increasing sequence of gambles $f_1 \leq f_2 \leq \cdots$ in $\mathcal{A}$, $\sup_k f_k$ will also be in $\mathcal{A}$.

This axiom is a version of the standard axiom of σ-additivity in measure-theoretic probability. It often simplifies proofs but on closer inspection turns out to be redundant; we will see examples in Sections 6.5 (the proof of Theorem 6.2) and 6.7 (the proof of Theorem 6.4). In the interests of readers who are willing to accept this axiom, in both cases we will first give a simpler argument using the axiom of continuity. However, none of the results stated in this chapter outside this section and Section 6.8 requires it.

It is widely accepted that the axiom of continuity is a purely formal requirement without much empirical substance. Kolmogorov's justification for it was merely ([405, § II.1]):

> *We limit ourselves, arbitrarily, to only those models which satisfy Axiom VI [the axiom of continuity].* This limitation has been found expedient in researches of the most diverse sort.

For further historical details see [583, especially §§ 3.3–3.4 and 5.1.1].

When AG1–AG5 are satisfied for a set $\mathcal{A}$ of available gambles, it is natural to extend $\mathcal{A}$ to the following set $\mathcal{A}'$ of functions $f : \Omega \to (-\infty, \infty]$: $f \in \mathcal{A}'$ if and only if f is bounded from below and $\min(f, C) \in \mathcal{A}$ for all $C \in \mathbb{R}$. It is easy to see that $\mathcal{A}'$ also satisfies Axioms AG1–AG5; moreover, it satisfies the following simplified form of AG5:

AG5′. For any increasing sequence of functions $f_1 \leq f_2 \leq \cdots$ in $\mathcal{A}'$, $\sup_k f_k$ will also be in $\mathcal{A}'$.

(Indeed, for any C, $\min(f_k, C)$ will be an increasing sequence of functions in $\mathcal{A}$, and so $\sup_k \min(f_k, C) = \min(\sup_k f_k, C) \in \mathcal{A}$; therefore, $\sup_k f_k \in \mathcal{A}'$.)

It is now easy to see that there is a natural bijection between the family $\mathbb{A}_5$ of all sets $\mathcal{A}$ of gambles satisfying Axioms AG1–AG5 and the family $\mathbb{A}'_5$ of all sets $\mathcal{A}'$ of functions in $(-\infty, \infty]^{\mathcal{X}}$ that are bounded from below and satisfy Axioms AG1–AG4 and AG5′: in one direction, the set $\mathcal{A}$ of all local gambles in $\mathcal{A}' \in \mathbb{A}'_5$ (which we call the *restriction* of $\mathcal{A}'$ to $\mathcal{L}(\mathcal{X})$) will be in $\mathbb{A}_5$; and in the other, for each $\mathcal{A} \in \mathbb{A}_5$ there is a unique $\mathcal{A}' \in \mathbb{A}'_5$ (defined as explained above) whose restriction to $\mathcal{L}(\mathcal{X})$ is $\mathcal{A}$.

In some cases Axiom AG5 is satisfied, or almost satisfied, automatically. At the end of the previous section we explained that Protocol 6.1 describes the situation where Forecaster always chooses a set $\mathcal{A}_p$ that consists of all bounded functions f such that $(\exists s)(\forall y) f(y) \leq s(y-p)$. For each $p \in (-C, C)$, this set satisfies AG5: this follows from the fact that, when $f_1 \leq f_2 \leq \cdots$ and $f_1, f_2, \ldots \in \mathcal{A}_p$, each $f_k(y)$ is bounded above by $s_k(y-p)$ for some monotonic and bounded sequence $s_k \in \mathbb{R}$; it is clear that $f(y)$, where $f := \lim_{k\to\infty} f_k$, will be bounded above by $s(y-p)$, where $s := \lim_{k\to\infty} s_k$. Notice that for $p \in \{-C, C\}$ the set $\mathcal{A}_p$ does not satisfy AG5 (e.g., for $p := -C$ the limit as $k \to \infty$ of the increasing sequence $f_k(y) := \min(k(y-p), 1)$ will be outside $\mathcal{A}_p$).

Let us now consider the case where the possibility space $\mathcal{X}$ is finite. It is clear that in this case we have:

AG5″. For any uniformly bounded and increasing sequence of gambles $f_1 \leq f_2 \leq \cdots$ in $\mathcal{A}$ and any $\epsilon > 0$, $\sup_k f_k - \epsilon$ will also be in $\mathcal{A}$.

In practice, there is very little difference between Axioms AG5 and AG5″.

In the previous section we defined null events. However, we cannot state Theorem 6.1 (and many other similar theorems given in [582]) in terms of null events without weakening it. For this, we need the following somewhat narrower definition. An event E is *strictly null* if there is a test supermartingale $\mathcal{S}$ such that $\mathcal{S}_t \to \infty$ on E. Theorem 6.1 says that the event (6.1) holds strictly almost surely, in the sense of its negation being strictly null.

It is obvious that any strictly null event is null. Under Axiom AG5, it is natural to replace $\mathbb{A}$ with $\mathbb{A}'_5$ in the definition of the sample space; a supermartingale is now a function $\mathcal{S}$ on the clearing situations taking values in $(-\infty, \infty]$ such that $\mathcal{S}(x\mathcal{A}\cdot) - \mathcal{S}(x) \in \mathcal{A}$ for all clearing situations x and all $\mathcal{A} \in \mathbb{A}'_5$. In this case there is no difference between null and strictly null events:

Lemma 6.2 *Suppose the sample space is $(\mathbb{A}'_5 \times \mathcal{X})^\infty$. An event is null if and only if it is strictly null.*

Proof. Suppose an event E is null. For each $k \in \mathbb{N}$ choose a test supermartingale $\mathcal{S}^{(k)}$ such that $\liminf_t \mathcal{S}_t^{(k)} \geq 2^{2k}$ on E. The test supermartingale $\mathcal{S} := \sum_{k=1}^{\infty} 2^{-k}\mathcal{S}^{(k)}$ witnesses that E is strictly null. (To see that $\mathcal{S}$ is a supermartingale notice that $\mathcal{S}(x\mathcal{A}\cdot) - \mathcal{S}(x) \in \mathcal{A}$ follows from $\mathcal{S}_K(x\mathcal{A}\cdot) - \mathcal{S}(x) \leq \mathcal{S}_K(x\mathcal{A}\cdot) - \mathcal{S}_K(x) \in \mathcal{A}$, where $\mathcal{S}_K := \sum_{k=1}^{K} 2^{-k}\mathcal{S}^{(k)}$, and Axiom AG5′.) □

Simplest examples show that without Axiom AG5 there are null events that are not strictly null: e.g., set $\mathcal{X} := [0, 1]$ and define $\mathcal{A}$ to be the set of all continuous functions f satisfying $\int_0^1 f(y)\, dy \leq 0$; the set of all sequences $0\mathcal{A}y_2\mathcal{A}y_3\mathcal{A}\ \ldots$, where $y_2, y_3, \ldots \in \mathcal{X}$, is null but not strictly null.

Let $\mu : \wp(\Omega) \to [0, \infty)$, where $\wp(\Omega)$ is the family of all subsets of Ω. Recall that μ is called an *outer measure* [102, 335] if:

O1. For any events E_1 and E_2,

$$E_1 \subseteq E_2 \Rightarrow \mu(E_1) \leq \mu(E_2). \tag{6.5}$$

O2. For any sequence of events $E_1, E_2, \ldots,$

$$\mu\left(\bigcup_{k=1}^{\infty} E_k\right) \leq \sum_{k=1}^{\infty} \mu(E_k).$$

O3. $\mu(\emptyset) = 0$.

The first and third conditions hold obviously when μ is game-theoretic upper probability, $\mu = \overline{\mathrm{P}}$. The following lemma establishes the property of countable subadditivity for conditional game-theoretic upper probability and so, in particular, implies that $\overline{\mathrm{P}}(\cdot|x)$ is an outer measure.

Lemma 6.3 *Suppose the sample space is* $(\mathbb{A}_5 \times \mathcal{X})^\infty$ *or* $(\mathbb{A}'_5 \times \mathcal{X})^\infty$. *For any sequence of events* $E_1, E_2, \dots$ *and any situation* x, *it is true that*

$$\overline{\mathrm{P}}\left(\bigcup_{k=1}^{\infty} E_k | x\right) \leq \sum_{k=1}^{\infty} \overline{\mathrm{P}}(E_k|x).$$

In particular, the union of a sequence of null events is null.

Proof. Assume, without loss of generality, that $x = \square$. Let $\epsilon > 0$ be arbitrarily small. For each $k \in \mathbb{N}$ choose a supermartingale $\mathcal{S}^{(k)}$ (automatically non-negative, by Lemma 6.1) such that $\liminf_t \mathcal{S}_t^{(k)} \geq I_{E_k}$ and $\mathcal{S}_0^{(k)} \leq \overline{\mathrm{P}}(E_k) + \epsilon/2^k$. It is easy to check that $\mathcal{S} := \min(1, \sum_{k=1}^{\infty} \mathcal{S}^{(k)})$ will be a supermartingale (cf. AG2 and AG5) that satisfies $\mathcal{S}_0 \leq \sum_{k=1}^{\infty} \overline{\mathrm{P}}(E_k) + \epsilon$ and $\liminf_t \mathcal{S}_t \geq I_{E_k}$ for all k. □

It is not difficult to check that Lemma 6.3 will continue to hold when the sample space is $(\mathbb{A}''_5 \times \mathcal{X})^\infty$, $\mathbb{A}''_5$ being the family of all sets of local gambles satisfying Axioms AG1–AG4 and AG5″.

Every outer measure gives rise to the important class of measurable sets in the sense of Carathéodory (see [102] or [335, § 11]). This class is a σ-algebra and the restriction of the outer measure to it is a measure. In particular, the restriction of game-theoretic upper probability (equivalently, of game-theoretic lower probability) to the σ-algebra of Carathéodory measurable sets is a probability measure. An interesting problem, largely unexplored, is to characterize the Carathéodory measurable sets for various instantiations of Protocol 6.3. For this class Kolmogorov's axioms of probability emerge in a natural way instead of being postulated. The main result of [664] suggests that the class of Carathéodory measurable sets becomes especially rich in the case of continuous time (this case will be discussed very briefly in Section 6.9).

6.5 Doob's argument

In this section we introduce Doob's argument, which is very useful in both measure-theoretic and game-theoretic probability. The form used in this section may be called additive; the multiplicative form will be used in Section 6.7.

For the sake of readers who have skipped the previous section, we define an event E to be *strictly null* if there is a test supermartingale $\mathcal{S}$ such that $\mathcal{S}_t \to \infty$ on E; an event holds *strictly almost surely* if its negation is strictly null. The following lemma is a game-theoretic version of Doob's famous martingale convergence theorem.

Theorem 6.2 *Any supermartingale* $\mathcal{S}$ *that is bounded from below is convergent in* $\mathbb{R}$ *strictly almost surely.*

Proof. We first assume Axiom AG5′; later we will get rid of this assumption. Namely, we consider the sample space $(\mathbb{A}'_5 \times \mathcal{X})^\infty$. Without loss of generality we assume that the supermartingale $\mathcal{S}$ is positive (we can always add a constant to it). It suffices to construct a positive supermartingale $\mathcal{T}_t$ (not necessarily starting from 1) that tends to ∞ on the paths $\omega \in \Omega$ on which $\mathcal{S}_t(\omega)$ is not convergent in $(-\infty, \infty]$: indeed, in this case $\mathcal{S}_t + \mathcal{T}_t$ will be a positive supermartingale that tends to ∞ when S_t is not convergent in $\mathbb{R}$. According to Lemma 6.3 (second statement) and Lemma 6.2, we can, without loss of generality, replace 'on which $S_t(\omega)$ is not convergent in $(-\infty, \infty]$' by

$$\liminf_{t\to\infty} S_t(\omega) < a < b < \limsup_{t\to\infty} S_t(\omega), \tag{6.6}$$

where a and b are given positive rational numbers such that $a < b$. (This is the place where we use Axiom AG5$'$.) The supermartingale $\mathcal{T}$ is defined as the capital process of the following strategy. Let $\omega \in \Omega$ be the sequence of moves chosen by Forecaster and Reality. Start with a monetary units. Wait until $S_t(\omega) \leq a$ (if this never happens, do nothing, i.e., always choose $f_t := 0$). As soon as this happens (perhaps for $t = 1$), maintain capital $S_t(\omega)$ until $S_t(\omega)$ reaches a value at least b (at which point Sceptic's capital is at least b). After that do nothing until $S_t(\omega) \leq a$. As soon as this happens, maintain capital $b - a + S_t(\omega)$ until $S_t(\omega)$ reaches a value at least b (at which point Sceptic's capital is at least $(b - a) + b$). After that do nothing until $S_t(\omega) \leq a$. As soon as this happens, maintain capital $2(b - a) + S_t(\omega)$ until $S_t(\omega)$ reaches a value at least b (at which point Sceptic's capital is at least $2(b - a) + b$). And so on. On the event (6.6) Sceptic's capital will tend to infinity.

It remains to replace the sample space $(\mathbb{A}_5' \times \mathcal{X})^\infty$ by $(\mathbb{A} \times \mathcal{X})^\infty$. Let $\mathcal{T}^{(a,b)}$ be the supermartingale constructed in the same way as $\mathcal{T}$ in the previous paragraph for a given pair (a, b) (so now we reflect formally the dependence on (a, b) in our notation). Let us check that $\mathcal{T} := \sum_{(a,b)} w_{(a,b)} \mathcal{T}^{(a,b)}$, where (a, b) ranges over the pairs of rational numbers such that $0 < a < b$ and $w_{(a,b)}$ are weights summing to 1, is a supermartingale. For each betting situation x, we have $\mathcal{T}(x\cdot) - \mathcal{T}(x) = \sum_{(a,b)} w_{(a,b)} (\mathcal{T}^{(a,b)}(x\cdot) - \mathcal{T}^{(a,b)}(x))$, and it remains to remember that $\mathcal{T}^{(a,b)}(x\cdot) - \mathcal{T}^{(a,b)}(x)$ is either $\mathcal{S}(x\cdot) - \mathcal{S}(x)$ or 0. □

The main application of Theorem 6.2 (more accurately, of its proof) will be the following theorem giving two equivalent definitions of game-theoretic upper expectation.

Theorem 6.3 *For all global gambles* $\xi : \Omega \to \mathbb{R}$ *and all situations* x,

$$\overline{\mathrm{E}}(\xi|x) = \inf\left\{ \mathcal{S}(x) : (\forall \omega \in \Gamma(x)) \limsup_{t\to\infty} \mathcal{S}_t(\omega) \geq \xi(\omega) \right\} \tag{6.7}$$

(i.e., on the right-hand side of (6.2), *we can replace* lim inf *by* lim sup*), where* $\mathcal{S}$ *ranges over the class of all supermartingales, and*

$$\overline{\mathrm{E}}(\xi|x) = \inf\left\{ \mathcal{S}(x) : (\forall \omega \in \Gamma(x)) \lim_{t\to\infty} \mathcal{S}_t(\omega) \geq \xi(\omega) \right\} \tag{6.8}$$

where $\mathcal{S}$ *ranges over the class* $\mathbb{L}$ *of all bounded supermartingales for which* $\lim_{t\to\infty} \mathcal{S}_t(\omega)$ *exists for all* $\omega \in \Omega$.

Proof. Fix ξ and x. It suffices to show that the right-hand side of (6.8) does not exceed the right-hand side of (6.7). Let $\epsilon > 0$. Take any non-negative supermartingale $\mathcal{S}$ such that $\mathcal{S}(x) < \overline{\mathrm{E}}(\xi|x) + \epsilon/2$ and $\limsup_{t\to\infty} \mathcal{S}_t \geq \xi$ on $\Gamma(x)$. It suffices to construct a supermartingale $\mathcal{S}^* \in \mathbb{L}$ such that $\mathcal{S}^*(x) < \overline{\mathrm{E}}(\xi|x) + \epsilon$ and $\lim_{t\to\infty} \mathcal{S}_t^* \geq \xi$ on $\Gamma(x)$. Let $\mathcal{T}$ be a non-negative supermartingale that tends to ∞ when $\mathcal{S}_t$ does not converge in $\mathbb{R}$ (e.g., the one constructed in the proof of Theorem 6.2); assume, without loss of generality, that

$\mathcal{T}(x) = \epsilon/2$. Set $\mathcal{S}^* := \min(\mathcal{S} + \mathcal{T}, \sup_{\omega\in\Gamma(x)}\xi(\omega))$. All the required conditions for $\mathcal{S}^*$ are obvious except for $\mathcal{S}^* \in \mathbb{L}$.

We only consider $\omega \in \Gamma(x)$. If $\mathcal{S}_t(\omega)$ is not convergent in $\mathbb{R}$ we have $\mathcal{T}_t(\omega) \to \infty$ and so $\mathcal{S}^*_t(\omega) = \sup_{\omega\in\Gamma(x)}\xi(\omega)$ from some t on. Therefore, we only need to consider the case where $\mathcal{S}_t$ converges in $\mathbb{R}$. In this case all $\mathcal{T}_t^{(a,b)}$ converge in $\mathbb{R}$. Suppose, without loss of generality, that $\sum_{(a,b)} w_{(a,b)} a < \infty$. Since $\mathcal{T}_{t_2}^{(a,b)} - \mathcal{T}_{t_1}^{(a,b)} \geq -a$ whenever $t_2 > t_1$, we can see that $\lim_{T\to\infty}\inf_{t_2>t_1\geq T}(\mathcal{T}_{t_2} - \mathcal{T}_{t_1}) \geq 0$. Therefore, $\mathcal{T}_t$ converges in $(-\infty, \infty]$, which implies the convergence of $\mathcal{S}^*_t$ in $\mathbb{R}$. □

Replacing the $\liminf_{t\to\infty}$ in (6.2) by $\inf_t$ or $\sup_t$ does change the definition. If we replace the $\liminf_{t\to\infty}$ by $\inf_t$, we will have $\overline{\mathrm{E}}(\xi|x) = \sup_{\omega\in\Gamma(x)}\xi(\omega)$. In the following example we consider replacing $\liminf_{t\to\infty}$ by $\sup_t$; for simplicity, we only consider $\overline{\mathrm{E}}(\xi|x)$ for $x = \square$.

Example 6.1 For each $\xi \in \mathcal{L}(\Omega)$, set

$$\overline{\mathrm{E}}_1(\xi) := \inf\left\{ \mathcal{S}(\square) : (\forall\omega \in \Omega) \sup_t \mathcal{S}_t(\omega) \geq \xi(\omega) \right\},$$

$\mathcal{S}$ ranging over the supermartingales. It is always true that $\overline{\mathrm{E}}_1(\xi) \leq \overline{\mathrm{E}}(\xi)$. Consider the coin-tossing protocol. For each $\epsilon \in (0, 1]$ there exists a non-negative global gamble $\xi \in \mathcal{L}(\Omega)$ such that $\overline{\mathrm{E}}(\xi) = 1$ and $\overline{\mathrm{E}}_1(\xi) = \epsilon$. ♦

Proof. Let us demonstrate the following equivalent statement: for any $C \geq 1$ there exists a non-negative global gamble ξ such that $\overline{\mathrm{E}}_1(\xi) = 1$ and $\overline{\mathrm{E}}(\xi) = C$. Fix such a C. Define $\Xi : \Omega \to [0, \infty)$ by the requirement $\Xi(\omega) := 2^n$ where n is the number of 1s at the beginning of ω: $n := \max\{i : \omega_1 = \cdots = \omega_i = 1\}$. It is obvious that $\overline{\mathrm{E}}_1(\Xi) = 1$ and $\overline{\mathrm{E}}(\Xi) = \infty$. However, Ξ is unbounded. We can always find $A \geq 1$ such that $\overline{\mathrm{E}}(\min(\Xi, A)) = C$ (as the function $a \mapsto \overline{\mathrm{E}}(\min(\Xi, a))$ is continuous). Since $\overline{\mathrm{E}}_1(\min(\Xi, A)) = 1$, we can set $\xi := \min(\Xi, A)$. □

Game-theoretic probability is a special case of game-theoretic expectation, and in this special case it is possible to replace $\liminf_{t\to\infty}$ not only by $\limsup_{t\to\infty}$ but also by $\sup_t$, provided we restrict our attention to non-negative supermartingales (simple examples show that this qualification is necessary).

Lemma 6.4 *The definition of upper probability will not change if we replace the* $\liminf_{t\to\infty}$ *in (6.4) by* $\limsup_{t\to\infty}$ *or by* $\sup_t$.

It is obvious that the definition will change if we replace the $\liminf_{t\to\infty}$ in (6.4) by $\inf_t$: in this case we will have

$$\overline{\mathrm{P}}(E|x) = \begin{cases} 0 & \text{if } E \cap \Gamma(x) = \emptyset \\ 1 & \text{otherwise.} \end{cases}$$

Proof. It suffices to prove that the definition will not change if we replace the $\liminf_{t\to\infty}$ in (6.4) by $\sup_t$. Consider a strategy for Sceptic resulting in a non-negative capital process. If this strategy ensures $\sup_t \mathcal{K}_t > 1$ when $\mathcal{A}_1 y_1 \mathcal{A}_2 y_2 \ldots \in E \cap \Gamma(x)$ (it is obvious that it does not matter whether we have $\geq$ or $>$ in (6.4)), Sceptic can also ensure $\liminf_{t\to\infty}\mathcal{K}_t > 1$

when $\mathcal{A}_1 y_1 \mathcal{A}_2 y_2 \ldots \in E \cap \Gamma(x)$ by stopping (i.e., always choosing $f_t := 0$) after his capital $\mathcal{K}_t$ exceeds 1. □

6.6 Limit theorems of probability

The main goal of this section is to give a constructive proof of Theorem 6.1, which says that event (6.1) holds strictly almost surely. The simple proof given here is inspired by [412] (proof of Theorem 3.1).

We say that a non-negative supermartingale $\mathcal{S}$ *forces* an event E if $\lim_{t\to\infty} \mathcal{S}_t(\omega) = \infty$ for every ω not in E. We are required to construct a supermartingale forcing (6.1), where we can set $p_t := 0$, as we mentioned in Section 6.2. This is not difficult given Theorem 6.2.

Lemma 6.5 *In Protocol 6.2, there is a game-theoretic supermartingale $\mathcal{S}$ such that the supermartingale constructed in the proof of Theorem 6.2 forces*

$$\lim_{T\to\infty} \frac{1}{T} \sum_{t=1}^{T} y_t = 0. \tag{6.9}$$

Proof. Set $\epsilon_t := 1/\max(2C, t)$. The protocol specifies that the initial capital is 1. Let Σ be the strategy $s_t := \epsilon_t \mathcal{K}_{t-1}$ of gambling the fraction ϵ_t of the current capital. Since Reality's move y_t is never less than $-C$, this strategy loses at most the fraction $\epsilon_t C \le 1/2$ of the current capital, and hence the capital process $\mathcal{K}^\Sigma$ is non-negative. It is given by $\mathcal{K}^\Sigma(\square) = 1$ and

$$\mathcal{K}^\Sigma(y_1 \ldots y_T) = \mathcal{K}^\Sigma(y_1 \ldots y_{T-1})(1 + \epsilon_T y_T) = \prod_{t=1}^{T} (1 + \epsilon_t y_t).$$

Let $\mathcal{T}$ be the supermartingale constructed in the proof of Theorem 6.2 for $\mathcal{S} := \mathcal{K}^\Sigma$. Consider two cases:

- If $\mathcal{K}^\Sigma(y_1 \ldots y_T)$ does not converge in $\mathbb{R}$ as $T \to \infty$, $\lim_{T\to\infty} \mathcal{T}(y_1 \ldots y_T) = \infty$.
- Now suppose $\mathcal{K}^\Sigma(y_1 \ldots y_T)$ does converge to a finite limit. Then $\sum_{t=1}^{T} \ln\,(1 + \epsilon_t y_t)$ converges to a finite limit. By Taylor's formula, $\ln(1 + u) = u + O(u^2)$, and so $\sum_{t=1}^{T} \epsilon_t y_t$ also converges to a finite limit. Kronecker's lemma now implies $\epsilon_T \sum_{t=1}^{T} y_t \to 0$, which is equivalent to (6.9). □

The proof of Lemma 6.5 shows that we can replace (6.1) by, e.g.,

$$\lim_{T\to\infty} \frac{1}{\sqrt{T} \ln T} \sum_{t=1}^{T} (y_t - p_t) = 0$$

in Theorem 6.1: indeed, the essential conditions for ϵ_t were $\epsilon_t < 1/C$ and $\sum_{t=1}^{\infty} \epsilon_t^2 < \infty$, and they permit any positive coefficient α_T in front of $\sum_{t=1}^{T}$ that satisfies $\sum_{T=1}^{\infty} \alpha_T^2 < \infty$. This is, however, still significantly weaker than the statement of the law of the iterated logarithm, which allows $\alpha_T = O(1/\sqrt{T \ln\ln T})$. The game-theoretic version of Kolmogorov's law of the iterated logarithm is stated and proved in [582, § 5]; for other game-theoretic laws of the iterated logarithm, see [474].

Another direction in which Theorem 6.1 can be strengthened is to remove the requirement that y_t should be bounded. This is done in the game-theoretic version of Kolmogorov's strong law of large numbers, which is the topic of [582, § 4] (for other game-theoretic strong laws of large numbers for unbounded y_t, see [413]). Both Kolmogorov's law of the iterated logarithm and Kolmogorov's strong law of large numbers ask more of Forecaster: he is asked to price not only y_t but also $(y_t - p_t)^2$ (or at least to provide an upper price for the latter).

The strong law of large numbers and the law of the iterated logarithm belong to strong laws of probability, which means, in our current context, that they do not require the notions of game-theoretic probability and expectation; for them the notion of forcing, as used in this section, is sufficient. Important examples of weak laws of probability, which do require them, are the weak law of large numbers and the central limit theorem. For a game-theoretic version of Lindeberg's central limit theorem, see [582, Chapter 7].

6.7 Lévy's zero-one law

Alongside the classical limit theorems, zero-one laws are among the most important results of probability theory. In this book we only prove a game-theoretic counterpart of Lévy's zero-one law [431, § 41], which implies most of the other well-known zero-one laws, such as Bártfai and Révész's [47] zero-one law and, via it, Kolmogorov's zero-one law ([405, Appendix]) and the ergodicity of Bernoulli shifts. For the derivation of these other laws, scc [585].

Theorem 6.4 *Let $\xi : \Omega \to \mathbb{R}$ be a global gamble. For strictly almost all $\omega \in \Omega$,*

$$\liminf_{t\to\infty} \overline{\mathrm{E}}(\xi|\omega^t) \geq \xi(\omega). \tag{6.10}$$

(Of course, by 'for strictly almost all $\omega \in \Omega$' we mean that (6.10) holds strictly almost surely.) This theorem is a game-theoretic version of Lévy's zero-one law. The name of this result might look puzzling, but it is justified by its connections with more standard zero-one phenomena mentioned above. The proof will be based on the multiplicative version of Doob's argument; for a much more detailed version of this proof, see [585].

Proof. As in the proof of Theorem 6.2, we will first assume Axiom AG5′, considering the sample space $(\mathbb{A}_5' \times \mathcal{X})^\infty$. It suffices to construct a test supermartingale that tends to ∞ on the paths $\omega \in \Omega$ for which (6.10) is not true. Without loss of generality we will assume ξ to be positive. According to Lemma 6.3 (second statement) and Lemma 6.2, we can, without loss of generality, replace 'for which (6.10) is not true' by

$$\liminf_{t\to\infty} \overline{\mathrm{E}}(\xi|\omega^t) < a < b < \xi(\omega) \tag{6.11}$$

where a and b are given positive rational numbers such that $a < b$. The supermartingale is defined as the capital process of the following strategy for Sceptic. Let $\omega \in \Omega$ be the sequence of moves chosen by Forecaster and Reality. Start with 1 monetary unit. Wait until $\overline{\mathrm{E}}(\xi|\omega^t) < a$ (if this never happens, do nothing). As soon as this happens, choose a non-negative supermartingale $\mathcal{S}^{(1)}$ starting from a, $\mathcal{S}^{(1)}(\omega^t) = a$, whose upper limit exceeds

ξ on $\Gamma(\omega^t)$. Maintain capital $(1/a)\mathcal{S}^{(1)}(\omega^t)$ until $\mathcal{S}^{(1)}(\omega^t)$ reaches a value $b_1 \geq b$ (at which point Sceptic's capital is $b_1/a \geq b/a$). After that do nothing until $\overline{\mathrm{E}}(\xi|\omega^t) < a$. As soon as this happens, choose a non-negative supermartingale $\mathcal{S}^{(2)}$ starting from a, $\mathcal{S}^{(2)}(\omega^t) = a$, whose upper limit exceeds ξ on $\Gamma(\omega^t)$. Maintain capital $(b/a^2)\mathcal{S}^{(2)}(\omega^t)$ until $\mathcal{S}^{(2)}(\omega^t)$ reaches a value $b_2 \geq b$ (at which point Sceptic's capital is $b_2 b/a^2 \geq (b/a)^2$). After that do nothing until $\overline{\mathrm{E}}(\xi|\omega^t) < a$. As soon as this happens, choose a non-negative supermartingale $\mathcal{S}^{(3)}$ starting from a whose upper limit exceeds ξ on $\Gamma(\omega^t)$. Maintain capital $(b^2/a^3)\mathcal{S}^{(3)}(\omega^t)$ until $\mathcal{S}^{(3)}(\omega^t)$ reaches a value $b_3 \geq b$ (at which point Sceptic's capital is $b_3 b^2/a^3 > (b/a)^3$). And so on. On the event (6.11) Sceptic's capital will tend to infinity.

It remains to replace the sample space $(\mathbb{A}'_5 \times \mathcal{X})^\infty$ by $(\mathbb{A} \times \mathcal{X})^\infty$. For each clearing situation x fix a non-negative supermartingale $\mathcal{S}^x$ such that $\mathcal{S}^x(x) = \overline{\mathrm{E}}(\xi|x) + 2^{-n}$, where n is the level of x, and $\limsup_{t\to\infty} \mathcal{S}^x_t \geq \xi$ on $\Gamma(x)$. The construction in the previous paragraph, for a given pair (a, b) with $a < b$, is now modified as follows. Start with 1 monetary unit. Wait until $\overline{\mathrm{E}}(\xi|\omega^t) < a$. Set $\mathcal{S}^{(1)} := \mathcal{S}^{\omega^t}$ and $a_1 := \max(a, \mathcal{S}^{(1)}(\omega^t))$. Maintain capital $(1/a_1)\mathcal{S}^{(1)}(\omega^t)$ until $\mathcal{S}^{(1)}(\omega^t)$ reaches a value $b_1 \geq b$ (at which point Sceptic's capital is at least $b_1/a_1 \geq b/a_1$). After that do nothing until $\overline{\mathrm{E}}(\xi|\omega^t) < a$. Set $\mathcal{S}^{(2)} := \mathcal{S}^{\omega^t}$ and $a_2 := \max(a, \mathcal{S}^{(2)}(\omega^t))$. Maintain capital $(b/a_1 a_2)\mathcal{S}^{(2)}(\omega^t)$ until $\mathcal{S}^{(2)}(\omega^t)$ reaches a value $b_2 \geq b$ (at which point Sceptic's capital is at least $b_2 b/a_1 a_2 \geq b^2/a_1 a_2$). After that do nothing until $\overline{\mathrm{E}}(\xi|\omega^t) < a$. Set $\mathcal{S}^{(3)} := \mathcal{S}^{\omega^t}$ and $a_3 := \max(a, \mathcal{S}^{(3)}(\omega^t))$. Maintain capital $(b^2/a_1 a_2 a_3)\mathcal{S}^{(3)}(\omega^t)$ until $\mathcal{S}^{(3)}(\omega^t)$ reaches a value $b_3 \geq b$ (at which point Sceptic's capital is at least $b_3 b^2/a_1 a_2 a_3 \geq b^3/a_1 a_2 a_3$). And so on. As $\limsup_{n\to\infty} a_n = a$, Sceptic's capital will still tend to infinity on the event (6.11).

Let $\mathcal{S}^{(a,b)}$ be the supermartingale constructed in the previous paragraph for a given pair (a, b). Let us check that $\mathcal{S} := \sum_{(a,b)} w_{(a,b)} \mathcal{S}^{(a,b)}$, where (a, b) ranges over the pairs of rational numbers such that $a < b$ and $w_{(a,b)}$ are positive weights summing to 1, is a supermartingale for a suitable choice of the weights (it is sufficient that $\sum_{(a,b)} w_{(a,b)} b^n/a^{n+1} < \infty$ for any $n \in \mathbb{N}_0$; e.g., we can take $w_{(a,b)}$ such that $\sum_{(a,b)} w_{(a,b)} e^{b/a}/a < \infty$). Fix a betting situation x. We have $\mathcal{S}(x\cdot) - \mathcal{S}(x) = \sum_{(a,b)} w_{(a,b)}(\mathcal{S}^{(a,b)}(x\cdot) - \mathcal{S}^{(a,b)}(x))$, and each $\mathcal{S}^{(a,b)}(x\cdot) - \mathcal{S}^{(a,b)}(x)$ is a linear combination of $\mathcal{S}^z(x\cdot) - \mathcal{S}^z(x)$ over the clearing situations $z \subseteq x$ ($z \subseteq x$ meaning that z is a prefix of x) with coefficients bounded by b^n/a^{n+1}, where n is the level of x. Therefore, $\mathcal{S}(x\cdot) - \mathcal{S}(x)$ is a linear combination of $\mathcal{S}^z(x\cdot) - \mathcal{S}^z(x)$; since there are finitely many $z \subseteq x$, we can deduce $\mathcal{S}(x\cdot) - \mathcal{S}(x) \in \mathcal{A}_n$ (where $\mathcal{A}_n$ is the last prediction in x), even without using Axiom AG5′. □

Specializing Theorem 6.4 to the indicators of events, we obtain:

Corollary 6.2 *Let E be any event. For strictly almost all $\omega \in E$,*

$$\overline{\mathrm{P}}(E|\omega^t) \to 1 \quad \text{as } t \to \infty. \tag{6.12}$$

6.8 The axiom of continuity revisited

Throughout this section we will assume Axiom AG5, and so Ω will stand for $(\mathbb{A}_5 \times \mathcal{X})^\infty$. To see more clearly the connection of Theorem 6.4 with Lévy's classical result, let us consider global gambles ξ satisfying $\overline{\mathrm{E}}(\xi) = \underline{\mathrm{E}}(\xi)$ and events E satisfying $\overline{\mathrm{P}}(E) = \underline{\mathrm{P}}(E)$.

Lemma 6.6 *Suppose a global gamble ξ satisfies $\overline{\mathrm{E}}(\xi) = \underline{\mathrm{E}}(\xi)$. Then it is almost certain that for all t it satisfies $\overline{\mathrm{E}}(\xi|\omega^t) = \underline{\mathrm{E}}(\xi|\omega^t)$.*

Proof. For any positive ϵ, there exist supermartingales $\mathcal{S}^{(1)}$ and $\mathcal{S}^{(2)}$ that start at $\overline{\mathrm{E}}(\xi) + \epsilon/2$ and $\overline{\mathrm{E}}(-\xi) + \epsilon/2$, respectively, and tend to ξ or more and to $-\xi$ or more, respectively. Set $\mathcal{S} := \mathcal{S}^{(1)} + \mathcal{S}^{(2)}$. The assumption $\overline{\mathrm{E}}(\xi) = \underline{\mathrm{E}}(\xi)$ can also be written $\overline{\mathrm{E}}(\xi) + \overline{\mathrm{E}}(-\xi) = 0$. So the non-negative (by Lemma 6.1) supermartingale $\mathcal{S}$ begins at ϵ and tends to 0 or more on all $\omega \in \Omega$.

Fix $\delta > 0$, and let A be the event that there exists t such that

$$\overline{\mathrm{E}}(\xi|\omega^t) + \overline{\mathrm{E}}(-\xi|\omega^t) > \delta.$$

By the definition of conditional upper expectation,

$$\mathcal{S}_t^{(1)}(\omega) \geq \overline{\mathrm{E}}(\xi|\omega^t) \quad \text{and} \quad \mathcal{S}_t^{(2)}(\omega) \geq \overline{\mathrm{E}}(-\xi|\omega^t).$$

So $\mathcal{S}_t$ exceeds δ for some t on A. So the upper probability of A is at most ϵ/δ. Since ϵ may be as small as we like for fixed δ, this shows that A has upper probability zero. Letting δ range over the positive rational numbers and applying the second part of Lemma 6.3 (thus using AG5), we obtain the statement of the lemma. □

Corollary 6.3 *Let ξ be a global gamble for which $\overline{\mathrm{E}}(\xi) = \underline{\mathrm{E}}(\xi)$. Then, almost surely, $\overline{\mathrm{E}}(\xi|\omega^t) = \underline{\mathrm{E}}(\xi|\omega^t) \to \xi(\omega)$ as $t \to \infty$.*

Proof. By Theorem 6.4,

$$\liminf_{t\to\infty} \overline{\mathrm{E}}(\xi|\omega^t) \geq \xi(\omega)$$

for almost all $\omega \in \Omega$ and (applying the theorem to $-\xi$)

$$\limsup_{t\to\infty} \underline{\mathrm{E}}(\xi|\omega^t) \leq \xi(\omega)$$

for almost all $\omega \in \Omega$. It remains to apply Lemma 6.6. □

Our definitions 6.3 make it easy to obtain the following corollaries for events.

Corollary 6.4 *Suppose an event E satisfies $\overline{\mathrm{P}}(E) = \underline{\mathrm{P}}(E)$. Then, almost surely, it also satisfies $\overline{\mathrm{P}}(E|\omega^t) = \underline{\mathrm{P}}(E|\omega^t)$ for all t.*

Corollary 6.5 *Let E be an event for which $\overline{\mathrm{P}}(E) = \underline{\mathrm{P}}(E)$. Then, almost surely, $\overline{\mathrm{P}}(E|\omega^t) = \underline{\mathrm{P}}(E|\omega^t) \to I_E(\omega)$ as $t \to \infty$.*

Lévy's zero-one law has an important implication for the foundations of game-theoretic probability theory. Let ξ be a global gamble, and let $x := \square$. We will obtain an equivalent definition of the upper expectation $\overline{\mathrm{E}}(\xi|x) = \overline{\mathrm{E}}(\xi)$ if we replace '$\forall\omega \in \Gamma(x)$' in (6.2) by 'for almost all $\omega \in \Omega$'. It turns out that if we do so, the infimum in (6.2) becomes attained;

namely, it is attained by the supermartingale $\mathcal{S}_t(\omega) := \overline{\mathrm{E}}(\xi|\omega^t)$. In view of Theorem 6.4, to prove this statement it suffices to check that $\mathcal{S}_t(\omega) := \overline{\mathrm{E}}(\xi|\omega^t)$ is indeed a supermartingale.

Theorem 6.5 *Let $\xi : \Omega \to \mathbb{R}$ be a global gamble. Then $\mathcal{S}_t(\omega) := \overline{\mathrm{E}}(\xi|\omega^t)$ is a supermartingale.*

Proof. As a first step, let us check that, for any $\epsilon > 0$, $\mathcal{S}_t^{(\epsilon)}(\omega) := \overline{\mathrm{E}}(\xi|\omega^t) + \epsilon 2^{-t}$ is a supermartingale, i.e., that

$$\overline{\mathrm{E}}(\xi|\omega^{t-1}\mathcal{A}\cdot) - \overline{\mathrm{E}}(\xi|\omega^{t-1}) - \epsilon 2^{-t} \in \mathcal{A}$$

for all $\omega \in \Omega$, $t \in \mathbb{N}$, and $\mathcal{A} \in \mathbb{A}_5$. The last inclusion follows from the existence of a supermartingale $\mathcal{T}$ that starts from $\overline{\mathrm{E}}(\xi|\omega^{t-1}) + \epsilon 2^{-t}$ in the situation ω^{t-1} and ensures $\liminf_t \mathcal{T}_t \geq \xi$ on $\Gamma(\omega^{t-1})$: it is clear that such $\mathcal{T}$ will satisfy $\mathcal{T}(\omega^{t-1}\mathcal{A}y) \geq \overline{\mathrm{E}}(\xi|\omega^{t-1}\mathcal{A}y)$ for all $y \in \mathcal{X}$.

It remains to notice that the infimum of a decreasing sequence of supermartingales is again a supermartingale (this step uses AG5) and that $\mathcal{S} = \inf_{\epsilon>0}\mathcal{S}^{(\epsilon)} = \inf_{n\in\mathbb{N}}\mathcal{S}^{(1/n)}$. □

A function $\mu : \wp(\Omega) \to [0, \infty)$ is called a *precapacity* if:

CC1. It satisfies $\mu(\emptyset) = 0$ and $\mu(\Omega) = 1$.

CC2. For any subsets E_1 and E_2 of Ω, the monotonicity property (6.5) holds.

CC3. For any nested increasing sequence $E_1 \subseteq E_2 \subseteq \cdots$ of arbitrary subsets of Ω,

$$\mu\left(\bigcup_{k=1}^{\infty} E_k\right) = \lim_{k\to\infty} \mu(E_k). \tag{6.13}$$

The definitions of this section agree with the definition of capacity in Section 4.2: on finite Ω every precapacity is already a capacity (as defined below) and item CC3 of the definition of a precapacity is redundant.

The following is an easy corollary of Lévy's zero-one law.

Theorem 6.6 *The set function $\overline{\mathrm{P}}$ is a precapacity.*

Proof. Let $\mu := \overline{\mathrm{P}}$. The conditions $\mu(\emptyset) = 0$, $\mu(\Omega) = 1$, and (6.5) are obvious.

To check (6.13), let $E_1, E_2, \ldots$ be a nested increasing sequence of events. The non-trivial inequality in (6.13) is $\leq$. By Theorem 6.5, for each E_k the process

$$\mathcal{S}^{(k)}(x) := \overline{\mathrm{P}}(E_k|x)$$

is a non-negative supermartingale. By Corollary 6.2, $\liminf_t \mathcal{S}^{(k)}(\omega^t) \geq 1$ for almost all $\omega \in E_k$. The sequence $\mathcal{S}^{(k)}$ is increasing, $\mathcal{S}^{(1)} \leq \mathcal{S}^{(2)} \leq \cdots$, so the limit $\mathcal{S} := \lim_{k\to\infty} \mathcal{S}^{(k)} = \sup_k \mathcal{S}^{(k)}$ exists and is (by AG5) a non-negative supermartingale such that $\mathcal{S}(\square) = \lim_{k\to\infty}\overline{\mathrm{P}}(E_k)$ and $\liminf_t \mathcal{S}(\omega^t) \geq 1$ for almost all $\omega \in \cup_k E_k$ (by Lemma 6.3). We can get rid of 'almost' by adding to $\mathcal{S}$ a non-negative supermartingale V that starts at $V(\square) < \epsilon$, for an arbitrarily small $\epsilon > 0$, and satisfies $\liminf_t V(\omega^t) \geq 1$ for all $\omega \in \cup_k E_k$ violating $\liminf_t \mathcal{S}(\omega^t) \geq 1$. □

A precapacity μ is a *capacity* (or, more fully, *Choquet capacity*) if:

CC4. For any nested decreasing sequence $K_1 \supseteq K_2 \supseteq \cdots$ of compact sets in Ω,

$$\mu(\cap_{k=1}^{\infty} K_k) = \lim_{k\to\infty} \mu(K_k). \tag{6.14}$$

An interesting question is whether $\overline{\mathrm{P}}$ is a capacity. For the binary prequential framework, where $\mathcal{X} = \{0, 1\}$ and $\mathcal{A}_t$ are maximally committal, it is answered in positive in [667]. In general, the answer will depend on the chosen topology on $\mathbb{A}_5$ (and at this time very little is known about suitable topologies on $\mathbb{A}_5$, or on the set of all coherent sets of desirable gambles on $\mathcal{X}$, which was denoted $\mathbb{D}(\mathcal{X})$ in Section 1.2.3).

The class of capacities that is especially important from the point of view of imprecise probability (see, e.g., [674]) is that of strongly subadditive (or 2-alternating) capacities. We say that a precapacity μ is *strongly subadditive* if

$$\mu(E_1 \cup E_2) + \mu(E_1 \cap E_2) \leq \mu(E_1) + \mu(E_2)$$

for all events E_1 and E_2. (The conjugate precapacity $\tilde{\mu}(E) := 1 - \mu(E^c)$ is then 2-monotone, as defined in Definition 4.2 and Section 4.3.) We can define $\overline{\mathrm{P}}(E)$ for subsets of $(\mathbb{A}_5 \times \mathcal{X})^t$ by (6.4) with $\liminf_t$ omitted and $x := \square$. This is an example of subsets E_1 and E_2 of $([0, 1] \times \{0, 1\})^2$ in the binary prequential protocol for which the condition of strong subadditivity is violated:

$$E_1 = \left\{\left(0, 0, \frac{1}{2}, 0\right), \left(\frac{1}{2}, 0, 0, 0\right)\right\}, \tag{6.15}$$

$$E_2 = \left\{\left(0, 0, \frac{1}{2}, 0\right), \left(\frac{1}{2}, 1, 0, 0\right)\right\}. \tag{6.16}$$

For these subsets we have

$$\overline{\mathrm{P}}(E_1 \cup E_2) + \overline{\mathrm{P}}(E_1 \cap E_2) = 1 + \frac{1}{2} > \frac{1}{2} + \frac{1}{2} = \overline{\mathrm{P}}(E_1) + \overline{\mathrm{P}}(E_2).$$

To obtain an example of subsets E_1 and E_2 of the full set Ω for which the condition of strong subadditivity is violated, it suffices to add $00\ldots$ at the end of each element of the sets E_1 and E_2 defined by (6.15) and (6.16).

In some cases Theorem 6.6 implies that $\overline{\mathrm{P}}$ admits an equivalent (although somewhat complicated) definition in terms of sets of probability measures, in the spirit of the lower envelope theorem (see Propositions 2.3 and 2.4, or [674, mathematical model 6 on p. 126]). In the binary prequential case this is demonstrated in [667].

6.9 Further reading

In this section we will give references to some of the work in game-theoretic probability that we could not cover in detail.

In this book we consider the case of discrete time only. For a review of recent work in game-theoretic probability in continuous time, see [584, § 4.4]. The paper [663], published since that review, proves that typical price paths of financial securities in idealized continuous-time financial markets cannot be more volatile than Brownian motion (formally, cannot have a variation index exceeding 2). The word 'typical' here means that there is a strategy for trading in that security that risks only one monetary unit and brings an infinite capital when the price path is too volatile. As usual in game-theoretic probability, no statistical assumptions are made about the price path; the only assumption is analytic: we assume that the price path is a right-continuous function. If we make the stronger assumption that price paths are continuous, we can prove more: typical price paths are either constant or as volatile as Brownian motion (formally, their variation index is either 0 or 2): see [662] (based on [619]) and [664].

We can interpret $\mathcal{K}_t$ as the evidence found by Sceptic against Forecaster. For connections of this game-theoretic measure of evidence with more standard statistical measures of evidence (namely, p-values and Bayes factors), see [581]. An unusual feature of $\mathcal{K}_t$ as measure of evidence is that evidence can be both gained and lost; various strategies for insuring against loss of evidence in game-theoretic probability are discussed in [180].

The method of 'defensive forecasting' uses game-theoretic laws of probability to obtain forecasting strategies satisfying various desiderata: see, e.g., [112, 668].

Game-theoretic probability theory studies protocols involving various players, such as Forecaster, Sceptic, and Reality. Defensive forecasting produces strategies for Forecaster. Classical laws of probability theory correspond to strategies for Sceptic. To obtain optimal strategies for Sceptic and to demonstrate that they are optimal, it often helps to construct matching strategies for Reality. For examples of such strategies for Reality see [473 § 4.3 and 5.2] and [665]. It is interesting that this kind of results does not have any counterparts in measure-theoretic probability theory.

Game-theoretic probability, as we have presented it here, takes on-line prediction as basic. In this setting, we observe a sequence of outcomes, predicting each one after observing the previous one. Most modern work in probability and statistics, and most modern work on imprecise probabilities, lies outside this setting. In some cases, the task is to predict only one outcome, and relevant prior experience is relatively unformalized. In some cases, past experience is formalized as a theory that gives probability predictions all at once for indefinitely many future observations. In some cases, the focus is on hypotheses or estimation, leaving prediction out of the discussion even if it is the ultimate goal.

We believe that on-line prediction is a good setting for modernizing the core of classical probability theory. It allows us to generalize elegantly the most classical theorems of probability theory, such as the law of large numbers, the law of the iterated logarithm, and the central limit theorem. Perhaps it is therefore also a reasonable starting point for imprecise probabilities. But we have not addressed in this book the question of how game-theoretic probability can be extended beyond on-line prediction to serve other applications.

In particular, we have not addressed the problem of assessing evidence, in which probabilities may need to be created or updated because of unanticipated new information, or in which competing bodies of previous experience need to be combined. However, the question of updating has been studied previously from a game-theoretic point of view [578–580]. In [578], Shafer argued that the use of conditional probability for updating is justified based on the willingness of a person (our Forecaster) to offer bets only if a protocol (such as our on-line prediction protocol) is given in advance for the information on which one conditions. In [580], this argument was extended to Walley's rule for updating imprecise probabilities, and it was also explained how both conditional probability and Walley's rule can be used to update on unanticipated information when we instead interpret the probabilities or upper and lower probabilities as betting rates that will not allow Sceptic to multiply the capital he risks by a large factor. The essential ingredient here is a judgement about what strategies can or cannot help Sceptic achieve this goal. In [579], this picture is boiled down to its essentials and extended to an explanation of the judgements of irrelevance underlying the rule of combination for Dempster-Shafer belief functions.

So far in this section we have been discussing recent mathematical work on game-theoretic probability. As we mentioned in Section 6.1, the game-theoretic approach to probability has a long history, and in the rest of the section we will give references to recent historical work. Two especially interesting periods in the history of both game-theoretic and measure-theoretic probability are the emergence of probability as a scientific discipline in the second half of the 17th century and the axiomatization of probability in the first third of the 20th century. It has been argued (see, e.g., [582, § 2.1], and [584, § 2.1]) that the game-theoretic approach goes back to Pascal's solution to the problem of points and the measure-theoretic approach to Fermat's. Whereas Pascal solved the problem by constructing a martingale, Fermat solved it by constructing a finite probability space equipped with the uniform probability measure. Of the two most well-known attempts at axiomatization of probability, one, proposed by Richard von Mises in 1919, was game-theoretic in spirit and the other, proposed by Andrei Kolmogorov in 1933, was thoroughly measure-theoretic. Richard von Mises's approach was made rigorous by Abraham Wald and Jean Ville, the latter introducing martingales into the foundations of probability *en route*. Joseph Doob translated Ville's idea into the measure-theoretic framework and greatly advanced the mathematical theory of martingales. This period is described in [582, §§ 2.2–2.4]. For further information about the history of martingales (mainly in measure-theoretic probability), see the special issue on martingales of the *Electronic Journal for History of Probability and Statistics* (June 2009).

Acknowledgements

The work on this chapter has been supported in part by EPSRC grant EP/F002998/1.

7

Statistical inference

Thomas Augustin[1], Gero Walter[1], and Frank P. A. Coolen[2]

[1]*Department of Statistics, Ludwig-Maximilians University Munich (LMU), Germany*

[2]*Department of Mathematical Sciences, Durham University, UK*

This chapter introduces the use of imprecise probabilities in statistical inference. In contrast to many other fields, in this area the development of methodology based on imprecise probabilities is still in its early stages. From the current perspective of statistical modelling, the procedures proposed so far are mostly rather basic, but even already there imprecise probabilities prove very powerful. By overcoming some fundamental limitations inherent in traditional statistics, imprecise probabilities are promising a comprehensive framework for reliable inference and data analysis.

With still so much to explore, statistical inference with imprecise probability offers excellent opportunities for research, and the authors hope to stimulate its advancement with this overview. To keep the material at an appropriate length, many choices had to be made in the presentation. Our major goal is to achieve an exemplary perception of the power of imprecise probability methods in the area of statistical inference, as well as to raise awareness of the many challenges still waiting. Therefore we often give up mathematical rigour in favour of an informal description aiming at an intuition of the basic concepts.[1]

This chapter is organized as follows: Taking the heterogeneous background of potential readers into account, we try to keep the text self-contained and recall at least those concepts

[1] In particular, we are very sloppy with respect to finite versus σ-additivity of underlying precise probabilities, and aspects of measurability. See in particular [323] for a mathematically rigorous exposition of the framework applied here.

Introduction to Imprecise Probabilities, First Edition.
Edited by Thomas Augustin, Frank P. A. Coolen, Gert de Cooman and Matthias C. M. Troffaes.

of traditional statistics which are fundamental for the generalizations to be discussed. The reader with a stronger background in traditional statistics may just browse through these parts (Sections 7.1, 7.4.1, and 7.5.1), mainly taking them as a guide to the notation used. Section 7.1 also includes a sketch of the most important different inference concepts, which are used to structure the later sections. Before discussing the specific extended frameworks, two introductory sections provide the common basis for the different approaches to statistical inference with imprecise probabilities. In Section 7.2 we work out different understandings and interpretations of imprecision in statistics, including an ideal typical distinction between model and data imprecision, and discuss some motives for the paradigmatic shift towards imprecise probabilities in statistics. Section 7.3 collects some basic models and general notions. We then first discuss the main inference concepts under model imprecision, namely generalized Bayesian inference (Section 7.4, where we mainly focus on a certain prototypical class of models), selected aspects of the frequentist approach (Section 7.5), the methodology of nonparametric predictive inference (Section 7.6), and also briefly mention in Section 7.7 some other approaches. Section 7.8 changes the perspective, first considering data imprecision and then more generally the concept of partial identification as a framework for handling observationally equivalent models. Some general challenges are discussed in Section 7.9, while Section 7.10 serves as a guide for further reading.

7.1 Background and introduction

7.1.1 What is statistical inference?

In short, statistical inference is all about learning from data.

In this, data are considered observable quantities from which we wish to learn. So what does 'learning' mean in the context of statistical inference? In a focused sense, statistical inference aims to generalize information from a set of observations (a *sample*) to a larger *population* from which the sample was drawn, e.g., to estimate the percentages of votes for candidates or parties in an election (population: all voters) from the percentages of votes obtained in an exit poll (sample: exit poll participants).[2]

Thus, statistical *inference* or inductive statistics is basically concerned with inductive reasoning: establishing a general rule from observations. It therefore goes beyond merely compressing (or summarizing) information, as in *descriptive statistics*, where, e.g., a number of observations is summarized by calculating the median or tabulating the frequency of different outcomes, mostly involving graphical representations like barplots, boxplots, etc. It also distinguishes itself from direct unstructured discovering of information (patterns) concealed in large quantities of data, as in *data mining* or *explorative data analysis* (where, e.g., in a merchant's database of customers, one strives to identify groups of customers with similar characteristics).

[2] In a broader sense, statistical inference aims to infer general laws not necessarily confined to the population the sample was drawn from. Consider, e.g., a medical study conducted to answer the question if some (new) medical treatment A works generally better than the (standard) treatment B. Such studies are based on the performance of treatments A and B in the participants of a clinical trial (the sample). Here, the goal is not to infer the performance of A and B in the population the study participants were drawn from (which is often confined to the populace that the participating hospitals are serving), but the potential benefit of these treatments for patients in general.

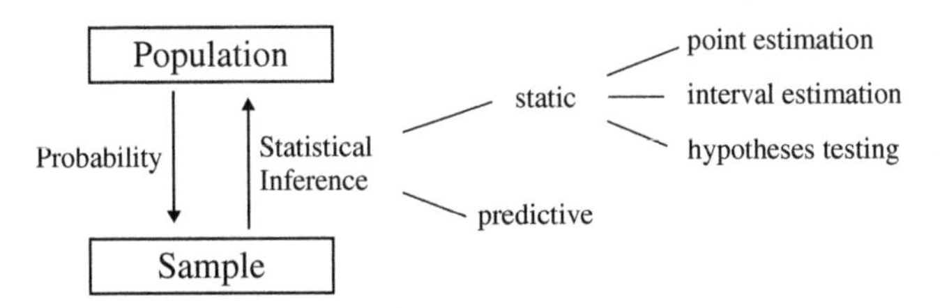

Figure 7.1 Schematic illustration of the relation between population and sample, motivating the role of statistical inference as 'inverting probability laws'.

As is long known as the problem of induction [371], it is impossible to justify inductive reasoning by pure reason, and therefore one cannot infer general statements (laws) with absolute truth from single observations. In this vein, every statement about properties of a population which is based on a sample is potentially incomplete and therefore possibly wrong, except in the trivial extreme case where the sample contains the whole population. The statistical remedy for this inevitable and fundamental dilemma of any type of inductive reasoning is (postulated, maybe virtual) *randomness* of the sampling process.[3] If, and only if, the sample is, or can be understood as, drawn randomly, probability theory allows to quantify the error of statistical propositions concluded from the sample.

To elaborate this in a more formal way, let X be a certain property of interest *(variable)*, like the gender or the age of the units in the population. By the sampling process, properties of the population units are transferred to random properties of the sample members: Before the sample of units $i = 1, \dots, n$ is drawn and analyzed, it is not known who is the i-th unit in the sample. Therefore also the concrete value x_i of the i-th unit in the sample is unknown, and thus has to be described by a random quantity X_i. In the very same way, the sampling process transfers frequencies in the population into probabilities. If, for instance, 60% of the population units are female, the probability – under the usually assumed simple random sample scheme[4] – that a randomly drawn unit i is female is 0.6. This induced probability governs the distribution of the single observations in the sample, and, by appropriate independence concepts, the joint distribution of the random sample $\mathbf{X} = (X_1, \dots, X_n)$ of size n.

7.1.2 (Parametric) statistical models and i.i.d. samples

The concept of probability builds a bridge between the population and a sample drawn from it. Statistical inference basically is the art to cross this bridge in the reverse direction, i.e. to learn from concrete observations in the sample about characteristics of the unknown (!) underlying distribution $P^*(\cdot)$ of the sample, and therefore about unknown characteristics of

[3] The term 'virtual randomness' is used here as a surrogate for 'uncertainty' in the subjective approaches to probability. (The term 'uncertainty' is avoided here because it is used in different meanings in various areas.)

[4] The common sampling procedure underlying statistical inference is simple random sampling, where every subset of units in the population of the same cardinality has equal probability to be drawn. Then, in particular, no unit in the population is favoured in the sampling process, all units have the same precise probability of being drawn. This common model of sampling is also assumed to underlie all the developments in this chapter; some ideas for more general, imprecise probability-based sampling procedures are sketched at the end of Section 7.2.5.

the population.[5] In traditional statistics, as discussed in the present section, this underlying distribution is always assumed to be precise.

To facilitate this reasoning, typically a *(classical) statistical model* is formulated. It is a tuple $(\mathcal{X}, \mathcal{Q})$, consisting of the sample space $\mathcal{X}$, i.e. the domain of the random quantity X under consideration, and a set $\mathcal{Q}$ of precise probability distributions,[6] collecting all precise probability distributions that are judged to be potential candidates for the precise distribution of X. In this setting $\mathcal{Q}$ is called *sampling model* and every element $P \in \mathcal{Q}$ *(potential) sampling distribution*[7]. The task is to learn the true element $P^* \in \mathcal{Q}$ from multiple observations of the random process producing X.

One typically distinguishes between two types of models with respect to $\mathcal{Q}$:[8]

- *Nonparametric models*, where $\mathcal{Q}$ is characterized by qualitative characteristics of its sampling distributions only, like symmetry or unimodality of the underlying densities.
- *Parametric models*. Here $\mathcal{Q}$ is parametrized by a parameter ϑ of finite dimension, assuming values in the so-called *parameter space* Θ, $\Theta \subseteq \mathbb{R}^q, q < \infty$, i.e. $\mathcal{Q} = (P_\vartheta)_{\vartheta \in \Theta}$. In such models the different sampling distributions P_ϑ are implicitly understood as belonging to a specific class of distributions (e.g. normal distributions, see the example in Section 7.1.5), the basic type of which is assumed to be known completely, and only some characteristics ϑ (e.g. the mean) of the distributions are unknown. Typically q is small in applications, often indeed $q = 1$.

In the sequel, we mainly focus on such parametric models (with a notable exception in Section 7.6), and throughout this chapter we call the tuple

$$(\mathcal{X}, (P_\vartheta)_{\vartheta \in \Theta}) \tag{7.1}$$

(classical) parametric statistical model for the random quantity X with *sampling model* $(P_\vartheta)_{\vartheta \in \Theta}$ and *parameter* $\vartheta \in \Theta \subseteq \mathbb{R}^q$. Single components of the vector ϑ are denoted by $\vartheta_{[\ell]}$, $\ell = 1, \ldots, q$. Throughout the chapter, we will assume (as is the case for almost all common applications) that the underlying candidate distributions P_ϑ of the random quantity X are either discrete or absolutely continuous with respect to the Lebesgue measure (see, e.g., [389, pp. 32f, 38] for some technical details) for every $\theta \in \Theta$. Then it is convenient to express every P_ϑ in the discrete case by its *mass function* f_ϑ, with $f_\vartheta(x) := P_\vartheta(X = x)$, $\forall x \in \mathcal{X}$, and in the continuous case by its *probability density function* (pdf) f_ϑ. Then, if X takes value on the real line, f_ϑ is such that $P_\vartheta(X \in [a, b]) = \int_a^b f_\vartheta(x)\mathrm{d}x$ for all $a, b \in \mathbb{R}$.

Learning now needs several, typically many, observations of the underlying process generating values of the variable X, i.e. a sample. More precisely, an *i.i.d. sample of size n*

[5] In this sense, statistical inference is in some disciplines entitled as 'inverse probability', providing a methodology for 'inverting probability laws', see Figure 7.1.

[6] Most models of statistical inference rely on σ-additive probability distributions. Therefore, technically, in addition an appropriate (σ-)field $\sigma(\mathcal{X})$, describing the domain of the underlying probability measure, has to be specified. In most applications there are straightforward canonical choices for $\sigma(\mathcal{X})$, and thus σ-fields are not explicitly discussed here, except when they are utilized to express coarsening aspects in Section 7.8.

[7] In this chapter we do not distinguish probability measures and the corresponding distributions in terminology.

[8] Note that this common distinction between parametric and nonparametric models is an informal one, based on heuristic modelling aspects. Mathematically the distinction cannot be maintained rigorously, because any probability distribution in $\mathbb{R}^q$ can be described by a real-valued parameter, e.g. [708, p. 5, Footnote 1], that parameterisation is however often too odd to be practically useful.

(where *i.i.d.* abbreviates independent, identically distributed) *based on the classical parametric statistical model* $(\mathcal{X}, (P_\vartheta)_{\vartheta\in\Theta})$ is then a vector

$$\mathbf{X} = (X_1, \ldots, X_n)^{\mathrm{T}}$$

of independent random quantities X_i with the same distribution P_ϑ. Then $\mathbf{X}$ is defined on $\mathcal{X}^n$ with probability distribution $P_\vartheta^{\otimes n}$ as the n-dimensional product measure describing the independent observations.[9] $P_\vartheta^{\otimes n}$ thus has the probability mass or density function

$$\prod_{i=1}^{n} f_\vartheta(x_i) =: f_\vartheta(x_1, \ldots, x_n).$$

The term *sample* is then also used for the concretely observed value(s) $\mathbf{x} = (x_1, \ldots, x_n)^T$.[10] In an analogous way, more generally an i.i.d. sample of size n is defined based on an arbitrary, not necessarily parametric classical statistical model $(\mathcal{X}, \mathcal{Q})$.

Current statistical applications mostly rely on extensions of these settings to *regression models*, where heterogeneity of the population with respect to certain individual characteristics $\mathbf{z}_i$ of unit $i, i = 1, \ldots, n$, *(covariates)* is allowed. Technically this is modelled in the parametric setting by allowing one component $\vartheta_{[\ell]}$ of ϑ to vary from unit to unit and to relate it to $\mathbf{z}_i$ by $\vartheta_{[\ell]}(i) = g(\mathbf{z}_i),\ i = 1, \ldots, n$, with some appropriate function g, which typically is parameterized again by some finite parameter vector β. These regression models have been further generalized to model also dependencies between observations, and thus can be used for repeated measurements at the same object, otherwise clustered observations and also for longitudinal or spatial data.[11]

7.1.3 Basic tasks and procedures of statistical inference

Different tasks and concepts of statistical inference can be classified by two fundamental dimensions:

1) The type of statement one wants to infer from the data, as well as

2) the methodological framework for deriving such statements in a proper way.

The different types of statements can furthermore be classified into

1a) static conclusions and

1b) predictive conclusions.[12]

---

[9] In the Bayesian setting, independence is often replaced by exchangeability (see, e.g., [70, § 4.2]).

[10] Throughout this chapter, random quantities are denoted by capital letters, their values by small letters.

[11] See, e.g., [269] for a monograph on regression models, and [525] for a series of short review articles on the application of statistical models in life and medical sciences, social science and business, physical sciences and engineering.

[12] Static conclusions are still more common in statistical applications. In the light of the success of procedures from machine learning in statistics, the predictive view, early propagated, e.g., by [308], has attracted increasing attention.

Static conclusions refer directly to properties of the sampling model, typically to its parameter(s). Common to the different inference schools described below are the following procedures (formulated in terms of parameters here):[13]

- *Point estimators*, where a certain parameter value is selected to describe the sample.
- *Interval estimators*, where the information is condensed in a certain subset of the parameter space Θ, typically in an interval when $\Theta \subseteq \mathbb{R}$.
- *Hypotheses tests*, where the information in the sample is only used to decide between two mutually exclusive statements ('hypotheses').

Predictive conclusions instead summarize the information by statements on properties of typical further units, either by describing the whole distribution or certain summary measures. This includes classification of further units, as discussed in Chapter 10.

Static and predictive conclusions can in fact all be formally understood as a special case of *decision making*, where, more generally, the conclusion may be to select certain utility maximizing or loss minimizing acts from a set of possible acts.[14]

7.1.4 Some methodological distinctions

The appropriate way to derive inference procedures has been a matter of intensive debate in traditional statistics, having led to different inference schools. The two most popular schools are *frequentist* approaches (sometimes also called classical approaches) and *Bayesian* approaches. The choice between both methodological approaches is predetermined by the interpretation of probability, and most notably with the crucial question whether probability statements on parameters are meaningful.

The Bayesian approach allows (subjective) knowledge on the parameter to be expressed by a probability distribution on[15] Θ, with probability mass or density function $\pi(\vartheta)$ called *prior (distribution)*. Interpreting the elements $f_\vartheta(x)$ of the sampling model as conditional distributions of the sample given the parameter, denoted by $f(\mathbf{x}|\vartheta)$ and called *likelihood*, turns the problem of statistical inference into a problem of probabilistic deduction, where the *posterior distribution*, i.e. the distribution of the parameter given the sample data, can be calculated by Bayes rule.[16] Thus, in the light of the sample $\mathbf{x} = (x_1, \ldots, x_n)$ the prior distribution is updated by Bayes' rule to obtain the posterior distribution with density or mass function

$$\pi(\vartheta|\mathbf{x}) \propto f(\mathbf{x}|\vartheta) \cdot \pi(\vartheta). \tag{7.2}$$

The posterior distribution is understood as comprising all the information from the sample and the prior knowledge. It therefore underlies all further inferences on the parameter ϑ, the static ones as well as the predictive ones, where it can in particular be used to calculate the posterior predictive distribution of further observations. (See also Section 7.4.1.)

---

[13] See also Figure 7.1.

[14] We do not pursue this point of view explicitly in this chapter, as decision making is treated as a separate chapter (see Chapter 8).

[15] We implicitly assume that also Θ is complemented by an appropriate σ-field $\sigma(\Theta)$.

[16] Gillies [313, 314] argues that Bayes Theorem was actually developed in order to confront the problem of induction as posed by Hume [371].

If, on the contrary, as in the *frequentist approaches*, the applicability of the concept probability is solely restricted to random phenomena, parameters are strictly treated as fixed, non-probabilistic quantities ('true parameter') and inference directly focuses on $(P_\vartheta^{\otimes n}(\mathbf{x}))_{\vartheta \in \Theta}$. Since here inference goes beyond mere deduction, additional, external criteria have to be introduced. As the name frequentist suggests, methods are evaluated in the 'long run', i.e. one aims at constructing inference procedures that behave well in the average of infinitely many virtual samples drawn from the population. Many frequentist estimators and tests again involve a notion of *likelihood* based on the sampling model, with a somewhat different interpretation compared to the Bayesian approach. In this setting, one defines after having (virtually) observed the sample $\mathbf{x}$ the likelihood of ϑ as

$$\text{Lik}(\vartheta || \mathbf{x}) := f_\vartheta(\mathbf{x}),$$

that means, the joint probability density or mass function of the sample $\mathbf{x}$ given the parameter ϑ is now, after having fixed $\mathbf{x}$, reinterpreted as a function of ϑ, describing how plausible it is to have observed the concrete sample $\mathbf{x}$ if ϑ is the true parameter.[17] The likelihood $\text{Lik}(\vartheta || \mathbf{x})$ is thus understood to measure how plausible it is that ϑ is the true parameter.

Alternatives to Bayes and frequentist approaches include approaches that are directly based on the likelihood and the concept of *fiducial probabilities*, which try to arrive at probability statements about the parameter after the sample without relying on prior distributions. On the one hand this concept is widely perceived with great scepticism, but on the other hand it has induced an intensive debate. Different related approaches have also been proposed under the name *logical probability*.[18]

7.1.5 Examples: Multinomial and normal distribution

We will now present two basic examples that will be repeatedly discussed in this chapter.

Example 7.1 (The multinomial distribution). The *multinomial distribution* is a common model for samples where only a limited number of distinct values can be observed, like in the data from an election poll, or from a pharmacological study asking about a certain set of adverse reactions. These distinct values are often named *categories* (hence the term *categorical data*), and are usually numbered from 1 to k, without imposing any natural ordering on these values. We have therefore a discrete distribution, giving the probability for observing certain category counts $(n_1, \ldots, n_k) = \mathbf{n}$ in a sample of n observations in total. Thus, $\sum_{j=1}^k n_j = n$.

We start the definition of the multinomial distribution by decomposing the collection of n observations into its constituents, single observations of either of the categories $1, \ldots, k$. Such a single observation, often named *multivariate Bernoulli observation*, can be encoded as a vector x_i of length k, where the j-th element, x_{ij}, equals 1 if category j has been observed, and all other elements being 0. Given the vectorial parameter θ of length k, where the component $\theta_{[j]}$ models the probability of observing category j in a single draw (therefore

[17] See also Section 5.2.2.

[18] See Section 7.7 for a brief sketch of such approaches in the light of imprecise probabilities.

$\sum_{j=1}^{k} \theta_{[j]} = 1$), the probability for observing x_i can be written as

$$f(x_i|\theta) = \prod_{j=1}^{k} \theta_{[j]}^{x_{ij}}.$$

Assuming independence, the probability for observing a certain sequence $\mathbf{x} = (x_1, \ldots, x_n)$ of n observations can thus be written as

$$f(\mathbf{x}|\theta) = \prod_{i=1}^{n} f(x_i|\theta) \propto \prod_{i=1}^{n} \prod_{j=1}^{k} \theta_{[j]}^{x_{ij}} = \prod_{j=1}^{k} \theta_{[j]}^{\sum_{i=1}^{n} x_{ij}} = \prod_{j=1}^{k} \theta_{[j]}^{n_j},$$

where $n_j = \sum_{i=1}^{n} x_{ij}$ tells us how often category j was observed in the sample.

For the probability to observe a certain category count $(n_1, \ldots, n_k) = \mathbf{n}$, we have to account for the different possible orderings in $\mathbf{x}$ leading to the same count vector $\mathbf{n}$. Therefore,

$$f(\mathbf{n}|\theta) = \binom{n}{n_1, \ldots, n_k} \prod_{j=1}^{k} \theta_{[j]}^{n_j} = \frac{n!}{n_1! \cdot \ldots \cdot n_k!} \prod_{j=1}^{k} \theta_{[j]}^{n_j} \propto \prod_{j=1}^{k} \theta_{[j]}^{n_j}. \tag{7.3}$$

As a shortcut, we write $\mathbf{n} \sim \text{Mn}(\theta)$. ◆

Example 7.2 (Sampling from a normal distribution). A common model for observations that in principle can assume any value on the real line is the *normal distribution* with parameters μ and σ^2, also called the *Gaussian distribution*. Typical examples for data of this kind are scores in intelligence testing, or technical measurements in general.[19]

For each variable X_i, $i = 1, \ldots, n$, the normal probability density is

$$f(x_i|\mu, \sigma^2) = \frac{1}{\sqrt{2\pi\sigma^2}} \exp\left\{-\frac{1}{2\sigma^2}(x_i - \mu)^2\right\},$$

with the two parameters $\mu \in \mathbb{R}$ and $\sigma^2 \in \mathbb{R}_{>0}$ being in fact the mean and the variance of (the distribution of) X_i, respectively. As a shortcut, we write $X_i \sim \text{N}(\mu, \sigma^2)$; a normal density with $\mu = 0$ and $\sigma^2 = 1$ is displayed in Figure 7.3, Page 167.

With the independence assumption, the density of $\mathbf{x} = (x_1, \ldots, x_n)$ amounts to

$$L(\mathbf{x}|\mu, \sigma^2) = \prod_{i=1}^{n} f(x_i|\mu, \sigma^2) = (2\pi\sigma^2)^{-\frac{n}{2}} \exp\left\{-\frac{1}{2\sigma^2} \sum_{i=1}^{n} (x_i - \mu)^2\right\}. \tag{7.4}$$

Later on, we restrict considerations to the case where the variance σ^2 is known to be σ_0^2, denoted by $X_i \sim \text{N}(\mu, \sigma_0^2)$. Inference may thus concern the parameter μ directly, or future observations $X_{n+1}, X_{n+2}, \ldots$ in a chain of i.i.d. observations. ◆

[19] The normal distribution is distinguished by the central limit theorem (see, e.g., [389, § 7.3], or [85, § 9]), stating that, under regularity conditions, the distribution of an appropriately scaled sum of n standardized random variables converges to a normal distribution for $n \to \infty$.

7.2 Imprecision in statistics, some general sources and motives

In this section we discuss some general aspects of imprecision in statistical inference. We start with explicitly introducing an ideal typical distinction between model and data imprecision and different interpretations of imprecision. Then we discuss some fundamental motives for using imprecise probabilities in statistics, and review some basic models.

7.2.1 Model and data imprecision; sensitivity analysis and ontological views on imprecision

Imprecision plays a crucial role in all components of the statistical inference process. Although the sources for imprecision are closely interrelated, it may be helpful to distinguish in an ideal typical way between two different sources:

- *Data imprecision*, addressing the common situation where (some of) the data are not observed in the resolution intended in the subject matter interpretation and application. Such data often are called *coarsened*; censored and missing values are included as extreme cases.
- *Model imprecision*, potentially occurring in all probability models involved, namely
  - in the sampling distributions,
  - in the prior distribution if a generalized Bayesian framework is used,
  - in the distribution summarizing the conclusions from the data, most notably in the posterior (predictive) distribution in the generalized Bayesian framework, as well as other types of predictive distributions or fiducial-like probabilities.

This chapter mainly focuses on model imprecision, data imprecision is only discussed in Section 7.8 in some detail. Imprecise probability will prove to be a powerful modelling tool for all the diverse interpretations of probability. Orthogonal to the interpretation of probability, there are also two ideal type ways to interpret the imprecision.

- The *sensitivity analysis* [672], *epistemic* [678] or *disjunctive* [692, p. 101] point of view understands lower and upper previsions as arising from a lack of knowledge only; they are interpreted as the best available bounds of the exact values of a true precise (i.e., linear) underlying prevision.
- The *ontological* [678] or *adjunctive* [692, p. 103] point of view, which is also at the core of Walley's book [672], understands lower and upper previsions, and also the corresponding credal sets, strictly as a holistic entity of its own, like a molecule, whose chemical properties are determined by the interplay of all its elementary particles.

This distinction, as well as the continuum between the ideal types, becomes particularly important when more complex composite models are considered, in statistical inference most notably models for product measures.[20]

[20] See Section 7.3.2.

In the coming subsections we discuss several motives for the paradigmatic shift to imprecise probabilities, which are partially overlapping. We start with two general aspects, namely the potentially enormous sensitivity of traditional statistical conclusions on assumptions on the sampling model or the prior, and the power of imprecision as an additional modelling dimension. Then we turn to the importance of imprecision to ensure credibility of inferences and we conclude with some further specific arguments for considering imprecise sampling models.

7.2.2 The robustness shock, sensitivity analysis

It is common practice (not only) in statistics 'to take models not too seriously' but to understand them as crude approximations to 'reality', and thus the choice of a certain class of models is often strongly influenced by mathematical convenience. Box and Draper's statement 'Essentially, all models are wrong, but some of them are useful' [84, p. 424] has become a frequently cited dictum, often understood as a guiding paradigm to statistical modelling. Such methodology implicitly assumes 'continuity' of statistical procedures in the sense that small perturbations of the underlying model do not destroy the substantive conclusions drawn from the data. Starting in the late 1960s, robust statistics[21] has destructed this myth, giving examples where standard statistical procedures that are optimal under a certain model behave in a disastrous way even under 'small deviations from that model'.[22] Thus, optimality, and even meaningfulness, of statistical procedures is often very sensitive to modelling assumptions. This discovery applies to the Bayesian as well as the frequentist paradigm.

Robust frequentist statistics is concerned with statistical methods which may not be fully optimal in the ideal model, but still lead to reasonable conclusions if the ideal model is only approximately true. Mathematically this can be achieved by considering as sampling distributions all probability distributions in a certain neighbourhood around the sampling distribution in the ideal model, and thus imprecise probabilities also provide a formal superstructure upon models used in frequentist robust statistics.[23]

In the Bayesian paradigm, this sample distribution issue is complemented by the issue of *sensitivity to the prior*. The potentially strong dependence of inferences on the prior has been an intensively debated topic. Especially in situations where data are scarce and contain little information with respect to the question at hand (i.e., a low signal-to-noise ratio), posterior-based conclusions may rely heavily on characteristics of the chosen prior distribution. Thus, in the Bayesian setting mostly the influence of different priors on the posterior conclusions has been investigated. Imprecise probabilities offer a formal, not casuistic framework for such *Bayesian sensitivity analysis*, also containing the so-called *robust Bayesian approach* (See, e.g., [538]; see also Section 7.4.2).

Robust Bayesians and robust frequentists typically ask slightly different questions. For the robust Bayesian the inference *procedure* is undebatable. Proceeding to sets of posterior distributions, the main focus is on the sensitivity of this part of the inference, i.e. on the

[21] See [369, 344] for monographs on robust statistics.

[22] Two examples are discussed in detail in Section 7.5.1.

[23] In the early development in the 1970s of the last century, this intimate relationship between imprecise probabilities and robust statistics was explicitly emphasized by Huber, one of its most influential pioneers, himself (see, in particulary [367, 368]), but then it has become forgotten again, and both areas developed in their own directions; see Section 7.5.

imprecision in the posterior set. Frequentist statistics typically is directly concerned with efficiency considerations, and thus strives for optimal *procedures*. These differences do, however, by far not justify the rather isolated development of both areas. Indeed it would be very important to intensify the exchange of technical results between both areas.

7.2.3 Imprecision as a modelling tool to express the quality of partial knowledge

The flexible, multidimensional perspective on uncertainty makes imprecise probabilities capable of mirroring the quality of knowledge. Only well supported knowledge is expressed by comparatively precise models, including the traditional concept of probability as the special case of perfect stochastic information, while highly imprecise (or even vacuous) models are used in the situation of scarce (or no) knowledge. This power to differentiate between different degrees of partial knowledge distinguishes imprecise probabilities as an attractive modelling tool in statistics. In particular, it allows to overcome severe limitations inherent to inference based in precise probabilities.

Here we already mention two aspects within the Bayesian framework, which will be considered in detail in Section 7.4.3. A first important issue is the proper modelling of no (or extremely vague) prior knowledge. In traditional statistics, so-called *noninformative priors* have been proposed, which, by different criteria, eventually all select a single traditional prior distribution, turning ignorance into a problem with rather precise probabilistic information.

Secondly, increasing imprecision to make the conclusions more cautious is the natural way to handle *prior-data conflict*, i.e. when outlier-free data are observed that nevertheless are rather unexpected under the prior assumptions. For instance, [267, p. 893] warns that if '[...] the observed data is surprising in the light of the sampling model and the prior, then we must be at least suspicious about the validity of inferences drawn.' While there is no way to express this caution ('being suspicious') in precise probabilistic models, imprecise probability models may be directly designed to take care of this issue (see Section 7.4.3) in an appropriate way. As a related topic, decreasing imprecision in sequential learning expresses naturally that the accumulating information is nonconflicting and stable.

7.2.4 The law of decreasing credibility

Originally formulated in the context of data imprecision and partial identification,[24] Manski's 'Law of Decreasing Credibility',

> The credibility of inferences decreases with the strength of the assumptions maintained. [447, p. 1]

can be understood more generally as a compelling motto for statistical inference with imprecise probabilities.[25] It brings credibility, or reliability, of conclusions as a new dimension explicitly into the argument. Manski's law addresses two aspects, so-to-say a cautionary one and an encouraging one.

---

[24] See Section 7.8.

[25] 'Credibility' is used here in an informal, nontechnical sense.

The first, obvious one cautions us about the fact that 'overprecision' by too rigorous assumptions may destroy the practical relevance of the results obtained. This does not only refer to the nonrobustness of traditional statistical inference with respect to assumptions on the sampling model and, if applicable, on the prior distribution as just discussed, but it also applies to other kinds of assumptions, like independence assumptions or assumptions on missingness processes in complex data sets. Imprecision is a natural way out, also framing a careful sensitivity analysis, and guiding the understanding of the specific role of certain assumptions.

The second one encourages us to impose justified assumptions only and reminds us that there may be a considerable gain in the results derived from them. These results may be rather imprecise, but, by their credibility, they nevertheless are of substantial interest for the scientific questions underlying the statistical analysis. Indeed, they are the best one can go for without adding restrictive assumptions. In a strict sense, imprecise probability-based results are thus more informative than conclusions based on the calculus of traditional probability. This becomes most evident when the knowledge at hand is too scarce to justify any precise solution even after imposing further assumptions. Then the traditional precise approach, captured in the binary choice between the extremes 'precise solution' or 'no solution', necessarily has to surrender, while imprecise probabilistic methodology still provides powerful quantitative insights into the underlying structure.

In this sense imprecise models are also quite powerful in situations where researchers disagree about the tenability of certain specific assumptions. Then imprecise probabilities make clear which conclusions can already be obtained from a consensus model, based solely on noncontroversial assumptions.

7.2.5 Imprecise sampling models: Typical models and motives

Carefully chosen imprecise sampling models are important in many applications and under both interpretations of imprecision in the sense of Section 7.2.1. From the ontological point of view on imprecise probabilities, sampling models are eo ipso imprecise. The work of Fine and students (e.g., [678, 503, 551]) is motivated by situations of *physical indeterminacy*, typically also resulting in *instable relative frequencies* of apparently stationary random processes.

But also from the epistemic point of view, imprecise sampling models prove to be quite powerful in statistical modelling. They are needed whenever the information is not sufficient to determine single precise probability distributions as potential sampling distributions. This includes the following situations:

First, as discussed above, neighbourhood models capturing all distributions 'close' to a certain 'ideal' model are the immediate way to formulate 'approximately true' models. They in particular address robustness concerns and can also be designed to cope with the presence of outliers. Then the measurements of small proportions of the data are allowed to deviate strongly from the remaining data, without completely spoiling the conclusions drawn from the data.

Secondly, imprecise sampling models also automatically arise from precise sampling models when the resolution of interest is finer than the domain of the sampling model specified. We will discuss this situation in the context of data imprecision in Section 7.8.

Thirdly, imprecise sampling models naturally account for *unobserved heterogeneity*, where not all units have exactly the same distribution. Instead the distribution varies with certain characteristics that are not observable in the study and thus cannot be incorporated into the modelling process (e.g., different genetic disposition in a medical study, different driver styles in models for premium calculation).

A fourth – up to now only rarely explicitly studied – source for imprecise sampling models are *imperfect randomization schemes* for drawing random samples, most notably simple random sampling surrogates with selection probability $p \in \left[\frac{1}{N} - \underline{\epsilon}, \frac{1}{N} + \overline{\epsilon}\right]$, slightly deviating from the uniform probability $\frac{1}{N}$ with 'small' $\underline{\epsilon}$ and $\overline{\epsilon}$. This in particular extends to the situation of symmetry-based selection schemes, where symmetry cannot be fully established.[26]

7.3 Some basic concepts of statistical models relying on imprecise probabilities

7.3.1 Most common classes of models and notation

As in other applications of imprecise probabilities, also in statistical inference, there are basically two particularly conveniently constructed, partially overlapping classes of imprecise probability models.

- The first class of models consists naturally of *(lower) envelopes* of a set of precise probabilistic models, either characterized by certain common properties (like sharing the same mean, for instance) or as *parametrically constructed models* as the envelope of traditional parametric models where the parameter ϑ varies in a certain set $\underline{\overline{\vartheta}} \subseteq \Theta$, typically an interval or cuboid. For instance, later on we will consider the credal set described by all scaled normals with a mean μ between $\underline{\mu}$ and $\overline{\mu}$. The latter models are particularly attractive to model unobserved heterogeneity, where one could imagine a latent variable, with unknown distribution in the background, that determines the concrete value $\vartheta \in \underline{\overline{\vartheta}}$.
- The second class of models are *neighbourhood models* (compare also Section 4.7 and, e.g., [37]). Here the credal set consists of all probability distributions that are 'close' in a certain sense to a so-called central distribution $\tilde{P}(\cdot)$. This proximity may be measured by probabilistic metrics in a rigorous sense, like the total-variation or the Kolmogorov distance, but also more informally, based on corridors for density or mass functions, or distribution functions (see, e.g., Section 4.6.4 on p-boxes). In an even more liberal sense, also parametric models from the first class could be understood as neighbourhood models, where the proximity is measured in the parameter space.
- In principle, it is also possible to combine both classes by considering neighbourhood models of imprecise probability distributions, for instance a contamination neighbourhood of an imprecise normal distribution with an interval-valued mean parameter. Such models are discussed, e.g., in [33, 680, 593, 594].

Throughout this chapter, we use the following notation and terminology. In the case of imprecise prior distributions, lower and upper prior previsions on the parameter space Θ are

[26] For some first ideas, see [690, 31].

denoted by $\underline{\Pi}(\cdot)$ and $\overline{\Pi}(\cdot)$ and the corresponding credal set by $\mathcal{M}$; generic precise elements are denoted by $\Pi(\cdot)$ with density or mass function $\pi(\cdot)$. The corresponding posterior distributions, summarizing the information after having evaluated the data $\mathbf{x}$, are then described by $\Pi(\cdot|\mathbf{x})$, $\underline{\Pi}(\cdot|\mathbf{x})$ and $\overline{\Pi}(\cdot|\mathbf{x})$, with corresponding credal set $\mathcal{M}_{|\mathbf{x}}$. To emphasize the role of the sample size n, we sometimes write $\mathcal{M}^{(n)}$ instead of $\mathcal{M}_{|\mathbf{x}}$, and then denote the prior credal set by $\mathcal{M}^{(0)}$. Lower and upper previsions of imprecise sampling models are denoted by $\underline{P}$ and $\overline{P}$, respectively, and the corresponding credal set by $\mathcal{P}$. All quantities will often be indexed by parameters, as discussed in the following subsection.

7.3.2 Imprecise parametric statistical models and corresponding i.i.d. samples

In the case of imprecise sampling models, the notion of a classical parametric statistical model in (7.1) is extended towards the concept of an *imprecise parametric statistical model*,

$$\left(\mathcal{X}, \left(\mathcal{P}_{\overline{\underline{\vartheta}}}\right)_{\vartheta \in \Theta}\right)$$

with the parameter $\overline{\underline{\vartheta}} \subseteq \Theta \subseteq \mathbb{R}^p$.[27] For every $\overline{\underline{\vartheta}} \subseteq \Theta$, the corresponding credal set is denoted by $\mathcal{P}_{\overline{\underline{\vartheta}}}$. Recalling the different ways just mentioned to construct imprecise probability models, typically two main cases are considered:

- The case of an *interval-parameter*, where

  $$\overline{\underline{\vartheta}} = \left[\underline{\vartheta}_{[1]}, \overline{\vartheta}_{[1]}\right] \times \ldots \times \left[\underline{\vartheta}_{[\ell]}, \overline{\vartheta}_{[\ell]}\right] \times \ldots \times \left[\underline{\vartheta}_{[k]}, \overline{\vartheta}_{[k]}\right]$$

  for real-valued components $\underline{\vartheta}_{[1]}, \ldots, \underline{\vartheta}_{[\ell]}, \ldots, \underline{\vartheta}_{[k]}; \overline{\vartheta}_{[1]}, \ldots, \overline{\vartheta}_{[\ell]}, \ldots, \overline{\vartheta}_{[k]}$ with $\underline{\vartheta}_{[\ell]} < \overline{\vartheta}_{[\ell]}$ for some ℓ and

  $$\mathcal{P}_{\overline{\underline{\vartheta}}} = \left\{P_\vartheta | \vartheta \in \overline{\underline{\vartheta}}\right\}, \tag{7.5}$$

  where P_ϑ are sampling distributions from a classical parametric sampling model $(\mathcal{P}_\vartheta)_{\vartheta \in \Theta}$.[28]

- The case of an *imprecise parametric statistical model with a precise parameter*, where $\overline{\underline{\vartheta}}$ is a singleton $\{\vartheta\}, \vartheta \in \Theta$. Whenever confusion with classical models can be excluded, we use in this situation ϑ as index instead of $\overline{\underline{\vartheta}}$. Typically the credal sets are then neighbourhoods of sampling distributions P_ϑ from a classical parametric statistical model.

Proceeding further along the lines of Section 7.1, the next step is to define the notion of an *i.i.d. sample of size n based on the imprecise parametric statistical model* $\left(\mathcal{X}, \left(P_{\overline{\underline{\vartheta}}}\right)_{\vartheta \in \Theta}\right)$. Still the sample is taken to be described by a random vector

$$\mathbf{X} = (X_1, \ldots, X_n)$$

[27] Again, we remain on a more or less intuitive level. See [323] for a more formal rigorous treatment.

[28] Sometimes the convex hull of the parametric sampling models $\{P_\vartheta | \vartheta \in \overline{\underline{\vartheta}}\}$ is considered as the imprecise statistical model; we comment on the question of convexity of credal sets in Section 7.4.3.

with values $\mathbf{x} = (x_1, \ldots, x_n)$ (see Section 7.1.2), but in the light of Chapter 3 there are different ways to define product measures $\mathcal{P}_{\underline{\vartheta}}^{\otimes n}$.[29] In fact this issue must be treated with care. It is a crucial part of the model building process, since these substantially different ways to understand the concept of an i.i.d. sample predetermine interpretation, optimality notions and suitability of statistical methods based on imprecise probabilities.

Literature has almost exclusively focused on strong-independence-like concepts based on underlying precise probabilities in the credal sets $\mathcal{P}_{\underline{\vartheta}}$. Still, there are two versions[30] to define the credal set $\mathcal{P}^{\otimes n}$. The first one

$$ {}^{(1)}\mathcal{P}_{\underline{\vartheta}}^{\otimes n} = \left\{ P^{\otimes n} \;\middle|\; P \in \mathcal{P}_{\underline{\vartheta}} \right\} \tag{7.6} $$

assumes that under independent repetition always the same precise probability generates the data. In contrast,

$$ {}^{(2)}\mathcal{P}_{\underline{\vartheta}}^{\otimes n} = \left\{ \bigotimes_{i=1}^{n} P_i \;\middle|\; P_i \in \mathcal{P}_{\underline{\vartheta}},\ i = 1, \ldots, n \right\} \tag{7.7} $$

allows to vary the underlying probability P_i with the repetition $i = 1, \ldots, n$, and considers the independent product $\otimes_{i=1}^{n} P_i$ of these probabilities.

${}^{(1)}\mathcal{P}_{\underline{\vartheta}}^{\otimes n}$, called *repetition independence* in [158], and type-2 product in [672, § 9.3],[31] is the most radical implementation of the sensitivity analysis point of view. There is one single, unique classical probability, and imprecision comes only from our incomplete ability to specify it precisely. Indeed, under ever increasing sample size, this probability can eventually be discovered by evaluating the empirical distribution. Consequently, in this setting often shrinking neighbourhood models, where the radius decreases with the sample size, are considered when studying large sample behaviour.[32]

A typical application of ${}^{(2)}\mathcal{P}^{\otimes n}$, called type-1 product with the same marginals in [672, p. 456], would be an imprecise probability based analysis of unobserved heterogeneity, leading to possibly different distributions in each repetition.

7.4 Generalized Bayesian inference

This section discusses the application of the general framework of coherent lower previsions (as established in Chapter 2) for generalizing Bayesian inference to imprecise probabilities.[33] We confine presentation on approaches with imprecise prior distributions but still precise sampling models. First, we briefly recall some fundamental results from traditional Bayesian statistics needed later on. After introducing the general Bayesian setting, we then focus on the prototypical important case of generalized conjugate inference in exponential families. There we consider first applications to formulate near-ignorance priors, and

[29] As a basis, see, in particular, the discussions in [678] and [158].

[30] In addition, Footnote 28 applies again.

[31] Note that, in spite of the similiar names, this Type-2 product is different in spirit from the Type-2 independence discussed in Section 3.2.3. and 3.5.1.

[32] See Section 7.5.3.

[33] See also Section 8.3.2.

secondly the situation of informative priors including their imprecise probability based opportunities in the context of prior-data-conflict. An outlook on some further aspects concludes this section.

7.4.1 Some selected results from traditional Bayesian statistics

7.4.1.1 Conjugate families of distributions

Traditional Bayesian inference is frequently based on so-called *conjugate priors*[34] (related to a specific likelihood). Such priors have the convenient property that the posterior resulting from (7.2) belongs to the same class of parametric distributions as the prior, and thus only the parameters have to be updated, which makes calculation of the posterior and thus the whole Bayesian inference easily tractable.

Fortunately, there are general results guiding the construction of conjugate priors for models used most frequently in practice, namely in the case where the sample distribution belongs to a so-called *regular canonical exponential family* [70, pp. 202 and 272f]. This indeed covers many sample distributions relevant in a statistician's everyday life, like Normal and Multinomial models, Poisson models, or Exponential and Gamma models. After presentation of the general framework, we will discuss its instantiation for the Multinomial[35] and the Normal sampling model, which were introduced in Section 7.1.5.

A sample distribution[36] is said to belong to the *regular canonical exponential family* if its density or mass function satisfies the decomposition

$$f(\mathbf{x}|\theta) \propto \exp\{\langle\psi, \tau(\mathbf{x})\rangle - n\mathbf{b}(\psi)\}, \tag{7.8}$$

where $\psi \in \Psi \subset \mathbb{R}^q$ is a transformation of the (possibly vectorial) parameter $\theta \in \Theta$, and $\mathbf{b}(\psi)$ a scalar function of ψ (or, in turn, of θ). $\tau(\mathbf{x})$ is a function of the sample $\mathbf{x}$ that fulfills $\tau(\mathbf{x}) = \sum_{i=1}^{n} \tau^*(x_i)$, with $\tau^*(x_i) \in \mathcal{T} \subset \mathbb{R}^q$, while $\langle\cdot,\cdot\rangle$ denotes the scalar product. From these ingredients, a conjugate prior density or mass function on ψ can be constructed as[37]

$$\pi(\psi|n^{(0)}, y^{(0)})d\psi \propto \exp\left\{n^{(0)}\left[\langle y^{(0)}, \psi\rangle - \mathbf{b}(\psi)\right]\right\} d\psi, \tag{7.9}$$

where $n^{(0)}$ and $y^{(0)}$ are now the parameters by which a certain prior can be specified. The domain of $y^{(0)}$ is $\mathcal{Y}$, the interior of the convex hull of $\mathcal{T}$, and $n^{(0)}$ must take strictly positive values for the prior to be proper (integrable to 1). The prior parameters $y^{(0)}$ and $n^{(0)}$ are updated to their posterior values $y^{(n)}$ and $n^{(n)}$ in the following way:

$$y^{(n)} = \frac{n^{(0)}}{n^{(0)}+n} \cdot y^{(0)} + \frac{n}{n^{(0)}+n} \cdot \frac{\tau(\mathbf{x})}{n}, \qquad n^{(n)} = n^{(0)} + n, \tag{7.10}$$

[34] Conjugacy in this sense is not to be confused with conjugacy of lower and upper previsions as introduced in Section 2.2, Page 31.

[35] Partly, as a special case, we turn turn to the Binomial distribution, as the calculations are then straightforward.

[36] In this section, we use this expression directly for the distribution of $P_\theta^{\otimes n}$ of an i.i.d. sample $\mathbf{x}$ of size n. Here, there are no specific issues regarding the construction of the product measure, because in the models discussed in Sections 7.4.2 and 7.4.3, the sampling distribution will be precise.

[37] In our notation, $^{(0)}$ denotes prior parameters; $^{(n)}$ posterior parameters.

such that the density or mass function of the posterior can be written as $\pi(\psi|n^{(n)}, y^{(n)}) := \pi(\psi|\mathbf{x}, n^{(0)}, y^{(0)})$.[38] In this setting $y^{(0)}$ and $y^{(n)}$ can be seen as the parameter describing the main characteristics of the prior and the posterior, and thus we will call them *main prior* and *main posterior parameter*, respectively. $y^{(0)}$ can also be understood as a prior guess for the random quantity $\tilde{\tau}(\mathbf{x}) := \tau(\mathbf{x})/n$ summarizing the sample, as $\mathbb{E}[\tilde{\tau}(\mathbf{x})|\psi] = \nabla\mathbf{b}(\psi)$, where in turn $\mathbb{E}[\nabla\mathbf{b}(\psi)|n^{(0)}, y^{(0)}] = y^{(0)}$. Characteristically, $y^{(n)}$ is a weighted average of this prior guess $y^{(0)}$ and the sample 'mean' $\tilde{\tau}(\mathbf{x})$, with weights $n^{(0)}$ and n, respectively.[39] Therefore, $n^{(0)}$ can be seen as 'prior strength' or 'pseudocounts', reflecting the weight one gives to the prior as compared to the sample size n.[40]

Besides the posterior $\pi(\psi|n^{(n)}, y^{(n)})$, being possibly transformed to a distribution on θ, also the posterior predictive distribution

$$f(\mathbf{x}^*|\mathbf{x}) = \int f(\mathbf{x}^*|\psi)\pi\left(\psi|n^{(n)}, y^{(n)}\right) \mathrm{d}\psi,$$

a distribution on future samples $\mathbf{x}^*$ after having seen a sample $\mathbf{x}$, forms the basis for the inference tasks as listed in Section 7.1.3. The concretion of this framework is demonstrated by means of the two examples below.

Example 7.3 (Beta-Binomial Model). As the special case of the multinomial model (7.3) with only two categories, we will consider the Binomial model

$$f(\mathbf{x}|\theta) = \binom{n}{s} \theta^s (1-\theta)^{n-s} \tag{7.11}$$

where $\mathbf{x}$, the vector of n observations, is composed of scalar x_i's being either 0 or 1 denoting 'failure' or 'success' in an experiment with these two outcomes. $s = \sum_{i=1}^{n} x_i$ is the number of successes, and the (unknown) parameter $\theta \in (0, 1)$ is the probability for 'success' in a single trial. (7.11) can be written in the canonical exponential family form (7.8):

$$f(\mathbf{x}|\theta) \propto \exp\left\{\log\left(\frac{\theta}{1-\theta}\right) s - n(-\log(1-\theta))\right\}$$

We have thus $\psi = \log(\theta/(1-\theta))$, $\mathbf{b}(\psi) = -\log(1-\theta)$, and $\tau(\mathbf{x}) = s$. From these ingredients, a conjugate prior density/mass function on ψ can be constructed along (7.9), leading here to

$$\pi\left(\log\left(\frac{\theta}{1-\theta}\right)\middle| n^{(0)}, y^{(0)}\right) \mathrm{d}\psi \propto \exp\left\{n^{(0)}\left[y^{(0)}\log\left(\frac{\theta}{1-\theta}\right) + \log(1-\theta)\right]\right\} \mathrm{d}\psi.$$

This prior, transformed to the parameter of interest θ,

$$\pi(\theta|n^{(0)}, y^{(0)})\, \mathrm{d}\theta \propto \theta^{n^{(0)}y^{(0)}-1}(1-\theta)^{n^{(0)}(1-y^{(0)})-1}\, \mathrm{d}\theta,$$

[38] In this section we verbally identify traditional prior and posterior distributions by their density or mass function.

[39] This weighted average property of Bayesian updating with conjugate priors is an important issue we comment on in Section 7.4.3.

[40] As noted in [684, p. 258], '$n^{(0)}$ can be interpreted as the size of an imaginary sample that corresponds to the trust on the prior information in the same way as the sample size of a real sample corresponds to the trust in conclusions based on such a real sample'.

is a Beta distribution with parameters $n^{(0)}y^{(0)}$ and $n^{(0)}(1 - y^{(0)})$, in short,

$$\theta \sim \text{Beta}(n^{(0)}y^{(0)}, n^{(0)}(1 - y^{(0)})).$$

The combination of a Binomial sampling model with this conjugate Beta prior is called *Beta-Binomial Model.* Here, $y^{(0)} = \mathbb{E}[\theta]$ can be interpreted as prior guess of θ, while $n^{(0)}$ governs the concentration of probability mass around $y^{(0)}$. Due to conjugacy, the posterior on θ is a $\text{Beta}(n^{(n)}y^{(n)}, n^{(n)}(1 - y^{(n)}))$, where the posterior parameters $n^{(n)}$ and $y^{(n)}$ are given by (7.10).

From the posterior distribution, a **point estimator** for θ can be extracted by choosing a *loss function* evaluating the distance of a point estimator $\hat{\theta}$ to θ (see also Section 8.1, Page 191, and Section 8.3.2, Pages 205f). The *quadratic loss function* leads to the posterior expectation as the Bayesian point estimator. Here, $\mathbb{E}[\theta|\mathbf{x}] = y^{(n)}$, and so the posterior expectation of θ is a weighted average of the prior expectation $\mathbb{E}[\theta] = y^{(0)}$ and the sample proportion s/n, with weights $n^{(0)}$ and n, respectively. Taking the *absolute loss function* $l(\hat{\theta}, \theta) = |\hat{\theta} - \theta|$ leads to the median of the posterior distribution as the point estimator. Here, the median $\text{med}(\theta|\mathbf{x})$ has no closed form solution, and must be determined numerically. The *indicator loss function* leads to the mode of the posterior, often abbreviated as MAP (maximum a posteriori) estimator. For a $\text{Beta}(n^{(n)}y^{(n)}, n^{(n)}(1 - y^{(n)}))$, the mode is

$$\text{mode } \pi(\theta|\mathbf{x}) = \frac{n^{(n)}y^{(n)} - 1}{n^{(n)} - 2} = \frac{n^{(0)}y^{(0)} - 1 + s}{n^{(0)} - 2 + n},$$

and thus is a weighted average of the prior mode $\frac{n^{(0)}y^{(0)}-1}{n^{(0)}-2}$ ($n^{(0)} > 2$) and the sample proportion s/n, with weights $n^{(0)} - 2$ and n, respectively.

In the Bayesian approach, **interval estimation** is rather simple, as the posterior distribution delivers a direct measure of probability for arbitrary subsets of the parameter space Θ. Mostly, so-called *highest posterior density* (HPD) intervals are considered, where for a given probability level γ the shortest interval covering this probability mass is calculated. For unimodal densities, this is equivalent to finding a threshold α such that the probability mass for all θ with $\pi(\theta|\mathbf{x}) \geq \alpha$ equals γ, hence the name. (See, e.g., [70, § 5.1.5, pp. 259f].) For the Beta posterior, a HPD interval for θ must be determined by numeric optimization. For approximately symmetric (around $\frac{1}{2}$) Beta posteriors, a good approximation is the symmetric credibility interval, delimited by the $\frac{1-\gamma}{2}$ and the $\frac{1+\gamma}{2}$ quantile of the posterior.

The **testing of hypotheses** concerning the parameter of interest θ can be done by comparing posterior probabilities of two subsets of the parameter space, which are usually denoted by H_0 and H_1. High values of the ratio of the posterior probabilities $\Pi(H_1|\mathbf{x})/\Pi(H_0|\mathbf{x})$ then indicate high plausibility of H_1 as compared to H_0.

The **posterior predictive** distribution, giving the probability for s^* successes in n^* future trials after having seen s successes in n trials, is

$$f(s^*|s) = \binom{n^*}{s^*} \frac{B(s^* + n^{(n)}y^{(n)}, n^* - s^* + n^{(n)}(1 - y^{(n)}))}{B(n^{(n)}y^{(n)}, n^{(n)}(1 - y^{(n)}))},$$

known as the *Beta-Binomial distribution.*

For the multinomial sampling model, the analogue to the Beta-Binomial model is the *Dirichlet-Multinomial model.* The conjugate prior to the multinomial sampling model constructed via (7.9) is the *Dirichlet* distribution,[41] [42] which can be seen as a generalization of the Beta distribution. Indeed, the Dirichlet prior on the now vectorial parameter θ, of which each element $\theta_{[j]}$ gives the probability for observing category j in a single draw, has the form

$$\pi(\theta) \propto \prod_{j=1}^{k} \theta_{[j]}^{n^{(0)} y_{[j]}^{(0)} - 1},$$

where the vectorial $y^{(0)}$ is an element of the interior of the unit simplex Δ, thus $y_{[j]}^{(0)} \in (0, 1), \forall j = 1, \ldots, k$ and $\sum_{j=1}^{k} y_{[j]}^{(0)} = 1$, in short $y^{(0)} \in \text{int}(\Delta)$. The main posterior parameter can be calculated as

$$y_{[j]}^{(n)} = \frac{n^{(0)}}{n^{(0)} + n} y_{[j]}^{(0)} + \frac{n}{n^{(0)} + n} \cdot \frac{n_j}{n} \quad j = 1, \ldots, k,$$

and is thus again a weighted average of the main prior parameter (which can be interpreted as prior guess for θ, as $\mathbb{E}[\theta] = y^{(0)}$) and the fractions of observations in each category, with weights $n^{(0)}$ and n, respectively. $n^{(0)}$ again governs the concentration of probability mass around $y^{(0)}$, with larger values of $n^{(0)}$ leading to higher concentrations. ♦

Example 7.4 (Normal-Normal Model). The normal density (7.4) also adheres to the exponential family form:

$$f(\mathbf{x}|\mu, \sigma_0^2) \propto \exp\left\{ \frac{\mu}{\sigma_0^2} \sum_{i=1}^{n} x_i - \frac{n\mu^2}{2\sigma_0^2} \right\}.$$

So we have here $\psi = \frac{\mu}{\sigma_0^2}$, $\mathbf{b}(\psi) = \frac{\mu^2}{2\sigma_0^2}$, and $\tau^*(x_i) = x_i$. From these ingredients, a conjugate prior can be constructed with (7.9), leading to

$$\pi\left(\frac{\mu}{\sigma_0^2} \middle| n^{(0)}, y^{(0)} \right) d\frac{\mu}{\sigma_0^2} \propto \exp\left\{ n^{(0)} \left(\langle y^{(0)}, \frac{\mu}{\sigma_0^2} \rangle - \frac{\mu^2}{2\sigma_0^2} \right) \right\} d\frac{\mu}{\sigma_0^2}.$$

This prior, transformed to the parameter of interest μ and with the square completed,

$$\pi(\mu|n^{(0)}, y^{(0)}) d\mu \propto \frac{1}{\sigma_0^2} \exp\left\{ -\frac{n^{(0)}}{2\sigma_0^2} (\mu - y^{(0)})^2 \right\} d\mu,$$

is a normal distribution with mean $y^{(0)}$ and variance $\frac{\sigma_0^2}{n^{(0)}}$, i.e. $\mu \sim \mathrm{N}(y^{(0)}, \frac{\sigma_0^2}{n^{(0)}})$.

With (7.10), the parameters for the posterior distribution are

$$y^{(n)} = \mathbb{E}[\mu|\mathbf{x}] = \frac{n^{(0)}}{n^{(0)} + n} \cdot y^{(0)} + \frac{n}{n^{(0)} + n} \cdot \bar{\mathbf{x}} \tag{7.12}$$

$$\frac{\sigma_0^2}{n^{(n)}} = \mathbb{V}(\mu|\mathbf{x}) = \frac{\sigma_0^2}{n^{(0)} + n}. \tag{7.13}$$

[41] See also the derivation of the naive Bayes classifier in Section 10.2.

[42] This result is not as straightforward to derive as for the Binomial distribution. The proof is made explicit, e.g., in [683].

The posterior expectation of μ thus is a weighted average of the prior expectation $y^{(0)}$ and the sample mean $\bar{\mathbf{x}}$, with weights $n^{(0)}$ and n, respectively. The effect of the update step on the variance is such that it decreases by the factor $n^{(0)}/(n^{(0)}+n)$.

Here, all of the three choices of loss functions mentioned in the Beta-Binomial example lead to the same **point estimator** $\hat{\mu} = y^{(n)}$, as in normal distributions mean, median, and mode coincide.

As **interval estimation**, the HPD interval can be calculated, due to symmetry of the normal posterior, as $[z^{(n)}_{\frac{1-\gamma}{2}}, z^{(n)}_{\frac{1+\gamma}{2}}]$, where, e.g., $z^{(n)}_{\frac{1-\gamma}{2}}$ is the $\frac{1-\gamma}{2}$-quantile of the normal distribution with mean $y^{(n)}$ and variance $\frac{\sigma_0^2}{n^{(n)}}$.

The **testing of hypotheses** about μ works again by comparing posterior probabilities of two disjunct subsets of the parameter space. Note that the frequentist analogue to such a test is the (one-sided) one-sample Z-test (or Gaussian test).

The **posterior predictive** distribution for n^* future observations denoted by $\mathbf{x}^*$ is again a normal distribution, $\mathbf{x}^*|\mathbf{x} \sim \mathrm{N}(y^{(n)}, \frac{\sigma_0^2}{n^{(n)}}(n^{(n)}+n^*))$, around the posterior mean with a variance increasing with the posterior variance and the number of observations to be predicted. ♦

7.4.2 Sets of precise prior distributions, robust Bayesian inference and the generalized Bayes rule

Walley [672, §§ 6, 7] has established a general framework for coherent statistical inference under imprecise probabilities. In essence, it allows to transfer the basic aspect of traditional Bayesian inference to the generalized setting. In particular, the fundamental paradigms of Bayesian inference as discussed in Section 7.1.4 are maintained. Prior knowledge on the parameter, expressed by a now imprecise prior distribution, i.e. a lower prevision $\underline{\Pi}(\cdot)$ on the parameter space, with credal set $\mathcal{M}$, is updated in the light of the observed sample $\mathbf{x}$ to the imprecise posterior, i.e. the lower prevision $\underline{\Pi}(\cdot|\mathbf{x})$, with the credal set $\mathcal{M}_{|\mathbf{x}}$, and this statistical inference is again understood as a deductive process, obtained directly by conditioning on the observed sample, now according to the *generalized Bayes rule* that ensures coherence of this inferential process (see Chapter 2.3 for general aspects). For practical implementation of the generalized Bayes rule, the lower envelope theorem for conditional previsions[43] is of particular relevance. In its spirit the prior credal set $\mathcal{M}$ consisting of linear previsions $\Pi(\cdot)$ is updated element by element to obtain the posterior credal set

$$\mathcal{M}_{|\mathbf{x}} = \{\Pi(\cdot|\mathbf{x}) \mid \Pi(\cdot) \in \mathcal{M}\}, \tag{7.14}$$

consisting of all posterior distributions obtained by traditional Bayesian updating of elements of the prior credal set.

7.4.2.1 Robust Bayes and imprecise probabilities

Walley's lower envelope theorem establishes a close connection to robust Bayesian approaches and Bayesian sensitivity analysis (see, e.g. [538, 548]) based on sets of distributions. The basic difference is in the interpretation of the underlying sets of probability

[43] See [672, § 6.4.2] and also Property ii) at Page 50.

distributions. While in the imprecise probability context the emphasis is on the whole credal set as an entity, the robust Bayesian approach focuses on the single elements in the set, and very often discusses the effects of deviations from a certain central element. Robust Bayesian inference and Bayesian sensitivity analysis understand robustness and insensitivity mostly as desirable properties, while the imprecise probability framework may use such behaviour actively in modelling, in particular in the context of prior-data conflict (see Section 7.4.3).

7.4.2.2 Some computational aspects

In the case of discrete prior distributions (with a support of size $k < \infty$) the computation of posterior credal sets is covered in Chapter 16 (via the alternative model formulation as conditional lower previsions). For large k, the algorithms considered there may easily become infeasible. In traditional statistics, absolutely continuous distributions are employed when inference using discrete distributions becomes too complex, typically approximating a nonparametric model with a parametric one.[44] The same path can be followed in generalized Bayesian inference, by considering sets of continuous distributions. Furthermore, before the advent of Markov Chain Monte Carlo (MCMC) techniques (see, e.g., [312]) to determine posterior distributions in complex settings by simulation, Bayesian inference was mostly confined to conjugate prior distributions (as introduced in Section 7.4.1), because posteriors derived from non-conjugate priors are very often intractable analytically.

7.4.3 A closer exemplary look at a popular class of models: The IDM and other models based on sets of conjugate priors in exponential families

7.4.3.1 The general model

Due to the computational issues as discussed above, models based on conjugate priors are of particular convenience also in the imprecise probability context. They have led to the so-called *Imprecise Dirichlet model* (IDM, see below) by [673], see also [66], and more generally to powerful imprecise probability models for inference based on i.i.d. exponential family sampling models including the approach by [521, 519], which was extended by [684] and [685] to allow in addition an elegant handling of prior-data conflict. We use this model class as a prototypic example to illustrate some characteristic modelling opportunities of generalized Bayesian inference.

The basic setting. Consider inference based on samples from a regular canonical exponential family (7.8) using the conjugate prior (7.9). One specifies a prior parameter set $\text{I}\Pi^{(0)}$ of $(n^{(0)}, y^{(0)})$ values and takes as imprecise prior – described via the credal set $\mathcal{M}^{(0)}$ – the

[44] Consider the test for independence in contingency tables. Fisher's exact test, a nonparametric test using a permutation argument (thus resulting in a discrete distribution), can become difficult to calculate for large samples. An alternative is then the chi-squared test that is based on the continuous, one-parametric $\chi^2(df)$ distribution, which, for large samples, is a good approximation of the distribution of the χ^2 test statistic then used to determine the test decision.

set of traditional priors with $(n^{(0)}, y^{(0)}) \in \Pi^{(0)}$. The credal set $\mathcal{M}^{(n)}$ of posterior distributions,[45] obtained by updating each element of $\mathcal{M}^{(0)}$ via Bayes rule, then can be described as the set of parametric distributions with parameters varying in the set of updated parameters $\Pi^{(n)} = \{(n^{(n)}, y^{(n)}) | (n^{(0)}, y^{(0)}) \in \Pi^{(0)}\}$.

Alternatively, $\mathcal{M}^{(0)}$ can be defined as the set of all convex mixtures of parametric priors with $(n^{(0)}, y^{(0)}) \in \Pi^{(0)}$. In this case, the set of priors corresponding to $\Pi^{(0)}$ considered above gives the set of extreme points for the actual convex set $\mathcal{M}^{(0)}$. Updating this convex prior credal set with the generalized Bayes rule results in a set $\mathcal{M}^{(n)}$ of posterior distributions that is again convex, and $\mathcal{M}^{(n)}$ conveniently can be obtained by taking the convex hull of the set of posteriors defined by the set of updated parameters $\Pi^{(n)}$.

For both cases, the relationship between the parameter sets $\Pi^{(0)}$ and $\Pi^{(n)}$ and the credal sets $\mathcal{M}^{(0)}$ and $\mathcal{M}^{(n)}$ will allow us to discuss different models $\mathcal{M}^{(0)}$ and $\mathcal{M}^{(n)}$ by focusing on the corresponding parameter sets $\Pi^{(0)}$ and $\Pi^{(n)}$.[46] As an example, in the precise Beta-Bernoulli model, the posterior predictive probability for the event that a future single draw is a success is equal to $y^{(n)}$, and so we get, for an imprecise model with $(n^{(0)}, y^{(0)}) \in \Pi^{(0)}$ after having observed s successes, the lower and upper probability

$$\underline{y}^{(n)} := \inf_{\Pi^{(n)}} y^{(n)} = \inf_{\Pi^{(0)}} \frac{n^{(0)}y^{(0)} + s}{n^{(0)} + n},$$

$$\overline{y}^{(n)} := \sup_{\Pi^{(n)}} y^{(n)} = \sup_{\Pi^{(0)}} \frac{n^{(0)}y^{(0)} + s}{n^{(0)} + n}.$$

Special imprecise probability models for learning in the case of an exponential family sampling model are now obtained by specific choices of $\Pi^{(0)}$. We distinguish the following types of models:

(a) $n^{(0)}$ is fixed, while $y^{(0)}$ varies in a set $\mathcal{Y}^{(0)}$.
The IDM [673], as well as its generalization to all sample distributions of the canonical exponential form (7.8) by [521] are of this type. The approach by [82] also belongs to this category, as she specifies bounds for $n^{(0)}y^{(0)}$ while holding $n^{(0)}$ constant (see [57, p. 1973]).

(b) $n^{(0)}$ varies in a set $\mathcal{N}^{(0)}$, while $y^{(0)}$ is fixed.
This type of model is rarely discussed in the literature, but is mentioned in [672, § 7.8.3 and also in § 1.1.4, Footnote 10]. Both instances assume the Normal-Normal model as described in Section 7.4.1, where the set of priors is spanned by normal distributions with a fixed mean $y^{(0)}$ and a range of variances $\sigma_0^2/n^{(0)}$.

(c) Both $n^{(0)}$ and $y^{(0)}$ vary in a set $\{(n^{(0)}, y^{(0)}) | n^{(0)} \in \mathcal{N}^{(0)}, y^{(0)} \in \mathcal{Y}^{(0)}\}$.
This type of model is first discussed in [672, § 5.4.3] for the Beta-Bernoulli model, and was later generalized by [684] to sample distributions of the canonical exponential form (7.8). It should be noted that while the prior parameter set is a Cartesian

[45] To emphasize the dependence of the sample size n, the posterior credal set is denoted by $\mathcal{M}^{(n)}$ instead of $\mathcal{M}_{|\mathbf{x}}$ in (7.14).

[46] Note that, although, by the general framework, the credal sets $\mathcal{M}^{(0)}$ and $\mathcal{M}^{(n)}$ may be defined as convex hulls, the parameter sets $\Pi^{(0)}$ and $\Pi^{(n)}$ generating them need not necessarily be so, and typically are not convex, indeed.

product of $\mathcal{N}^{(0)} = [\underline{n}^{(0)}, \overline{n}^{(0)}]$ and $\mathcal{Y}^{(0)}$, the posterior parameter set is not of such a form. This is due to (7.10), which results in different ranges for $y^{(n)}$ depending on the value of $n^{(0)}$ used in the update step.

(d) Both $n^{(0)}$ and $y^{(0)}$ vary in other sets $\Pi^{(0)} \subset (\mathbb{R}_{>0} \times \mathcal{Y})$.
In this type, also in the prior parameter set the range of $y^{(0)}$ may depend on $n^{(0)}$ as in [685, § 2.3], or, vice versa, the range of $n^{(0)}$ may depend on the value of $y^{(0)}$, as in [57].

Properties and criteria regarding the models. Before discussing the approaches mentioned above in some detail, we will describe some properties that they have in common. The first three of the four properties below are due to the update mechanism (7.10) for $n^{(0)}$ and $y^{(0)}$ and the resulting size and position of $\Pi^{(n)}$, being a direct consequence of the (generalized) Bayes rule in the setting of canonical exponential families. Remember that $n^{(0)}$ is incremented with n, while $y^{(n)}$ is a weighted average of $y^{(0)}$ and the sample statistic $\tilde{\tau}(\mathbf{x})$, with weights $n^{(0)}$ and n, respectively. Thus, while the (absolute) stretch of $\Pi^{(n)}$ in the $n^{(0)}$ resp. $n^{(n)}$ dimension will not change during updating, the stretch in the $y^{(0)}$ resp. $y^{(n)}$ dimension will do so. When speaking of the size of $\Pi^{(n)}$, we will thus refer to the stretch in the main parameter dimension, also denoted by $\Delta_y(\Pi^{(n)})$.

- The larger n relative to (values in the range of) $n^{(0)}$, ceteris paribus (c.p.) the smaller $\Pi^{(n)}$, i.e. the more precise the inferences. Vice versa, the larger the $n^{(0)}$ value(s) as compared to n, c.p. the larger $\Pi^{(n)}$, and the more imprecise the inferences. Thus, a high weight on the imprecise prior $\mathcal{M}^{(0)}$ will lead to a more imprecise posterior $\mathcal{M}^{(n)}$.[47]
- In particular, for $n \to \infty$, the stretch of $y^{(n)}$ in $\Pi^{(n)}$ will converge towards zero, i.e. $\Delta_y(\Pi^{(n)}) \to 0$, with the limit located at $\tilde{\tau}(\mathbf{x})$. For inferences based mainly on $y^{(n)}$, this leads to a convergence towards the 'correct' conclusions. This applies, e.g., to point estimators like the posterior mean or median, which converge to the 'true' parameter, to interval estimates (HPD intervals) contracting around this point estimate (length of interval $\to 0$), and to the probability that a test decides for the 'right' hypothesis, which converges to 1.[48] This property, that holds also for traditional (precise) Bayesian inferences, is similar to the consistency property often employed in frequentist statistics.
- The larger $\Delta_y(\Pi^{(0)})$, the larger c.p. $\Delta_y(\Pi^{(n)})$; a more imprecise prior $\mathcal{M}^{(0)}$ will naturally lead to a more imprecise posterior $\mathcal{M}^{(n)}$, which carries over to the inferences.
- When the credal set $\mathcal{M}^{(0)}$ is taken to contain convex combinations (i.e., finite mixtures) of parametric priors, $\mathcal{M}^{(0)}$ is very flexible and contains a wide variety of priors. Nevertheless, maximization and minimization over $\mathcal{M}^{(n)}$ is quite feasible for inferences that are linear in the parametric posteriors contained in $\mathcal{M}^{(n)}$, as then we can be sure that suprema and infima are attained at the extreme points of $\mathcal{M}^{(n)}$,

[47] For model types with fixed $n^{(0)}$, if $n^{(0)} = n$, then $\Delta_y(\Pi^{(n)}) = \Delta_y(\Pi^{(0)})/2$, i.e. the width of the posterior expectation interval is half the width of the prior interval. This fact may also guide the choice of $n^{(0)}$.

[48] In the binomial and normal example, the posteriors in $\mathcal{M}^{(n)}$ will concentrate all their probability mass at $y^{(n)} \to \tilde{\tau}(\mathbf{x})$, and as $\tilde{\tau} \to \theta$ in probability, all these inference properties follow.

which are the parametric distributions generated by $\mathrm{I\!I}^{(n)}$, enabling us to search over $\mathrm{I\!I}^{(n)}$ only. For other inferences, suprema and infima over $\mathcal{M}^{(n)}$ may be very difficult to obtain, or the model could even be useless for them (because the model may give too imprecise, or even vacuous, bounds).

Apart from the above properties that are guaranteed in all of the four model types (a) – (d), one might want the models to adhere to (either of) the following additional criteria:

- **Prior-data conflict sensitivity.** In order to mirror the state of posterior information, posterior inferences should, all other things equal, be more imprecise in the case of prior-data conflict. Reciprocally, if prior information and data coincide especially well, an additional gain in posterior precision may be warranted. Such models could deliver (relatively) precise answers when the data confirm prior assumptions, while rendering much more cautionary answers in the case of prior-data conflict, thus leading to cautious inferences if, and only if, caution is needed.
  Most statisticians using precise priors would probably expect a more diffuse posterior in case of prior-data conflict. However, in the canonical conjugate setting (7.9), which is often used when data are scarce and only strong prior beliefs allow for a reasonably precise inference answer, this is usually not the case. E.g., for the Normal-Normal model, the posterior variance (7.13) is not sensitive to the location of $\bar{\mathbf{x}}$, and decreases by the factor $n^{(0)}/(n^{(0)}+n)$ for any $\bar{\mathbf{x}}$, thus giving a false certainty in posterior inferences in case of prior-data conflict. In the Beta-Binomial model, the posterior variance $y^{(n)}(1-y^{(n)})/(n^{(0)}+n)$ depends on the location of s/n, but in the very same way as the prior variance $y^{(0)}(1-y^{(0)})/n^{(0)}$ does, mirroring only the fact that Beta distributions centred at the margins of the unit interval are constrained in their spread. Thus, there is no systematic reaction to prior-data conflict also in this case.
  In the imprecise Bayesian framework as discussed here, prior-data conflict sensitivity translates into having a larger $\mathrm{I\!I}^{(n)}$ (leading to a larger $\mathcal{M}^{(n)}$) in case of prior-data conflict, and, mutatis mutandis, into a smaller $\mathrm{I\!I}^{(n)}$ if prior and data coincide especially well.

- **Possibility of weakly or noninformative priors.** When only very weak or (almost) no prior information is available on the parameter(s) one wishes to learn about, it should be possible to model this situation adequately. The traditional Bayesian approach to this problem, so-called *noninformative priors*, are, due to their nature as single, i.e. precise, probability distributions, not expressive enough; a precise probability distribution necessarily produces a precise value for $\Pi[\theta \in A]$ for any $A \subseteq \Theta$, which seems incompatible with the notion of prior ignorance about θ. Furthermore, in the literature there are several, often mutually incompatible, approaches to construct noninformative priors, such as Laplace's prior, Jeffreys' prior, or reference priors [70, § 5.6.2]. Most of these precise priors seem to convey mainly a state of indifference instead of ignorance [547, p. 271].[49]

[49] E.g., in the designation of the uniform prior as a noninformative prior by the principle of insufficient reason (i.e., taking Laplace's prior), it is argued that there is no reason to favour one parameter value over another, and thus, all of them should get the same probability resp. density value. For analysts restricting themselves to precise priors, this argument leads necessarily to the uniform prior. When considering imprecise priors, however, the principle of insufficient reason does not uniquely determine a certain precise prior. It only states that

The approaches mentioned in (a) – (d) are now discussed in detail, sorted according to the two basic scenarios regarding the intended use: When no prior information on θ is available (or nearly so), so-called *near-ignorance priors* are used to model the state of prior ignorance. When, in contrast, there is substantial prior information available, the challenge is to model this information adequately in the prior, while ensuring easy handling and prior-data conflict sensitivity.[50]

7.4.3.2 The IDM and other prior near-ignorance models

The *Imprecise Dirichlet model* (IDM) was developed by Walley [673] as a model for inferences from multinomial data when no prior information is available.[51] As indicated by its name, it uses as imprecise prior a (near-) noninformative set of Dirichlet priors, which is obtained by choosing $\mathcal{Y}^{(0)}$ as the whole interior of the unit simplex Δ, with $n^{(0)}$ fixed,[52]

$$\mathcal{M}^{(0)} = \left\{ \Pi\left(\theta \,\middle|\, n^{(0)}, y^{(0)}\right) \,\middle|\, 0 < y^{(0)}_{[j]} < 1, j = 1, \ldots, k, \sum_{j=1}^{k} y^{(0)}_{[j]} = 1 \right\}.$$

The prior credal set is thus determined by the choice of $n^{(0)}$. [673] argues for choosing $n^{(0)} = 1$ or $n^{(0)} = 2$, where, in the latter case, inferences from the IDM encompass both frequentist and Bayesian results based on standard choices for noninformative priors. For any choice of $n^{(0)}$, however, this imprecise prior expresses a state of ignorance about θ, as, for all $j = 1, \ldots, k$,

$$\underline{\Pi}[I_{\theta_{[j]}}] = \underline{y}^{(0)} = \inf_{y^{(0)} \in \mathcal{Y}^{(0)}} y^{(0)} = 0,$$

$$\overline{\Pi}[I_{\theta_{[j]}}] = \overline{y}^{(0)} = \sup_{y^{(0)} \in \mathcal{Y}^{(0)}} y^{(0)} = 1,$$

and probabilities for events regarding $\theta_{[j]}$ are vacuous, that is $\underline{\Pi}(\vartheta_{[j]} \in A) = 0$ and $\overline{\Pi}(\vartheta_{[j]} \in A) = 1$ for any subset A of $[0, 1]$.[53]

---

the probability resp. density *interval* should be equal for all parameter values, but that interval may be any interval $\subseteq [0, 1]$. We may thus realize that the principle of insufficient reason actually implies *indifference* between parameter values only, and that other considerations are needed to distinguish a certain imprecise prior as (nearly) noninformative; usually, it is postulated that (a certain class of) inferences based on the prior should be (nearly) vacuous, and specific information (e.g., symmetry of parameter values) would be needed to reduce the size of the prior credal set. See, in particular, [692, § 4.3] formulating two symmetry principles extending the principle of insufficient reason, and the closely related discussion of weak and strong invariance in Section 3.3.

For a critique on noninformative priors from an imprecise probability viewpoint see, e.g., [672, § 5.5], their partition dependence is also discussed in the context of elicitation (see Page 326).

[50] In situations when data is revealed to the analyst sequentially in distinct batches, it might also be useful if the model is able to resonate unusual patterns or extreme differences between the batches. This actually effects to doubting the i.i.d. assumptions on which these models are founded. This could be useful in the area of *statistical surveillance*, where, e.g., the number of cases of a certain infectious disease is continuously monitored, with the aim to detect epidemic outbreaks in their early stages.

[51] The imprecise Beta-Binomial model from [672, § 5.3.2] can be seen as a precursor to the IDM, covering the special case of two categories.

[52] Our notation relates to [673] as $t_j \leftrightarrow y^{(0)}_{[j]}$, $s \leftrightarrow n^{(0)}$, $t^*_j \leftrightarrow y^{(n)}_{[j]}$.

[53] However, the IDM may give nonvacuous prior probabilities for some more elaborate events. An example [673, p. 14] is the event (A_J, A_K) that the next observation belongs to a category subset $J \subseteq \{1, \ldots, k\}$, and the observation following that belongs to a category subset K, where $J \cap K = \emptyset$ and $|n(A_J) - n(A_K)| < n^{(0)}$.

The posterior credal set $\mathcal{M}^{(n)}$ is then the set of all Dirichlet distributions with parameters $n^{(n)}$ and $y^{(n)}$ obtained by (7.10),

$$\mathcal{M}^{(n)} = \left\{ \Pi\left(\theta \,\middle|\, n^{(n)}, y^{(n)}\right) \,\middle|\, 0 < y^{(0)}_{[j]} < 1, j = 1, \ldots, k, \sum_{j=1}^{k} y^{(0)}_{[j]} = 1 \right\}.$$

For any event A_J that the next observation belongs to a subset J of the categories, $J \subseteq \{1, \ldots, k\}$, the posterior lower and upper probabilities are

$$\underline{\Pi}(A_J|\mathbf{x}) = \frac{n(A_J)}{n^{(0)} + n} \qquad \overline{\Pi}(A_J|\mathbf{x}) = \frac{n^{(0)} + n(A_J)}{n^{(0)} + n},$$

where $n(A_J) = \sum_J n_j$ is the number of observations from the category subset J.

The IDM is motivated by a number of inference principles put forward in [673, § 1], most notably the *representation invariance principle* (RIP, [673, § 2.9]): Inferences based on the IDM are invariant under different numbers of categories considered in the sample space.[54] The usefulness of the RIP has been controversially discussed (see, e.g., the discussion to [673]), and alternative imprecise probability models that do not rely on it have been developed.[55] [56]

Due to its tractability, the IDM has been applied in a number of situations, including the following:

In [673] Walley offers an application to data from medical studies; [65] details an extension of the IDM to contingency table data that was briefly covered in [673]. In 2009, a special issue of the *International Journal of Approximate Reasoning* [66] was devoted to the IDM. Since its introduction, the IDM has found applications in, e.g., reliability analysis (e.g., [128, 651, 652, 434]), and operations research (e.g., [642]); however, the IDM has had an especially strong impact in the area of artificial intelligence, namely in the construction of classification methods (including, e.g., pattern recognition) and in inference based on graphical models, as covered in the Chapters 9 and 10 of this book. These IDM-based methods, in turn, are used in a vast variety of tasks from all kinds of subjects, such as medicine (e.g., [723]), agriculture (e.g., [717]), or geology (e.g., [22]). The IDM can be regarded as the most influential imprecise probability model so far.

In the IDM, satisfying near-ignorance for the prior and still having nonvacuous posterior probabilities is possible because the domain of the prior main parameter $y^{(0)}$ is bounded ($\mathcal{Y} = \text{int}(\Delta)$). For most conjugate priors to exponential family distributions, $\mathcal{Y}$ is not bounded, and thus, trying to reach prior ignorance in the same way as in the IDM, by taking $\mathcal{Y}^{(0)} = \mathcal{Y}$ for a fixed $n^{(0)}$, would lead to vacuous posterior probabilities.[57] Instead, as was shown by [57], for conjugate priors to one-parameter exponential family

[54] In the example discussed in [673], where coloured marbles are drawn from a bag, it does not matter, e.g., for prior and posterior probabilities for 'red' as the next draw, whether one considers the categorization {red, other} or {red, blue, other}.

[55] The IDM can be shown to lead to ∞-monotone (see Section 4.2) lower probabilities, and its credal set is already completely determined by the lower and upper probabilities of the singletons, i.e. the corresponding lower and upper probabilities constitute a probability interval (see Section 4.4).

[56] See also [133, 134] and the brief discussion at the end of Section 7.6.1 of an alternative based on the NPI approach.

[57] From (7.10), it follows that for $y^{(0)} \to \infty$ we get $y^{(n)} \to \infty$ if $n^{(0)}$ is fixed.

distributions, one needs to vary $n^{(0)}$ in conjunction with $y^{(0)}$ to get both prior near-ignorance and nonvacuous posterior probabilities. In essence, the term $n^{(0)}y^{(0)}$ appearing in (7.10) must be bounded while letting $\mathcal{Y}^{(0)} = \mathcal{Y}$, which effects to a prior parameter set $\Pi^{(0)}$ where the range of $n^{(0)}$ depends on $y^{(0)}$.

To summarize, imprecise probability methods allow for a much more adequate modelling of prior ignorance than noninformative priors, the traditional Bayesian approach to this problem, can deliver. Instead of the somehow awkward choice of a certain noninformativeness approach, to define an imprecise noninformative prior the analyst just needs to specify one parameter (or two, as partly in [57]) determining the learning speed of the model, namely $n^{(0)}$ for the IDM.

7.4.3.3 Substantial prior information and sensitivity to prior-data conflict

Models intended specifically for use in situations with substantial prior information are presented in [672, Footnote 10 in §1.1.4, and §7.8.3], [521], and [684]; the IDM can be modified by not taking $\mathcal{Y}^{(0)} = \text{int}(\Delta)$, but a smaller set $\mathcal{Y}^{(0)}$ fitting the prior information. Generalizing the IDM approach to conjugate priors for sample distributions of the canonical form (7.8), [521] proposed an imprecise prior $\mathcal{M}^{(0)}$ based on $\Pi^{(0)} = \mathcal{Y}^{(0)} \times n^{(0)}$. E.g., in the Normal-Normal model as described in Section 7.4.1, one can take as imprecise prior all convex mixtures of normals with mean in $\mathcal{Y}^{(0)} = [\underline{y}^{(0)}, \overline{y}^{(0)}]$ and a fixed variance $\sigma_0^2/n^{(0)}$. $\mathcal{Y}^{(n)}$, the posterior set of expectations (or modes, or medians) of μ, is then bounded by

$$\underline{y}^{(n)} = \inf_{\Pi^{(0)}} y^{(n)} = \inf_{\mathcal{Y}^{(0)}} \frac{n^{(0)}y^{(0)} + n\bar{\mathbf{x}}}{n^{(0)} + n} = \frac{n^{(0)}\underline{y}^{(0)} + n\bar{\mathbf{x}}}{n^{(0)} + n} \tag{7.15}$$

$$\overline{y}^{(n)} = \sup_{\Pi^{(0)}} y^{(n)} = \sup_{\mathcal{Y}^{(0)}} \frac{n^{(0)}y^{(0)} + n\bar{\mathbf{x}}}{n^{(0)} + n} = \frac{n^{(0)}\overline{y}^{(0)} + n\bar{\mathbf{x}}}{n^{(0)} + n}. \tag{7.16}$$

The lower (upper) posterior expectation of μ is thus a weighted average of the lower (upper) prior expectation and the sample mean, with weights $n^{(0)}$ and n, respectively. As mentioned above, [521] model for Bernoulli or multinomial data leads to the IDM; because $\mathcal{Y}$, the domain of $y^{(0)}$, is not bounded in general, the model is normally used to express substantial prior information.[58]

More generally in case of a one-parameter exponential family, $\Pi^{(0)}$ is fully described by the three real parameters $\underline{y}^{(0)}$, $\overline{y}^{(0)}$, and $n^{(0)}$, which are straightforward to elicit; furthermore, also $\Pi^{(n)}$ is fully described by $\underline{y}^{(n)}$, $\overline{y}^{(n)}$, and $n^{(n)}$, and many inferences will be expressible in terms of these three parameters only. Models of this kind allow for a simple yet powerful imprecise inference calculus, where ambiguity in the prior information can be represented by the choice of the set $\mathcal{Y}^{(0)}$, with $n^{(0)}$ determining the learning speed.

The downside of this easily manageable model is that it is insensitive to prior-data conflict, as the imprecision for the main posterior parameter,

$$\Delta_y(\Pi^{(n)}) = \overline{y}^{(n)} - \underline{y}^{(n)} = \frac{n^{(0)}(\overline{y}^{(0)} - \underline{y}^{(0)})}{n^{(0)} + n}, \tag{7.17}$$

[58] However, it could be used for near-ignorance prior situations in case of other sampling models where $\mathcal{Y}^{(0)}$ can encompass the whole domain without causing posterior vacuousness. This applies, e.g., to circular distributions like the von Mises distribution, where the mean direction angle μ has the domain $(-\pi, \pi]$, see [519, § B.1.4] and [450].

does not depend on the sample $\mathbf{x}$. Imprecision is thus the same for any sample $\mathbf{x}$ of size n, whenever prior information about μ as encoded in $\mathcal{Y}^{(0)}$ is in accordance with data information $\tilde{\tau}(\mathbf{x})$ or not. The relation of $\Delta_y(\Pi^{(n)})$ (which determines the precision of posterior inferences) to the inferential situation at hand is loosened, as possible conflict between prior and data is not reflected by increased imprecision. In that sense, the IDM with prior information and the model by [521] do not utilize the full expressive power of imprecise probability models, behaving, with regard to this aspect, similar to precise conjugate models by basically ignoring prior-data conflict.

To counter this behaviour, [684] suggested that imprecise priors to canonical sample distributions (7.8) should be based on parameter sets $\Pi^{(0)} = [\underline{n}^{(0)}, \overline{n}^{(0)}] \times \mathcal{Y}^{(0)}$, as [672, § 5.4.3] had already implemented for the binomial distribution. Then, (7.15) and (7.16) become

$$\underline{y}^{(n)} = \begin{cases} \dfrac{\overline{n}^{(0)}\underline{y}^{(0)} + n\bar{\mathbf{x}}}{\overline{n}^{(0)} + n} & \bar{\mathbf{x}} \geq \underline{y}^{(0)} \\[2ex] \dfrac{\underline{n}^{(0)}\underline{y}^{(0)} + n\bar{\mathbf{x}}}{\underline{n}^{(0)} + n} & \bar{\mathbf{x}} < \underline{y}^{(0)} \end{cases}, \qquad \overline{y}^{(n)} = \begin{cases} \dfrac{\overline{n}^{(0)}\overline{y}^{(0)} + n\bar{\mathbf{x}}}{\overline{n}^{(0)} + n} & \bar{\mathbf{x}} \leq \overline{y}^{(0)} \\[2ex] \dfrac{\underline{n}^{(0)}\overline{y}^{(0)} + n\bar{\mathbf{x}}}{\underline{n}^{(0)} + n} & \bar{\mathbf{x}} > \overline{y}^{(0)} \end{cases}.$$

If $\underline{y}^{(0)} < \bar{\mathbf{x}} < \overline{y}^{(0)}$, both $\underline{y}^{(n)}$ and $\overline{y}^{(n)}$ are calculated using $\overline{n}^{(0)}$; when $\bar{\mathbf{x}}$ falls into $[\underline{y}^{(0)}, \overline{y}^{(0)}]$, the range of prior expectations for the mean, prior information gets maximal weight $\overline{n}^{(0)}$ in the update step (7.10), leading to the same results as for a model with fixed $n^{(0)} = \overline{n}^{(0)}$. If, however, $\bar{\mathbf{x}} < \underline{y}^{(0)}$, then $\underline{y}^{(n)}$ is calculated using $\underline{n}^{(0)}$, giving less weight to the prior information that turned out to be in conflict with the data. Thus, as $\underline{y}^{(n)}$ is a weighted average of $\underline{y}^{(0)}$ and $\bar{\mathbf{x}}$, with weights $n^{(0)}$ and n, respectively, $\underline{y}^{(n)}$ will be lower (nearer towards $\bar{\mathbf{x}}$) as compared to an update using $\overline{n}^{(0)}$, resulting in increased imprecision $\Delta_y(\Pi^{(n)})$ compared to the situation with $\underline{y}^{(0)} < \bar{\mathbf{x}} < \overline{y}^{(0)}$. In the same way, there is additional imprecision $\Delta_y(\Pi^{(n)})$ if $\bar{\mathbf{x}} > \overline{y}^{(0)}$.[59] Indeed, (7.17) can be written as

$$\Delta_y(\Pi^{(n)}) = \frac{\overline{n}^{(0)}(\overline{y}^{(0)} - \underline{y}^{(0)})}{\overline{n}^{(0)} + n} + \inf_{y^{(0)} \in \mathcal{Y}^{(0)}} |\tilde{\tau}(\mathbf{x}) - y^{(0)}| \frac{n(\overline{n}^{(0)} - \underline{n}^{(0)})}{(\underline{n}^{(0)} + n)(\overline{n}^{(0)} + n)},$$

such that we have the same $\Delta_y(\Pi^{(n)})$ as for a model with $\Pi^{(0)} = \mathcal{Y}^{(0)} \times \overline{n}^{(0)}$ when $\tilde{\tau}(\mathbf{x}) \in \mathcal{Y}^{(0)}$, whereas $\Delta_y(\Pi^{(n)})$ increases if $\tilde{\tau}(\mathbf{x}) \notin \mathcal{Y}^{(0)}$, the increase depending on the distance of $\tilde{\tau}(\mathbf{x})$ to $\mathcal{Y}^{(0)}$, as well as on $\underline{n}^{(0)}$, $\overline{n}^{(0)}$, and n. This model is illustrated in Figure 7.2 for the Normal-Normal model.

The model with $\Pi^{(0)} = \mathcal{N}^{(0)} \times \mathcal{Y}^{(0)}$ is sensitive to prior-data conflict, where prior-data conflict is operationalized as $\tilde{\tau}(\mathbf{x}) \notin \mathcal{Y}^{(0)}$. There is no such direct mechanism for a gain in precision when prior and data information coincide especially well. However, $\mathcal{Y}^{(0)}$ could be chosen relatively small such that it mirrors this situation, considering as the neutral situation $\tilde{\tau}(\mathbf{x})$ being not too far away from $\mathcal{Y}^{(0)}$, and taking as prior-data conflict situations when $\tilde{\tau}(\mathbf{x})$ is in a greater distance to $\mathcal{Y}^{(0)}$.

[59] In the above, $\bar{\mathbf{x}}$ can be replaced by $\tilde{\tau}(\mathbf{x})$ to hold for canonical priors (7.9) in general.

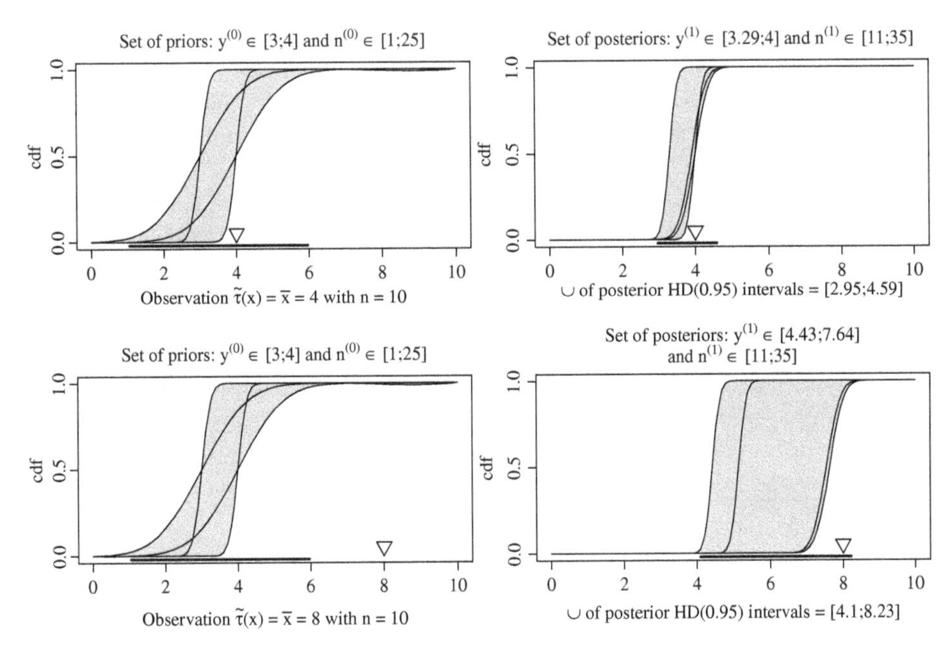

Figure 7.2 Illustration of prior-data conflict sensitivity in the normal-normal model. Displayed are sets of normal cdfs, where the graphs in the upper row show the update step if $\bar{\mathbf{x}} \in [\underline{y}^{(0)}, \overline{y}^{(0)}]$, and the graphs in the lower row show a situation with prior-data conflict. The black bars below the cdfs denote the unions of prior resp. posterior highest density intervals, while the position of $\bar{\mathbf{x}}$ used in the update step is indicated by the lower point of the triangle. Note how prior-data conflict in the example depicted in the lower row leads to increased imprecision, as reflected by the larger set of posteriors and the longer union of HPD intervals. Graph taken from [684, p. 267]. Reprinted by permission of Taylor & Francis Ltd, http://www.tandf.co.uk/journals.

As mentioned above, the parameter set $\mathrm{I\!\Pi}^{(n)}$ resulting from the update step (7.10) is not a Cartesian product of $\mathcal{Y}^{(n)}$ and $\mathcal{N}^{(n)}$, i.e., a *rectangle* set in case of one-dimensional $y^{(0)}$, as was the case for $\mathrm{I\!\Pi}^{(0)}$. It might therefore be necessary to minimize and maximize over $\mathrm{I\!\Pi}^{(n)}$ itself if inferences depend on $y^{(n)}$ and $n^{(n)}$ simultaneously. If, e.g., $n^{(n)}y^{(n)}$ must be minimized to determine the posterior lower bound of a characteristic of interest, $\min_{\mathrm{I\!\Pi}^{(n)}} n^{(n)}y^{(n)}$ may not be found at $\underline{y}^{(0)}$, i.e., $\min_{\mathrm{I\!\Pi}^{(n)}} n^{(n)}y^{(n)} \neq \min_{\mathcal{N}^{(0)}} n^{(n)}\underline{y}^{(n)}$.

The model briefly mentioned at [672, Footnote 10 in §1.1.4, and §7.8.3], where $\mathrm{I\!\Pi}^{(0)} = \mathcal{N}^{(0)} \times y^{(0)}$, also leads to a more complex description of $\mathrm{I\!\Pi}^{(n)}$ as compared to the models with $\mathrm{I\!\Pi}^{(0)} = n^{(0)} \times \mathcal{Y}^{(0)}$.

In principle, $\mathrm{I\!\Pi}^{(0)}$ could have any form fitting the prior information at hand. On close inspection, a rectangular shape for $\mathrm{I\!\Pi}^{(0)}$ may not be appropriate in many situations. One could, e.g., argue that the $y^{(0)}$ interval should be narrower at $\underline{n}^{(0)}$ than at $\overline{n}^{(0)}$, because we might be able to give quite a precise $y^{(0)}$ interval for a low prior strength $\underline{n}^{(0)}$, whereas for a high prior strength $\overline{n}^{(0)}$, we should be more cautious with our elicitation of $y^{(0)}$ and thus give a wider interval; interestingly, one could also argue the other way round based on similarly convincing arguments. To fully specify $\Pi^{(0)}$ along these lines, lower and upper bounds for $y^{(0)}$ must be given for all intermediate values of $n^{(0)}$ between $\underline{n}^{(0)}$ and $\overline{n}^{(0)}$, e.g., by some functional form $\underline{y}^{(0)}(n^{(0)})$ and $\overline{y}^{(0)}(n^{(0)})$. The choice of such general forms is not straightforward, as it may influence the posterior inferences considerably (see [685]), and it may be very difficult to elicit as a whole. One such choice is discussed in [685] for the Binomial case, developed with the intention to create a smoother reaction to prior-data conflict than in the model with rectangle $\Pi^{(0)}$. However, there is a trade-off between easy description and handling of $\Pi^{(0)}$, and modelling accuracy and fulfillment of desired properties:

(i) The model in [521], which takes $\Pi^{(0)} = n^{(0)} \times \mathcal{Y}^{(0)}$, is very easy to handle, as the posterior parameter set $\Pi^{(n)} = n^{(n)} \times \mathcal{Y}^{(n)}$ has the same form as $\Pi^{(0)}$, and it often suffices to consider the two elements $(n^{(0)}, \underline{y}^{(0)})$ and $(n^{(0)}, \overline{y}^{(0)})$ to find posterior bounds for inferences. It is, however, insensitive to prior-data conflict.

(ii) The model by [684] is sensitive to prior-data conflict, but this advantage is payed for by a more complex description of $\Pi^{(n)}$.

(iii) More general set shapes $\Pi^{(0)} \subset \mathbb{R}_{>0} \times \mathcal{Y}$ are possible, but may be difficult to elicit and complex to handle.

To conclude, when substantial prior information is available, that, however, is not permitting the identification of a single prior distribution, imprecise models allow for adequate modelling of partial information and prior-data conflict sensitivity, and will ultimately result in more reliable inferences.

7.4.4 Some further comments and a brief look at other models for generalized Bayesian inference

In this subsection, we will first consider direct extensions of the models discussed in Section 7.4.3, then briefly turn to other approaches based on sets of priors, and conclude with some general remarks on updating and the generalized Bayes rule.

As a direct extension of the models discussed in Section 7.4.3, it may be stressed that their central properties rely on the specific form of the update step (7.10) only. Thus, they can be generalized to settings that are not based on i.i.d. observations from canonical exponential family distributions, but nevertheless follow the same update step. This is elaborated in [684] by the definition of so-called LUCK-models, which were already used in [686] to generalize Bayesian linear regression.

So far, we have strongly focused on a specific class of models to illustrate in an exemplary way the main ideas and potential of generalized Bayesian inference. Other work on exponential family sampling models includes [124], who use a different type of imprecise conjugate prior densities, and [75], who suggest the logit-normal model as an alternative to the IDM.

Many other models have been investigated and applied, in particular neighbourhood models (see, e.g. the surveys [63] and [548]). By studying credibility intervals, [512] gives a neat overview on a number of approaches based on sets of priors; [537] is a very instructive example of a study on elicitation of the so-called *interval of measures* priors (cf. [229]), also called *density ratio class*.

For many of these models, computation is a crucial issue. [688] shows that linearization techniques directly related to the generalized Bayes rule can be applied in a large variety of models. [72] suggests a powerful algorithm for handling the important case when the prior is specified by a finite number of lower previsions.

Most models in generalized Bayesian inference rely on a precise sampling model, but the general framework also allows for imprecise sampling models (see, in particular, [672, § 8.5]), also discussed under the notion of 'likelihood robustness' (e.g., [589]) in the robust Bayesian framework.

In the literature the seemingly self-evident character of the generalized Bayes rule has also been questioned.[60] Firstly, it has been shown that the decision theoretic justification of the Bayes rule as producing prior risk minimizing decision functions does not extend to the case of sets of priors. Updating by generalized Bayes rule does no longer necessarily lead to optimal decision functions and thus, as one could argue, also not to optimal inference procedures; see, in particular, [34] for a detailed discussion and [496, 35] for algorithms to calculate optimal decision functions. Secondly, it has been debated that – intuitively – in the case of no prior-data conflict the resulting interval limits may be too wide, and in this sense the inference is too imprecise. Such possibly counterintuitive bounds result from elements in the prior credal set under which the observed sample is rather unlikely. In order to maintain the understanding of updating as conditioning on the sample, several authors therefore suggested to reduce the prior sets in dependence on the sample. A radical way to achieve this is to consider in the conditioning process only those priors under which the observed sample has highest probability, e.g., [355], see also [678], resulting in a procedure that is related to Dempster's rule of conditioning (see Section 5.3.2). Other ways to address this issue are updating rules where imprecision is updated in accordance with an information measure ([124, 126], see also [125]) or hierarchical models, like the model suggested by [104] in the (profile) likelihood context (see Section 7.7), a variant of which also allows to incorporate prior information by using a so-called prior likelihood.

7.5 Frequentist statistics with imprecise probabilities

Now we discuss frequentist methods for statistical inference with imprecise probabilities. In this setting an imprecise sampling model, with the parameter understood as a fixed, non-probabilistic quantity, is considered. Methods currently available have almost exclusively been developed in the context of robust statistics, and thus presentation here also focuses on this area, highlighting those results from robust statistics where the relation to, and the (potential) impact on, imprecise probabilities is most striking.

---

[60] Indeed, Walley himself notes [672, p. 335] that, although the theory of coherence suggests that [... the generalized Bayes rule (GBR)] is a reasonable updating strategy, [...] there is nothing in the theory that requires You to [...] construct conditional previsions [...] through the GBR' and to '[...] adopt [... them] as Your new unconditional prevision'. He is also very clear about the fact (see [672, p. 334]) that 'there is a role for other updating strategies, not because the updated beliefs constructed through the GBR are unjustified, but because they are often indeterminate'. Indeed, he [672, § 6.11.1] lists twelve items summarizing '[...] the reasons for which the GBR may fail to be applicable or useful as an updating strategy.'

Influenced by robust statistics, the methodology is typically relying on the sensitivity analysis point of view in the sense of Section 7.2.1. Moreover, aiming at robustness, imprecision is mainly used as a device to safeguard against potentially devastating consequences of slight deviations from an underlying traditional model.[61] Thus in robust statistics the models considered are usually neighbourhood models, but many of the concepts, results and methods developed there are also of high relevance for general imprecise probability models. This in particular applies to hypothesis testing.

We start with describing the background, giving two prototypical examples of nonrobustness of traditional statistical procedures. Then, in the second subsection, we review the comparatively widely developed theory of statistical hypothesis testing based on the so-called Huber-Strassen theorem. In the third subsection we turn to estimation theory, where the connection between robust statistics and imprecise probabilities has been loosened. Finally, we speculate on further developments by identifying some issues for further research in that field.

7.5.1 The nonrobustness of classical frequentist methods

Example 7.5 (Nonrobustness of standard estimators). As a simple but striking example, consider an i.i.d. sample $\mathbf{X} = (X_1, \ldots, X_n)$, where the probability density function of the population underlying the sample can be described by 'some regularly looking, symmetric, bell-shaped curve'. Of course, there are several parametric models fitting to such an informal description, including the normal and the Cauchy distribution (see Figure 7.3), which indeed look very similar. Now, let μ, the centre of symmetry, be the quantity to be estimated and assume – to simplify the presentation of the argument – the second parameter of the distribution (σ^2 resp. τ) to be known. Whether, however, one assumes the underlying distribution to be normal ($X_i \sim \mathrm{N}(\mu, \sigma^2)$) or to be Cauchy ($X_i \sim C(\mu, \tau)$) makes a substantial difference indeed. The behaviour of the sample mean $\bar{X} := \frac{1}{n} \sum_{i=1}^{n} X_i$ as the optimal standard procedure under normality is completely different in the Cauchy case. One obtains

$$\bar{X} \sim N\left(\mu, \frac{\sigma^2}{n}\right) \qquad \text{and} \qquad \bar{X} \sim C(\mu, \tau),$$

respectively. Both estimators are distributed around the true centre and therefore are not systematically biased. Under normality, the variance decreases with the sample size n; the larger n, the more the variability in the estimator decreases, and learning about μ from samples is possible in a straightforward way. Indeed, in this situation $\bar{X}$ can even be shown to be the optimal procedure.[62] However, assuming a Cauchy distribution, the shape parameter is again τ, independent of the sample size n; thus, in this setting, learning from the arithmetic mean is impossible. Irrespectively of the number of items sampled, the variability of that estimator is the same. In other words, in the Cauchy case, evaluating even thousands of observations by the arithmetic mean, the optimal procedure under normality, does not provide any further information beyond evaluating a single unit. ♦

[61] A vivid interpretation compares robust statistics with an insurance contract. One pays a premium (some loss of power in the ideal model) to be protected in the case of an accident (unfortunate observations deviating from the ideal model), compare, in the situation of outlier rejection to [17, p. 127f.])

[62] UMVU, uniformly minimum variance unbiased, compare (7.26) and (7.27).

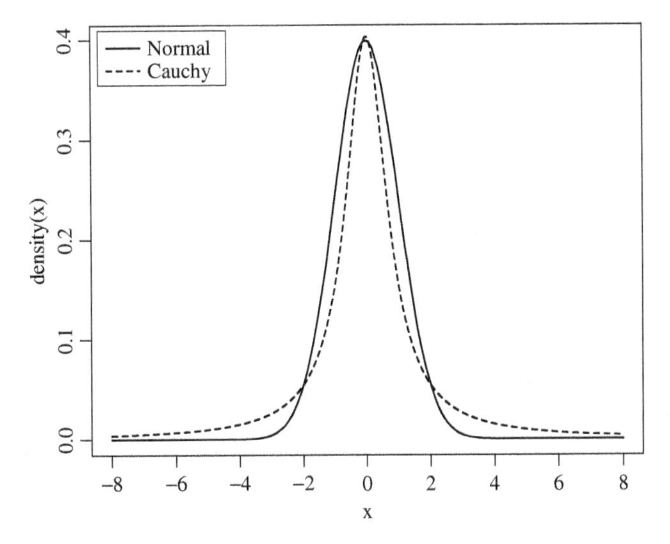

Figure 7.3 Densities of the Normal(0, 1) and the Cauchy(0, 0.79) distribution.

Robustness is also an important issue in hypothesis testing. As a preparation for the result reported in Section 7.5.2, we briefly recall the technical basis of the traditional frequentist approach here.

In its basic form, statistical hypothesis testing makes a data-based decision between two scenarios. It decides whether every observation $x_i, i = 1, \ldots, n$, of an i.i.d. sample $\mathbf{x} = (x_1, \ldots, x_n)$ is distributed according to a distribution P_0 with probability density or mass function $f_0(x_i)$ (*null hypothesis* H_0) or a distribution P_1 with probability density or mass function $f_1(x_i)$ (*alternative hypothesis* H_1), i.e. whether the distribution generating $\mathbf{x}$ has density or mass function $\tilde{f}_0(\mathbf{x}) := \prod_{i=1}^{n} f_0(x_i)$ or $\tilde{f}_1(\mathbf{x}) := \prod_{i=1}^{n} f_1(x_i)$.

Technically, (frequentist) statistical hypothesis testing can be described by randomized decision functions, which can be represented in this setting by a single random variable $\varphi(\cdot)$ (*test function*), where $\varphi(\mathbf{x})$ describes the probability of rejecting the null hypothesis after having observed the sample $\mathbf{x}$, and $A^{\varphi} := \{\mathbf{x} | \varphi(\mathbf{x}) = 1\}$, as the set of all samples where $\varphi(\cdot)$ takes the value 1, is called *rejection region*. Typically, for continuous sample spaces, $\varphi(\cdot)$ simply takes values in $\{0, 1\}$; in particular in the discrete case, randomization is often needed, i.e. $\varphi(\mathbf{x})$ means that after having observed $\mathbf{x}$, a (virtual) binary experiment with precise probability $\varphi(\mathbf{x})$ of success is performed, and H_0 is rejected if and only if success is observed.

Relying on test functions allows to describe and construct test decisions in the simple testing problem

$$H_0 : \text{`}P_0 \text{ is true'} \qquad \text{against} \qquad H_1 : \text{`}P_1 \text{ is true'}$$

by expectations[63]. If one writes $\mathbb{E}_\ell$ for the expectation with respect to $P_\ell^{\otimes n}, \ell = 0, 1$, then $\mathbb{E}_0\varphi$ describes the total *error probability of the first kind* (or Type I Error probability), i.e. of

[63] In this section we join the convention in the corresponding literature and use the term 'expectations' instead of 'previsions'.

falsely rejecting H_0 although it is true. Its counterpart is $\mathbb{E}_1(1-\varphi)$, the total *error probability of the second kind* (or Type II Error probability), i.e. of not rejecting H_0 although H_1 is true, which is complementary to the *power* $\mathbb{E}_1(\varphi)$ giving the total probability of correctly discovering H_1.

The main approach, Neyman-Pearson theory, treats the hypotheses H_0 and H_1 asymmetrically. Starting with an external given *level of significance* α constraining the error probability of the first kind, only *level-α tests* are admissible, i.e. those tests where the error probability of the first kind does not exceed α. Among all those tests, a test is defined as optimal if and only if it maximizes the *power*, i.e. minimizes the error probability of second kind compared to all other level-α tests. Very often, tests are considered where both f_0 and f_1 are from the same parametric family, so $f_0 = f_{\vartheta_0}$ and $f_1 = f_{\vartheta_1}$, with parameters $\vartheta_0 \neq \vartheta_1$. Intuitively, H_0 has then to be rejected in the light of the data $\mathbf{x} = (x_1, \ldots, x_n)$ if and only if ϑ_1 has high plausibility compared to ϑ_0, given $\mathbf{x}$. This suggests to compare the likelihoods under the hypotheses. Indeed, it can be shown by the so-called Neyman-Pearson fundamental lemma[64] that optimal tests can be constructed by considering the likelihood ratio

$$\mathrm{LR}(\mathbf{x}) := \frac{\mathrm{Lik}(\vartheta_1 \| x_1, \ldots, x_n)}{\mathrm{Lik}(\vartheta_0 \| x_1, \ldots, x_n)} = \prod_{i=1}^{n} \frac{f_1(x_i)}{f_0(x_i)}.$$

Then every so-called *level-α likelihood ratio test*, i.e. every test with

$$\varphi^*(\mathbf{x}) = \begin{cases} 1 & LR(\mathbf{x}) > \kappa \\ 0 & LR(\mathbf{x}) < \kappa \end{cases}, \tag{7.18}$$

such that $\mathbb{E}_0\varphi^* = \alpha$, is optimal.[65]

Note, however, that this straightforward construction via the likelihood ratio leads to highly nonrobust tests. As a product of ratios, the likelihood ratio may be determined completely by a single factor coming from a single outlying observation x_{i_0}, where $f_{\vartheta_\ell}(x_{i_0})$ is very close to zero for one $\ell \in \{0, 1\}$, and thus the whole ratio tends to infinity or zero, irrespective of the other observations. In other words, similar to estimation theory, also in testing theory the optimal procedure under the ideal model may show disastrous behaviour under small deviations from the ideal model. Intuitively, the test may be robustified by restricting the influence of single observations, i.e. by bounding very large or very small factors in the product forming the likelihood ratio. Indeed, this idea from [365] can be justified rigorously, and has proven successful in many circumstances, initiating the development of a general testing theory between credal sets arising from hypotheses described by certain lower previsions, which will be presented in the following subsection, discussing a generalized Neyman-Pearson setting and optimal tests arising from so-called least favourable pairs.

[64] See, e.g., [429, Theorem 3.2.1].

[65] It can be shown that under distributions with so-called monotone likelihood ratio, for instance for exponential families in canonical form, the resulting test is also optimal (among all so-called unbiased level-α tests) for $H_0 : \vartheta \in (-\infty, \vartheta_0]$ versus $H_1 : \vartheta \in (\check{\vartheta}, \infty)$, with $\check{\vartheta} \geq \vartheta_0$ as a fixed value separating the hypotheses.

7.5.2 (Frequentist) hypothesis testing under imprecise probability: Huber-Strassen theory and extensions

7.5.2.1 A generalized Neyman-Pearson setting: Basic concepts

To extend the testing problem just discussed to imprecise probabilities, the hypotheses H_0 and H_1 are now described by coherent lower previsions $\underline{P}_0$ and $\underline{P}_1$ with corresponding credal sets $\mathcal{P}_0$ and $\mathcal{P}_1$ with $\mathcal{P}_0 \cap \mathcal{P}_1 = \emptyset$, respectively, i.e. the testing problem becomes

$$H_0 \; : \; \text{`}\underline{P}_0 \text{ is true'} \quad \text{versus} \quad H_1 \; : \; \text{`}\underline{P}_1 \text{ is true'}. \tag{7.19}$$

For technical reasons, the formulation of the general setting here uses sample size 1, and we denote the sample space by $\mathcal{Y}$. Of course, situations with sample size n can be included by considering independent products according to the different concepts (e.g., [158] and Chapter 3, see also Section 7.3.2) for the lower prevision fields describing the hypotheses.

In the Gamma-maximin spirit (see Chapter 8) typical for robust statistics, the literature has mainly concentrated on the case where only the upper error probabilities are taken into account. Then the Neyman-Pearson principle 'Minimize the probability of the error of the second kind (i.e. not recognizing H_1 although it is true) while controlling for the error probability of the first kind (i.e. falsely rejecting H_0) constrained by the level of significance α,' leads to a complex, nonparametric (not 'easily parameterizable') maximin[66] problem between the credal sets $\mathcal{P}_0$ and $\mathcal{P}_1$. Then, for $\alpha \in (0, 1)$ given, a test φ^* is called *level-α maximin test* or *minimax test* if φ^* respects the level of significance, i.e.

$$\sup_{P_0 \in \mathcal{P}_0} \mathbb{E}_{P_0} \varphi = \overline{P}_0 \varphi \leq \alpha, \tag{7.20}$$

and φ^* has maximal power among all tests under consideration, i.e.

$$\underline{P}_1 \varphi^* = \inf_{P_1 \in \mathcal{P}_1} \mathbb{E}_{P_1} \varphi^* \geq \inf_{P_1 \in \mathcal{P}_1} \mathbb{E}_{P_1} \varphi = \underline{P}_1 \varphi \tag{7.21}$$

for every test φ with $\overline{P}_0 \varphi \leq \alpha$.

7.5.2.2 Globally least favourable pairs and Huber-Strassen theory

Apart from trivial special cases, it seems to be impossible to solve the complex testing problem between the credal set directly by constructing a level-α-maximum test right away. A way out would be to find an easily tractable auxiliary testing problem between two classical probabilities Q_0 and Q_1 that is representative for the testing problem between the whole credal sets. For that purpose, Q_0 and Q_1 have to be *least favourable* in the sense that testing between them is difficult enough to hope that a test for this reduced problem is also admissible for the testing between the whole credal sets.

[66] See, for instance, [429] for maximin (minimax) tests in classical statistics.

Huber and Strassen [370] have formalized this intuition in the following way: A pair

$$(Q_0, Q_1) \in \mathcal{P}_0 \times \mathcal{P}_1 \tag{7.22}$$

of classical probabilities is called *a (globally) least favourable pair (for testing $\underline{P}_0$ versus $\underline{P}_1$ or $\mathcal{P}_0$ versus $\mathcal{P}_1$)* if the likelihood ratio LR(·) of Q_0 and Q_1 satisfies for all $t \geq 0$ the condition

$$Q_0(\{y \in \mathcal{Y} \mid \mathrm{LR}(y) > t\}) = \overline{P}_0(\{y \in \mathcal{Y} \mid \mathrm{LR}(y) > t\}) \\ = \max_{P_0 \in \mathcal{P}_0} P_0(\{y \in \mathcal{Y} \mid \mathrm{LR}(y) > t\}), \tag{7.23}$$

$$Q_1(\{y \in \mathcal{Y} \mid \mathrm{LR}(y) > t\}) = \underline{P}_1(\{y \in \mathcal{Y} \mid \mathrm{LR}(y) > t\}) \\ = \min_{P_1 \in \mathcal{P}_1} P_1(\{y \in \mathcal{Y} \mid \mathrm{LR}(y) > t\}). \tag{7.24}$$

The heuristics given above about least favourable positions for testing can be justified rigorously. It is straightforward to prove that every level-α likelihood ratio test based on a globally least favourable pair indeed is a level-α maximin test, and, in this way, an optimal test has been found.

Illustrating this in the case where all the level-α maximin tests are unrandomized, then, with $A_t := \{y \in \mathcal{Y} \mid \mathrm{LR}(y) > t\}$ as the so-to-say 'candidates' for rejection regions, the level-α maximin test is[67] $\varphi^* = A_{t^*}$, with t^* such that $\overline{P}_0(A_{t^*}) = \alpha$. This test then has power $\underline{P}_1(A_{t^*})$, and A_{t^*} is just the rejection region of the conventional optimal level-α test of Q_0 versus Q_1. Indeed, by (7.23) and (7.24), Q_0 and Q_1 are least favourable in a literal sense; the candidates A_t for rejection regions have maximal error probabilities over $\mathcal{P}_0$ and $\mathcal{P}_1$ at Q_0 and Q_1.

Remarkably, by the condition $\forall t \geq 0$ in (7.23) and (7.24), the property of being globally least favourable does not depend on the underlying level of significance. This also makes globally least favourable pairs independent of the sample size, i.e., if $(Q_0(\cdot), Q_1(\cdot))$ is least favourable for testing $\mathcal{P}_0$ versus $\mathcal{P}_1$ then $(Q_0^{\otimes n}(\cdot), Q_1^{\otimes n}(\cdot))$ is least favourable for testing $\mathcal{P}_0^{\otimes n}$ versus $\mathcal{P}_1^{\otimes n}$, if $\mathcal{P}_i^{\otimes n}$ is defined according to (7.6) or (7.7).[68] This ensures that the calculation of least favourable pairs remains computationally tractable even for large-scale problems. For every sample size n, for the construction of a globally least favourable pair of distributions on $\mathcal{X}^{\otimes n}$ it is sufficient to consider the basic sample space $\mathcal{X}$ only instead of $\mathcal{X}^{\otimes n}$, which would be of extremely high dimension in typical applications.

[370, Theorem 4.1, p. 257] had been able to formulate an existence result, which became famous under the name *Huber-Strassen theorem*. With an extension by [92], the existence of a globally least favourable pair is assured under the additional condition that the lower previsions $\underline{P}_0(\cdot)$ and $\underline{P}_1(\cdot)$ describing the hypotheses are two-monotone capacities (see Section 4.3) satisfying some mild regularity conditions.[69] Since the usual neighbourhood models of precise probabilities discussed in Chapter 4.7 indeed satisfy these conditions, this theorem has become a milestone for frequentist approaches to imprecise probabilities,

---

[67] Again indicator functions and events are identified: $I_{A_{t^*}} \cong A_{t^*}$.

[68] For technical details see [708, p. 237f., Satz 2.57], and in particular the discussions in [678, § 5] and [32, p. 221–227], who also show that this result holds for both independence concepts described in (7.6) and (7.7).

[69] In particular, the continuity conditions (6.13 and 6.14), automatically satisfied on finite sample spaces, are sufficient.

the more so as soon afterwards methods for concretely constructing least favourable pairs have been found. [533] presented a solution for (ε, δ)-contamination models containing ε-contamination, i.e. linear-vacuous mixtures, and total variation models as special cases.[70] [327] gives an overview of further results that have been achieved by working with the generalized risk function. [51, p. 402f] was able to show that, as intuitively argued in the motivation in Section 7.5.1, indeed the likelihood ratio of the least favourable pair is often a monotone transformation of the likelihood ratio of the central distributions, which is robustified by 'flattening' its factors.

While according to the Huber-Strassen theorem, two-monotonicity is (together with some mild regularity conditions) a sufficient condition for the existence of globally least favourable pairs, it not necessary. An important class of examples of general imprecise probability models where a globally least favourable pair exists ([33, 558]) are imprecise parametric statistical models with an interval-parameter where the classical statistical model has a monotone density ratio (for instance, 'imprecise normal distributions') in the parameter that is assumed to vary in an interval.[71] As shown by [33, pp. 164–169] the Huber-Strassen theory, including its construction methods, can, under regularity conditions, also be extended to certain classes of generalized neighbourhood models with an interval-valued central distribution. Conveniently enough, there the calculation of globally least favourable pairs can be separated into two steps. First the testing problem between the generalized neighbourhood models is reduced to the testing between 'least favourable conventional neighbourhood models'. For these two neighbourhoods of precise probability measures, least favourable pairs can be calculated in the usual way.

7.5.3 Towards a frequentist estimation theory under imprecise probabilities – some basic criteria and first results

While, driven by the Huber-Strassen theory, frequentist testing theory for imprecise probabilities is – at least within the restricted focus just discussed – fairly developed, the more complex area of frequentist estimation theory under imprecise probabilities is still in a very early stage, and only a few basic results are available up to now. In contrast to testing theory, where results achieved in robust statistics proved very helpful as a catalyzing starting point, the impact of robust estimation theory on imprecise probabilities is more limited. There is a variety of reasons why borrowing from robust statistics is so difficult in this setting. One important reason is the fact that most of the results in – traditional as well as in robust – estimation theory are of asymptotic nature, i.e. they refer to the limiting case of infinite sample size, whereas the corresponding generalizations of limit theorems to imprecise probabilities are still in development. In addition, in recent years many robust estimation procedures have frequently been developed without an explicit reference to an underlying imprecise sampling model, and often robustness is even perceived as a simple add-on to precise statistical procedures, where some nonrobust features are alleviated. Moreover, even if proper neighbourhood models are considered, robust statistics mostly uses so-called shrinking neighbourhoods, where the radius decreases with increasing sample (see below).

Developing a neat frequentist estimation theory for imprecise probabilities is thus still a big challenge. The following paragraphs mainly aim at an exposition of the background

[70] See, also, e.g., [429, Theorem 8.3.1] for the solution in the case of an ε-contamination model.

[71] Compare (7.5).

of some of the fundamental issues. Some finite-sample criteria for estimators that could be of relevance are discussed, and some first results already achieved in the literature are reported. Then a brief flavour of the development in the asymptotic setting is given.

7.5.3.1 Basic traditional criteria

Being outside the Bayesian paradigm with its inherently given inference procedure, the first step towards a solid frequentist estimation theory lies in the formulation of appropriate criteria for assessing the properties of estimators. We reflect here on this still demanding task, and touch on some of the most important results achieved.

Basically, an estimator is a statistical procedure how to deduce ('estimate') a parameter value from a sample (cp. Section 7.1). In the classical setting, given an i.i.d. sample of size n based on the classical parametric statistical model $(\mathcal{X}, (P_\vartheta)_{\vartheta \in \Theta})$, an estimator is therefore a mapping[72]

$$T : \mathcal{X}^n \to \Theta.$$

Under imprecise probability, this definition can be maintained. In the case of an interval-parameter, a point estimator may have some first glance similarities with interval estimators in the traditional parametric setting, but its aim and semantic interpretation are different. Here, we mainly focus on a single real valued parameter.

The construction of finite-sample optimal estimators is demanding even in traditional statistics. On a more fundamental basis, it is already a challenging question how to generalize the common traditional finite-sample optimality criteria based on expectations $\mathbb{E}_{P_\vartheta}(\cdot)$ and variances $\mathbb{V}_{P_\vartheta}(\cdot)$, like *uniform minimum variance unbiasedness (UMVU)* to imprecise probabilities. To repeat as a preparation the basic traditional concepts, an estimator T of the parameter ϑ of a classical statistical model is called *unbiased* if there is no systematic deviation, i.e. if

$$\mathbb{E}_{P_\vartheta}(T) = \vartheta, \ \forall \vartheta \in \Theta, \tag{7.25}$$

or equivalently, by introducing the term *bias* to measure the extent of the systematic deviation, $\text{Bias}(T, \vartheta) := | \ \mathbb{E}_{P_\vartheta}(T) - \vartheta \ |$, unbiasedness means that

$$\text{Bias}(T, \vartheta) = 0, \quad \forall \vartheta \in \Theta.$$

An unbiased estimator T^* is said to be *UMVU* if

$$\mathbb{V}_{P_\vartheta}(T^*) \leq \mathbb{V}_{P_\vartheta}(T), \quad \forall T \in \mathcal{U}, \tag{7.26}$$

with

$$\mathcal{U} := \{ T : \mathcal{X}^n \to \theta | \ \text{Bias}(T, \vartheta) = 0, \ \forall \vartheta \in \Theta \} \tag{7.27}$$

denoting the set of all unbiased estimators.[73]

[72] Again, measurability issues with respect to appropriate underlying σ-fields are neglected in this exposition.

[73] When comparing variances, it is indeed crucial to confine consideration in (7.26) to all unbiased estimators, i.e. elements of $\mathcal{U}$ satisfying (7.27), in order to exclude 'obviously silly estimators', like constant estimators ignoring the sample having variance 0.

7.5.3.2 Generalized unbiasedness

Generalizing (7.25) in the usual way to an imprecise statistical model with a precise parameter $\vartheta \in \mathbb{R}$ would lead either to a strong version

$$\mathbb{E}_P(T) = \vartheta, \quad \forall \vartheta \in \Theta, \forall P \in \mathcal{P}_\vartheta, \tag{7.28}$$

or to a weak version,

$$\forall \vartheta \in \Theta, \exists P \in \mathcal{P}_\vartheta : \mathbb{E}_P(T) = \vartheta. \tag{7.29}$$

After adapting (7.27) appropriately, (7.26) could be generalized by looking at the upper variance or using criteria in the spirit of E-admissibility. Generally, (7.28) seems to be rather strong in practical applications. A relevant example, however, where it is satisfied are credal sets $\mathcal{P}_\vartheta$ produced by normal distributions with fixed mean ϑ and a variance varying in a known interval. On the other hand, (7.29) may be judged as rather weak, it does not exclude high biases $\mid \mathbb{E}_P(T) - \vartheta \mid$ for other elements $P \in \mathcal{P}_\vartheta$. This motivates to use as a criterion the maximal bias over $\mathcal{P}_\vartheta$

$$\overline{\text{Bias}}(T, \vartheta) := \overline{\mathbb{E}}_{\mathcal{P}_\vartheta}(|T - \vartheta|) := \sup_{P \in \mathcal{P}_\vartheta} (|\mathbb{E}_P(T - \vartheta)|). \tag{7.30}$$

This concept indeed has been considered in robust statistics in the case where $\mathcal{P}_\vartheta$ is a ε-contamination neighbourhood model with radius ε. Then the *max bias curve* $\overline{\text{Bias}}(\varepsilon, T, \vartheta)$ describes the maximal bias in the dependence of ε. A related concept is the (finite-sample) *breakdown point*, the proportion of outlying observations needed to spoil an estimator completely, e.g. [542]. For instance, the arithmetic mean $\bar{X}_n$ has finite sample breakdown point $\frac{1}{n}$; indeed one single outlying observation may be enough to produce arbitrarily high bias. In contrast, for the median the breakdown point is 0.5.

Another criterion, motivated by the aim to protect oneself against huge biases, tries to ensure that the probability of large deviations of the estimator from the true parameter ϑ is small. Then, for a given constant $\tau \in (0, \infty)$ specifying the maximally tolerable bias, one seeks an estimator T^{**} minimizing

$$\sup_{Q \in \mathcal{P}^{\otimes n}} \max\{Q(T > \theta + \tau), Q(T < \theta - \tau)\}. \tag{7.31}$$

This is one of the very few criteria where finite-sample results have been achieved, at least for certain neighbourhood models. For the case where $\mathcal{P}^{\otimes n}$ is defined by (7.6), this has been solved by [366]; see also [534] for an extension to simple linear regression.

7.5.3.3 Minimum distance estimation

A general concept to construct estimators in traditional statistics is *minimum distance estimation*. There, for an i.i.d. sample $\mathbf{X} = (X_1, \ldots, X_n)$, with X_i taking values in $\mathbb{R}$, the empirical measure $P_n(\mathbf{X})$ consisting of point masses with weights $\frac{1}{n}$ in the observed values, or the empirical distribution function $F_n(\mathbf{X})$, is compared to the probabilities $(P_\vartheta)_{\vartheta \in \Theta}$, or the corresponding distribution functions $F_\vartheta(x)$, respectively. From this, an estimator $\hat{\vartheta}$ is determined such that $\text{dist}(P_n(\mathbf{X}), P_\vartheta)$ or $\text{dist}(F_n(\mathbf{X}), F_\vartheta(x))$ is minimal at $\hat{\vartheta}$ for some appropriate distance function dist, like the χ^2-distance. The resulting estimators typically do not satisfy the common optimality criteria, but are known to behave reasonably well in applications.

[324, 325] was the first to generalize this construction method to imprecise probabilities under repetition independence. He considers imprecise parametric statistical models $(\mathcal{X}, (\mathcal{P}_\vartheta)_{\vartheta \in \Theta})$ with a single-valued precise parameter ϑ, where it is assumed that for every ϑ the credal set $\mathcal{P}_\vartheta$ can already be described by upper or lower bounds on the probabilities of a finite number of events only. The resulting estimator $\hat{\vartheta}$ is then a minimizer of the distance between the empirical measure and the credal set $\mathcal{P}_\vartheta$, where the distance is measured in the total variation norm. Practically important, based on an appropriate discretization, $\hat{\vartheta}$ can be calculated by linear optimization.[74]

7.5.3.4 Asymptotics

Most current results on optimal robust procedures are of asymptotic nature, addressing the case of large samples. Rieder [535] has developed a general framework of asymptotic optimality for asymptotically linear estimators. If neighbourhood models are considered in this setting, it is typically assumed that their radius is shrinking with increasing sample size.[75] [402, 403] has elaborated and implemented Rieder's framework for the important case of shrinking neighbourhood models where the ideal central distribution belongs to an exponential family of full rank.

7.5.4 A brief outlook on frequentist methods

In the previous subsection, some basic aspects guiding the development of a frequentist estimation theory have already been briefly discussed. There is still a long way to go until frequentist estimation and testing theory fully exploits the richness of imprecise probability. Up to now most work is, explicitly or implicitly, strongly influenced by some robustness considerations, in particular in so far as exclusively Gamma-Maximin-type (see Chapter 8) criteria have been used, evaluating procedures only in worst case scenarios. Other criteria, like Hurwicz-based criteria, or even criteria in the spirit of E-admissibility or maximality producing incomplete orderings have not yet been considered. Up to now the independence concept underlying the i.i.d. samples has been confined to the strong independence-based notions (7.6) and (7.7); in particular epistemic irrelevance (see Section 3.2.1) deserves detailed investigation.

Working with neighbourhood models, the choice of the radius is often a crucial modelling issue. In hypothesis testing with neighbourhood models, a sensitivity analysis of the final test decision in dependence of the radius of the neighbourhood would provide valuable insights. In the estimation context it could also be of interest to work with a sort of least favourable radius (see [536]).

Speaking of frequentist methods, also a radically different approach has to be mentioned at least briefly. The so-called *chaotic models*, that are based on the direct frequentist understanding of imprecise probabilities developed by Fine with some students

[74] The estimator has been implemented in the statistical programming language **R**, and is publicly available as **R** package `imprProbEst`; see [322].

[75] Here it is important to recall once more that robust statistics strictly relies on the sensitivity analysis point of view (see Section 7.2.1). The main argument is then that, given i.i.d. replications following a single distribution in the credal set, goodness-of-fit tests could detect decreasing deviations with increasing sample size (see, e.g., [546]).

and colleagues [287–289, 529], provide a different perspective on the estimation of credal sets.[76]

7.6 Nonparametric predictive inference

7.6.1 Overview

Nonparametric predictive inference (NPI) is a statistical method based on Hill's assumption $A_{(n)}$ [361], which gives a direct conditional probability $P(\cdot)$ for a future observable random quantity X_{n+1}, based on observed values $x_1, \ldots, x_n$ of related random quantities $X_1, \ldots, X_n$, see, in particular, [36, 130]. Suppose that $X_1, \ldots, X_n, X_{n+1}$ are continuous and exchangeable random quantities. Let the ordered observed values of $X_1, \ldots, X_n$ be denoted by $x_{(1)} < x_{(2)} < \ldots < x_{(n)}$, and let $x_{(0)} = -\infty$ and $x_{(n+1)} = \infty$ for ease of notation. For a future observation X_{n+1}, based on n observations, the assumption $A_{(n)}$ [361] is

$$P(X_{n+1} \in (x_{(j-1)}, x_{(j)})) = \frac{1}{n+1} \text{ for } j = 1, 2, \ldots, n+1.$$

$A_{(n)}$ does not assume anything else, and is a post-data assumption related to exchangeability. The frequentist nature of $A_{(n)}$ can easily be understood by considering a simple case like $n = 2$, for which $A_{(2)}$ states that the third observation has equal probability to be the minimum, median or maximum of all three observations, no matter the values x_1 and x_2. For repeated applications in situations with exchangeable random quantities, this post-data assumption is clearly seen to hold with a frequency interpretation of probability. For one-off applications, such an inference can be considered reasonable if one has no information at all about the location of X_3 relative to x_1 and x_2, or if one explicitly does not wish to use any such information. $A_{(n)}$ is also closely related to simple random sampling, for this case with $n = 2$ it just implies that the minimum, mean and maximum of the three random quantities each have the same probability to be X_3. Hill [362] discusses $A_{(n)}$ in detail.

Inferences based on $A_{(n)}$ are predictive and nonparametric, and avoid the use of so-called counterfactuals, i.e. the reference to what else could have been observed, which play an important role in classical inferences like hypothesis testing and which is often considered a main disadvantage of frequentist statistical methods. $A_{(n)}$ is not sufficient to derive precise probabilities for many events of interest, but it provides optimal bounds for probabilities for all events of interest involving X_{n+1}. These bounds are lower and upper probabilities and are coherent [36]. NPI is a framework of statistical theory and methods that use these $A_{(n)}$-based lower and upper probabilities, and also considers several variations of $A_{(n)}$ which are suitable for different inferences. For example, NPI has been presented for Bernoulli data, multinomial and lifetime data with right-censored observations. NPI provides a solution to some explicit goals formulated for objective (Bayesian) inference, which cannot be obtained when using precise probabilities [130]. NPI is also exactly calibrated [425], which is a strong consistency property, and it never leads to results that are in conflict with inferences based on empirical probabilities.

[76] See in particular [287, § 6] for an illustrative application of the estimation of chaotic probability models and their simulation from exchange rate data.

NPI enables inference for multiple future observations [28], with their interdependence explicitly taken into account. If we are interested in $m \geq 1$ future observations X_{n+i}, for $i = 1, \ldots, m$, we can link the data and future observations by sequentially assuming $A_{(n)}, \ldots, A_{(n+m-1)}$. Let $S_j = \#\{X_{n+i} \in (x_{(j-1)}, x_{(j)}),\ i = 1, \ldots, m\}$, then inferences about these m future observations under these assumptions can be based on the following probabilities, for any $(s_1, \ldots, s_{n+1})$ with non-negative integers s_j with $\sum_{j=1}^{n+1} s_j = m$

$$P(\bigcap_{j=1}^{n+1}\{S_j = s_j\}) = \binom{n+m}{n}^{-1}$$

So, all possible orderings of the m future observations among the n data observations are equally likely, again this leads to lower and upper probabilities for many events of interest, and can be related to simple random sampling of the m observations out of $n + m$ which are not revealed as data, assuring the exact calibration of such inferences from frequentist statistics perspective. Problems where NPI for multiple future real-valued observations has been applied include predicting system reliability following component testing [144] and development of nonparametric control charts [27]. This also immediately enables inference for future order statistics, for example the minimum or maximum of the m future observations.

For data including right-censored observations, as often occur in lifetime data analysis, NPI is based on a variation of $A_{(n)}$ which effectively uses a similar exchangeability assumption for the future lifetime of a right-censored unit at its moment of censoring [140]. This method provides an attractive predictive alternative to the well-known Kaplan-Meier estimator for such data, with the lower survival function decreasing at times of both actual events and right-censorings, but the upper survival function only decreasing at an actual event time, reflecting that a right-censoring does not provide actual evidence against survival beyond that moment, yet reduces the evidence in favour of surviving. An interesting application of NPI for right-censored data has been presented for situations with multiple failure modes, also called 'competing risks' [454]. It is shown that the NPI lower and upper probabilities for the event that the next unit fails due to a specific failure mode depends on possible grouping of failure modes, with possibly less (but not more) imprecision if more failure modes are put together into a single group. This reflects that, if more failure modes are distinguished, there are more aspects of the problem about which one has limited information, leading possibly to more imprecision for some events of interest.

NPI for Bernoulli random quantities [129] is based on a latent variable representation of Bernoulli data as real-valued outcomes of an experiment in which there is a completely unknown threshold value, such that outcomes to one side of the threshold are successes and to the other side failures. The use of $A_{(n)}$ together with lower and upper probabilities enables inference without a prior distribution on the unobservable threshold value as is needed in Bayesian statistics, where this threshold value is typically represented by a parameter. Suppose that there is a sequence of $n + m$ exchangeable Bernoulli trials, each with 'success' and 'failure' as possible outcomes, and data consisting of s successes in n trials. Let Y_1^n denote the random number of successes in trials 1 to n, then a sufficient representation of the data for NPI is $Y_1^n = s$, due to the assumed exchangeability of all trials. Let Y_{n+1}^{n+m} denote the random number of successes in trials $n + 1$ to $n + m$. Let $R_t = \{r_1, \ldots, r_t\}$, with $1 \leq t \leq m + 1$ and $0 \leq r_1 < r_2 < \ldots < r_t \leq m$, and, for ease of notation, define $\binom{s + r_0}{s} = 0$. Then the NPI upper probability for the event $Y_{n+1}^{n+m} \in R_t$, given data $Y_1^n = s$, for $s \in \{0, \ldots, n\}$, is

$$\overline{P}(Y_{n+1}^{n+m} \in R_t | Y_1^n = s) = \binom{n+m}{n}^{-1} \sum_{j=1}^{t} \left[\binom{s+r_j}{s} - \binom{s+r_{j-1}}{s} \right] \binom{n-s+m-r_j}{n-s}$$

The corresponding NPI lower probability is derived via the conjugacy property

$$\underline{P}(Y_{n+1}^{n+m} \in R_t | Y_1^n = s) = 1 - \overline{P}(Y_{n+1}^{n+m} \in R_t^c | Y_1^n = s)$$

where $R_t^c = \{0, 1, \ldots, m\} \backslash R_t$.

For multinomial data, a latent variable representation via segments of a probability wheel has been presented, together with a corresponding adaptation of $A_{(n)}$ [134]. This provides an alternative to the popular imprecise Dirichlet model (IDM) by Walley [673],[77] resolving some perceived shortcomings of the latter as raised by Walley and discussants to his paper, also deliberately violating the representation invariance principle. While for applications with substantial data NPI and the IDM lead to very similar results, both of course also close to empirical probabilities, the NPI approach distinguishes between defined or undefined categories with zero observations, as can be important in reliability and risk analysis [132]. Typically, if outcome categories have not occurred yet, the NPI lower probability of the next observation falling in such a category is zero, but the corresponding NPI upper probability is positive and depends on whether or not the category is explicitly defined, on the total number of categories or whether this number is unknown, and on the number of categories observed so far. NPI enables inference if the number of categories is unknown, and the results can depend on the specific grouping of categories which may seem unnatural when considered from a precise probabilistic perspective, but there is no strong argument why imprecision should not be affected by different grouping of categories. Indeed, imprecision offers a way to reflect the different information contained in count data on such different groupings of categories. Study of the relation between imprecision and data representation is a major topic for research on foundations of statistical methods using imprecise probabilities.

7.6.2 Applications and challenges

Many applications of NPI to problems in statistics, operations research, and reliability and risk have been presented in the literature.[78] In statistics, NPI methods for comparisons of groups of real-valued data are attractive for situations where such comparisons are naturally formulated in terms of comparison of future observations from the different groups [127, 138]. For example, Table 7.1 gives birth weights (in grams) for 12 male and 12 female babies. In classical statistics, such data are often compared by testing a hypothesis, for example that both data sets are randomly drawn from the same underlying population. In NPI, the comparison is necessarily predictive, for example by comparing the random birth weights of a single further male (M_{13}) and a single further female baby (F_{13}), with the idea that these were exchangeable with the 12 observed birth weights in their respective groups, and assuming $A_{(12)}$ for each group. This leads to the NPI lower and upper probabilities $\underline{P}(M_{13} > F_{13}) = \frac{86}{169} = 0.509$ and $\overline{P}(M_{13} > F_{13}) = \frac{111}{169} = 0.657$. These values

[77] See Section 7.4.3.

[78] For applications in reliability and risk see Section 14.4.

Table 7.1 Birth weights data.

Male	2625	2628	2795	2847	2925	2968
	2975	3163	3176	3292	3421	3473
Female	2412	2539	2729	2754	2817	2875
	2935	2991	3126	3210	3231	3317

result by counting, of the 169 equally likely combinations of intervals created by the data per group, those for which $M_{13} > F_{13}$ is necessarily or possibly true, leading to the NPI lower and upper probability, respectively. In many applications, such inferences may be more naturally suited to the problems of actual interest than testing of hypotheses.

A nice example of the general advantage of the use of imprecise probabilities, namely that one does not need to add assumptions to data which one feels are not justified, occurs in precedence testing, where experiments to compare different groups may be terminated early in order to save costs or time [147]. In such cases, the NPI lower and upper probabilities are the sharpest bounds corresponding to all possible orderings of the not-fully observed data. There are several similarities between NPI and methods of robust statistics, for example NPI pairwise or multiple comparisons of groups are based on the ranks of the observations, and the influence of possible outliers can be reduced in NPI by terminating tails, which may be motivated similarly to tails truncation as used in robust methods [453]. NPI provides an attractive framework for decision support in a wide range of problems where the focus is naturally on a future observation. For example, NPI methods for replacement decisions of technical units are powerful and fully adaptive to process data [142].

NPI has been applied for comparisons of multiple groups of proportions data [136], where the number m of future observations per group plays an interesting role in the inferences. Effectively, if m increases the inferences tend to become more imprecise, while imprecision tends to decrease if the number of observations in the data set increases. NPI for Bernoulli data has also been implemented for system reliability, with particularly attractive algorithms for optimal redundancy allocation [144, 145]. Some further examples of NPI applications are presented in Chapter 14.

Development of NPI is gathering momentum; inferential problems for which NPI solutions have recently been presented or are being developed include aspects of medical diagnosis with the use of ROC curves, robust classification, inference on competing risks, quality control and acceptance sampling. Main research challenges for NPI include its generalization for multi-dimensional data, which is similarly challenging for NPI as for general nonparametric methods due to the lack of a unique natural ordering of the data. NPI theory and methods that enable information from covariates to be taken into account also provide interesting and challenging research opportunities.[79]

7.7 A brief sketch of some further approaches and aspects

Beyond the Bayesian and the frequentist paradigm, there is a variety of further approaches to statistical inference, see, e.g., [45]. At least three of them have had an inspiring influence on imprecise probability work and shall be discussed here briefly.

[79] Further information is available from www.npi-statistics.com.

The first to mention is the *likelihood* paradigm. Beyond its use as a technical tool in frequentist statistics, the notion of likelihood can also be understood as the basis of a whole inference theory (see, e.g., [257, 504]). [104] suggests to use a slightly extended notion of likelihood as relative plausibility to construct a hierarchical model. It consists of a credal set $\mathcal{P}$ and of the likelihood function lik:$\mathcal{P} \to \mathbb{R}_0^+$ as a second level, describing the plausibility of its elements based on the sample. Inference on certain deduced quantities $\vartheta = g(P), P \in \mathcal{P}$, then can be based on the so-called *profile likelihood*

$$\text{proflik}\,(\vartheta) = \sup_{\{P \in \mathcal{P} | g(P) = \vartheta\}} \text{lik}\,(P). \tag{7.32}$$

The approach has up to now not often been used, but shows, as already mentioned in Section 7.4.4, some promising behaviour as an updating rule, and as a basis for a direct likelihood based decision theory [103, 106][80]. Moreover, it has been successfully applied in graphical models (e.g., [20], see also Section 10.4.3), and in the context of data imprecision (see Section 7.8.4). Walley [675] uses another notion of normalized likelihood for inferences, using this as a showcase where frequentist properties and the likelihood principle can be reconciled.

Surely the most intensively debated paradigm of statistical inference is *fiducial inference* as proposed by R. A. Fisher [293, 294], who also had introduced the notion of likelihood. Summarizing it in a simplified way, fiducial inference is an attempt to deduce probability statements on the parameter space without relying on a prior distribution. Although this way of reasoning is widely judged as problematic and rarely applied in practice, the reflection and discussion of fiducial probabilities has inspired important work also in the area of imprecise probabilities. Several prominent authors have been arguing that the methodology of imprecise probabilities is the right framework for attempts to reformulate, save or extend some of Fisher's fundamental ideas. This in particular applies to Dempster's work in the 1960s (see, in particular, [222, 223])[81], but also includes for instance Hampel's work on the foundation of statistics [340, 341, 342].

Finally work on *logical* or *epistemiological probability* with a focus on statistical inference should be briefly mentioned. Both Levi and Kyburg (see, in particular, [430] and [418], as well as [78]) directly use imprecise probabilities as a basic entity underlying their reasoning. In his so-called *symmetrical theory*, Weichselberger [693, 694] rigorously understands interval-valued logical probability $P(A\|B)$ as the probability with which a proposition A (understood as conclusion) can be derived from a proposition B (understood as premise). He sets out to build a framework in which A and B can exchange their role ('symmetry'), and thus the interval-valued probability $P(B\|A)$ can be deduced. Specializing A to 'observations' and B to 'parameters', this directly yields a general framework for statistical inference, which also has some relation to fundamental ideas of fiducial probabilities.

7.8 Data imprecision, partial identification

In this section we discuss some basic aspects of data imprecision and then briefly introduce the notion of partial identification, which provides a general framework for handling observationally equivalent models, including data imprecision as a special case.

---

[80] See also Section 5.2.2.

[81] See also, e.g., [692, p. 70ff.] and [562, 563], trying to elaborate this relation more explicitly.

7.8.1 Data imprecision

7.8.1.1 The basic setting

Data imprecision is used as an umbrella term to denote all situations where at least some of the data are *coarse(ned)*, i.e. they are not observed in the resolution intended in the subject matter context. This means, there is a certain true precise value $x \in \mathcal{X}$ of a generic variable X of material interest, but we only observe a set $A \supseteq \{x\}$. A typical example in the case of continuous X is constituted by interval data of exact monthly income values. In the discrete setting coarsening occurs, for instance, when participants of an opinion poll are only asked about their preferences for certain coalitions of parties.[82] In technical terms, one would be materially interested in data in a space $\mathcal{X}$ ($\mathbb{R}^+$ for income values or $\{1, \ldots, k\}$ for k voting alternatives), but the observations can consist of nonsingleton-subsets of $\mathcal{X}$, i.e. they may be elements A of the power set of $\mathcal{X}$ with $|A| > 1$. An extreme special case of coarse data are *missing data*, where the missingness of value x_i of unit i can be interpreted as having observed the whole sample space $\mathcal{X}$.

Before turning to the formal representation, some issues have to be clarified. First of all, it must be stressed that the term 'coarse' is a relative term. Whether data are coarse or not depends on the specified sample space, reflecting the subject matter context to be investigated. If, for instance, the sample space is taken to consist of some a priori specified ranges for income data, and that is all what is needed, then data are not coarse, while if precise income values are of interest, the data are coarse. Indeed even unions of intervals may constitute precise observations, for instance as the response to the question 'When did you live in Munich?', measured in years. Then, e. g., $\{[1986; 1991] \cup [1997; 2000]\}$ is a precise observation in the sample space of all finite unions of closed intervals $[a, b]$ with $a, b \in \mathbb{N}$.[83]

Secondly, it is important to note that coarse data typically are not just the result of sloppy research, like an insufficient study design or improper data handling, and therefore could be avoided by careful planning of the study. On the contrary, they are an integral part of data collection, in particular in social surveys. Interval data arise naturally from the use of categories in order to avoid refusals in the case of sensitive questions, and are a means to model roughly rounded responses (e.g., [449]). Coarsened categorical data are, for instance, produced by matching data sets with not fully overlapping categories. They can be a direct result of data protection by some anonymization techniques for disclosure control (see, e.g. [235, 236]), or are arising naturally from the combination of spaces with given marginals by so-called Fréchet bounds (see, e.g., [417]). Another prototypic setting is the case of systematically missing data, arising from treatment evaluations in nonrandomized designs as in observational studies. To evaluate effects of treatment or intervention A compared to treatment B, in principle, it would be necessary to have information from a parallel universe, so-to-say, i.e. to know in addition how the units treated with A *would* have reacted *if* they had been given the treatment B and vice versa.[84]

---

[82] Compare also Example 4.1.

[83] See in particular the distinction between *conjunctive* and *disjunctive* random sets in [239, § 1.4], from which also this example is adopted.

[84] See also the references in Subsection 7.8.4. A typical, very instructive example is the case study in [615]. There the propensity of recidivism of young offenders depending on whether or not they had been assigned to residential facilities after their prior offence. Of course, judges will assign 'treatments' not randomly, and in this situation thus missingness of the counterfactual outcomes is most probably not ignorable.

7.8.1.2 Basic coarsening model

To formalize coarsened data, it is helpful to consider two layers, namely the 'ideal level', in which the 'true', but unobservable i.i.d. sample $\mathbf{X} = (X_1, \ldots, X_n)$ with values $\mathbf{x} = (x_1, \ldots, x_n)$ based on a classical statistical model $(\mathcal{X}, \mathcal{Q})$ exists, and an 'observation level', where the *coarsened sample* $\mathfrak{x} = (\mathfrak{x}_1, \ldots, \mathfrak{x}_n)$ with $\mathfrak{x}_i \subseteq \mathcal{X}, i = 1, \ldots, n$, is observed.[85] Throughout this section, $\mathfrak{x}_i, i = 1, \ldots, n$, is assumed to arise from x_i by an unknown coarsening process such that

$$x_i \in \mathfrak{x}_i, \quad \forall i = 1, \ldots, n.^{85} \tag{7.33}$$

The aim is inference about the true underlying probability distribution $P^* \in \mathcal{Q}$ based on the observed data $\mathfrak{x}$.

7.8.1.3 Coarse data in traditional statistics

Traditional approaches to cope with coarse data are forced to try to escape the imprecision in the data. An immediate way is to represent the set-valued observations $\mathfrak{x}_i, i = 1, \ldots, n$, by certain precise values $\mathring{x}_i \in \mathfrak{x}_i$ and then to proceed with a standard analysis based on the fictitious sample $\mathring{\mathbf{x}} = (\mathring{x}_1, \ldots, \mathring{x}_n)$. A typical example in that vein in the case of interval data $\mathfrak{x}_i = [\underline{x}_i, \overline{x}_i]$ is to represent the interval by its center $\mathring{x}_i = (\underline{x}_i + \overline{x}_i)/2$.

More sophisticated approaches add complex and typically untestable assumptions either to model explicitly the coarsening process by a precise model, or to characterize situations where the coarsening can be included in standard likelihood and Bayesian inference without biasing the analysis systematically. Most prominent is here Little and Rubin's [438] classification, distinguishing situations of *missingness completely at random (MCAR)* or *missing at random (MAR)* from *missing not at random (MNAR)* settings, where a systematic bias has to be expected, see also Sections 9.5.2 and 10.3.3. This classification has been extended to coarsening by [354].

A very popular approach, in particular for missing data, is *multiple imputation* (see, e.g., [544]), where for every missing observation several substituting values are drawn randomly, based on a model for the data and the missingness mechanism. Then separate standard analyses are performed and eventually merged together.

7.8.2 Cautious data completion

In recent years, awareness has grown that these traditional modelling approaches build on rather strong assumptions that are often hardly justifiable by substantive arguments, and thus the price for the seemingly precise result of the statistical analysis is a lack of credibility of the conclusions.[87] In the light of this, it is of particular interest to develop approaches

[85] By doing this, and naively ignoring aspects of measurability, we implicitly introduced set-valued random variables $\mathfrak{X}_i$ and a corresponding sample $\mathfrak{X} = (\mathfrak{X}_1, \ldots, \mathfrak{X}_n)$ underlying the observed sample $\mathfrak{x} = (\mathfrak{x}_1, \ldots, \mathfrak{x}_n)$; see, in particular, Section 5.2.1 and [61] for a proper description of coarse data as random sets, and also Section 7.8.3. Note that with set-valued variables, elements of the powerset of $\mathcal{X}$ now play a double-role, namely as events in terms of X_i and as atoms in terms of $\mathfrak{X}_i, i = 1, \ldots, n$.

[86] This means that we implicitly assume here the coarsening process to be error free. It always gives values $\mathfrak{x}_i$ in which the true value x_i is contained.

[87] See Manski's Law of Decreasing Credibility in Section 7.2.4.

that reflect the underlying imprecision properly by considering all possible data on the ideal level compatible with the information on the observation level. Related approaches in this vein have been developed almost independently in different settings, ranging from reliable computing and interval analysis in engineering (e.g., [493]) and extensions of generalized Bayesian inference [215, 721] applied in graphical models and classification (see also Sections 9.5.2 and 10.3.3) to reliable descriptive statistics in social sciences ([539, § 17f., 517]). Moreover, the methodology is closely related to the concept of partial identification, in particular propagated by Manski (e.g., [447]) in econometrics, and to systematic sensitivity analysis (e.g. [654], [539, § 17f., 517]) in biometrics (see the next subsection for more general settings in that framework).

To grasp the common basic idea technically, it is again helpful to take rigorously the point of view that statistical procedures can simply be described by random quantities summarizing the random sample. This not only applies to estimators and tests, but under certain regularity conditions also to (generalized) Bayesian inference via the posterior distribution $\pi(\vartheta|\mathbf{X})$, which indeed may be understood as a function-valued random quantity before the sample $\mathbf{X}$ is realized. In the light of this, a quite natural way to cope with coarse data is to select a standard statistical procedure $T(\mathbf{X})$ and then to take the lower and upper envelope over all possible data completions, i.e. over all potential samples $\mathbf{x}$ on the ideal level compatible with the concrete data $\mathfrak{x}$ on the observation level, i.e. satisfying (7.33). We will use the term *cautious data completion* for this way to proceed.

Definition 7.1 (Cautious data completion). Consider a classical statistical model $(\mathcal{X}, \mathcal{Q})$, and let $T(\mathbf{X})$ be a statistical procedure that would be applied if a corresponding i.i.d. sample $\mathbf{X} = (X_1, \ldots, X_n)$ was available. Given a coarse sample $\mathfrak{x} = (\mathfrak{x}_1, \ldots, \mathfrak{x}_n)$,

$$T(\mathfrak{x}) := \{T(\mathbf{x})|\mathbf{x} \in C(\mathfrak{x})\}$$

with

$$C(\mathfrak{x}) := \{\mathbf{x}|\mathbf{x} = (x_1, \ldots, x_n), x_i \in \mathfrak{x}_i, i = 1, \ldots, n\}$$

will be called *cautious data completion of* $T(\mathbf{X})$ *based on* $\mathfrak{x}$.

Indeed, $C(\mathfrak{x})$ is the set of all possible values compatible with the observed data. This application of the evaluation procedure $T(\mathbf{x})$ to all elements of $C(\mathfrak{x})$ can be seen as an optimal nonparametric procedure. It is the best one can learn without adding further assumptions on the coarsening mechanism. In the case of interval data, continuity of T guarantees that $T(\mathfrak{x})$ is connected, and thus – assuming one-dimensional interval data – leads to an interval-valued estimator. In the more general situations where $T(\mathfrak{x})$ is not connected, it is often reasonable to consider a connected superset or to pass over to the convex hull: $\tilde{T}(\mathfrak{x}) = \text{conv}(T(\mathfrak{x}))$.

Obviously, cautious data completion reflects the imprecision in a natural way. If one imagines two different levels of coarsening of a sample $\mathbf{x} = (x_1, \ldots, x_n)$ into $\mathfrak{x}_\ell = (\mathfrak{x}_{\ell 1}, \ldots, \mathfrak{x}_{\ell n}), \ell = 1, 2,$ with $\mathfrak{x}_{1i} \subseteq \mathfrak{x}_{2i}, i = 1, \ldots, n,$ then $T(\mathfrak{x}_1) \subseteq T(\mathfrak{x}_2)$. Moreover, for sequences $(\mathfrak{x}^{(j)})_{j\in\mathbb{N}}$ with ever decreasing imprecision, i.e. $\mathfrak{x}^{(j)} \supseteq \mathfrak{x}^{(j+1)}$, with $\cap_{j=1}^{\infty} \mathfrak{x}^{(j)} = \{x\}$, the sequence $T(\mathfrak{x}^{(j)})_{j\in\mathbb{N}}$ tends to $T(\mathbf{x})$.

Example 7.6 (Cautious data completion). As an illustrative example, consider a sample $\mathfrak{x}$ of interval data $\mathfrak{x}_1, \ldots, \mathfrak{x}_n$ with

$$\mathfrak{x}_i = [\underline{x}_i, \overline{x}_i], \quad i = 1, \ldots, n, \quad \underline{x}_i \le \overline{x}_i \in \mathbb{R}$$

from a generic variable X with unknown distribution p^* and existing expectation μ and variance σ^2. Thinking of an underlying ideal sample $\mathbf{X} = (X_1, \ldots, X_n)$ and assuming no outliers to be present, the most common estimator for μ is the arithmetic mean: $T(\mathbf{X}) = \frac{1}{n}\sum_{i=1}^{n} X_i$. Cautious data completion of $T(\mathbf{X})$ based on $\mathfrak{x}$ then leads to

$$T(\mathfrak{x}) = \{T(\mathbf{x})|\mathbf{x} \in C(\mathfrak{x})\} = \left[\frac{1}{n}\sum_{i=1}^{n} \underline{x}_i, \frac{1}{n}\sum_{i=1}^{n} \overline{x}_i\right].$$

This simple calculation of cautious data completion by just plugging in lower and upper interval limits is more generally successful for procedures $T(\mathbf{X})$ that are monotone in $\mathbf{X}$ (understood component-wise)[88], but cannot be extended naively to other situations. Already for the standard estimator of the variance $S^2 = \frac{1}{n}\sum_{i=1}^{n}\left(X_i - \frac{1}{n}\sum_{i=1}^{n} X_i\right)^2 = \frac{1}{n}\sum_{i=1}^{n} X_i^2 - \left(\frac{1}{n}\sum_{i=1}^{n} X_i\right)^2$ such elementary arithmetics based on the interval limits only is no longer valid. The simplest way to illustrate this is to consider the case where $\cap_{i=1}^{n} \mathfrak{x}_i \neq \emptyset$, and thus[89]

$$\min_{\{x \in C(\mathfrak{x})\}} S^2(\mathbf{x}) = 0. \qquad \blacklozenge$$

In the discrete case applying cautious data completion to relative frequencies of the events can be shown to correspond directly to constructing Dempster-Shafer structures from relative frequencies as described in Section 5.2.1. Thus, cautious data completion can be understood as a natural generalization of the Dempster-Shafer like construction of empirical distributions from coarse discrete data.

7.8.3 Partial identification and observationally equivalent models

7.8.3.1 (Partial) Identifiability, identification regions

In the previous subsection, methods for cautious handling of data imprecision were discussed. They lead immediately to imprecise models, being serious about the fact that under data imprecision even the information of infinitely many observations may not allow for the reliable selection of a single distribution; there are several distributions 'corresponding to the data'. Indeed, the presence of such *observationally equivalent models* is a frequent general problem, where data imprecision is just one, yet prominent, instance. Mainly in econometrics, concepts of *partial identification*[90] (see, e.g., [447], and Section 7.8.4 for selected further references) are gathering strong momentum, providing a general framework for handling such situations.

The basic concept to prepare a somewhat more detailed discussion of this concept is the notion of identifiability. A precise statistical model is called *identifiable*[91] from the data if it is always possible to learn the true underlying distribution from an infinitely large amount of

[88] An important class of models is the class of exponential family-based models as intensively discussed in Section 7.4, see [640, 641, 645] for an investigation of the IDM under coarse observations, cf. also Page 311.

[89] Calculating $\max_{\{x \in C(\mathfrak{x})\}} S^2(\mathbf{x})$ is much more complex, see, e.g., [410] and [616].

[90] A closely related development has taken place in biometrics, typically called *systematic sensitivity analysis*, see, e.g., [654].

[91] See, e.g., [320, p. 645].

data. In the case of a classical parametric statistical model $(\mathcal{X}, (P_\vartheta)_{\vartheta\in\Theta})$, this means that any two distinct values $\vartheta_1 \in \Theta$ and $\vartheta_2 \in \Theta$ lead to different distributions of the corresponding generic random variable X, and thus an infinitely large i.i.d. sample is able to discriminate between the different parameter values.

The issue of identifiability is fundamental whenever a statistical model is applied to data. In traditional statistics, identifiability is often presupposed and always perceived as a yes/no question. Either there is identifiability, and thus traditional statistical procedures can be applied, or not. In the latter case, statistical analysis is understood as impossible, unless further assumptions are added until identifiability is eventually enforced; these assumptions come often, however, at the price of loosing credibility of subject matter conclusions (see Section 7.8.1 above).

In contrast, *partial identification* understands identifiability as a continuum. The models that are observationally equivalent given the data form a set of models or parameters, called *identification set* or *identification region*, respectively. If this set is a singleton, then *point identification* is given, i.e. identification in the traditional sense. Typically, this set contains more than one element, but is still a proper subset of the set of all models. Then the identification is *partial* in the literal meaning. The fascinating insight, corroborated by a variety of applications mainly in econometrics (see also Section 7.8.4 for some references), is that in many studies this still is enough to answer important substantive science questions, and if not, the scientist is alerted that strong conclusions drawn from the data may be mere artefacts.

The methodology of partial identification can be understood as a general framework for the statistical analysis of complex observations, where it is necessary to distinguish between an observational layer with observable generic variable $\mathfrak{X}$ and a latent layer with a, partially or completely unobservable, generic variable X. This includes missingness or coarsening as discussed above, but also covers error-prone observations, like misclassification of discrete variables[92] or measurement error of continuous variables.

Again the sensitivity analysis point of view on imprecision is taken, assuming that there is a true underlying precise distribution. Neglecting issues of sampling variability for a moment, the (possibly imprecise) distribution $P_{\mathfrak{X}}$ of $\mathfrak{X}$ can eventually be learned from an i.i.d. sample and can be taken as given for the reasoning. For inference about the distribution P_X of X, an additional model, called *observation model* here, has to be formulated. It specifies the relationship between $\mathfrak{X}$ and X, in terms of the typically imprecise distributions of (some components of) $\mathfrak{X}$ given X or of X given $\mathfrak{X}$, or the joint distribution of $(X^\mathsf{T}, \mathfrak{X}^\mathsf{T})^\mathsf{T}$. Solving the system arising from $P_{\mathfrak{X}}$, the observation model, and the unknown probability P_X with respect to P_X, and taking possibly further, for instance parametric, assumptions on P_X into account, yields the identification region for P_X. Concrete estimators are then obtained by replacing $P_{\mathfrak{X}}$ by appropriate estimators $\widehat{P_{\mathfrak{X}}}$, and then sample variability is taken additionally into account by appropriate confidence or credibility intervals.[93]

The naturally arising imprecision in the identification region is essentially influenced by assumptions made in specifying the observation model, and thus the framework allows also the systematic study of the consequences of different assumptions, including a sensitivity analysis of deviations from certain point identifying assumptions.

---

[92] See also Section 7.8.3.2.

[93] See Section 7.8.4 for references.

As an illustration, consider coarse discrete data. Let X be a generic variable taking values in $\mathcal{X} = \{1, \ldots, k\}$. Now let only coarse observations as a sample $\mathfrak{X}$ of the variable $\mathfrak{X}$ be available. This allows to identify $p(\mathfrak{X} = \mathfrak{x}), \mathfrak{x} \subseteq \{1, \ldots, k\}$. The relationship between X and $\mathfrak{X}$ is described by an observation model $(P(X \in A|\mathfrak{X} = \mathfrak{x}))_{A, \mathfrak{x} \subseteq \{1, \ldots, k\}}$. As again error-freeness of the coarsening is assumed, for all A, the conditional probability $P(X \in A|\mathfrak{X} = \mathfrak{x})$ equals 1 if $A \supseteq \mathfrak{x}$, and 0 if $A \cap \mathfrak{x} = \emptyset$. Then, by the elementary theorem of complete probability, for any event $A \subseteq \{1, \ldots, k\}$,

$$
\begin{aligned}
P(X \in A) &= \sum_{\mathfrak{x}:\mathfrak{x} \subseteq \mathcal{X}} P(X \in A|\mathfrak{X} = \mathfrak{x}) \cdot P(\mathfrak{X} = \mathfrak{x}) \\
&= \sum_{\mathfrak{x}:\mathfrak{x} \subseteq A} 1 \cdot P(\mathfrak{X} = \mathfrak{x}) + \sum_{\mathfrak{x}:\mathfrak{x} \cap A = \emptyset} 0 \cdot P(\mathfrak{X} = \mathfrak{x}) + \sum_{\mathfrak{x} \in \mathfrak{M}_A} P(X \in A|\mathfrak{X} = \mathfrak{x}) \cdot P(\mathfrak{X} = \mathfrak{x}) \\
&= \sum_{\mathfrak{x}:\mathfrak{x} \subseteq A} p(\mathfrak{X} = \mathfrak{x}) + \sum_{\mathfrak{x} \in \mathfrak{M}_A} \left(\sum_{x \in \mathfrak{x} \cap A} P(X = x|\mathfrak{X} = \mathfrak{x}) \cdot P(\mathfrak{X} = \mathfrak{x}) \right),
\end{aligned}
$$

with $\mathfrak{M}_A := \{\mathfrak{x} | \mathfrak{x} \cap A \neq \emptyset \text{ and } \mathfrak{x} \cap A^C \neq \emptyset\}$. If no additional assumptions are made, for all $\mathfrak{x}$, the conditional probability masses $P(X = x|\mathfrak{X} = \mathfrak{x})$ can vary freely in the probability simplex, resulting, with $\mathfrak{M}_A \cup \{\mathfrak{x} : \mathfrak{x} \subseteq A\} = \{\mathfrak{x} : \mathfrak{x} \cap A \neq \emptyset\}$, in

$$
P(X \in A) \in \left[\sum_{\mathfrak{x}:\mathfrak{x} \subseteq A} P(\mathfrak{X} = \mathfrak{x});\ \sum_{\mathfrak{x}:\mathfrak{x} \cap A \neq \emptyset} P(\mathfrak{X} = \mathfrak{x}) \right]. \tag{7.34}
$$

Estimating these probabilities by relative frequencies gives again the well-known Dempster-Shafer-like structure and is equivalent to cautious data completion as discussed in Section 7.8.2, and thus to the construction in Section 5.2.1.

7.8.3.2 Partial identification and misclassification

The simplest situation to illustrate the methodology of partial identification in the context of error-ridden data is the case of misclassification of a binary variable X taking the values 0 and 1 only, where $P(X = 1)$ has to be estimated from potentially misclassified data $\mathfrak{X} \in \{0, 1\}$. The problem is omnipresent in, but by far not limited to, biometrics, where typically X is the true disease status with *prevalence* $P(X = 1)$, and $\mathfrak{X}$ describes the outcome of some diagnostic test. The observation model is fully specified by the probabilities $P(\mathfrak{X} = 1|X = 1)$ and $P(\mathfrak{X} = 0|X = 0)$ of correct classification, called *sensitivity* and *specificity* in the biometric context. By elementary probability calculus,

$$
\begin{aligned}
P(\mathfrak{X} = 1) &= P(\mathfrak{X} = 1|X = 1) \cdot P(X = 1) + P(\mathfrak{X} = 1|X = 0) \cdot P(X = 0) \\
&= P(\mathfrak{X} = 1|X = 1) \cdot P(X = 1) + (1 - P(\mathfrak{X} = 0|X = 0)) \cdot (1 - P(X = 1)).
\end{aligned} \tag{7.35}
$$

Solving for $P(X = 1)$, and respecting the trivial bounds $0 \leq P(X = 1) \leq 1$, yields directly a corrected estimator of the true prevalence.

However, exact knowledge of $P(\mathfrak{X} = 1|X = 1)$ and $P(\mathfrak{X} = 0|X = 0)$ is quite rare, and validation data sets allowing their reliable estimation are typically not available. Partial identification provides a welcome alternative to process the background information on the measurement process at hand in a reliable way. [477] discusses direct correction based on (7.35) for interval-valued misclassification probabilities, and its generalization for the case of more than two categories. [411] use the so-called *kappa coefficient* of inter-rater agreement based on two measurements to derive identification regions, with an illustrative example of estimating the prevalence of caries among children.

7.8.4 A brief outlook on some further aspects

7.8.4.1 Partial identification and other concepts of imprecise probabilities

It is challenging to elaborate precisely the relationship between the work of the protagonists of partial identification and the most commonly applied other imprecise probability methods. As a first answer the following aspects may be mentioned. Both areas have been developing rather isolated from each other, and cross-references are quite rare (see, however, [616], for one of the exceptions), but both approaches clearly share the appreciation of the expressive power of imprecision, supporting reliability of conclusions.

From the technical point of view, in the case of data imprecision there is the close relationship via the concept of random sets (see, in particular, [61], and also Equation (7.34)). Moreover, as Manski [447, p.3] emphasises, '... an enormous amount about identification can be learned from judicious application of the Law of Total Probability and Bayes Theorem'. Such construction can be seen as being closely related to formulating probabilistic constraints on conditional previsions or desirability statements and applying natural extension, as discussed in Chapters 1 and 2. The exact concrete mutual relationships, however, have still to be worked out. This also applies to the role of additional assumptions, which may be interpreted as structural judgements (see Chapter 3 and [672, § 9]). As a main fundamental difference, at least up to now, the term 'partial identification' is understood to suggest the existence of an underlying not-fully known and only partially characterized *precise* probability. Thus, partial identification-based approaches up to now exclusively rely on a strict sensitivity analysis point of view of imprecision, assuming a precise probabilistic model in the background.[94] Further differences include the clear focus on econometric models as main applications of partial identification, including a much more prominent role of regression models, and, at least up to now, a clear dominance of methods based on the frequentist paradigm.

7.8.4.2 A brief outlook on further developments in the literature

The discussion in the previous subsection gives some flavour of the power of partial identification in statistical models under complex data structures. Current research in that area goes beyond the situation considered so far mainly in two important directions, often considered simultaneously. First, of course, sound methods are needed to incorporate sampling uncertainty in addition by constructing appropriate inference procedures (e.g., confidence regions) for the partially identified parameter and the identification region.[95]

[94] See, in particular, [672, § 9.7.5], on the partial incompatibility of some imprecise probability-based structural judgements and sensitivity analysis based on structural judgements for precise probabilities.

[95] In the light of this, the methods described so far are mostly descriptive/empirical, as they would lead to the paradoxical situation that, under a constant proportion of coarse to precise data, overall imprecision of the conclusions would increase with more data.

Secondly, typically regression models are considered, modelling the relationship between explanatory and dependent variables (cf. also the remarks).

The basic ideas behind the technical constructions of inference procedures in the context of partial identification is explained in [620, § 4]. [725] and [107] show that coarse data can be powerfully and reliably handled in the profile likelihood context, the later work also explicitly develops a robust regression method in this framework. Further insights are also provided by recent discussions of observational data from a Bayesian point of view (see, in particular, [321] and the references therein).

In the regression context the additional question arises whether the underlying regression relation is understood as the true structure governing the data process or whether it is just taken as an approximative, simplified description ('model') of the data. While in the traditional statistical setting both views are not distinguishable in the results, the analysis may differ substantially when data imprecision is taken into account properly. The first point of view is usually technically framed by so-called moment inequality models (see, for example, [16, 91, 95, 435] for recent developments), while the later one indeed leads to corresponding cautious data completion procedures of regression analysis (e.g., [516, 564]).

Selected practical applications of partial identification include the study by [615] (see also Footnote 84), [426] investigating the German reform of unemployment compensation based on register data, [478] for an analysis of treatment effects in observational studies, illustrated by a longitudinal youth survey, and [590] reanalysis of a randomized social experiment with imperfect compliance.

7.9 Some general further reading

Due to the heterogeneity of the discussed approaches, references on specific further topics have directly been included in the corresponding Sections 7.4.4, 7.5.4, 7.6.2, and 7.8.4, and thus the present section mainly guides the reader briefly to material on the background and to relevant surveys of general aspects.

A wide-range introduction to statistics in general is, for instance, [687]. Many issues of regression modelling are presented, among others, in [269], while [351] is an influential book on the relationship between statistics and machine learning. Classical monographs on the notion of probability in statistics include [290] and, focusing on subjective probabilities, [421], for more recent work see also, e.g., [314]. The monographs [45] and [552] discuss different inference schools from a comparative point of view. An important basis for Bayesian methodology is [70]; classical work on frequentist estimation theory and hypothesis testing includes [428], and [429], respectively. The first monographs on traditional robust statistics are the first edition of [369] and [344]; see, e.g., [451] for a current introduction to the field, and [613] for a discussion of the historical development of this area.

The general theory of coherent inference based on the generalized Bayes rule has been founded by Walley [672, §§ 7–9], the state-of-the-art progress in the related concept of robust Bayesian analysis is surveyed by [63] and [548]. Some aspects of the relation between frequentist inference with imprecise probabilities and robust statistics are discussed in [37], and [493] summarizes current results in reliable computing of statistical measures, providing a technical basic for treating data imprecision.

General introductory surveys on partial identification include the books [447, 448], and the review by [620]. The latter also shows that related ideas have quite a long history in econometrics. The history of nonadditive probabilities in statistics is traced in [343].

7.10 Some general challenges

Brief specific outlooks have already been given at the end of the different sections, devoted to the various inference paradigms. Thus this section is confined to some general aspects.

The almost exclusive focus on independence concepts relying on the sensitivity analysis point of view, as discussed in the frequentist context at the end of Section 7.5, is a strong restriction in general. A careful study of other independence concepts in the context of imprecise sampling models is an important topic for further research, since the specific independence concept may have a strong impact on the statistical procedure to be chosen. Other issues to be emphasized again in the general context are a thorough investigation of the subtle relation between data representation and inference (see the brief discussion at the end of Section 7.6.1) and the unexploited potential generalized sampling designs offer (see the comment at the end of Section 7.2.5). As the natural generalization of the typically assumed simple random sampling (SRS), imprecise selection probabilities would be able to model not fully established symmetry between units. Moreover, they have great prospects in the context of not completely assignable probabilities in complex multi-stage designs.

A major reason why imprecise probability approaches are often perceived with reservation by applied statisticians is most probably the rather limited role regression models play up to now in statistical methodology based on imprecise probabilities. While many applied statisticians understand the different facets of regression modeling as the core part of statistical analysis, explicit imprecise probability based work on regression models is mostly restricted to data imprecision or other situations of partial identification. Application of generalized Bayesian methods and intrinsic imprecise probability based frequentist methods to regression models is rather rare up to now, and mostly also limited to comparatively simple models (e.g., [686]), see however recent work by [48] and [643] providing important first steps towards the handling of generalized linear models and support vector machines, respectively. Not only in this context, the development of an appropriate framework for simulating imprecise probability models is a crucial but demanding task, given the prominent role simulations play in current statistical methodology, in order to study the properties of advanced procedures in complex models, and also as a computational device. First attempts may utilize the one-to-one correspondence of belief-functions with traditional probabilities on the power set, or the lower envelope theorem (Proposition 2.3). A direct, genuine imprecise probability based approach could build on the genuine frequentist imprecise probability approach by Fine and co-authors ([287–289, 529]).

From a more general perspective, one could argue that imprecise probabilities in statistics have focused almost exclusively on extending traditional statistical models by taking a so-to-say set-based perspective. The shift to imprecise probability typically means then to introduce an additional layer, in order to take imprecision into account. This way of thinking makes imprecise probability-based methods necessarily at least as complex as traditional procedures, and indeed often the high computational effort needed makes a successful application of imprecise probabilities in large-scale applications cumbersome. The additional development of so-to-say 'direct methods' that use imprecise probabilities immediately at the core of the subject matter problem, without the detouring reference to a set of traditional models, would make a further substantial contribution to a sophisticated statistical analysis under complex uncertainty, but research in that direction is still in its infancy.

Acknowledgements

We are very grateful to Erik Quaeghebeur for inspiring discussions about the background of Section 7.4.

8

Decision making

Nathan Huntley[1], **Robert Hable**[2], **and Matthias C. M. Troffaes**[1]
[1]*Department of Mathematical Sciences, Durham University, UK*
[2]*Department of Mathematics, University of Bayreuth, Germany*

8.1 Non-sequential decision problems

In this section, we consider a single choice under uncertainty: the subject must choose one of a number of possible actions, each of which leads to an uncertain reward. The reward depends on the act and the true state of nature. How does a subject use her beliefs about the state of nature and her relative preferences over rewards to choose her act?

We use the following terminology. An *act* is something the subject can choose to do. A *choice* or *decision* refers to the choosing of a single act from a set. A *reward* is whatever the subject can receive once the state of nature is revealed.

Example 8.1 Suppose we can bet £1 that a six-sided die will score 4 or higher. The possible acts are {bet, do not bet}. The possibility space consists of the scores shown on the die: $\mathcal{X} = \{1, 2, 3, 4, 5, 6\}$. Three rewards are possible: losing £1, gaining £1, and receiving nothing – so rewards can also include negative results and also the status quo.

We want to employ coherent lower previsions (Section 2.2) for decision making. To do so, we need to express the decision problem in terms of gambles. Suppose the subject can choose among a set of possible acts D. For a particular act $d \in D$, if $x \in \mathcal{X}$ is the true state of nature, then the subject receives a reward $R(d, x)$. The reward could be anything; however, following the previous chapters (see for instance Section 1.1, Page 2 in particular), we assume that it is a real number corresponding to the subject's utility $U(d, x)$ for that reward. Note that decision theory literature often expresses rewards as losses $L(d, x)$ instead

Introduction to Imprecise Probabilities, First Edition.
Edited by Thomas Augustin, Frank P. A. Coolen, Gert de Cooman and Matthias C. M. Troffaes.

of utility $U(d,x)$, but the two are equivalent through $L(d,x) = -U(d,x)$. We use utility for consistency with the previous chapters.

As a simple example, suppose we are interested in a parameter X, which can take any value in $\mathcal{X}$. We must choose a point estimate $\hat{x}$. In this case, $\hat{x}$ is an act, and our set of acts is $D = \mathcal{X}$. A typical reward in such circumstances is quadratic loss: if x is the true value of the parameter and $\hat{x}$ is our estimate (act), then the reward is $U(x,\hat{x}) = -(x-\hat{x})^2$.

Whence, each act defines a gamble. Indeed, for a particular act d, if the true state of nature is x, then the subject receives $U(d,x)$. So, choosing d corresponds to receiving the gamble f_d defined by $f_d(x) = U(d,x)$. For example, in the dice game described earlier, the act of betting corresponds to the gamble

$$f_{\text{bet}}(x) = \begin{cases} -1 & \text{if } x \in \{1,2,3\}, \\ 1 & \text{if } x \in \{4,5,6\}. \end{cases}$$

In the estimation example, the act $\hat{x}$ corresponds to the gamble $f_{\hat{x}}(x) = -(x-\hat{x})^2$. ♦

Clearly, a subject should take into account her beliefs about the true state of nature. If she can express her beliefs through a probability measure, or equivalently, through a linear prevision P (see Section 2.2.2), then a common solution is for her to choose the act d that maximizes her expected utility $P(f_d)$. More generally, as argued in previous chapters, when little information is available, it may be more reasonable for the subject to express her beliefs through a more general uncertainty model. In this chapter we consider beliefs modelled by a coherent lower prevision (see Section 2.2).

In Section 8.1.1, we examine the decision problem as a choice between gambles. Section 8.1.2 then studies how we can choose between gambles using a coherent lower prevision.

8.1.1 Choosing from a set of gambles

Ideally, given a set of gambles where each gamble corresponds to an act, one wants to use one's knowledge to select a single optimal choice (or at least a set of equivalent optimal choices), for instance by maximizing expected utility. In some cases – for example, when betting on dice – this may be reasonable. However, as has been argued, there are occasions when our information cannot be represented through a linear prevision, but instead by a more general uncertainty model, say a coherent lower prevision. In such circumstances, is it possible to identify a best act in any choice?

For example, suppose that one must choose between two gambles f and g, where $\underline{P}(f) = 0$, $\overline{P}(f) = 3$, $\underline{P}(g) = 1$, and $\overline{P}(g) = 2$. Can there really be any well-motivated means of determining which should be chosen? Those more comfortable interpreting a lower prevision through the set of dominating linear previsions $\mathcal{M}(\underline{P})$ (see Equation (2.5)) can view the same question thus: if for some $P \in \mathcal{M}$, f maximizes expected utility, but for a different $Q \in \mathcal{M}$, g maximizes expected utility, can we claim one should be preferred to the other?

Dealing with such situations requires a language for discussing preference that is more general than, say, a weak order[1] of gambles. For this purpose, we use the concept of a *choice function*. Although we may not be able to identify our preferred gamble from a particular set, we can at least imagine being able to specify gambles that we *do not* want

[1] We use standard terminology from the decision making literature, and take a weak order to be a total preorder.

to choose. For some sets of gambles this may be nothing, for other sets this may be all but one gamble, and so on. The *optimal set* of gambles is what remains after all unacceptable gambles are eliminated. The interpretation of this set is that we have no further means to eliminate gambles, and we will just have to pick one of the remaining ones.

Definition 8.1 A choice function C is a function that maps sets of gambles to non-empty subsets: for any non-empty set of gambles $\mathcal{K}$,

$$\emptyset \neq C(\mathcal{K}) \subseteq \mathcal{K}.$$

Observe that weak orders, such as maximizing expected utility, are also special cases of choice functions (ones that usually return a single gamble), and so we lose no generality by adopting this interpretation.

In Section 2.3, upper and lower previsions conditional on an event A were introduced. One can do the same with choice functions: $C(\mathcal{K}|A)$ can be defined just as above, and given the same called-off interpretation. In the next section, when we look at choice functions for a coherent lower prevision $\underline{P}$, we give the definitions for the unconditional case. The corresponding conditional choice functions will simply use $\underline{P}(\cdot|A)$ instead of $\underline{P}(\cdot)$.

Beware that choice functions only work under *act-state independence*. If the true state of nature depends on the act – for instance, if for each act, we have a different coherent lower prevision – then it is not possible to compare the gambles using a choice function. This situation is called act-state dependence, and poses significant difficulties in decision making. Although it is usually possible to extend the possibility space $\mathcal{X}$ to convert the problem into an act-state independent one, such a transformation can make $\mathcal{X}$ extremely large, complicated, and arguably unintuitive. Problems of act-state dependence are outside the scope of this chapter.

Choice functions provide a means of analyzing simple decision problems without requiring a weak order of gambles. This is particularly useful in the context of coherent lower previsions, since modelling indecision is a principal motivation for their use. In the following section, we examine the most popular choice functions for coherent lower previsions, and discuss their advantages, drawbacks and relationships.

8.1.2 Choice functions for coherent lower previsions

Throughout this section, $\underline{P}$ is a coherent lower prevision, and $\mathcal{M}$ is its corresponding credal set (see Section 1.6.2 and Equation (2.5)). We must choose from the set of gambles $\mathcal{K}$, conditional on some event A – for simplicity, we assume $A = \mathcal{X}$, but the general case follows trivially by conditioning everywhere.

Although we argued that choice functions should generally be less determinate than a weak order, there are nonetheless choice functions corresponding to weak orders that are quite popular. Reasons for using these include:

- a desire for determinacy, because a decision theory telling us what to do in all situations is convenient, and
- computational ease, because weak orders are the easiest to compute of all in this section.

A very popular one is Γ-maximin, which maximizes the lower prevision of gambles. This is a conservative, risk-averse choice, as can also be seen from a credal set interpretation: whatever gamble f we choose, assume that the true $P \in \mathcal{M}$ is the worst case one, that is, the one that minimizes $P(f)$. Under this assumption, we choose the gamble for which its minimum is greatest, that is, we employ Γ-maximin.

Definition 8.2 (Γ-maximin). The choice function corresponding to maximizing the lower prevision is called Γ-maximin, and is defined by

$$C_{\underline{P}}(\mathcal{K}) := \{f \in \mathcal{K} : (\forall g \in \mathcal{K})(\underline{P}(f) \geq \underline{P}(g))\}.$$

Γ-maximin has roots in *robust Bayesian statistics* (examined later in this chapter), and is discussed by Berger [62, § 4.7.6]. Γ-maximin has been criticized for being too conservative. The related choice function Γ-maximax is similar to Γ-maximin, but maximizes the upper expectation. Of course, Γ-maximax can be criticized for being too bold, and often conservatism is considered more appealing than boldness. This is analogous to the discussion of the classical maximin and maximax criteria.

Definition 8.3 (Γ-maximax). The choice function corresponding to maximizing upper prevision is called Γ-maximax, and is defined by

$$C_{\overline{P}}(\mathcal{K}) := \{f \in \mathcal{K} : (\forall g \in \mathcal{K})(\overline{P}(f) \geq \overline{P}(g))\}.$$

In the case of complete ignorance, a choice function commonly called the Hurwicz criterion involves maximizing a combination of the best and worst possible outcomes [26]. A similar approach is possible with coherent lower previsions, creating a choice function also usually referred to as Hurwicz [379].

Definition 8.4 (Hurwicz). The choice function corresponding to maximizing a combination of upper and lower prevision is called Hurwicz, and is defined by

$$C_{\beta}(\mathcal{K}) := \{f \in \mathcal{K} : (\forall g \in \mathcal{K})(\beta\underline{P}(f) + (1-\beta)\overline{P}(f) \geq \beta\underline{P}(g) + (1-\beta)\overline{P}(g))\}.$$

for some $\beta \in [0, 1]$.

These three choice functions do not reflect the indecision that coherent lower previsions represent. Consider a gamble f for which we have assessed $\underline{P}(f) = 0$ and $\overline{P}(f) = 1$. Recall the interpretation of these numbers: we would buy f for any price less than 0, and sell it for any price above 1, but have stated nothing about our behaviour between 0 and 1. A choice function that reflects this should have, for instance, $C(\{f - 0.5, 0\}) = \{f - 0.5, 0\}$. But we have $C_{\underline{P}}(\{f - 0.5, 0\}) = \{0\}$ and $C_{\overline{P}}(\{f - 0.5, 0\}) = \{f - 0.5\}$. In other words, these choice functions are more deterministic than our original statement of beliefs.

These criteria are to be used with caution, particularly given their well-documented problems when applied to sequential decision problems (see for instance Seidenfeld [570] or Berger [62, § 5]). However, usually, they do give a single answer, which is useful in practice, and, as we shall see, the single answer they give is usually a reasonable choice. In any case, can we find a choice function that better represents our imprecision? This must be one that generally returns a set of gambles. The obvious first try is *interval dominance*, as proposed by for instance Satia and Lave [554], Kyburg [419], and many others. If the upper prevision of f is lower than the lower prevision of g, then if f and g are among our options we should never choose f. The motivation for this follows easily from behavioural interpretation of lower and upper previsions. Suppose one chooses f. Then one would sell f for $\overline{P}(f) + \epsilon$ if offered. One would then also buy g for $\underline{P}(g) - \epsilon$, thus one would have $\overline{P}(f) - \underline{P}(g) + g < g$. It would have been more sensible to have chosen g in the first place.

Definition 8.5 (Interval dominance). Let $\sqsupset_{\underline{P}}$ be the partial order defined by

$$f \sqsupset_{\underline{P}} g \quad \text{if } \underline{P}(f) > \overline{P}(g).$$

Let $C_{\sqsupset_{\underline{P}}}$ be the choice function corresponding to this partial order, called interval dominance, and given by

$$C_{\sqsupset_{\underline{P}}}(\mathcal{K}) := \{f \in \mathcal{K} : (\forall g \in \mathcal{K})(g \not\sqsupset_{\underline{P}} f)\}.$$

This can be equivalently written

$$C_{\sqsupset_{\underline{P}}}(\mathcal{K}) = \{f \in \mathcal{K} : (\forall g \in \mathcal{K})(\overline{P}(f) \geq \underline{P}(g))\}.$$

Giving interval dominance a credal set interpretation is perhaps helpful only to see its shortcomings. A gamble f is optimal if there is a $P \in \mathcal{M}$ such that, for each gamble g there is a $Q \in \mathcal{M}$ such that $P(f) \geq Q(g)$. Note that the Q can be different for different g.

Interval dominance is a good start, but generally retains too many gambles.

Example 8.2 Consider a particularly extreme case: $\mathcal{K} = \{f, f - \epsilon\}$. If ϵ is small enough and $\overline{P}(f) \neq \underline{P}(f)$, then $C_{\sqsupset_{\underline{P}}}(\mathcal{K}) = \mathcal{K}$. It seems highly unreasonable to consider choosing $f - \epsilon$ when f is available. ♦

Interval dominance could be refined to immediately eliminate any pointwise dominated gambles, such as $f - \epsilon$ in this example, and this seems a sensible refinement, but this would not remove the fundamental problem with interval dominance.

Example 8.3 Suppose that $\mathcal{K} = \{f, g\}$, $\underline{P}(f - g) > 0.$ and $\overline{P}(g) \geq \underline{P}(f)$. Then, $C_{\sqsupset_{\underline{P}}}(\mathcal{K}) = \{f, g\}$. Suppose one chooses g. Then one would pay $\underline{P}(f - g) - \epsilon$ to buy the gambles $f - g$, leaving us with $g + f - g - \underline{P}(f - g) + \epsilon = f - \underline{P}(f - g) + \epsilon$. Since $\underline{P}(f - g) > 0$, one can find an ϵ such that $f - \underline{P}(f - g) + \epsilon < f$, that is, one would have been better off choosing f initially. Since after choosing g every action taken was a obligatory consequence of the supremum buying price assessment, choosing g is not to be advised. ♦

So, if there are ever gambles f and g in $\mathcal{K}$, and $\underline{P}(f - g) > 0$, then choosing g is never advisable. Therefore we should augment interval dominance with this new criterion, which is called *maximality* [672, p. 161]. It turns out, however, that if for a gamble g there is no $f \in \mathcal{K}$ such that $\underline{P}(f - g) > 0$, then g is optimal under interval dominance anyway, so *all* we need is maximality.

Definition 8.6 (Maximality). Let $>_{\underline{P}}$ be the partial order defined by $f>_{\underline{P}}g$ if

$$\underline{P}(f-g)>0$$

or

$$f(x)\geq g(x) \quad \text{for all } x\in\mathcal{X} \quad \text{and } f\neq g.$$

Let $C_{>_{\underline{P}}}$ be the choice function corresponding to this partial order, called maximality, and given by

$$C_{>_{\underline{P}}}(\mathcal{K}):=\{f\in\mathcal{K}:(\forall g\in\mathcal{K})(g\not>_{\underline{P}}f)\}.$$

Why was interval dominance introduced if it is implied by maximality? As well as for motivation (interval dominance is an obvious place to start), interval dominance may be useful when there are very many gambles in $\mathcal{K}$: it is linear in number of gambles, whereas maximality is quadratic. Since calculating lower previsions using natural extension can be computationally expensive anyway (see Chapter 16), this difference can be important. In particular, one can apply interval dominance first and then maximality second.

Maximality also has an intuitive credal set interpretation. Consider a gamble $f\in\mathcal{K}$. If, for each gamble $g\in\mathcal{K}$, there is a $P\in\mathcal{M}$ such that $P(f)\geq P(g)$, then f is maximal. In other words, f is maximal if there is no $g\in\mathcal{K}$ that dominates f everywhere in the credal set. This interpretation also suggests a refinement of maximality: consider a gamble f to be optimal if there is a $P\in\mathcal{M}$ such that, for each gamble $g\in\mathcal{K}$, $P(f)\geq P(g)$. In other words, f maximizes expected utility for at least one element of the credal set. This is stronger than maximality, because with maximality it is possible that for all the elements of $\mathcal{M}$ where f is not dominated by g, there is an $h\in\mathcal{K}$ that does dominate f, so f maximizes expected utility nowhere. This stronger criterion is called E-admissibility and was proposed by Levi [430].

Definition 8.7 (E-admissibility). Let $C_{\mathcal{M}}$ be the choice function, called E-admissibility, defined by

$$C_{\mathcal{M}}(\mathcal{K}):=\{f\in\mathcal{K}:(\exists P\in\mathcal{M})(\forall g\in\mathcal{K},P(f)\geq P(g))\}.$$

In other words, there is a linear prevision in the credal set under which f maximizes expected utility. E-admissibility can be equivalently expressed as

$$C_{\mathcal{M}}(\mathcal{K})=\bigcup_{P\in\mathcal{M}}C_{>_P}(\mathcal{K}).$$

The lower prevision interpretation of E-admissibility is rather more complicated, and requires the concept of *randomized acts*. If f and g are two possible choices of gambles, we could consider tossing a fair coin to decide which one to choose. This is an example of a randomized act. More generally, for any collection of gambles f_1 to f_n, a randomized act is one that chooses gamble f_1 with probability p_1, f_2 with probability p_2, and so on. If $\mathcal{K}$

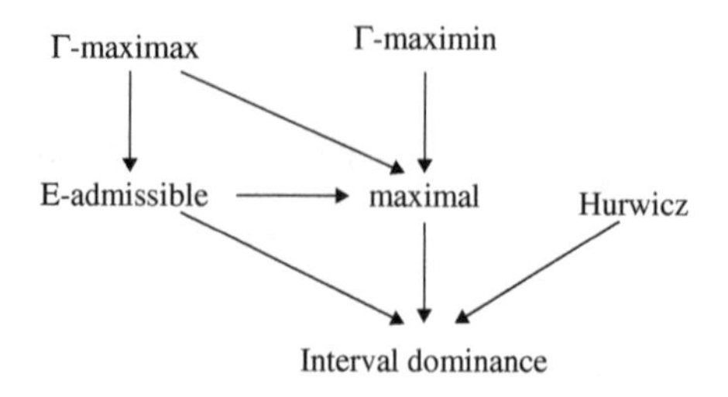

Figure 8.1 Relationships between the choice functions.

includes all possible randomized acts, then maximality and E-admissibility coincide [672, § 3.9.5].

Proposition 8.1 *Let $\mathcal{K}$ be a set of gambles, and $\mathcal{H}$ be the convex hull of $\mathcal{K}$ (that is, the set of all randomized gambles from $\mathcal{K}$). Then $f \in C_{\mathcal{M}}(\mathcal{K})$ if and only if $f \in C_{>_{\underline{P}}}(\mathcal{H}) \cap \mathcal{K}$. In other words,*

$$C_{\mathcal{M}}(\mathcal{K}) = C_{>_{\underline{P}}}(\mathcal{H}) \cap \mathcal{K}.$$

As indicated at several points in this section, some of the choice functions are related. For instance, a maximal gamble is also an intervally dominant one. Figure 8.1 shows all such links. Any links not present in the diagram are not generally true (although in special cases they can be). Proofs of most of these relationships are given by Troffaes [630], although Hurwicz is not considered. It is easy to show that Hurwicz must be intervally dominant.

Perhaps the most surprising result is that optimality under Γ-maximin does not imply E-admissibility. A simple example (as seen for instance in Seidenfeld [570]) is instructive.

Example 8.4 Let $\mathcal{X} = \{x_1, x_2\}$, and let f, g, and h be the gambles in Table 8.1. Suppose that we are completely ignorant about which of x_1, x_2 is true, and so assign the vacuous lower prevision. Then, $C_{\underline{P}}(\{f, g, h\}) = \{h\}$ since its lower prevision is 0.4 rather than 0. But consider a $P \in \mathcal{M}$. If $P(x_1) \geq 0.5$, then $P(g) \geq 0.5 > P(h)$. And if $P(x_1) \leq 0.5$, then $P(f) \geq 0.5 > P(h)$. So there is no $P \in \mathcal{M}$ such that h maximizes expected utility, and so $h \notin C_{\mathcal{M}}(\{f, g, h\}) = \{f, g\}$.

This can be easily understood by considering the randomized gamble that chooses f with probability $1/2$ and g with probability $1/2$. For any $P \in \mathcal{M}$, this randomized gamble has expected value 0.5, and so dominates h everywhere in $\mathcal{M}$. That is, $\underline{P}(1/2f + 1/2g - h) = 0.1 > 0$, so in the set of all randomized gambles of $\{f, g, h\}$, h is not maximal, and is therefore not E-admissible in $\{f, g, h\}$. ♦

Table 8.1 Gambles for Example 8.4.

	x_1	x_2
f	0	1
g	1	0
h	0.4	0.4

This section has mainly considered the theoretical aspects of the choice functions for coherent lower previsions. An important practical consideration is computation of the choice functions; this is discussed in detail in Section 16.3.

8.2 Sequential decision problems

In this section we consider problems where the subject may have to make more than one decision, at different times. It must be noted that this is a very wide class of problems, for which many ideas and methods of solutions exist, and about which there is often disagreement over the best approach. Therefore, we provide only an introductory overview of some simple approaches that can be applied using coherent lower previsions.

Consider for instance the classic oil wildcatter example from Raiffa and Schlaifer ([526, Example 1.4.3]). An oil wildcatter can drill for oil at a particular location (act d_1), or sell his rights to it (act d_2). Before making his decision, he has the option to spend money performing an experiment (act e_1), which can give three possible results B_1, B_2, B_3, where each result suggests different chances of oil being present. He does not have to perform this experiment (this act is called e_2). Finally, there will either be much oil (A_1) or not (A_2). This problem is displayed on a *decision tree*[2] in Figure 8.2. Branches emerging from square nodes represent acts, and branches emerging from circular nodes represent events. At the end of each path is a number representing the utility reward of that particular combination of acts and events.

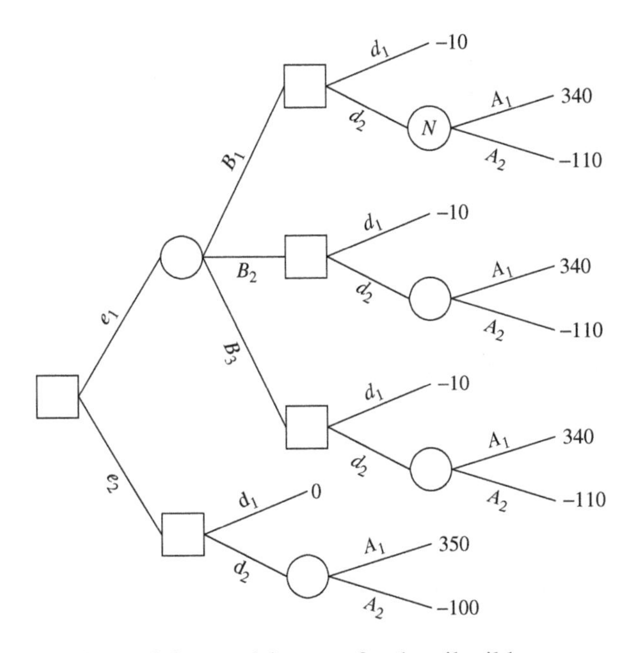

Figure 8.2 Decision tree for the oil wildcatter.

[2] Here, we use the term 'decision tree' to describe a sequential decision problem. However, sometimes the term is also used for classification trees (see Section 10.5), which are quite different.

How does one solve such a problem? It is instructive to look first at the solution for a linear prevision, since this is easy to understand and well-documented. So, suppose that each event arc is assigned a (possibly conditional) probability. A natural way to analyze the problem goes as follows. At ultimate decision nodes (the ones furthest to the right), each act can be immediately represented by a gamble. For example, at the node N, the corresponding gamble is $340I_{A_1} - 110I_{A_2}$ (recall, I_{A_1} is the indicator gamble of the event A_1; see Page 2). Therefore, the (conditional) expected utility of each final chance node can be computed, and then expected utility can be maximized at the final decision nodes. Record for each node which act (or acts) led to that expected utility, and then *replace* the decision node with the maximum expected utility. Now the penultimate layer of decision nodes have become ultimate decision nodes, and the same procedure can be applied. Eventually, the root node will be reached, yielding an optimal act (or acts) at each node and the corresponding expected utility of this strategy.

There is a more tedious but also intuitively reasonable approach available. Consider choosing a single act at each decision node. This sets up a plan for all eventualities. Observe we can turn this into a gamble. For instance, the plan 'choose e_1, and then choose d_2 unless B_3 is observed' has corresponding gamble $(I_{B_1} + I_{B_2})(340I_{A_1} - 110I_{A_2}) - 10I_{B_3}$. We can do this for every possible plan, obtaining a (large) set of gambles. Then, expected utility can be maximized over these gambles, and the plan yielding the maximal gamble can be chosen. It is easy to show that the two methods yield the same solution when using expected utility. This is not true for general choice functions (see for instance Seidenfeld [570], Jaffray [378], LaValle and Wapman [424], or Machina [444]). Indeed, for the more complicated choice functions that we have seen earlier, the first method (usually called the extensive form) is not obvious to extend: take maximality for example; what can we replace decision nodes with when there is no expected utility? The second method (usually called the normal form) is always straightforward for any choice function on gambles, but often impractical and perhaps not philosophically motivated. We now consider the two approaches in more detail.

8.2.1 Static sequential solutions: Normal form

The interpretation of the normal form method above is quite natural: initially, we state exactly what action we will take at every decision node. Having adopted this strategy, we carry out the mandated act upon reaching a decision node. So, the problem is effectively reduced to a simple static problem from before: one could transform the initial decision tree into a new one with only a single decision node, at the root. This would then be a choice between acts, albeit very complicated acts, and we know in principle how to solve this problem.

Definition 8.8 A *plan* (also called strategy, pure strategy, policy, or normal form decision) is a set of acts, one for every relevant decision node of a tree (there is no need for a plan to specify an act for a decision node that cannot be reached due to earlier decisions). Equivalently, it is a subtree of the original tree in which all but one arc is removed from every decision node, and every node not connected to the root is removed. A plan has an associated gamble, because once a particular plan is chosen, the reward received is determined entirely by the state of nature.

Figure 8.3 is an example of a plan represented in decision tree format. It is the plan 'choose e_1, and then choose d_2 unless B_3 is observed' for the oil wildcatter example.

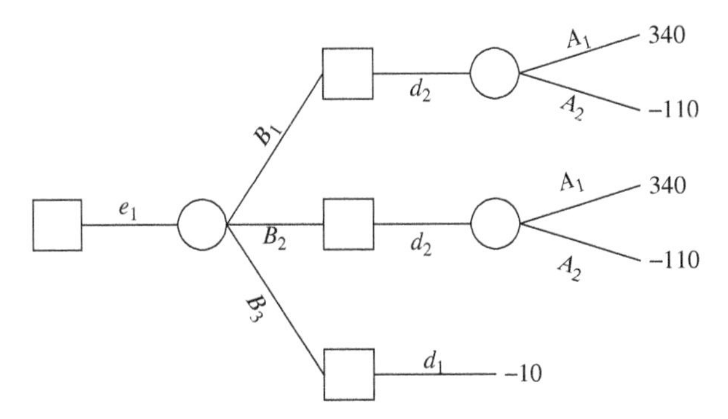

Figure 8.3 An example of a normal form plan for the oil wildcatter problem.

Definition 8.9 For any choice function C and any decision tree, the *normal form solution induced by C* is obtained as follows. Find all plans associated with the tree. Find the set of gambles $\mathcal{K}$ induced by the plans ($f \in \mathcal{K}$ if f is induced by at least one of the plans). Find $C(\mathcal{K})$. The normal form solution is then the set of plans that induce a gamble in $C(\mathcal{K})$.

The interpretation of the normal form solution is simple: if one has to choose a plan, then one should choose a plan that is in the normal form solution.

There are of course many ways to define an optimal set of plans, of which applying the choice function to all gambles is only one (admittedly the most popular and perhaps most intuitive). Investigation of more general types of normal form solution is quite limited, but some attempts have been made, such as using backward induction to find an optimal set of plans [395, 372, 373, 588, 634]. This contrasts with the more traditional 'dynamic' backward induction discussed in the introduction to this section, to which we now turn our attention.

8.2.2 Dynamic sequential solutions: Extensive form

The normal form solution is perhaps an unnatural way of approaching a sequential problem. It requires all actions to be specified in advance, but the structure of the problems are such that the subject does not have to choose what to do at a particular decision node until that node is actually reached. Therefore we want to consider solutions that specify only what the subject may do at each decision node, and defer the final choice to the moment it must really be made. We call this type of solution an *extensive form solution*, although it is more general than the typical use of this term (see for instance [441, 526], instead corresponding to Hammond's *behaviour norms* [338]).

Definition 8.10 An *extensive form solution* of a decision tree is a specification for each decision node in the tree of a non-empty subset of the node's decision arcs. This corresponds to a subgraph of the original tree in which some decision arcs are removed.

An example of an extensive form solution for the oil wildcatter problem is given in Figure 8.4. In this solution, the subject will definitely choose e_1 initially, and will choose d_2 should B_1 or B_2 obtain, but may choose d_1 or d_2 should B_3 obtain. Since this is an extensive

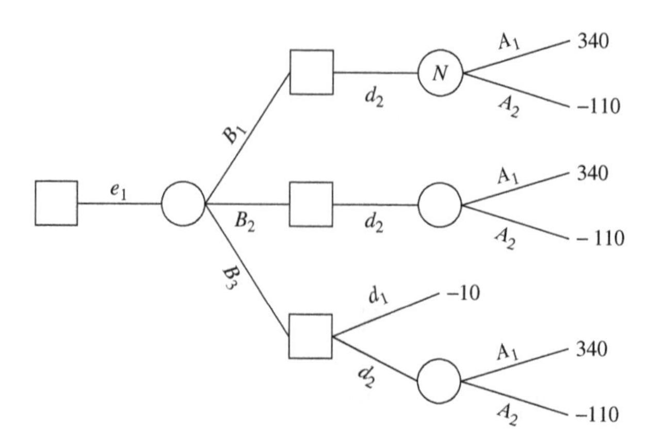

Figure 8.4 An extensive form solution for the oil wildcatter.

form solution, the subject need not decide whether d_1 or d_2 will be chosen unless B_3 actually obtains.

When the extensive form is discussed in most of the literature, it is usually more specific than just any behaviour norm: there is a particular one in mind, obtained by backward induction. The idea of backward induction is straightforward when using probability and expected utility, but difficult to generalize for other choice functions. It is useful to consider a specific example. We use a modification of a decision tree discussed in Seidenfeld [570, Sequential Example 1], using Γ-maximin. The example also highlights some challenges in sequential decision making. Following this example, we will use backward induction and extensive form solution as equivalent terms, though this is only because we do not explore more general theories where they are not equivalent.

Example 8.5 Consider two coins. One is known to be fair (with probability $1/2$ of landing heads, and $1/2$ of landing tails). Nothing is known about the other, so it has lower probability 0 of landing heads, and lower probability 0 of landing tails. We do however assess that the two coins are epistemically independent (see Section 3.2.2; in short, observing one coin's result does not change our buying and selling prices for gambling on the other coin). We will first toss the fair coin, and then the mysterious coin. Consider the following gamble, f: the subject receives 1 if both coins land heads, 1 if both coins land tails, and 0 otherwise.

This corresponds to the following situation: the subject can bet on the mysterious coin showing heads, and receive 1 if it does – call this f_1. Similarly, she can bet on the mysterious coin showing tails, and receive 1 if it does – call this f_2. The gamble f is equivalent to our subject tossing the fair coin, and choosing the first gamble if the fair coin lands head, and the second gamble if the fair coin lands tails.

This randomization serves to eliminate the imprecision. The gambles f_1 and f_2 have lower prevision 0 and upper prevision 1. In order to win the 1 from the gamble f, all that is necessary is that the outcome of the fair coin coincides with the outcome of the the mysterious coin, and this happens with probability $1/2$ (if this is not obvious, consider tossing the mysterious coin first, then trying to match it with the fair coin: clearly this has probability $1/2$, and the order of tossing should be irrelevant). Therefore the expected value of f is 0.5.

Now suppose the subject is offered the sequential problem in Figure 8.5. She chooses among two acts initially: to pay 0.4 for the randomization f, or to be given 0.05 to observe the fair coin. After observing the fair coin, she again must decide whether or not to buy f for

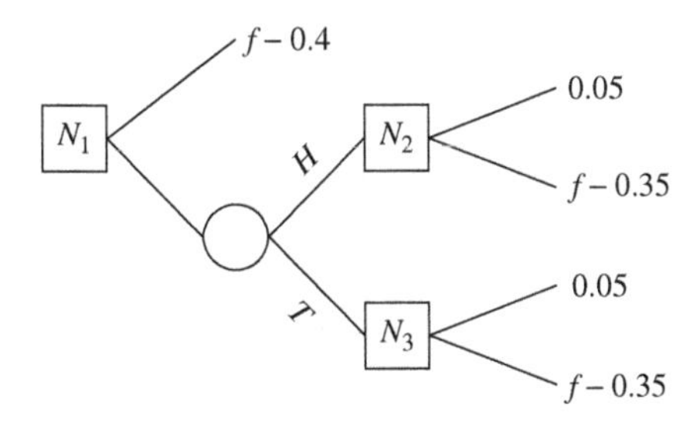

Figure 8.5 An example for backward induction using Γ-maximin.

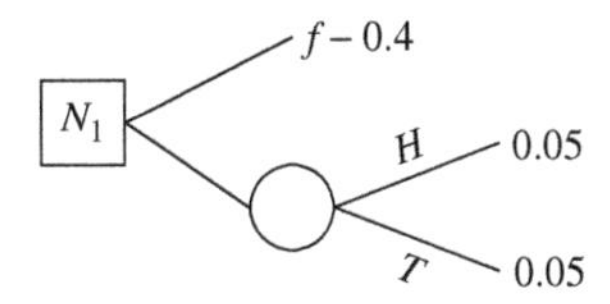

Figure 8.6 Second step of backward induction.

0.4. Of course, at this point f is no longer a randomization because the fair coin's result is known. Having observed H, f is equivalent to f_1, and having observed T, f is equivalent to f_2.

To solve this by backward induction, choose one of the terminal decision nodes, say N_2 first. There are two gambles available at this node, 0.05 and $f - 0.35$, The lower prevision of the former is obvious. For the latter, we have observed H, so f is now just f_1, and $\underline{P}(f|H) = \underline{P}(f_1) = 0$, so $\underline{P}(f - 0.35|H) < 0.05$ and rejecting f is optimal according to Γ-maximin. The same is true at N_3. So now we remove those rejected arcs, giving us the tree in Figure 8.6 (where the previously terminal decision nodes have been removed).

Now the root node has become a terminal decision node, with gambles $f - 0.4$ and 0.05. The lower prevision of the former is 0.1, and so the extensive form solution is simply to pay for f at the start. ♦

Two points for discussion arise from the example. The first is that the solution found by backward induction does not match that of the normal form, and it does not seem to be a particularly good solution. The subject has the option of *being paid* to receive information relevant to f, but rejects it. Also, she ends up choosing $f - 0.4$, whereas using the normal form would leave her with $f - 0.35$. This illustrates that care must be taken when trying to use the extensive form for Γ-maximin and indeed any other choice function corresponding to a weak order.

The second point is that our method above only works if we find a uniquely optimal gamble at each decision node. Suppose we were to use E-admissibility rather than Γ-maximin. Then we would not be able to remove any of the arcs at N_2 and N_3, and so the second step of the backward induction would not work at all. Since one of the main motivations for using a coherent lower prevision is to deal with situations of indecision, we need a more advanced form of backward induction for choice functions that are not weak orders.

We follow a method suggested by Seidenfeld [569], using Figure 8.5 as an example. We know that there is indecision between the two acts at N_2 and also at N_3. Using the extensive form interpretation, the subject will only have to make the choice if she reaches one of these nodes. When the subject is trying to decide what to do at N_1, all she knows is that if she is to

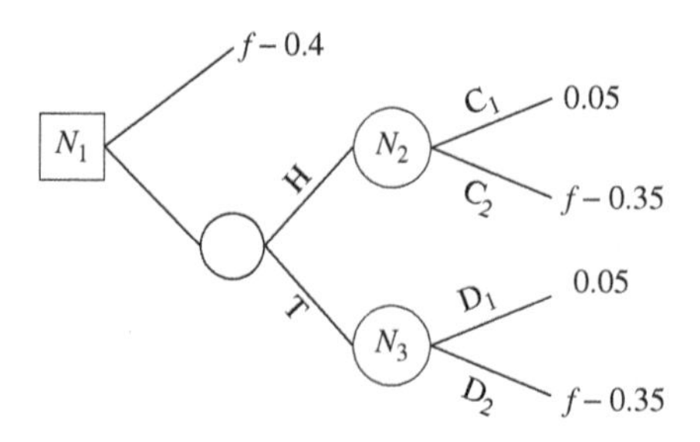

Figure 8.7 Backward induction using Seidenfeld's method.

reach N_2 or N_3, she will take one of the optimal acts there, but knows nothing about which one it is. Therefore it makes sense to model N_2 and N_3 as *chance nodes*, with completely vacuous probabilities on the arcs. This yields the decision tree in Figure 8.7, where C_1 is the event 'the subject chooses 0.05 if she reaches N_2', and so on. Observe that each decision arc at N_1 now corresponds to a gamble, so E-admissibility can be applied to find the optimal acts at the root (it turns out that both acts are optimal).

So, the general backward induction method is as follows. At terminal decision nodes, apply the choice function and eliminate all nonoptimal arcs. Turn all terminal decision nodes into chance nodes and place vacuous events on their arcs. This creates a new tree where the previously penultimate layer of decision nodes (N_1 in the example) have now become ultimate decision nodes. Repeat the process until all decision nodes have been dealt with. All the chance nodes that used to be decision nodes can be turned back into decision nodes, and an extensive form solution has been found.

Note that, in the example, the arc corresponding to $f - 0.4$ was considered acceptable by this method, even though there is a plan corresponding to $f - 0.35$ available (and so $f - 0.4$ would not appear in the normal form solution). This is because the subject cannot guarantee that she will implement the $f - 0.35$ plan, and so cannot argue that choosing $f - 0.4$ is a bad idea. Which of extensive form and normal form solutions an individual prefers largely depends on whether she finds this argument reasonable or not.

Using this method with E-admissibility and maximality will always give reasonable answers (provided one is convinced by the above argument). Using the method for Γ-maximin and related choice functions is risky and should be done with care, as we see from the first example. Interval dominance is less clear, but it should not be too bad: the solution will include everything in the maximal solution and occasionally some extra arcs, as is usual for interval dominance.

8.3 Examples and applications

8.3.1 Ellsberg's paradox

Ellsberg's paradox is a simple example that shows that the presence of complex uncertainty, i.e. uncertainty about the true probabilities, leads to situations where reasonable behaviour can violate the classical axioms of subjective expected utility, namely the so-called *sure-thing principle*.

To this end, consider the following situation. There is an urn with coloured balls. You know that $\frac{1}{3}$ of the balls are red and that the remaining balls can only be yellow or black.

So, you know that the yellow and the black balls together account for $\frac{2}{3}$ of the balls but you do not know the fraction of black balls. That is, up to $\frac{2}{3}$ of the balls can be black but there might even be no black ball in the urn. First, imagine you are offered the following Game A: a ball is randomly drawn from the urn and you may bet either on red or on black. If you are right, you get a prize – that is, you cannot get a prize if the randomly drawn ball turns out to be yellow. Do you bet on red or on black?

Secondly, imagine you are offered the slightly different Game B: again, a ball is randomly drawn from the urn, you may again bet either on red or on black and, if you are right, you get a prize. However, this time you also get a prize if the randomly drawn ball is yellow, irrespective of your bet. Do you bet on red or on black now?

In order to analyze the favourability of the bets, let us first take a closer look at the probabilities of getting the prize.

- Game A: If you bet on red, you know the exact probability of getting the prize; the probability is $\frac{1}{3}$. If you bet on black, you do not know the probability exactly but you know that it is somewhere between 0 and $\frac{2}{3}$.
- Game B: Now, you know the exact probability of getting the prize if you bet on black; the probability is $\frac{2}{3}$. If you bet on red you do not know the probability exactly but you know that it is somewhere between $\frac{1}{3}$ and 1.

Since we have two games and, in each game, two different bets are possible, there are four possible combinations of bets: (r, r), (r, b), (b, r) and (b, b) where, e.g., (r, b) means betting on red in Game A and betting on black in Game B. It turns out that most people choose the 'safe' combination (r, b). The attractiveness of this combination is due to the fact that this is the only combination where the probabilities of getting the prize are known exactly. Other people choose the 'risky' combination (b, r) where the probabilities of getting the prize are potentially very high but might also be very small.

Though these two combinations are indeed reasonable, they are in contradiction with the axioms of subjective expected utility. The so-called sure-thing principle says that if you prefer red to black in Game A, you should also prefer red to black in Game B because Game B just adds the prize for drawing a yellow ball to Game A.

The fact that a combination such as (r, b) violates the axioms of subjective expected utility can also be seen in a different way. A preference relation $\succ$ over decisions fulfills the axioms of subjective expected utility if and only if there is a (subjective) linear prevision P such that

$$d_1 \succ d_2 \iff P(f_{d_1}) > P(f_{d_2}). \tag{8.1}$$

Let us investigate the implications of Equation (8.1) to the two games that we introduced earlier.

Assume, without loss of generality, that the utility of getting the prize is 1, and the utility of not getting the prize is 0. Then, evidently, on $\mathcal{X} = \{r, b, y\}$, we have the gambles:

$$f_r^A = I_{\{r\}} \qquad f_b^A = I_{\{b\}}$$

$$f_r^B = I_{\{r,y\}} \qquad f_b^B = I_{\{b,y\}}$$

where f_r^A is the gamble corresponding to the act to bet on red in Game A, and so on.

So, if you strictly prefer red to black in Game A, then the axioms of subjective expected utility imply that your subjective linear prevision P fulfills

$$P(I_{\{r\}}) = P(f_r^A) > P(f_b^A) = P(I_{\{b\}})$$

because, otherwise, Equation (8.1) would be violated. But the above inequality also implies that

$$P(f_r^B) = P(I_{\{r,y\}}) = P(I_{\{r\}}) + P(I_{\{y\}}) > P(I_{\{b\}}) + P(I_{\{y\}}) = P(I_{\{b,y\}}) = P(f_b^B).$$

That is, you have to prefer red to black also in Game B according to the axioms of subjective expected utility. This is the so-called Ellsberg's paradox; it is a paradox because the betting combinations (r, b) and (b, r) violate the axioms of subjective expected utility even though these are reasonable combinations.

This paradox can nicely be solved by use of imprecise probabilities in decision theory. The information given in Game A and Game B leads to a coherent lower prevision $\underline{P}$ with credal set

$$\mathcal{M}(\underline{P}) := \left\{P \in \mathbb{P}(\mathcal{X}) \colon P(I_{\{r\}}) = \tfrac{1}{3} \text{ and } 0 \leq P(I_{\{b\}}) \leq \tfrac{2}{3}\right\}$$

on $\mathcal{X} = \{r, b, y\}$. Say that the utility of getting the prize is 1 and the utility of not getting the prize is 0. Then, we get the following lower and the upper previsions for the utility of betting on red or on black:

$$\underline{P}(f_r^A) = \frac{1}{3} \qquad \underline{P}(f_b^A) = 0$$

$$\underline{P}(f_r^B) = \frac{1}{3} \qquad \underline{P}(f_b^B) = \frac{2}{3}$$

and

$$\overline{P}(f_r^A) = \frac{1}{3} \qquad \overline{P}(f_b^A) = \frac{2}{3}$$

$$\overline{P}(f_r^B) = 1 \qquad \overline{P}(f_b^B) = \frac{2}{3}.$$

To see how we arrived at these numbers, consider for instance the upper prevision for the utility of act red in Game B:

$$\begin{aligned}\overline{P}(f_r^B) &= \max_{P \in \mathcal{M}(\underline{P})} P(f_r^B) = \max_{P \in \mathcal{M}(\underline{P})} P(I_{\{r,y\}}) \\ &= \max_{P \in \mathcal{M}(\underline{P})} \left[P(I_{\{r\}}) + P(I_{\{y\}})\right] = 1 \cdot \frac{1}{3} + 1 \cdot \frac{2}{3} = 1.\end{aligned}$$

Now, the betting combinations (r, b) and (b, r) can easily be explained by the theory presented in the previous sections. A decision maker who uses the (pessimistic) Γ-maximin choice function considers betting combination (r, b) as being optimal, a decision maker who uses the (optimistic) Γ-maximax choice function prefers the betting combination (b, r).

The so-called Ellsberg's paradox was given by D. Ellsberg in [264] and has influenced a large amount of subsequent articles; see e.g. [339] for a collection of many such articles. In the context of imprecise probabilities (interval probabilities), Ellsberg's paradox is investigated in [695]. A classical reference for subjective expected utility is [555].

8.3.2 Robust Bayesian statistics

It is common to formalize Bayesian statistical problems by use of decision theory. In order to increase the reliability of statistical conclusions, a theory of *robust* Bayesian statistics has been developed (see also Section 7.4). In this subsection, it is shown that the theory presented in the previous sections of this chapter covers large parts of robust Bayesian statistics as a special case.

In a Bayesian estimation problem, the goal is to estimate a parameter $\theta \in \Theta$. In order to evaluate the quality of an estimate $\hat{\theta}$, a loss function

$$L : \Theta \times \Theta \to \mathbb{R}, \quad (\theta, \hat{\theta}) \mapsto L(\theta, \hat{\theta})$$

is used – the least-squares loss

$$L(\theta, \hat{\theta}) = (\theta - \hat{\theta})^2$$

for $\Theta \subseteq \mathbb{R}$ is a very popular choice of loss function. In terms of decision theory, the parameter θ is the state of nature and the estimate $\hat{\theta}$ is the act. That is, the set of possible acts is equal to the possibility space ($D = \Theta$), and the utility of an act $d = \hat{\theta}$ is given by

$$U(\theta, d) = -L(\theta, \hat{\theta}).$$

In Bayesian statistics (see also Section 7.1.4) a precise prior distribution on the parameter space Θ is assumed. A precise prior distribution is a linear prevision

$$P : \mathcal{L}(\Theta) \to \mathbb{R}.$$

Then, after observing the data y, the prior distribution is updated to a conditional linear prevision $P(\cdot|y)$, the so-called posterior distribution.

An optimal estimate is an estimate that minimizes the posterior risk:

$$\hat{\theta}^* = \arg\min_{\hat{\theta}\in\Theta} P(L(\cdot, \hat{\theta})|y) = \arg\min_{\hat{\theta}\in\Theta} \int L(\theta, \hat{\theta})\; P(\mathrm{d}\theta|y),$$

or equivalently, in terms of decision theory, an act $d^* \in D = \Theta$ is optimal if it maximizes the posterior expected utility:

$$d^* = \arg\max_{d\in D} P(U(\cdot, d)|y) = \arg\max_{d\in D} \int U(\theta, d)\; P(\mathrm{d}\theta|y).$$

This means that, in Bayesian statistics, an estimator is chosen according to the axioms of subjective expected utility.

Just as the assumption of a known linear prevision on the states of nature in decision theory, assuming a known precise prior distribution in Bayesian statistics has been criticized because this requires a very detailed prior knowledge which is often not available in practice and, even worse, it has been noticed that the results can be very sensitive to the choice of the prior distribution. That is, even if two prior distributions are arbitrarily close to each other, the corresponding results from analysing the posterior distributions can differ massively (see also Section 7.2.2).

For this reason, the single precise prior distribution is replaced by a set $\mathcal{M}$ of prior distributions in robust Bayesian statistics. Accordingly, after observing the data y, the set of priors is updated to the set of posteriors

$$\mathcal{M}_{\cdot|y} := \{P(\cdot|y) : P \in \mathcal{M}\}$$

and an estimate is called optimal if it minimizes the upper posterior risk:

$$\hat{\theta}^* = \arg\min_{\hat{\theta}\in\Theta} \sup_{P(\cdot|y)\in\mathcal{M}_{\cdot|y}} P(L(\cdot,\hat{\theta})|y).$$

Focusing on the *upper* posterior risk corresponds to a worst-case consideration, which is most common in robust statistics. Equivalently, in terms of decision theory, an act $d^* \in D = \Theta$ is optimal if it maximizes the lower expected utility:

$$d^* = \arg\max_{d\in D} \inf_{P(\cdot|y)\in\mathcal{M}_{\cdot|y}} P(U(\cdot,d)|y).$$

Under some regularity assumptions, including coherence, the calculation of $\mathcal{M}_{\cdot|y}$ is equivalent to applying the generalized Bayes rule to the lower prevision defined by

$$\underline{P}(f) = \inf_{P\in\mathcal{M}} P(f)$$

for all $f \in \mathcal{L}(\Theta)$; see Sections 2.3.3 and 2.3.4 and following for further details. Hence, an act $d^* \in D = \Theta$ is optimal if it maximizes the conditional lower prevision of the utility:

$$d^* = \arg\max_{d\in D} \underline{P}(U(\cdot,d)|y).$$

This means that optimal estimates in robust Bayesian statistics are exactly Γ-maximin optimal acts in the theory presented in the previous sections.

More details about how robust Bayesian statistics fits into decision theory under imprecise probabilities can be found in [34]. An introductory textbook in Bayesian statistics is [62], which also contains a large chapter on robust Bayesian statistics. A review on the connections between imprecise probabilities and robust statistics is given in [37].

9

Probabilistic graphical models

Alessandro Antonucci, Cassio P. de Campos, and Marco Zaffalon
Instituto Dalle Molle di Studi sull'Intelligenza Artificiale (IDSIA), Manno-Lugano, Switzerland

9.1 Introduction

In the previous chapters of this book the reader has been introduced to a number of powerful tools for modelling uncertain knowledge with imprecise probabilities. These have been formalized in terms of sets of desirable gambles (Chapter 1), lower previsions (Chapter 2), and sets of linear previsions (Section 1.6.2), while their relations with other uncertainty models have been also described (Chapter 4). In the discrete multivariate case, a direct specification of models of this kind might be expensive because of too a high number of joint states, this being exponential in the number of variables. Yet, a compact specification can be achieved if the model displays particular invariance (Chapter 3) or *composition* properties. The latter is exactly the focus of this chapter: defining a model over its whole set of variables by the composition of a number of sub-models each involving only fewer variables. More specifically, we focus on the kind of composition induced by *independence* relations among the variables. Graphs are particularly suitable for the modelling of such independencies, so we formalize our discussion within the framework of probabilistic graphical models. Following these ideas, we introduce a class of probabilistic graphical models with imprecision based on directed graphs called *credal networks*.[1] The example below is used to guide the reader step-by-step through the application of the ideas introduced in this chapter.

[1] This chapter mostly discusses credal networks. Motivations for this choice and a short outline on other imprecise probabilistic models is reported in Section 9.6.

Introduction to Imprecise Probabilities, First Edition.
Edited by Thomas Augustin, Frank P. A. Coolen, Gert de Cooman and Matthias C. M. Troffaes.

Example 9.1 (Lung health diagnostic). Assume the lung health status of a patient can be inferred from a probabilistic model over the binary variables: lung cancer (C), bronchitis (B), smoker (S), dyspnoea (D), and abnormal X-rays (R).[2] An *imprecise* specification of this model can be equivalently achieved by assessing a coherent set of desirable gambles, a coherent lower prevision or a credal set, all over the joint variable $\mathbf{X} := (C, B, S, D, R)$. This could be demanding because of the exponentially large number of states of the joint variable to be considered (namely, 2^5 in this example). ♦

Among the different formalisms which can be used to model imprecise probabilistic models, in this chapter we choose credal sets as they appear relatively easy to understand for readers used to work with standard (precise) probabilistic models, in particular for credal sets induced from intervals (see Chapter 15).[3] The next section reports some background information and notation about them.

9.2 Credal sets

9.2.1 Definition and relation with lower previsions

We define a *credal set* (CS) $\mathcal{M}(X)$ over a categorical variable X as a closed convex set of probability mass functions over X.[4] An *extreme point* (or vertex) of a CS is an element of this set which cannot be expressed as a convex combination of other elements. Notation $\mathrm{ext}[\mathcal{M}(X)]$ is used for the set of extreme points of $\mathcal{M}(X)$. We focus on finitely-generated CSs, i.e., sets with a finite number of extreme points. Geometrically speaking, a CS of this kind is a polytope on the probability simplex, which can be equivalently specified in terms of linear constraints to be satisfied by the probabilities of the different outcomes of X (e.g., see Figure 9.1).[5] As an example, the *vacuous* CS $\mathcal{M}_0(X)$ is defined as the whole set of probability mass functions over X:

$$\mathcal{M}_0(X) := \left\{ P(X) \;\middle|\; \begin{array}{l} P(x) \geq 0, \forall x \in \mathcal{X}, \\ \sum_{x \in \mathcal{X}} P(x) = 1 \end{array} \right\}. \tag{9.1}$$

The vacuous CS is clearly the largest (and hence least informative) CS we can consider. Any other CS $\mathcal{M}(X)$ over X is defined by imposing additional constraints to $\mathcal{M}_0(X)$.

A single probability mass function $P(X)$ can be regarded as a 'precise' CS made of a single element. Given a real-valued function f of X (which, following the language of the previous chapters, can be also regarded as a *gamble*), its expectation is, in this precise case, $E_P(f) := \sum_{x \in \mathcal{X}} P(x) \cdot f(x)$. This provides a one-to-one correspondence between probability

---

[2] These variables are referred to the patient under diagnosis and supposed to be self-explanatory. For more insights refer to the *Asia network* [423], which can be regarded as an extension of the model presented here.

[3] There are also historical reasons for this choice: this chapter is mainly devoted to credal networks with strong independence, which have been described in terms of credal sets from their first formalization (see Fabio Cozman's paper [160] and earlier works by Serafín Moral, like for instance [96]).

[4] In Section 1.6.2 CSs have been defined as sets of linear previsions instead of probability mass functions. Yet, the one-to-one correspondence between linear previsions and probability mass functions makes the distinction irrelevant. Note also that, in this chapter, we focus on discrete variables. A discussion about extensions to continuous variables is given in Section 9.6.

[5] Standard algorithms can be used to move from the enumeration of the extreme points to the linear constraints generating the CS and *vice versa* (e.g., [40]).

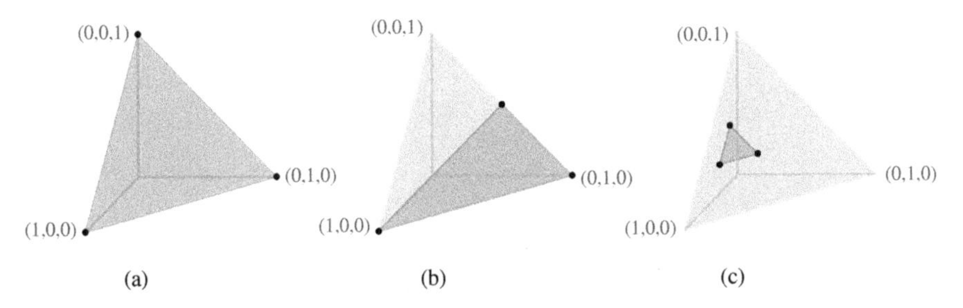

Figure 9.1 Geometrical representation of CSs over a variable X with $\mathcal{X} = \{x', x'', x'''\}$ in the three-dimensional space with coordinates $[P(x'), P(x''), P(x''')]^T$. Dark grey polytopes represent respectively: (a) the vacuous CS as in (9.1); (b) a CS defined by constraint $P(x''') \geq P(x'')$; (c) a CS $\mathcal{M}(X)$ such that $\text{ext}[\mathcal{M}(X)] = \{[.1, .3, .6]^T, [.3, .3, .4]^T, [.1, .5, .4]^T\}$. The extreme points are in black.

mass functions and *linear* previsions. Given a generic CS $\mathcal{M}(X)$, we can evaluate the lower expectation $\underline{E}_{\mathcal{M}}(f) := \min_{P(X)\in\mathcal{M}(X)} P(f)$ (and similarly for the upper). This defines a coherent lower prevision as a lower envelope of a set of linear previsions (see Section 2.2.2). As an example, the vacuous CS $\mathcal{M}_0(X)$ in (9.1) defines the (vacuous) coherent lower prevision, i.e., $\underline{E}_{\mathcal{M}_0}(f) = \min_{x\in\mathcal{X}} f(x)$. Note that a set and its convex closure have the same lower envelope, and hence a set of distributions and its convex closure define the same coherent lower prevision. This means that, when computing expectations, there is no lack of generality in defining CSs only as closed convex sets of probability mass functions, and the correspondence with coherent lower previsions is bijective [672, § 3.6.1]. Note also that the optimization task associated to the above definition of $\underline{E}_{\mathcal{M}}(f)$ is an LP optimization problem, whose solution can be equivalently obtained by considering only the extreme points of the CS [170], i.e.,

$$\underline{E}_{\mathcal{M}}(f) = \min_{P(X)\in\text{ext}[\mathcal{M}(X)]} E_P(f). \tag{9.2}$$

The above discussion also describes how inference with CSs is intended. Note that the complexity of computations as in (9.2) is linear in the number of extreme points, this number being unbounded for general CSs.[6]

A notable exception is the Boolean case: a CS over a binary variable[7] cannot have more than two extreme points. This simply follows from the fact that the probability simplex (i.e., the vacuous CS) is a one-dimensional object.[8] As a consequence of that, any CS over a binary variable can be specified by simply requiring the probability of a single outcome to belong to an interval. E.g., if $\mathcal{M}(X) := \{P(X) \in \mathcal{M}_0(X) | .4 \leq P(X = x) \leq .7\}$, then $\text{ext}[\mathcal{M}(X)] = \{[.4, .6]^T, [.7, .3]^T\}$ (see also Figure 9.2).

[6] Some special classes of CSs with bounded number of extreme points are the vacuous ones as in (9.1) and those corresponding to linear-vacuous mixtures as in Section 4.7.3 (for which the number of the extreme points cannot exceed the cardinality of $\mathcal{X}$). Yet, these theoretical bounds are not particularly binding for (joint) variables with high dimensionality.

[7] If X is a binary variable, its two states are denoted by x and $\neg x$.

[8] Convex sets on one-dimensional varieties are isomorphic to intervals on the real axis, whose extreme points are the lower and upper bounds.

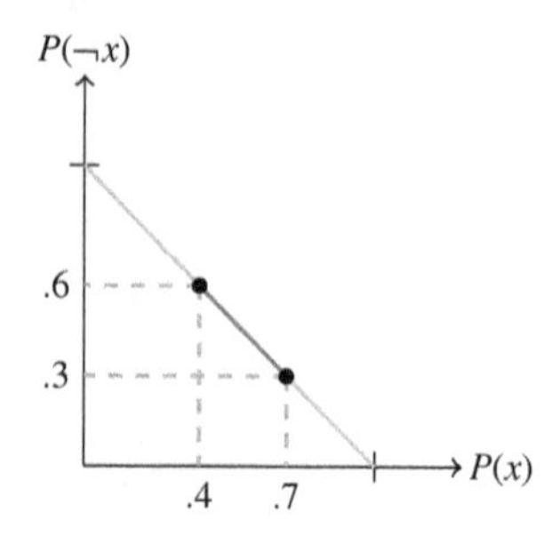

Figure 9.2 Geometrical representation (in dark gray) of a CS over a binary variable X in the two-dimensional space with coordinates $[P(x), P(\neg x)]^T$. The two extreme points are in black, while the probability simplex, which corresponds to the vacuous CS, is in light grey.

9.2.2 Marginalization and conditioning

Given a joint CS $\mathcal{M}(X, Y)$, the corresponding *marginal* CS $\mathcal{M}(X)$ contains all the probability mass functions $P(X)$ which are obtained by marginalizing out Y from $P(X, Y)$, for each $P(X, Y) \in \mathcal{M}(X, Y)$. Notably, the marginal CS can be equivalently obtained by only considering its extreme points, i.e.,

$$\mathcal{M}(X) = \mathrm{CH}\left\{ P(X) \;\middle|\; \begin{array}{l} P(x) := \sum_{y \in \mathcal{Y}} P(x, y), \forall x \in \mathcal{X}, \\ \forall P(X, Y) \in \mathrm{ext}[\mathcal{M}(X, Y)] \end{array} \right\}, \tag{9.3}$$

where CH denotes the convex hull operation.[9]

We similarly proceed for *conditioning*. For each $y \in \mathcal{Y}$, the conditional CS $\mathcal{M}(X|y)$ is made of all the conditional mass functions $P(X|y)$ obtained from $P(X, Y)$ by Bayes rule, for each $P(X, Y) \in \mathcal{M}(X, Y)$ (this can be done under the assumption $P(y) > 0$ for each mass function $P(X, Y) \in \mathcal{M}(X, Y)$, i.e., $\underline{P}(y) > 0$).[10] This corresponds to the *generalized Bayes rule* described in Section 2.3.3. As in the case of marginalization, conditional CSs can be obtained by only considering the extreme points of the joint CS, i.e.,

$$\mathcal{M}(X|y) = \mathrm{CH}\left\{ P(X|y) \;\middle|\; \begin{array}{l} P(x|y) := \frac{P(x,y)}{\sum_{x \in \mathcal{X}} P(x,y)}, \forall x \in \mathcal{X}, \\ \forall P(X, Y) \in \mathrm{ext}[\mathcal{M}(X, Y)] \end{array} \right\}. \tag{9.4}$$

The following notation is used as a shortcut for the collection of conditional CSs associated to all the possible values of the conditioning variable: $\mathcal{M}(X|Y) := \{\mathcal{M}(X|y)\}_{y \in \mathcal{Y}}$.

Example 9.2 In the medical diagnosis setup of Example 9.1, consider only variables lung cancer (C) and smoker (S). The available knowledge about the joint states of these two

[9] In order to prove that the CS in (9.3) is consistent with the definition of marginal CS, it is sufficient to check that any extreme point of $\mathcal{M}(X)$ is obtained marginalizing out Y from an extreme point of $\mathcal{M}(X, Y)$. If that would not be true, we could express an extreme point of $\mathcal{M}(X)$ as the marginalization of a convex combination of two or more extreme points of $\mathcal{M}(X, Y)$, and hence as a convex combination of two or more probability mass functions over X. This is against the original assumptions.

[10] See Section 2.3.4 for a discussion about conditioning on events for which this assumption is not satisfied.

Table 9.1 The eight extreme points of the joint CS $\mathcal{M}(C,S) = \mathrm{CH}\{P_j(C,S)\}_{j=1}^{8}$. Linear algebra techniques (e.g., see [40], even for a software implementation) can be used to check that none of these distributions belong to the convex hull of the remaining seven.

j	1	2	3	4	5	6	7	8
$P_j(c,s)$	$\frac{1}{16}$	$\frac{1}{8}$	$\frac{3}{16}$	$\frac{3}{8}$	$\frac{1}{16}$	$\frac{1}{8}$	$\frac{3}{16}$	$\frac{3}{8}$
$P_j(\neg c,s)$	$\frac{3}{16}$	$\frac{3}{8}$	$\frac{1}{16}$	$\frac{1}{8}$	$\frac{3}{16}$	$\frac{3}{8}$	$\frac{1}{16}$	$\frac{1}{8}$
$P_j(c,\neg s)$	$\frac{3}{8}$	$\frac{1}{4}$	$\frac{3}{8}$	$\frac{1}{4}$	$\frac{9}{16}$	$\frac{3}{8}$	$\frac{9}{16}$	$\frac{3}{8}$
$P_j(\neg c,\neg s)$	$\frac{3}{8}$	$\frac{1}{4}$	$\frac{3}{8}$	$\frac{1}{4}$	$\frac{3}{16}$	$\frac{1}{8}$	$\frac{3}{16}$	$\frac{1}{8}$

variables is modelled by a CS $\mathcal{M}(C,S) = \mathrm{CH}\{P_j(C,S)\}_{j=1}^{8}$, whose eight extreme points are those reported in Table 9.1 and depicted in Figure 9.3. It is indeed straightforward to compute the marginal CS for variable S as in (9.3):

$$\mathcal{M}(S) = \mathrm{CH}\left\{ \begin{bmatrix} \frac{1}{4} \\ \frac{3}{4} \end{bmatrix}, \begin{bmatrix} \frac{1}{2} \\ \frac{1}{2} \end{bmatrix} \right\}. \tag{9.5}$$

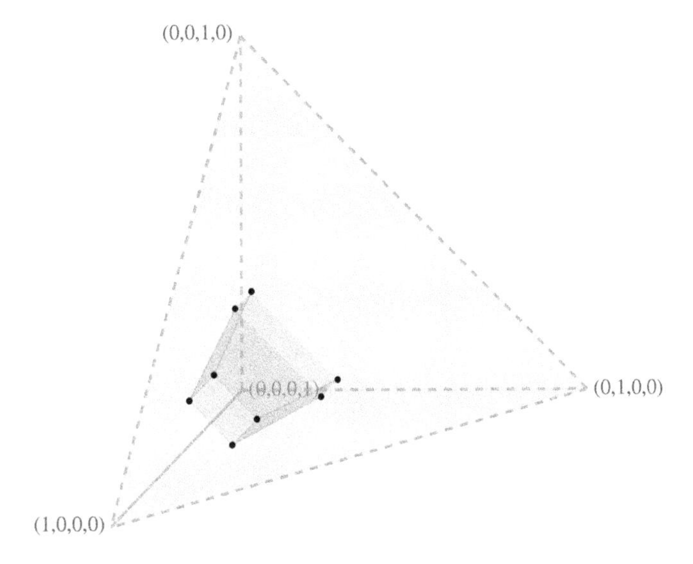

Figure 9.3 Geometrical representation of the CS over a (joint) quaternary variable (C,S), with both C and S binary, as quantified in Table 9.1. The CS is the dark grey polyhedron (and its eight extreme points are in black), while the probability simplex is the light grey tetrahedron. The representation is in the three-dimensional space with coordinates $[P(c,s), P(\neg c,s), P(c,\neg s)]^T$, which are the barycentric coordinates of the four-dimensional probability simplex.

Similarly, the conditional CSs for variable C as in (9.4) given the two values of S are:

$$\mathcal{M}(C|s) = \text{CH}\left\{\begin{bmatrix}\frac{1}{4}\\ \frac{3}{4}\end{bmatrix}, \begin{bmatrix}\frac{3}{4}\\ \frac{1}{4}\end{bmatrix}\right\} \quad , \quad \mathcal{M}(C|\neg s) = \text{CH}\left\{\begin{bmatrix}\frac{1}{2}\\ \frac{1}{2}\end{bmatrix}, \begin{bmatrix}\frac{3}{4}\\ \frac{1}{4}\end{bmatrix}\right\}. \tag{9.6}$$

♦

9.2.3 Composition

Let us define a *composition* operator in the imprecise-probabilistic framework. Given a collection of conditional CSs $\mathcal{M}(X|Y)$ and a marginal CS $\mathcal{M}(Y)$, the marginal extension introduced in Chapter 2 within the language of coherent lower previsions, corresponds to the following specification of a joint CS as a composition of $\mathcal{M}(Y)$ and $\mathcal{M}(X|Y)$:

$$\mathcal{M}(X,Y) := \text{CH}\left\{ P(X,Y) \,\middle|\, P(x,y) := P(x|y)\cdot P(y) \begin{array}{l} \forall x \in \mathcal{X}, \forall y \in \mathcal{Y}, \\ \forall P(Y) \in \mathcal{M}(Y), \\ \forall P(X|y) \in \mathcal{M}(X|y) \end{array} \right\}. \tag{9.7}$$

Notation $\mathcal{M}(X|Y) \otimes \mathcal{M}(Y)$ will be used in the following as a shortcut for the right-hand side of (9.7). As usual, in practice, the joint CS in (9.7) can be equivalently obtained by considering only the extreme points, i.e.,

$$\mathcal{M}(X|Y) \otimes \mathcal{M}(Y) = \text{CH}\left\{ P(X,Y) \,\middle|\, P(x,y) := P(x|y)\cdot P(y) \begin{array}{l} \forall x \in \mathcal{X}, \forall y \in \mathcal{Y}, \\ \forall P(Y) \in \text{ext}[\mathcal{M}(Y)], \\ \forall P(X|y) \in \text{ext}[\mathcal{M}(X|y)] \end{array} \right\}. \tag{9.8}$$

Example 9.3 As an exercise, compute by means of (9.8) the composition of $\mathcal{M}(C|S) \otimes \mathcal{M}(S)$ of the unconditional CS in (9.5) and the conditional CSs in (9.6). In this particular case, check that the so obtained CS $\mathcal{M}(C,S)$ coincides with that in Table 9.1. ♦

Example 9.4 As a consequence of (9.7), we may define a joint CS over the variables in Example 9.1 by means of the following composition:

$$\mathcal{M}(D,R,B,C,S) = \mathcal{M}(D|R,B,C,S) \otimes \mathcal{M}(R,B,C,S),$$

and then, iterating,[11]

$$\mathcal{M}(D,R,B,C,S) = \mathcal{M}(D|R,B,C,S) \otimes \mathcal{M}(R|B,C,S) \otimes \mathcal{M}(B|C,S) \otimes \mathcal{M}(C|S) \otimes \mathcal{M}(S) \tag{9.9}$$

♦

Note that (9.9) does not make the specification of the joint CS less demanding (the number of probabilistic assessments we should make for the CS on the left-hand side is

[11] Brackets setting the composition ordering in (9.9) are omitted because of the associativity of the composition operator $\otimes$.

almost the same required by the first on the right-hand side). In next section, we show how independence can make the specification of these multivariate models more compact.

9.3 Independence

First, let us formalize the notion of independence in the precise probabilistic framework. Consider variables X and Y, and assume that a (precise) joint probability mass function $P(X, Y)$ models the knowledge about their joint configurations. We say that X and Y are *stochastically* independent if $P(x, y) = P(x) \cdot P(y)$ for each $x \in \mathcal{X}$ and $y \in \mathcal{Y}$, where $P(X)$ and $P(Y)$ are obtained from $P(X, Y)$ by marginalization.

The concept might be easily extended to the imprecise probabilistic framework by the notion of *strong independence*.[12] Given a joint CS $\mathcal{M}(X, Y)$, X and Y are *strongly* independent if, for all $P(X, Y) \in \text{ext}[\mathcal{M}(X, Y)]$, X and Y are stochastically independent, i.e., $P(x, y) = P(x) \cdot P(y)$, for each $x \in \mathcal{X}$, $y \in \mathcal{Y}$. The concept admits also a formulation in the conditional case. Variables X and Y are strongly independent given Z if, for each $z \in \mathcal{Z}$, every $P(X, Y|z) \in \text{ext}[K(X, Y|z)]$ factorizes as $P(x, y|z) = P(x|z) \cdot P(y|z)$, for each $x \in \mathcal{X}$ and $y \in \mathcal{Y}$.

Example 9.5 In Example 9.1, consider only variables C, B and S. According to (9.8):

$$\mathcal{M}(C, B, S) = \mathcal{M}(C, B|S) \otimes \mathcal{M}(S).$$

Assume that, once you know whether or not the patient is a smoker, there is no relation between the fact that he could have lung cancer and bronchitis. This can be regarded as a conditional independence statement regarding C and B given S. In particular, we consider the notion of strong independence as above. This implies the factorization $P(c, b|s) = P(c|s) \cdot P(b|s)$ for each possible value of the variables and each extreme point of the relative CSs. Expressing that as a composition, we have:[13]

$$\mathcal{M}(C, B, S) = \mathcal{M}(C|S) \otimes \mathcal{M}(B|S) \otimes \mathcal{M}(S). \tag{9.10}$$

◆

In this particular example, the composition in (9.10) is not providing a significantly more compact specification of the joint CS $\mathcal{M}(C, B, S)$. Yet, for models with more variables and more independence relations, this kind of approach leads to a substantial reduction of the number of states to be considered for the specification of a joint model. In the rest of this section, we generalize these ideas to more complex situations where a number of conditional independence assessments is provided over a possibly large number of variables. In

---

[12] Strong independence is not the only independence concept proposed within the imprecise-probabilistic framework. See Section 3.2 for a discussion on this topic, and Section 9.6 for pointers on imprecise probabilistic graphical models based on other concepts.

[13] In (9.10) the composition operator has been extended to settings more general than (9.8). With marginal CSs, a joint CS $\mathcal{M}(X, Y) := \mathcal{M}(X) \otimes \mathcal{M}(Y)$ can be obtained by taking all the possible combinations of the extreme points of the marginal CSs (and then taking the convex hull). Thus, $\mathcal{M}(X, Y|Z := \mathcal{M}(X|Z) \otimes \mathcal{M}(Y|Z)$ is just the conditional version of the same relation. Similarly, $\mathcal{M}(X, Z|Y, W) := \mathcal{M}(X|Y) \otimes \mathcal{M}(Z|W)$. Notably, even in these extended settings, the composition operator remains associative.

order to do that, we need a compact language to describe conditional independence among variables. This is typically achieved in the framework of probabilistic graphical models, by assuming a one-to-one correspondence between the variables under consideration and the nodes of a directed acyclic[14] graph and then by assuming the so-called strong *Markov condition*: 'any variable is strongly independent of its non-descendant non-parents given its parents'. Assuming a one-to-one correspondence between a set of variables and the nodes of a directed graph, the *parents* of a variable are the variables corresponding to the immediate predecessors. Analogously, we define the *children* and, by iteration, the *descendants* of a node/variable. We point the reader to [175] for an axiomatic approach to the modelling of probabilistic independence concepts by means of directed (and undirected) graphs. Here, in order to clarify the semantics of this condition, we consider the following example.

Example 9.6 Assume a one-to-one correspondence between the five binary variables in Example 9.1 and the nodes of the directed acyclic graph in Figure 9.4. The strong Markov condition for this graph implies the following conditional independence statements:

- given smoker, lung cancer and bronchitis are strongly independent;
- given smoker, bronchitis and abnormal X-rays are strongly independent;
- given lung cancer, abnormal X-rays and dyspnoea are strongly independent, and abnormal X-rays and smoker are strongly independent;
- given lung cancer and bronchitis, dyspnoea and smoker are strongly independent.

The above independence statements can be used to generate further independencies by means of the axioms in [175]. ♦

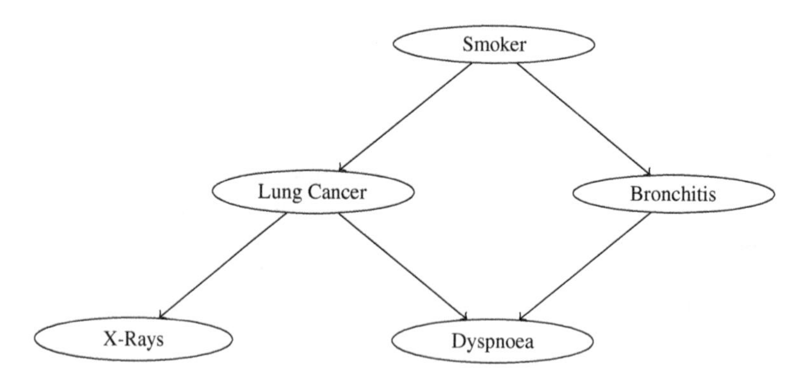

Figure 9.4 A directed graph over the variables in Example 9.1.

[14] A cycle in a directed graph is a directed path connecting a node with itself. A directed graph is acyclic if no cycles are present in it.

9.4 Credal networks

Let us introduce the definition of credal network by means of the following example.

Example 9.7 Consider the variables in Example 9.1 associated to the graph in Figure 9.4. Assume that, for each variable, conditional CSs given any possible value of the parents have been assessed. This means that $\mathcal{M}(S)$, $\mathcal{M}(C|S)$, $\mathcal{M}(B|S)$, $\mathcal{M}(R|C)$, and $\mathcal{M}(D|C,B)$ are available. A joint CS can be defined by means of the following composition, based on the operator introduced in (9.8):

$$\mathcal{M}(D,R,B,C,S) := \mathcal{M}(D|C,B) \otimes \mathcal{M}(R|C) \otimes \mathcal{M}(B|S) \otimes \mathcal{M}(C|S) \otimes \mathcal{M}(S). \quad (9.11)$$

♦

In general situations, we aim at specifying a probabilistic graphical model over a collection of categorical variables $\mathbf{X} := (X_1, \ldots, X_n)$, which are in one-to-one correspondence with the nodes of a directed acyclic graph $\mathcal{G}$. The notation $\mathrm{Pa}(X_i)$ is used for the variables corresponding to the parents of X_i according to the graph $\mathcal{G}$ (e.g., in Figure 9.4, $\mathrm{Pa}(D) = (C, B)$). Similarly, $\mathrm{pa}(X_i)$ and $Pa(X_i)$ are the generic value and possibility space of $\mathrm{Pa}(X_i)$. Assume the variables in $\mathbf{X}$ to be in a topological ordering.[15] Then, by analogy with what we did in Example 9.7, we can define a joint CS as follows:

$$\mathcal{M}(\mathbf{X}) := \bigotimes_{i=1}^{n} \mathcal{M}(X_i|\mathrm{Pa}(X_i)). \quad (9.12)$$

This leads to the following formal definition.

Definition 9.1 A credal network (CN) over a set of variables $\mathbf{X} := (X_1, \ldots, X_n)$ is a pair $\langle \mathcal{G}, \mathbb{M} \rangle$, where $\mathcal{G}$ is a directed acyclic graph whose nodes are associated to $\mathbf{X}$, and $\mathbb{M}$ is a collection of conditional CSs $\{\mathcal{M}(X_i|\mathrm{Pa}(X_i))\}_{i=1,\ldots,n}$, where, as usual, $\mathcal{M}(X_i|\mathrm{Pa}(X_i)) := \{\mathcal{M}(X_i|\mathrm{pa}(X_i))\}_{\mathrm{pa}(X_i) \in Pa(X_i)}$. The joint CS $\mathcal{M}(\mathbf{X})$ in (9.12) is called the *strong extension* of the CN.

A characterization of the extreme points of the strong extension $\mathcal{M}(\mathbf{X})$ as in (9.12) is provided by the following proposition [25].

Proposition 9.1 *Let* $\{P_j(\mathbf{X})\}_{j=1}^{v}$ *denote the extreme points of the strong extension* $\mathcal{M}(\mathbf{X})$ *of a CN, i.e.,* $\mathrm{ext}[\mathcal{M}(\mathbf{X})] = \{P_j(\mathbf{X})\}_{j=1}^{v}$. *Then, for each* $j = 1, \ldots, v$, $P_j(\mathbf{X})$ *is a joint mass functions obtained as the product of extreme points of the conditional CSs, i.e.,* $\forall \mathbf{x} \in \mathcal{X}$:

$$P_j(\mathbf{x}) = \prod_{i=1}^{n} P_j(x_i|\mathrm{pa}(X_i)), \quad (9.13)$$

where, for each $i = 1, \ldots, n$ *and* $\mathrm{pa}(X_i) \in Pa(X_i)$, $P_j(X_i|\mathrm{pa}(X_i)) \in \mathrm{ext}[\mathcal{M}(X_i|\mathrm{pa}(X_i))]$.

[15] A topological ordering for the nodes of a directed acyclic graph is an ordering in which each node comes before all nodes to which it has outbound arcs. As an example, (S, C, B, R, D) is a topological ordering for the nodes of the graph in Figure 9.4. Note that every directed acyclic graph has one or more topological orderings.

According to Proposition 9.1, the extreme points of the strong extension of a CN can be obtained by combining the extreme points of the conditional CSs involved in its specification. Note that this can make the number of extreme points of the strong extension exponential in the input size.

Example 9.8 The CS in (9.11) can be regarded as the strong extension of a CN. According to Proposition 9.1, each vertex of it factorizes as follows:

$$P(d, r, b, c, s) = P(d|c, b)P(r|c)P(b|s)P(c|s)P(s).$$

It is a simple exercise to verify that this joint distribution satisfies the conditional independence statements following from the Markov condition (intended with the notion of stochastic instead of strong independence). ♦

The above result can be easily generalized to the strong extension of any CN. Thus, if any extreme point of the strong extension obeys the Markov condition with stochastic independence, the strong extension satisfies the Markov condition with strong independence.

An example of CN specification and its strong extension are reported in the following.

Example 9.9 Given the five binary variables introduced in Example 9.1, associated with the directed acyclic graph in Figure 9.4, consider the following specification of the (collections of conditional) CSs $\mathcal{M}(S)$, $\mathcal{M}(C|S)$, $\mathcal{M}(B|S)$, $\mathcal{M}(R|C)$, $\mathcal{M}(D|C, B)$, implicitly defined by the following constraints:

$$\begin{cases} .25 \leq P(s) \leq .50 \\ \\ .05 \leq P(c|\neg s) \leq .10 \\ .15 \leq P(c|s) \leq .40 \\ \\ .20 \leq P(b|\neg s) \leq .30 \\ .30 \leq P(b|s) \leq .55 \\ \\ .01 \leq P(r|\neg c) \leq .05 \\ .90 \leq P(r|c) \leq .99 \\ \\ .10 \leq P(d|\neg c, \neg b) \leq .20 \\ .80 \leq P(d|\neg c, b) \leq .90 \\ .60 \leq P(d|c, \neg b) \leq .80 \\ .90 \leq P(d|c, b) \leq .99. \end{cases}$$

The strong extension of this CN is a CS $\mathcal{M}(D, R, B, C, S)$ defined as in (9.12). As a consequence of Proposition 9.1, the extreme points of $\mathcal{M}(D, R, B, C, S)$ are combinations of the extreme points of the local CSs, and can therefore be up to 2^{11} if no combination lies in the convex hull of the others. As a simple exercise let us compute the lower probability for the joint state where all the variables are in the state true, i.e., $\underline{P}(d, r, b, c, s)$.

This lower probability should be intended as the minimum, with respect to the strong extension $\mathcal{M}(D, R, B, C, S)$, of the joint probability $P(d, r, b, c, s)$. Thus, we have:

$$\min_{P(D,R,B,C,S)\in\mathcal{M}(D,R,B,C,S)} P(d, r, b, c, s) = \min_{P(D,R,B,C,S)\in\text{ext}[\mathcal{M}(D,R,B,C,S)]} P(d, r, b, c, s)$$

$$= \min_{\bullet} P(s)P(c|s)P(b|s)P(r|c)P(d|c, s) = \underline{P}(s)\underline{P}(c|s)\underline{P}(b|s)\underline{P}(r|c)\underline{P}(d|c, s),$$

$$\text{with } \bullet \hat{=} \left\{ \begin{array}{c} P(S) \in \text{ext}[\mathcal{M}(S)] \\ P(C|s) \in \text{ext}[\mathcal{M}(C|s)] \\ P(B|s) \in \text{ext}[\mathcal{M}(B|s)] \\ P(R|c) \in \text{ext}[\mathcal{M}(R|c)] \\ P(D|c) \in \text{ext}[\mathcal{M}(D|c, s)] \end{array} \right\}$$

with the first step because of (9.2), the second because of Proposition 9.1, and the last because each conditional distribution takes its values independently of the others. This result, together with analogous for upper probability, gives $P(d, r, b, c, s) \in [.0091125, .1078110]$. ♦

The above computation can be regarded as a simple example of inference based on the strong extension of a CN. More challenging problems based on more sophisticated algorithmic techniques are described in Section 9.5.3.

Overall, we introduced CNs as a well-defined class of probabilistic graphical models with imprecision. Note that exactly as a single probability mass function can be regarded as a special CS with a single extreme point, we can consider a special class of CNs, whose conditional CSs are made of a single probability mass function each. This kind of CNs are called *Bayesian networks* [505] and their strong extension is a single joint probability mass function, which factorizes according to the (stochastic) conditional independence relations depicted by its graph, i.e., as in (9.13). In this sense, CNs can be regarded as a generalization to imprecise probabilities of Bayesian networks. With respect to these precise probabilistic graphical models, CNs should be regarded as a more expressive class of models.

9.4.1 Nonseparately specified credal networks

In the definition of strong extension as in (9.12), each conditional probability mass function is free to vary in (the set of extreme points of) its conditional CS independently of the others. In order to emphasize this feature, CNs of this kind are said to be defined with separately specified CSs, or simply *separately specified.* Separately specified CNs are the most commonly used type of CN, but it is possible to consider CNs whose strong extension cannot be formulated as in (9.12). This corresponds to having relationships between the different specifications of the conditional CSs, which means that the choice for a given conditional mass function can be affected by that of some other conditional mass functions. A CN of this kind is simply called *nonseparately specified.*

As an example, some authors considered so-called *extensive* specifications, where instead of a separate specification for each conditional mass function associated to X_i, the *probability table* $P(X_i|\text{Pa}(X_i))$, i.e., a function of both X_i and $\text{Pa}(X_i)$, is defined to belong to a finite set (of tables). This corresponds to assuming constraints between the

Table 9.2 A data set about three of the five binary variables of Example 9.1; '$*$' denotes a missing observation.

S	C	R
s	c	r
$\neg s$	$\neg c$	r
s	c	$\neg r$
s	$*$	r

specification of the conditional CSs $\mathcal{M}(X_i|\text{pa}(X_i))$ corresponding to the different values of $\text{pa}(X_i) \in Pa(X_i)$. The strong extension of an extensive CN is obtained as in (9.12), by simply replacing the separate requirements for each single conditional mass function with extensive requirements about the tables which take values in the corresponding finite set.

Example 9.10 (extensive specification). Consider the CN defined in Example 9.9 over the graph in Figure 9.4. Keep the same specification of the conditional CSs, but this time use extensive constraints for the CSs of B. According to Definition 9.1, in the joint specifications of the two CSs of B, all the four possible combinations of the extreme points of $\mathcal{M}(B|s)$ with those of $\mathcal{M}(B|\neg s)$ appear. An example of extensive specification for this variable would imply that only the following two tables can be considered:

$$P(B|S) \in \left\{ \begin{bmatrix} .20 & .30 \\ .80 & .70 \end{bmatrix}, \begin{bmatrix} .30 & .55 \\ .70 & .45 \end{bmatrix} \right\}. \tag{9.14}$$

♦

Extensive specifications are not the only kind of nonseparate specification we consider for CNs. In fact, we can also consider constraints between the specification of conditional CSs corresponding to different variables. This is a typical situation when the quantification of the conditional CSs in a CN is obtained from a data set. A simple example is illustrated below.

Example 9.11 (learning from incomplete data). Among the five variables of Example 9.1, consider only S, C, and R. Following Example 9.6, S and R are strongly independent given C. Accordingly, let us define a joint $\mathcal{M}(S, C, R)$ as a CN associated to a graph corresponding to a chain of three nodes, with S first (parentless) node and R last (childless) node of the chain.

Assume that we learn the model probabilities from the incomplete data set in Table 9.2, assuming no information about the process making the observation of C missing in the last instance of the data set. A possible approach is to learn two distinct probabilities from the two complete data set corresponding to the possible values of the missing observation, and use them to specify the extreme points of the conditional CSs of a CN.

To make things simple we compute the probabilities for the joint states by means of the relative frequencies in the complete data sets. Let $P_1(S, C, R)$ and $P_2(S, C, R)$ be the joint mass functions obtained in this way, from which we obtain the same conditional mass functions for

$$P_1(s) = P_2(s) = \tfrac{3}{4}$$
$$P_1(c|\neg s) = P_2(c|\neg s) = 0$$
$$P_1(r|\neg c) = P_2(r|\neg c) = 1;$$

and different conditional mass functions for

$$\begin{array}{ll} P_1(c|s) = 1 & P_2(c|s) = \tfrac{2}{3} \\ P_1(r|c) = \tfrac{2}{3} & P_2(r|c) = \tfrac{1}{2}. \end{array} \tag{9.15}$$

We have therefore obtained two, partially distinct, specifications for the local models over variables S, C and R. The conditional probability mass functions of these networks are the extreme points of the conditional CSs for the CN we consider. Such a CN is nonseparately specified. To see that, just note that if the CN would be separately specified the values $P(c|s) = 1$ and $P(r|c) = \frac{1}{2}$ could be regarded as a possible instantiation of the conditional probabilities, despite the fact that there are no complete data sets leading to this combination of values. ♦

Despite their importance in the modelling various problems, nonseparate CNs have received relatively small attention in the literature. Most of the inference algorithms for CNs are in fact designed for separately specified models. However, two important exceptions are the credal classifiers presented in Chapter 10: the naive credal classifier and the credal TAN (see Section 10.3) are credal networks whose credal sets are nonseparately specified for reasons similar to those described in Example 9.11. Another exception is the inference technique proposed in [456], where an extensive specification is assumed for each variable of the network.[16]

Furthermore, it has been shown that nonseparate CNs can be equivalently described as separately specified CNs augmented by a number of auxiliary parent nodes enumerating only the possible combinations for the constrained specifications of the conditional CSs. This can be described by the following example.

Example 9.12 ('separating' a nonseparately specified CN). Consider the extensively specified CN in Example 9.10. Augment this network with an auxiliary node A, which is used to model the constraints between the two, nonseparately specified, conditional CSs $\mathcal{M}(B|s)$ and $\mathcal{M}(B|\neg s)$. Node A is therefore defined as a parent of B, and the resulting graph becomes that in Figure 9.5. The states of A are indexing the possible specifications of the table $P(B|S)$. So, A should be a binary variable such that $P(B|S, a)$ and $P(B|S, \neg a)$ are the two tables in (9.14). Finally, specify $\mathcal{M}(A)$ as a vacuous CS. Overall, we obtain a separately specified CN whose strong extension coincides with that of the CN in Example 9.10.[17] ♦

This procedure can be easily applied to any nonseparate specification of a CN. We point the reader to [25] for details.

[16] This is not preventing the application to separately specified CNs: a separate specification of the conditional credal sets of a variable can be equivalently formulated as an extensive specification obtained by considering all the possible combinations of the vertices of the conditional credal sets.

[17] Once rather than the auxiliary variable A is marginalized out.

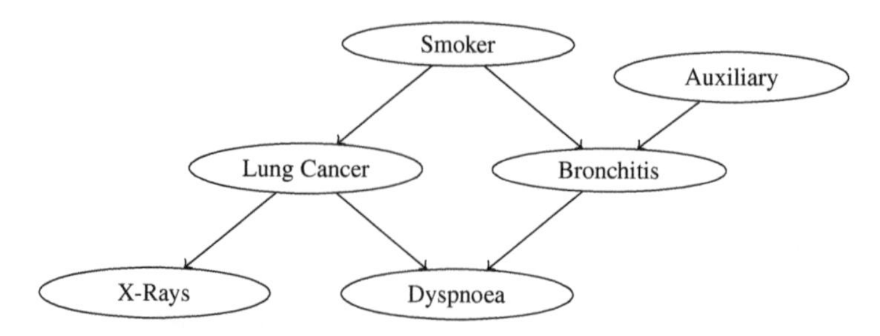

Figure 9.5 The network in Figure 9.4 with an auxiliary node indexing the tables providing the extensive specification of $\mathcal{M}(B|S)$.

9.5 Computing with credal networks

9.5.1 Credal networks updating

In the previous sections we have shown how a CN can model imprecise knowledge over a joint set of variables. Once this modelling phase has been achieved, it is possible to interact with the model through *inference* algorithms. This corresponds, for instance, to query a CN in order to gather probabilistic information about a variable of interest X_q given evidence x_E about some other variables X_E. This task is called *updating* and consists in the computation of the lower (and upper) posterior probability $P(x_q|x_E)$ with respect to the network strong extension $\mathcal{M}(\mathbf{X})$. For this specific problem, (9.2) rewrites as follows:

$$\underline{P}(x_q|x_E) = \min_{P(\mathbf{X})\in\mathcal{M}(\mathbf{X})} P(x_q|x_E) = \min_{j=1,\dots,v} \frac{\sum_{x_M} \prod_{i=1}^{n} P_j(x_i|\mathrm{pa}(X_i))}{\sum_{x_M,x_q} \prod_{i=1}^{n} P_j(x_i|\mathrm{pa}(X_i))}, \tag{9.16}$$

where $\{P_j(\mathbf{X})\}_{j=1}^{v}$ are the extreme points of the strong extension, $X_M = \mathbf{X}\setminus(\{X_q\}\cup X_E)$, and in the second step we exploit the result in Proposition 9.1. A similar expression with a maximum replacing the minimum defines upper probabilities $\overline{P}(x_q|x_E)$. Note that, for each $j = 1, \dots, v$, $P_j(x_q|x_E)$ is a posterior probability for a Bayesian network over the same graph. In principle, updating could be therefore solved by simply iterating standard Bayesian network algorithms. Yet, according to Proposition 9.1, the number v of extreme points of the strong extension might be exponential in the input size, and (9.16) can be hardly solved by such an exhaustive approach. In fact, exact updating displays higher complexity in CNs rather than Bayesian networks: CNs updating is NP-complete for polytrees[18] (while polynomial-time algorithms exist for Bayesian networks with the same topology [505]), and NP^{PP}-complete for general CNs [186] (while updating of general Bayesian networks is PP-complete [173]). Yet, a number of exact and approximate algorithms for CNs updating has been developed. A summary about the state of the art in this field is reported in Section 9.5.3.

[18] A credal (or a Bayesian) network is said to be a *polytree* if its underlying graph is singly connected, i.e., if given two nodes there is at most a single undirected path connecting them. A *tree* is a polytree whose nodes cannot have more than a single parent.

Algorithms of this kind can compute, given the available evidence x_E, the lower and upper probabilities for the different outcomes of the queried variable X_q, i.e., the set of probability intervals $\{[\underline{P}(x_q|x_E), \overline{P}(x_q|x_E)]\}_{x_q \in \mathcal{X}_q}$. In order to identify the most probable outcome for X_q, a simple *interval dominance*[19] criterion can be adopted. The idea is to *reject* a value of X_q if its upper probability is smaller than the lower probability of some other outcome. Clearly, this criterion is not always intended to return a single value as the most probable for X_q. In general, after updating, the posterior knowledge about the state of X_q is described by the set $\mathcal{X}_q^* \subseteq \mathcal{X}_q$, defined as follows:

$$\mathcal{X}_q^* := \left\{ x_q \in \mathcal{X}_q \;\middle|\; \nexists\, x_q' \in \mathcal{X}_q \text{ s.t. } \overline{P}(x_q|x_E) < \underline{P}(x_q'|x_E) \right\}. \tag{9.17}$$

Criteria other than interval dominance have been proposed in the literature and formalized in the more general framework of decision making with imprecise probabilities (see Chapter 8 for a deeper discussion on this topic). As an example, the set of nondominated outcomes $\mathcal{X}_q^{**}$ according to the *maximality* criterion [672, § 3.9] is obtained by rejecting the outcomes whose probabilities are dominated by those of some other outcome, for any distribution in the posterior CS, i.e., underlying (9.17)

$$\mathcal{X}_q^{**} := \left\{ x_q \in \mathcal{X}_q \;\middle|\; \nexists x_q' \in \mathcal{X}_q \text{ s.t. } P(x_q|x_E) < P(x_q'|x_E) \forall P(X_q|x_E) \in \text{ext}[\mathcal{M}(X_q|x_E)] \right\}. \tag{9.18}$$

Maximality is in general more informative than interval dominance, i.e., $\mathcal{X}_q^{**} \subseteq \mathcal{X}_q^*$. As a matter of fact, most of the algorithms for CNs are designed to compute the posterior probabilities as in (9.16), this allowing for the application of interval dominance, but apparently preventing the application of maximality. Yet, the inequalities to be discussed for the computation of (9.18) can be reduced to standard updating tasks by simply adding to the queried node an auxiliary child with an appropriate quantification of its conditional probabilities.

9.5.2 Modelling and updating with missing data

The updating problem in (9.16) refers to a situation where the actual values x_E of the variables X_E are available, while those of the variables in X_M are missing. The latter variables are simply marginalized out. This corresponds to the most popular approach to missing data in the literature and in the statistical practice: the so-called *missing at random* assumption MAR, [438], see also Section 7.8.1.3 and 10.3.3, which allows missing data to be neglected, thus turning the incomplete data problem into one of complete data. In particular, MAR implies that the probability of a certain value to be missing does not depend on the value itself, neither on other nonobserved values. Yet, MAR is not realistic in many cases, as shown for instance in the following example.

Example 9.13 *Consider the variable* smoker (*S*) *in Example* 9.1. *For a given patient, we may want to 'observe' S by simply asking him about that. The outcome of this observation is missing when the patient refuses to answer. MAR corresponds to a situation where the probability that the patient would not answer is independent of whether or not he actually*

[19] Compare Definition 8.5.

smokes. Yet, it could be realistic to assume that, for instance, the patient is more reluctant to answer when he is a smoker.

If MAR does not appear tenable, more conservative approaches than simply ignoring missing data are necessary in order to avoid misleading conclusions. De Cooman and Zaffalon have developed an inference rule based on much weaker assumptions than MAR, which deals with near-ignorance about the missingness process [215]. This result has been extended [721] to the case of mixed knowledge about the missingness process: for some variables the process is assumed to be nearly unknown, while it is assumed to be MAR for the others. The resulting updating rule is called *conservative inference rule*[20] (CIR).

To show how CIR-based updating works, we partition the variables in $\mathbf{X}$ in four classes: (i) the queried variable X_q, (ii) the observed variables X_E, (iii) the unobserved MAR variables X_M, and (iv) the variables X_I made missing by a process that we basically ignore. CIR leads to the following CS as our updated beliefs about queried variable:[21]

$$\mathcal{M}(X_q||^{X_I}x_E) := \mathrm{CH}\{P_j(X_q|x_E, x_I)\}_{x_I\in\mathcal{X}_I, j=1,\dots,v}, \tag{9.19}$$

where the superscript on the double conditioning bar is used to denote beliefs updated with CIR and to specify the set of missing variables X_I assumed to be non-MAR, and $P_j(X_q|x_E, x_I) = \sum_{x_M} P_j(X_q, x_M|x_E, x_I)$. The insight there is that, as we do not know the actual values of the variables in X_I and we cannot ignore them, we consider all their possible explanations. In particular, when computing lower probabilities, (9.19) implies:

$$\underline{P}(X_q||^{X_I}x_E) = \min_{x_I\in\mathcal{X}_I} \underline{P}(X_q|x_E, x_I). \tag{9.20}$$

When coping only with the MAR variables (i.e., if X_I is empty), (9.20) becomes a standard updating task to be solved by one of the algorithms we will list in Section 9.5.3. Although these algorithms cannot be directly applied if X_I is not empty, a procedure to map a CIR task as in (9.19) into a standard updating task as in (9.16) for a CN defined over a wider domain has been developed [24].[22] The transformation is particularly simple and consists in the augmentation of the original CN with an auxiliary child for each non-missing-at-random variable, as described by the following example.

Example 9.14 (CIR-based updating by standard algorithms). Consider the CN in Example 9.9. In order to evaluate the probability of a patient having lung cancer (i.e., $C = c$), you perform an X-rays test, and you ask the patient whether he smokes. The X-rays are abnormal (i.e., $R = r$), while, regarding S, the patient refuses to answer (i.e., $S = *$). Following the discussion in Example 9.13, we do not assume MAR for this missing variable. Yet, we do not formulate any particular hypothesis about the reasons preventing the observation of S, so we do CIR-based updating. The problem can be equivalently

---

[20] Compare also Section 7.8.2.

[21] This updating rule can be applied also to the case of *incomplete* observations, where the outcome of the observation of X_I is missing according to a non-missing-at-random process, but after the observation some of the possible outcomes can be excluded. If $\mathcal{X}'_I \subset \mathcal{X}_I$ is the set of the remaining outcomes, we simply rewrite Equation (9.19), with $\mathcal{X}'_I$ instead of $\mathcal{X}_I$.

[22] An exhaustive approach to the computation of (9.20) consisting in the computation of all the lower probabilities on the right-hand side is clearly exponential in the number of variables in X_I.

solved by augmenting the CN with an auxiliary (binary) child O_S of S, such that $\mathcal{M}(O_S|s)$ is a vacuous CS for each $s \in \mathcal{S}$. It is easy to prove that $\underline{P}(c||^S d) = \underline{P}(c|d, o_s)$, where the latter inference can be computed by standard algorithms in the augmented CN. ♦

9.5.3 Algorithms for credal networks updating

Despite the hardness of the problem, a number of algorithms for exact updating of CNs have been proposed. Most of these methods generalize existing techniques for Bayesian networks. Concerning Pearl's algorithm for efficient updating on polytree-shaped Bayesian networks [505], a direct extension to CNs is not possible unless all variables are binary. The reason is that a CS over a binary variable has at most two extreme points (see Section 9.2.1) and it can therefore be identified with an interval. This enables an efficient extension of Pearl's propagation scheme. The result is an exact algorithm for binary polytree-shaped separately specified CNs, called *2-Updating* (or simply 2U), whose computational complexity is linear in the input size.[23]

Another exception exists if one works with a CN under the concept of *epistemic* irrelevance (see Section 3.2.1). In this case, updating can be performed in polynomial time if the topology is a tree [204]. A similar result cannot be obtained for strong independence, for which updating is NP-hard [186], even in trees [183]. Yet, the problem has been shown to be approximable in polynomial time for general topologies under very reasonable assumptions [456].[24] Hence, even if inference cannot be processed by an exact method, there are various approximate algorithms to be considered instead [23, 96, 99, 100, 165, 183, 275, 276, 375].

Other approaches to exact inference are also based on generalizations of the most known algorithms for Bayesian networks. For instance, the *variable elimination* technique of Bayesian networks [219] corresponds, in a credal setting, to a *symbolic* variable elimination, where each elimination step defines linear constraints among the different conditional probabilities where the variable to be eliminated appears. The elimination is said to be symbolic because numerical calculation are not performed, instead constraints are generated to later be treated by a specialized (nonlinear) optimization tool. Overall, this corresponds to a mapping between CNs updating as in (9.16) and *multilinear programming* [194]. Similarly, an approach analogous to the *recursive conditioning* technique of Bayesian networks [172] can be used to transform the problem into an *integer linear programming* problem [188]. Other exact inference algorithms examine potential extreme points of the strong extension according to different strategies in order to produce the required lower/upper values [96, 160], but are limited in the size of networks that they can handle.

Concerning approximate inference, there are three types: (i) inner approximations, where a (possibly local optimal) solution is returned; (ii) outer approximations of the bounds of the objective value (which are not associated to a feasible solution), and (iii) other methods that cannot guarantee to be inner or outer. Some of these algorithms emphasize enumeration of extreme points, while others resort to nonlinear optimization techniques. Possible techniques include branch-and-bound methods [184, 188], relaxation of probability values [100], or relaxation of the constraints that define the optimization

[23] This algorithm has been extended to the case of extensive specifications in [25], while a modified version to update more efficiently binary polytrees with *imprecise noisy-OR gates* is in [18].

[24] These assumptions being a bound on the treewidth and on the number of states for each variable.

problem [100]. Inner approximation methods search through the feasible region of the problem using well-known techniques, such as genetic programming [99], simulated annealing [96], hill-climbing [97], or even specialized multilinear optimization methods [163, 165]. Finally, there are methods that cannot guarantee the quality of the result, but usually perform very well in practice. For instance, *loopy propagation* is a popular technique that applies Pearl's propagation to multiply connected Bayesian networks [488]: propagation is iterated until probabilities converge or for a fixed number of iterations. In [374], Ide and Cozman extend these ideas to belief updating on CNs, by developing a loopy variant of 2U that makes the algorithm usable for multiply connected binary CNs. This idea has been further exploited by the *generalized loopy 2U*, which transforms a generic CN into an equivalent binary CN, which is indeed updated by the loopy version of 2U [23]. Another promising approach is in [456], where it is shown that a Bayesian network inference algorithm which propagates sets instead of single potentials, enhanced with pruning techniques, can lead to a competitive exact method for belief updating in credal networks, and can serve as basis for efficient approximation methods.

9.5.4 Inference on credal networks as a multilinear programming task

In this section we describe by examples two ideas to perform exact inference with CNs: symbolic variable elimination and conditioning. In both cases, the procedure generates a multilinear programming problem, which must be later solved by a specialized software. Multilinear problems are composed of multivariate polynomial constraints and objective where the exponents of optimization variables are either zero or one, that is, each nonlinear term is formed by a product of distinct optimization variables.

Example 9.15 (multilinear programming). Consider the task of computing $\underline{P}(s|\neg d)$ in the CN of Example 9.9. In order to perform this calculation, we write a collection of multilinear constraints to be later processed by a multilinear programming solver. The objective function is defined as

$$\min P(s|\neg d) \tag{9.21}$$

subject to

$$P(s|\neg d) \cdot P(\neg d) = P(s, \neg d), \tag{9.22}$$

and then two symbolic variable eliminations are used, one with query $\{s, \neg d\}$ and another with query $\{\neg d\}$, to build the constraints that define the probability values appearing in Expression (9.22). Note that the probability values $P(s|\neg d)$, $P(\neg d)$, and $P(s, \neg d)$ are viewed as optimization variables such that the multilinear programming solver will find the best configuration to minimize the objective function that respects all constraints. Strictly speaking, the minimization is over all the Ps that appear in the multilinear programming problem. Expression (9.22) ensures that the desired minimum value for $P(s|\neg d)$ is indeed computed as long as the constraints that specify $P(\neg d)$ and $P(s, \neg d)$ represent exactly what is encoded by the CN. The idea is to produce the bounds for $P(s|\neg d)$ without having to explicitly compute the extension of the network.

The symbolic variable elimination that is executed to write the constraints depends on the variable elimination order, just as the bucket elimination in Bayesian networks [219].

In this example, we use the elimination order S, B, C. For the computation of the probability of the event $\{s, \neg d\}$ using that order, the variable elimination produces the following list of computations, which in our case are stored as constraints for the multilinear programming problem (details on variable elimination can be found in [219, 404]):

- Bucket of S: $\forall c', \forall b'$: $\mathbf{P(s, c', b')} = P(s) \cdot P(c'|s) \cdot P(b|s')$.[25] No variable is summed out in this step because S is part of the query. Still, new intermediate values (marked in bold) are defined and will be processed in the next step. In a usual variable elimination, the values $P(s, c', b')$, for every c', b', would be computed and propagated to the next bucket. Here instead optimization variables $P(s, c', b')$ are included in the multilinear programming problem.

- Bucket of B: $\forall c'$: $\mathbf{P(s, c', \neg d)} = \sum_b \mathbf{P(s, c', b)} \cdot P(\neg d|c', b)$. In this bucket, B is summed out (eliminated) in the probabilistic interpretation. New intermediate values $P(s, c', \neg d)$ (for every c') appear and will be dealt in the next bucket (again, in a usual variable elimination, they would be the propagated values).

- Bucket of C: $\mathbf{P(s, \neg d)} = \sum_{c'} \mathbf{P(s, c', \neg d)}$. By this equation C is summed out, obtaining the desired result. In the multilinear interpretation, the optimization variables $P(s, c', \neg d)$ are employed to form a constraint that defines $P(s, \neg d)$.

In bold we highlight the intermediate probability values that are not part of the CN specification, so they are solely tied by the equations just presented to the probability values that are part of the input (those not in bold). Overall, these equations define the value of $P(s, \neg d)$ in terms of the input values of the problem.

The very same idea is employed in the symbolic computation of the probability of $\neg d$:

- Bucket of S: $\forall c', \forall b'$: $\mathbf{P(c', b')} = \sum_{s'} P(s') \cdot P(c'|s') \cdot P(b'|s')$. Now S is not part of the query, so it is summed out. Four constraints (one for each joint configuration c', b') define the values $P(c', b')$ in terms of the probability values involving S.

- Bucket of B: $\forall c'$: $\mathbf{P(c', \neg d)} = \sum_{b'} \mathbf{P(c', b')} \cdot P(\neg d|c', b')$. Here B is summed out, and the (symbolic) result is $P(c', \neg d)$, for each c'.

- Bucket of C: $\mathbf{P(\neg d)} = \sum_{c'} \mathbf{P(c', \neg d)}$. This final step produces the glue between values $P(c', \neg d)$ and $P(\neg d)$ by summing out C.

Finally, the constraints generated by the symbolic variable elimination are put together with those forcing the probability mass functions to lie inside their local CSs, which are simply those specified in Example 9.9. All these constraints are doing is to formalize the dependence relations between the variables in the network, as well as the local CS specifications. Clearly, a different elimination ordering would produce a different multilinear program leading to the same solution. The chosen elimination order has generated terms with up to three factors, that is, polynomials of order three. ♦

We illustrate another idea to generate constraints based on multilinear programming, which uses the idea of conditioning. Instead of a variable elimination, we keep an active

[25] Lowercase letters with a prime are used to indicate a generic state of a binary variable.

set of conditioning variables that cut the network graph, in a similar manner as done by the recursive conditioning method for Bayesian networks.

Example 9.16 (Conditioning). Let us evaluate $P(\neg d)$ and write the constraints that define it, we have:

- Cut-set $\{S\}$: $P(\neg d) = \sum_{s'} P(s') \cdot \mathbf{P}(\neg \mathbf{d}|\mathbf{s}')$. In this step, the probability of D is conditioned on S by one constraint. New intermediate probability values arise (in bold) that are not part of the network specification. Next step takes $P(\neg d|s')$ to be processed and defined through other constraints.
- Cut-set $\{S, C\}$: $\forall s'$: $P(\neg d|s') = \sum_{c'} P(c'|s') \cdot \mathbf{P}(\neg \mathbf{d}|\mathbf{s}', \mathbf{c}')$. The probability of $D|S$ is conditioned on C by two constraints (one for each value of S). Again, new intermediate values appear in bold, which are going to be treated in the next step.
- Cut-set $\{C, B\}$: $\forall s', \forall c'$: $P(\neg d|s', c') = \sum_{b'} P(b'|s') \cdot P(\neg d|c', b')$. Here $D|S, C$ is further conditioned on B by using four constraints (one for each value of S and C), which leaves only C and B in the cut-set (D is independent of S given C, B). The probability values that appear are all part of the network specification, and thus we may stop creating constraints. $P(\neg d)$ is completely written as a set of constraints over the input values.

As in the previous example, the cut-set constraints are later put together with the local constraints of the CSs, as well as the constraints to specify $P(s, \neg d)$ (in this example, this last term is easily defined by a single constraint: $P(s, \neg d) = P(s) \cdot P(\neg d|s)$, because the latter element had already appeared in the cut-set constraints for the cut $\{S, C\}$ and thus is already well-defined by previous constraints). ◆

The construction of the multilinear programming problem just described takes time proportional to an inference in the corresponding precise Bayesian network. The great difference is that, after running such transformation, we still have to optimize the multilinear problem, while in the precise case the result would already be available. Hence, to complete the example, we have run an optimizer over the problems constructed here to obtain $\underline{P}(s|\neg d) = 0.1283$. Replacing the minimization by a maximization, we get $\overline{P}(s|\neg d) = 0.4936$. Performing the same transformation for d instead of $\neg d$, we obtain $P(s|d) \in [0.2559, 0.7074]$, that is, the probability of smoking given dyspnoea is between one fourth and 70 per cent, while smoking given not dyspnoea is between 12 and 49 per cent. The source code of the optimization problems that are used here, together with other tools for inference based on credal networks, is available online in the website http://ipg.idsia.ch. The reader is invited to try them out.

9.6 Further reading

We conclude this chapter by surveying a number of challenges, open problems and alternative models to those we have presented here. We start with a discussion on probabilistic graphical models with imprecision other than credal networks with strong independence. As noted in Section 9.5.3, the literature has recently started exploring an alternative definition of credal network where strong independence is replaced by the weaker concept of *epistemic irrelevance* [204]. This change in the notion of independence used by a credal network affects the results of the inferences [204, § 8] even if the probabilistic information with which one starts is the same in both cases (in particular the inferences made under strong independence will be never less precise, and typically more precise, than those obtained under irrelevance). This means that it is very important to choose the appropriate notion of independence for the domain under consideration.

Yet, deciding which one is the 'right' concept for a particular problem is not always clear. A justification for using strong independence may rely on a sensitivity analysis interpretation of imprecise probabilities: one assumes that some 'ideal' precise probability satisfying stochastic independence exists, and that, due to the lack of time or other resources, can only be partially specified or assessed, thus giving rise to sets of models that satisfy stochastic independence. Although this seems to be a useful interpretation in a number of problems, it is not always applicable. For instance, it is questionable that expert knowledge should comply with the sensitivity analysis interpretation. Epistemic irrelevance has naturally a broader scope, as it only requires that some variables are judged not to influence other variables in a model. For this reason, research on epistemic irrelevance is definitely a very important topic in the area of credal networks. On the other hand, at the moment we know relatively little about how practical is using epistemic irrelevance. A recent paper [182] shows that the maximal sequences of an imprecise hidden Markov model based on irrelevance (which can be regarded as a very simple tree-shaped credal network) can be computed very easily. Similarly, the above mentioned paper [204] shows that trees with irrelevance can be updated efficiently. This, for instance, is not the case under strong independence.[26] But that paper also shows that irrelevance can frequently give rise to dilation [573] in a way that may not always be desirable. This might be avoided using the stronger, symmetrized, version of irrelevance called *epistemic independence* [210, § 4] (some earlier analysis was also done in [187]). But the hope to obtain efficient algorithms under this stronger notion is much less that under irrelevance. Also, epistemic irrelevance and independence have been shown to make some of the graphoid axioms fail [167], which is an indication that the situation on the front of efficient algorithms could become complicated on some occasions. In this sense, the situation of credal networks under strong independence is obviously much more consolidated, as research on this topic has been, and still is, intense, and has been going on for longer.[27] Moreover, the mathematical properties of strong independence make it particularly simple to represent a credal network as a collection of Bayesian networks, and this makes quite natural to (try to) extend algorithms originally developed for Bayesian networks into the credal setting. This makes it easier, at the present time, to address applications using credal networks under strong independence. In summary, we believe that it is too early to make any strong claims on the relative benefits of the two concepts, and moreover we see the introduction of models based on epistemic irrelevance as an exciting and important new avenue for research on credal networks.

Another challenge concerns the development of credal networks with continuous variables. Benavoli et al. [59] have proposed an imprecise hidden Markov model with continuous variables using Gaussian distributions, which produces a reliable Kalman filter algorithm. This can be regarded as a first example of credal network with continuous variables

[26] The hardness of updating credal trees with strong independence follows from a result in [183].

[27] There are also historical reasons for this focus, which are basically those reported in Footnote 3 of this chapter.

over a tree topology. Similarly, the framework for the fusion of imprecise probabilistic knowledge proposed in [56] corresponds to a credal network with continuous variables over the naive topology (see Section 10.2). These works use coherent lower previsions (Chapter 2) for the general inference algorithms, and also provide a specialized version for linear-vacuous mixtures. The use of continuous variables within credal networks is an interesting topic and deserves future attention.

Decision trees have been also explored [372, 382, 395, 588, 634]. Usually in an imprecise decision tree, the decision nodes and utilities are treated in the same way as in their precise counterpart, while chance nodes are filled with imprecise probabilistic assessments. The most common task is to find the expected utility of a decision or find the decisions (or *strategies*) that maximize the expected utility. However, the imprecision leads to imprecise expected utilities, and distinct decision criteria (see Chapter 8) can be used to select the best strategy (or set of strategies). By some (reasonably simple) modifications of the tree structures, it is possible to obtain a credal network (which is not necessarily a tree) whose inference is equivalent to that of the decision tree. In a different approach, where the focus is on classification, imprecise decision trees have been used to build more reliable *classifiers* [3] (see Chapter 10).

Qualitative and semi-qualitative networks, which are Bayesian networks extended with qualitative assessments about the probability of events, are also a type of credal network. The (semi-)qualitative networks share most of the characteristics of a credal network: the set of variables, the directed acyclic graph with a Markov condition, and the local specification of conditional probability mass functions in accordance with the graph. However, these networks admit only some types of constraints to specify the local credal sets. For example, qualitative *influences* define that the probability of a state s of a variable is greater given one parent instantiation than given another, which indicates that a given observed parent state implies a greater chance of seeing s. Other qualitative relations are *additive* and *multiplicative* synergies. The latter are nonlinear constraints and can be put within the framework of credal sets by extending some assumptions (for example, we cannot work only with finitely many extreme points). Qualitative networks have only qualitative constraints, while semi-qualitative networks also allow mass functions to be numerically defined. Some inferences in qualitative networks can be processed by fast specialized algorithms, while inferences in semi-qualitative networks (mixing qualitative and quantitative probabilistic assessments) are as hard as inferences in general credal networks [185, 191].

Markov decision processes have received considerable attention under the theory of imprecise probability [376, 587]. In fact, the framework of Markov decision process has been evolved to deal with deterministic, nondeterministic and probabilistic planning [316]. This has happened in parallel with the development of imprecise probability, and recently it has been shown that Markov decision processes with imprecise probability can have precise and imprecise probabilistic transitions, as well as set-valued transitions, thus encompassing all those planning paradigms [626]. Algorithms to efficient deal with Markov decision processes with imprecise probabilities have been developed [220, 376, 395, 587, 628].

Other well-known problems in precise models have been translated to inferences in credal networks in order to exploit the ideas of the latter to solve the former. For instance, the problem of strategy selection in influence diagrams and in decision networks was mapped to a query in credal networks [190, 455]. Most probable explanations and maximum a posteriori problems of Bayesian networks can be also easily translated into credal networks inferences [186]. Inferences in probabilistic logic, when augmented by stochastic irrelevance/

independence concepts, naturally become credal network inferences [164, 166, 189]. Other extensions are possible, but still to be done. For example, dynamic credal networks have been mentioned in the past [221], but are not completely formalized and widely used. Still, hidden Markov models are a type of dynamic Bayesian networks, so the same relation exists in the credal setting [21]. Besides imprecise hidden Markov models, dynamic credal networks have appeared to model the probabilistic relations of decision trees and Markov decision processes. Moreover, undirected probabilistic graphical models (such as Markov random fields) can clearly be extended to imprecise probabilities. Markov random fields are described by an undirected graph where local functions are defined over the variables that belong to a same clique. These functions are not constrained to be probability distribution, as the whole network is later normalized by the so called partition function. Hence, the local functions can be made imprecise in order to build an imprecise Markov random field, on which inferences would be more reliable.

Overall, a number of probabilistic graphical models with imprecision other than credal networks has been proposed in the literature. We devoted most of this chapter to credal networks because their theoretical development is already quite mature, thus making it possible to show the expressive power (as well as the computational challenges) of approaches based on imprecise probabilities. Furthermore, credal networks have been already applied in a number of real-world problems for the implementation of knowledge-based expert systems (see [19, 22, 191] for some examples, and [514] for a tutorial on implementing these applications). Applications to classification will be considered in the next chapter.

Acknowledgements

This work was supported by the Swiss NSF grants nos. 200020_134759 / 1, 200020_137680 / 1, by the Hasler foundation grant n. 10030, and by the Armasuisse Competence Sector for Science and Technology Management.

10

Classification

Giorgio Corani[1], Joaquín Abellán[2], Andrés Masegosa[2], Serafin Moral[2], and Marco Zaffalon[1]

[1]*Instituto Dalle Molle di Studi sull'Intelligenza Artificiale (IDSIA), Manno-Lugano, Switzerland*

[2]*Department of Computer Science and Artificial Intelligence, University of Granada, Spain*

10.1 Introduction

Classification is the problem of predicting the *class* of a given instance, on the basis of some attributes (*features*) of it. In the Bayesian framework,[1] a classifier is learned from data by updating a *prior* density, which represents the beliefs before analyzing the data and which is usually assumed uniform, with the *likelihood*, which models the evidence coming from the data; this yields a *posterior* joint probability over classes and features. Once the classifier has been learned from data, it can classify novel instances; under 0-1 loss, it returns the most probable class after conditioning on the value of the features of the instance to be classified.

Yet, Bayesian classifiers can happen to issue *prior-dependent* classifications, namely the most probable class changes under different priors. This might be acceptable if the prior has been carefully elicited and represents domain knowledge; however, in general the uniform prior is taken as default, without further investigation.

This consideration has lead to the development of *credal classifiers*, which extend Bayesian classifiers to imprecise probabilities. The term 'credal classifier' was firstly

[1] Cp. Section 7.1.4

Introduction to Imprecise Probabilities, First Edition.
Edited by Thomas Augustin, Frank P. A. Coolen, Gert de Cooman and Matthias C. M. Troffaes.

used in [715], when introducing the naive credal classifier (NCC). NCC generalizes naive Bayes, representing a condition of prior ignorance[2] through the Imprecise Dirichlet Model (IDM) [673], namely adopting a set of priors instead of a single prior.[3] The set of priors is then updated by element-wise application of Bayes rule, yielding a set of posteriors.

From this starting idea, imprecise-probability classification has been developed following two main directions: (a) extensions to imprecise probabilities of models based on Bayesian networks classifiers and (b) extension to imprecise probabilities of models based on classification trees. The common trait of such approaches is that they return a set of classes rather than a single class when there is not enough evidence for safely classifying an instance by a single class. In this way, credal classifiers produce reliable classifications also on small data sets. Traditional classifiers generally undergo a sharp decrease of accuracy on the instances indeterminately classified by credal classifiers.

In Section 10.2 we present the traditional naive Bayes; in Section 10.3 we show how it has been extended to imprecise probability, yielding the naive credal classifier (NCC); in Section 10.4 we review how NCC has been evolved yielding more sophisticated classifiers and in Section 10.5 we present the credal classifiers based on classification trees. In Section 10.6, we discuss the metrics suitable for scoring credal classifiers and present some experiments.

10.2 Naive Bayes

The naive Bayes classifier (NBC) [237] 'naively' assumes the features to be independent given the class; this introduces a severe bias in the estimate of probabilities, as the real data generation mechanism does *not* generally satisfy such condition. As a consequence of such an assumption, NBC tends to assign an unrealistic high probability to the most probable class, being too confident in its predictions [345]; this phenomenon is emphasized when there are highly redundant features or when there is a large number of features.

Yet, NBC performs well under the 0-1 loss [237, 345]. A first reason is that the bias of the probability estimates may not matter under 0-1 loss: given two classes c_1 and c_2 (c_1 being the correct one), even severe bias in the estimate of $P(c_1)$ will not matter, provided that $P(c_1) > P(c_2)$ [302]. The success of NBC can be further explained by looking at the bias-variance decomposition of the misclassification error: NBC has in fact high bias but also low variance, which is the key factor on small data sets. In this way, NBC can be competitive in small and medium data sets, while it is generally outperformed by more complex classifiers on larger data sets.

Moreover, under careful feature selection NBC can become competitive [263, 295] even against state-of-the-art algorithms. A further strength of NBC is computational speed.

There has been a huge research aimed at improving the performance of naive Bayes; comprehensive references can be found in [345, 237] and [363]. The quantity of such attempts demonstrates the interest raised by NBC, which is recognized as one of the top 10 data mining algorithms [710]. The extension of NBC to imprecise probability is called *naive credal classifier* (NCC) [715].

---

[2] Actually, prior *near-ignorance*: full ignorance is not compatible with learning [672, § 7.3.7], see also Page 159.

[3] Cp. Section 7.4.3

10.2.1 Derivation of naive Bayes

Let us denote by C the classification variable (taking values in $\mathcal{C}$) and as $A_1, \ldots, A_k$ the k *discrete* feature variables (taking values from the sets $\mathcal{A}_1, \ldots, \mathcal{A}_k$). Values of class and features are denoted by lower-case letters, such as c and $\mathbf{a} = (a_1, \ldots, a_k)$. For a variable X (be it the class variable or a feature variable), $P(X)$ denotes a probability mass function over all the states $x \in \mathcal{X}$, while $P(x)$ denotes the probability of $X = x$. Both $P(X)$ and $P(x)$ are *epistemic* probabilities; they represent subjective probabilities, obtained by updating prior beliefs with the evidence of the observations.

Instead, the *physical* probability of the actual data generation mechanism is called *chance*; it is unknown and we aim at making inference about it. We denote by $\theta_{c,\mathbf{a}}$ the chance that $(C, A_1, \ldots, A_k) = (c, \mathbf{a})$, by $\theta_{a_i|c}$ the chance that $A_i = a_i$ conditional on $C = c$, by $\theta_{\mathbf{a}|c}$ the chance that $A_1, \ldots, A_k = (a_1, \ldots, a_k)$ conditional on $C = c$.

The naive assumption of independence of the features given the class can be expressed as:

$$\theta_{\mathbf{a}|c} = \prod_{i=1}^{k} \theta_{a_i|c}. \tag{10.1}$$

We assume the data to be complete; the treatment of missing data will be discussed later in Section 10.3 and thus we can proceed by a standard conjugate Bayesian analysis, see also Section 7.4.1. We denote by n the total number of instances; by $n(c)$ the counts of class c and by $n(a_i, c)$ the counts of instances where $A_i = a_i$ and $C = c$; by $\mathbf{n}$ the vector of all counts of type $n(c)$ and $n(a_i, c)$. The *multinomial* likelihood can be expressed as a product of powers of the theta-parameters:

$$L(\theta|\mathbf{n}) \propto \prod_{c \in \mathcal{C}} \left[\theta_c^{n(c)} \prod_{i=1}^{k} \prod_{a_i \in \mathcal{A}_i} \theta_{a_i|c}^{n(a_i,c)} \right]. \tag{10.2}$$

The prior conjugate to the multinomial likelihood is constituted by a product of Dirichlet distributions; in fact, it is analogous to the likelihood, with the counts $n(\cdot)$ replaced everywhere by $st(\cdot) - 1$. The parameter $s > 0$ is the *equivalent sample size* and can be thought of as a number of *hidden instances*, thus interpreting conjugate Bayesian priors as additional samples. The parameters $t(\cdot)$ can be interpreted as the proportion of hidden instances of a given type; for instance, $t(c_1)$ is the expected proportion of hidden instances for which $C = c_1$.

A so-called *non-informative* prior is given by:

$$t(c) = \frac{1}{|\mathcal{C}|}; \quad t(a_i, c) = \frac{1}{|\mathcal{C}||\mathcal{A}_i|}, \tag{10.3}$$

which is also known as the BDe prior in the literature of Bayesian networks [404, § 18]. We denote by $\mathbf{t}$ the vector collecting all the parameters of type $t(c)$ and $t(a_i, c)$ which characterize the joint prior.

Alternatively, the Laplace prior is commonly adopted; it initializes to 1 all counts $n(c)$ and $n(a_i, c)$ before analyzing the data; such prior is also referred to as K2 [404, § 18] in the literature of Bayesian networks. The Laplace prior is obtained by setting $s = |\mathcal{C}|$ and $t(c) = 1/|\mathcal{C}|$ for the marginal probability of the class and $s = |\mathcal{A}_i|$ and $t(c) = 1/|\mathcal{A}_i|$ for the conditional probability of feature $\mathcal{A}_i$ given the class. As can be seen, the Laplace prior does not generally correspond to a joint prior over features and class; moreover, the equivalent sample size cannot be interpreted as the number of hidden instances, since s can be different for each variable.

By multiplying the joint Dirichlet prior and the multinomial likelihood, we obtain a posterior density for $\theta_{c,\mathbf{a}}$, which is again a product of Dirichlet densities; compared to the likelihood (10.2), the parameters $n(\cdot)$ are replaced by $n(\cdot) + st(\cdot)$:

$$P(\theta_{c,\mathbf{a}}|\mathbf{n}, \mathbf{t}, s) \propto \prod_{c \in \mathcal{C}} \left[\theta_c^{n(c)+st(c)-1} \prod_{i=1}^{k} \prod_{a_i \in \mathcal{A}_i} \theta_{a_i|c}^{n(a_i,c)+st(a_i,c)-1} \right]. \tag{10.4}$$

Once $P(\theta_{c,\mathbf{a}}|\mathbf{n}, \mathbf{t})$ has been estimated, the learning step has been accomplished.

At this point, NBC can classify instances, once the assignment $\mathbf{a} = (a_1, \ldots, a_k)$ of the features is known. Considering that $P(c|\mathbf{a}, \mathbf{n}, \mathbf{t}) \propto P(c, \mathbf{a}|\mathbf{n}, \mathbf{t})$, we focus on the latter expression, obtaining:

$$P(c, \mathbf{a}|\mathbf{n}, \mathbf{t}) = P(c|\mathbf{n}, \mathbf{t}) \prod_{i=1}^{k} P(a_i|c, \mathbf{n}, \mathbf{t}), \tag{10.5}$$

and by taking expectations we get:

$$P(c|\mathbf{n}, \mathbf{t}) = \frac{n(c) + st(c)}{n + s} \tag{10.6}$$

$$P(a_i|c, \mathbf{n}, \mathbf{t}) = \frac{n(a_i, c) + st(a_i, c)}{n(c) + st(c)}. \tag{10.7}$$

A problem of NBC, and of Bayesian classifiers in general, is that sometimes the classification is *prior-dependent*, namely the most probable class varies under different **t**. This can be acceptable if the prior is carefully elicited and thus models domain knowledge; yet, this situation is not common and thus in general prior-dependent classifications are unreliable. It can be moreover argued ([672, § 5.5.1]) that the uniform prior is a model of prior *indifference*, which does not properly represent a condition of prior ignorance. To address such concerns NCC extends NBC to imprecise probability, substituting the non-informative prior by the Imprecise Dirichlet Model [673].

10.3 Naive credal classifier (NCC)

NCC adopts a *joint* credal set, which is modelled by the IDM to represent prior *near-ignorance* [672, § 4.6.9]. As a consequence, **t** ranges within polytope $\mathcal{T}$, rather than being fixed. The polytope contains all the densities for which **t** satisfy the constraints, $\forall(i, c)$: $\sum_c t(c) = 1$, $\sum_{a_i} t(a_i|c) = t(c)$, $0 < t(a_i|c) < t(c)$, $0 < t(c) < 1$.

10.3.1 Checking Credal-dominance

To classify an instance under 0-1 loss, traditional classifiers return the most probable class from $P(C|\mathbf{a}, \mathbf{n}, \mathbf{t})$. Instead, NCC identifies the *non-dominated* classes using the criterion of *maximality*[4]: c_1 dominates c_2 if $P(c_1, \mathbf{a}|\mathbf{n}, \mathbf{t}) > P(c_2, \mathbf{a}|\mathbf{n}, \mathbf{t})$ $\forall$ $\mathbf{t} \in \mathcal{T}$. Maximality does not

[4] See also Definition 8.6.

always determine a total order on the possible classes; as a consequence, the decision rule can sometimes select a set of classes rather than a single class.

The test of dominance for NCC under maximality has been designed in [715]. In particular, c_1 dominates c_2 iff:

$$\inf_{\mathbf{t}\in\mathcal{T}} \frac{P(c_1, \mathbf{a}|\mathbf{n}, \mathbf{t})}{P(c_2, \mathbf{a}|\mathbf{n}, \mathbf{t})} > 1 \tag{10.8}$$

subject to:

$$\begin{aligned} & 0 \leq t(c) \leq 1 \quad \forall c \\ & \textstyle\sum_c t_c = 1 \\ & 0 < t(a_i, c) < t(c) \quad \forall (a_i, c) \\ & \textstyle\sum_{a_i} t(a_i, c) = t(c) \quad \forall (c). \end{aligned} \tag{10.9}$$

Considering Equation (10.6 – 10.7), the function (10.8) to be minimized becomes:

$$\inf_{\mathbf{t}\in\mathcal{T}} \left\{ \left[\frac{n(c_2) + st(c_2)}{n(c_1) + st(c_1)}\right]^{k-1} \prod_i \frac{n(a_i, c_1) + st(a_i, c_1)}{n(a_i, c_2) + st(a_i, c_2)} \right\} \tag{10.10}$$

subject to the same constraints.

The infimum of problem (10.10) is obtained by setting, $\forall a_i$, $t(a_i|c_1) \to 0$ and $t(a_i|c_2) \to t(c_2)$; moreover, the infimum is achieved when $t(c_1) + t(c_2) = 1$, which allows to express $t(c_2)$ as $1 - t(c_1)$. The obtained function depends only on $t(c_1)$ and, being convex in $t(c_1)$, can be exactly minimized. More details are given in [715].

NCC identifies the *non-dominated* classes by running many pairwise comparisons between classes, as shown in Figure 10.1. The classification is *determinate* or *indeterminate* if there are respectively one or more non-dominated classes. The set of non-dominated classes contains the most probable class identified by NBC,[5], induced with any joint Dirichlet prior[6]; thus NCC, when determinate, returns the same class of NBC. If there are more non-dominated classes, the most probable class is *prior-dependent*, namely changes under different **t**. The non-dominated classes are *incomparable*, namely there is no information to rank them. In fact, NCC drops the dominated classes as sub-optimal and expresses indecision about the optimal class by yielding the remaining set of non-dominated classes.

The indeterminacy (% of instances indeterminately classified) of NCC decreases with the size of the learning set because the prior becomes less influential as more data are available. For similarly sized data sets, the determinacy of NCC generally decreases with increasing number of classes or values of the feature, which fragment more the data set.

NBC is generally not accurate on the instances indeterminately classified by NCC; in [152], an average drop of some 30 points of accuracy for NBC is reported, between the instances determinately and indeterminately classified by NCC. The overall performance of NBC is therefore constituted by good accuracy on the *non*-prior-dependent instances and a much lower accuracy on the prior-dependent ones. Instead, indeterminate classification

[5] Assuming the s used to learn NBC to be no larger than the s used to learn NCC.

[6] This is not guaranteed for instance under the Laplace prior, although even in this case the statement is most often satisfied.

IDENTIFICATION OF NON-DOMINATED CLASSES
Input: The feature values **a**
Output: the non-dominated classes.

1. set Non-Dominated Classes := ;
2. **for** class $c_1 \in$
   - **for** class $c_2 \in \ , c_2 \neq c_1$
     - **if** $\min_{\mathbf{t} \in} \frac{P(c_1|\mathbf{a},\mathbf{n},\mathbf{t})}{P(c_2|\mathbf{a},\mathbf{n},\mathbf{t})} > 1$, drop c_2 from Non-Dominated Classes;
3. **return** Non-Dominated Classes.

Figure 10.1 Identification of non-dominated classes via pairwise comparisons.

preserves the reliability of NCC on prior-dependent instances; they can moreover be informative, when for instance only few classes are returned, out of many possible ones.

Applications of NCC to real-world case studies include diagnosis of dementia [723] and prediction of presence of parasites in crops [717].

10.3.2 Particular behaviours of NCC

Consider again problem (10.10), namely the minimization of $\frac{P(c_1,\mathbf{a}|\mathbf{n},\mathbf{t})}{P(c_2,\mathbf{a}|\mathbf{n},\mathbf{t})}$. The infimum is obtained by setting $t(a_i, c_1) \to 0\ \forall i$; in the particular case of $n(a_i, c_1) = 0$, we have $n(a_i, c_1) + st(a_i, c_1) = 0$. In this case, $P(a_i|c_1, \mathbf{n}, \mathbf{t})$ goes to sharp 0 when solving the minimization problem (10.10), causing $P(c_1, \mathbf{a}|\mathbf{n}, \mathbf{t})$ to be 0 as well and therefore preventing c_1 from dominating any class. Thus, the presence of a single count $n(a_i, c_1) = 0$ prevents c_1 from dominating any class, regardless the information coming from the remaining features. This phenomenon is called the *feature problem* in [148]; experiments show [148] that NBC can be quite accurate on the instances which are indeterminately classified because of the feature problem. This can be explained by considering that NCC is in fact ignoring the information from all the remaining features.

A further counter-intuitive behaviour of NCC is called the *class problem* in [148] and was already observed in [715, § 6]; the point is that it is very difficult to dominate a class which is never observed (or very rarely observed) in the data. Let us consider what happens when such class is at the denominator of Equation (10.8), namely it corresponds to c_2. For any value a_i of the feature, there are no data for estimating $P(a_i|c_2, \mathbf{n}, \mathbf{t})$, which therefore only depends on the prior; under the IDM, $P(a_i|c_2, \mathbf{n}, \mathbf{t})$ varies between 0 and 1 and it is set to 1 to minimize the ratio of probabilities. As this behaviour repeats for each feature, $P(\mathbf{a}|c_2, \mathbf{n}, \mathbf{t}) \gg P(\mathbf{a}|c_1, \mathbf{n}, \mathbf{t})$. Recalling that $P(c|\mathbf{a}, \mathbf{n}, \mathbf{t}) \propto P(c|\mathbf{n}, \mathbf{t})P(\mathbf{a}|c, \mathbf{n}, \mathbf{t})$, one can see that the class problem often prevents c_2 from being dominated; thus, c_2 is often returned as non-dominated. When indeterminacy is due to the class problem, the accuracy of NBC generally does not drop on the instances indeterminately classified.

Both the feature and the class problem can be addressed by restricting the set of priors of the IDM [148], for instance by considering an ϵ-contamination of the credal set of the IDM with the uniform prior. Yet, restricting the credal set expresses preferences among possible values of the parameters; after the restriction, the prior credal set becomes at some extent informative and does not satisfy some properties of the original IDM, such as the representation invariance principle.

10.3.3 NCC2: Conservative treatment of missing data

Very often, real data sets are subject to missingness: this is the case when some values of the variables are not present in the data set.[7] We call these data sets *incomplete*.[8]

Dealing with incomplete data sets rests on the assumptions done about the process responsible for the missingness. This process can be regarded as one that takes in input a set of complete data, which is generally not accessible for learning, and that outputs an incomplete data set, obtained by turning some values into missing. Learning about the missingness process' behaviour is usually not possible by only using the incomplete data. This fundamental limitation explains why the assumptions about the missingness process play such an important role in affecting classifiers' predictions. Moreover, a missingness process may also be such that the empirical analysis of classifiers is doomed to provide misleading evidence about their actual predictive performance, and hence, indirectly, about the quality of the assumptions done about the missingness process. This point in particular has been discussed in [152, § 4.6] and [721, § 5.3.2]. For these reasons, assumptions about the missingness process should be stated with some care.

In the vast majority of cases, common classifiers deal with missing values (sometimes implicitly) assuming that the values are *missing at random* [543] (MAR), see also Sections 7.8.1.3 and 9.5.2. This assumption makes it possible to deal with incomplete data through the tools used for complete data. In the case of NBC, MAR justifies dropping the missing values from the learning set, as well as those in the instance to classify. Unfortunately, MAR entails the idea that the missingness is generated in a non-selective way (for example in an unintentional way), and this is largely recognized to substantially lead to a narrow scope of MAR in applications. In classification, this may mean that assuming MAR decreases the predictive performance, sometimes in a way that is not easy (or possible) to assess empirically.

As a result, the imprecision introduced by missing data leads to an increase in the indeterminacy of the NCC, which is related to the amount of missingness. In other words, the NCC copes with the weak knowledge about the missingness process by weakening the answers it gives, thus maintaining reliability.

Later work [152] has extended the way NCC deals with missing values in two directions: it has enabled NCC to deal conservatively with missing values also in the instance to classify; and it has given the option to declare variables of two types: those subject to a MAR process and those to a process whose behaviour is basically unknown. The first group is treated according to MAR; the second in a conservative way. The resulting classifier is called NCC2.

Distinguishing variables that are subject to the two types of processes is important because treating MAR variables in a conservative way leads to an excess of indeterminacy in the output that is not justified. In fact, the experimental results of NCC2 show that the indeterminacy originated from missing data is compatible with informative conclusions provided that the variables treated in a conservative way are kept to a reasonable number (feature selection can help in this respect, too). Moreover, they show that the classifiers that assume MAR for all the variables are often substantially unreliable when NCC2 is indeterminate.

---

[7] This may happen also to the class variable, but we do not consider such a case in the present chapter.

[8] See also Section 7.8 on data imprecision.

Formal justifications of the rule NCC2 uses to deal with missing values can be found in [152]. This work discusses also, more generally, the problem of incompleteness for uncertain reasoning. An open source implementation of NCC2 is available [151].

10.4 Extensions and developments of the naive credal classifier

10.4.1 Lazy naive credal classifier

Two aspects of NCC which can be improved are (a) the high bias due to the naive assumption and (b) a sometimes high indeterminacy, not only because of the feature or the class problem. The lazy NCC (LNCC) [153] addresses both problems by combining NCC and *lazy learning*.

The lazy learning algorithms defer the training, until it has to classify an instance (*query*). In order to classify an instance, a lazy algorithm:

1. ranks the instances of the training set according to the distance from the query;
2. trains a local classifier on the k instances nearest to the query and returns the classification using the local classifier;
3. discards the locally trained classifier and keeps the training set in memory in order to answer new queries.

Lazy classifiers are *local*, as they get trained on the subset of instances which are nearest to the query. The parameter k (*bandwidth*) controls the bias-variance trade-off for lazy learning. In particular, a smaller bandwidth implies a smaller bias (even a simple model can fit a complex function on a small subset of data) at a cost of a larger variance (as there are less data for estimating the parameters). Therefore, learning locally NBC can be a winning strategy as it allows reducing the bias; moreover, working locally reduces the chance of encountering strong dependencies between features [300]. The good performance of local NBC is studied for instance in [300].

An important problem dealing with lazy learning is how to select the bandwidth k. The simplest approach is to empirically choose k (for instance, by cross-validation on the training set) and to then use the same k to answer all queries. However, the performance of lazy learning can significantly improve if the bandwidth is adapted query-by-query, as shown in [80] in the case of regression.

LNCC tunes the bandwidth query-by-query using a criterion based on imprecise probability. After having ranked the instances according to their distance from the query, a local NCC is induced on the k_{min} closest instances (for instance, $k_{min} = 25$) and classifies the instance. The classification is accepted if determinate; otherwise, the local NCC is updated by adding a set of further k_{upd} instances (we set $k_{upd} = 20$) to its training set. The procedure continues until either the classification is determinate or all instances have been added to the training of the local NCC. Therefore, the bandwidth is increased until the locally collected data smooth the effect of the choice of the prior. The naive architecture makes it especially easy updating LNCC with the k_{upd} instances; it only requires to update the vector $\mathbf{n}$ of counts, which is internally stored by LNCC.

By design LNCC is generally more determinate than NCC; in the worst case, LNCC is as determinate as NCC. Thus, LNCC addresses both the bias and the indeterminacy problem mentioned at the beginning of this section.

Extensive experiments [153] show that LNCC generally is more accurate or as accurate as NCC, but generates a less indeterminate output; it can be therefore regarded as delivering higher performance than NCC. This is especially apparent in data sets containing thousands of instances, where LNCC generally uses a bandwidth of a few hundreds; this allows a considerable bias reduction and results in generally better classifications.

10.4.2 Credal model averaging

Model uncertainty is the problem of having multiple models which provide a good explanation of the data, but lead to different answers when used to make inference. In this case, selecting a single model while discarding the competing ones underestimates uncertainty. *Bayesian model averaging* (BMA) [364] addresses model uncertainty by averaging over the set of candidate models, assigning to each model a weight corresponding to its posterior probability.

In case of NBC, given k features, there are 2^k possible NBCs, each characterized by a different subset of features; we denote by $\mathcal{G}$ the set of such models and by g a generic model of the set (g can be seen as standing for *graph.*). Using BMA, the posterior probability $P(c, \mathbf{a}|\mathbf{n}, \mathbf{t})$ is computed by averaging over *all* the 2^k different NBCs, namely by marginalizing g out:

$$P(c, \mathbf{a}|\mathbf{n}, \mathbf{t}) \propto \sum_{g \in \mathcal{G}} P(c, \mathbf{a}|\mathbf{n}, \mathbf{t}, g) P(\mathbf{n}|g) P(g), \tag{10.11}$$

where $P(g)$ is the prior probability of model g and $P(\mathbf{n}|g) = \int P(\mathbf{n}|g, \boldsymbol{\theta}) P(\boldsymbol{\theta}|g) d\boldsymbol{\theta}$ is its marginal likelihood. The posterior probability of model g is $P(g|\mathbf{n}) \propto P(\mathbf{n}|g)P(g)$. For a Bayesian network learned with the BDe prior, the marginal likelihood corresponds to the BDeu score [353].

BMA implies two main challenges [116]: the computation of the exhaustive sum of Equation (10.11) and the choice of the prior distribution over the models. The computation of BMA is difficult, because the sum of Equation (10.11) is often intractable; in fact, BMA is often computed via algorithms which are both approximate and time-consuming. However, Dash and Cooper [174] provide an exact and efficient algorithm to compute BMA over all the 2^k NBCs.

As for the choice of the prior, a common choice is to assign equal probability to all models; in the following it is understood that BMA is induced with the uniform prior over the models. However, the uniform prior has the drawbacks already discussed in Section 10.2; its adoption is questioned also within the literature of BMA (see the rejoinder of [364]). The idea of *credal model averaging* (CMA) is thus to substitute such uniform prior by a credal set.

Within BMA, the prior probability of model g is generally expressed as:

$$P(g) = \pi_{A_i \in g} p_i \pi_{A_i \notin g} (1 - p_i), \tag{10.12}$$

where p_i is the prior probability of feature A_i to be present in the true model; moreover, $A_i \in g$ and $A_i \notin g$ index the features which, within the naive model g, are respectively linked to the class or isolated. The uniform prior is obtained by setting $p_i := 0.5$ for all i.

CMA models a condition close to ignorance about the prior probability of the 2^k NBC_s, by letting vary each p_i within the interval $\epsilon < p_i < 1 - \epsilon$, with $\epsilon > 0$ and $\epsilon \ll 1$. This defines a credal set $K(G)$ of mass functions $P(G)$. The introduction of ϵ is necessary to allow learning from the data; otherwise, the posterior probability of the models will keep ranging between 0 and 1, even after having observed the data.

The test of maximality for CMA is:

$$\inf_{P(G)\in K(G)} \frac{\sum_{g\in\mathcal{G}} P(c_1, \mathbf{a}|\mathbf{n}, \mathbf{t}, g)P(\mathbf{n}|g)P(g)}{\sum_{g\in\mathcal{G}} P(c_2, \mathbf{a}|\mathbf{n}, \mathbf{t}, g)P(\mathbf{n}|g)P(g)} > 1. \tag{10.13}$$

The dominance test [150] is computed by extending to imprecise probability the BMA algorithm by [174]. The set of non-dominated classes identified by CMA includes the most probable class identified by BMA; thus CMA, when determinate, returns the same class of BMA.

The experiments of [150] show that the accuracy of BMA sharply drops on the instances where CMA gets indeterminate. The finding that a Bayesian classifier is little accurate on the instances indeterminately classified by its counterpart based on imprecise probabilities is indeed consistent across the various credal classifiers developed so far. Future research could involve CMA for NCC, namely imprecise averaging of credal classifiers.

10.4.3 Profile-likelihood classifiers

A way for extending graphical models to imprecise probability, alternative to coping with multiple priors via the IDM, is adopting a profile likelihood approach.[9] The profile likelihood [504] approach extends the paradigm of maximum likelihood by retaining a set of parameter estimates: all the estimates yielding a likelihood above a certain threshold.

Consider the *credal set* $\mathbf{P}$, collecting probability distributions P_θ indexed by the parameter ϑ, which takes values in a set Θ. Given the available data $\mathcal{D}$, the *normalized likelihood* [504] is:

$$lik(\theta) := \frac{P_\theta(\mathcal{D})}{\sup_{\theta'\in\Theta} P_{\theta'}(\mathcal{D})}. \tag{10.14}$$

We then remove from $\mathbf{P}$ the distributions whose profile likelihood is below a threshold $\alpha \in [0, 1]$:

$$\mathbf{P}_\alpha := \{P_\theta\}_{\theta\in\Theta | lik(\theta)\geq\alpha}. \tag{10.15}$$

In [105, 20], it is discussed how to perform inference and classification using the profile likelihood with naive structures: the resulting model is called *naive hierarchical credal classifier* (HNCC) and corresponds to a collection of NBCs, each characterized by a different parameter estimate with profile likelihood $lik \geq \alpha$. The classifier applies instance by instance the criterion of *maximality* in order to determine the class (or the set of classes) to be returned. Further investigation could be necessary to set up a principled criterion for defining the value of α, which controls the degree of imprecision of HNCC; however, in [20], preliminary experiments are performed using $\alpha = 0.75$ or 0.95. For this choice of the parameter, the behaviour of HNCC shows some similarity with that of NCC.

---

[9] See also the sketch of further inference approaches at Page 179.

10.4.4 Tree-augmented networks (TAN)

A way to reduce the bias of NBC is to relax the independence assumption using TAN (tree-augmented naive Bayes) [303]. In particular, TAN is a Bayesian network where each feature node has one parent (the class) and possibly also a second one (a feature); yet, a feature node cannot have more than two parents. In fact, TAN has been shown to outperform both general Bayesian networks and NBC [303, 446].

The test of dominance for a credal TAN takes in principle a form similar to Equation (10.8) for NCC; yet, the TAN structure generates a set of additional constraints w.r.t. to the NCC case, so that the minimization problem becomes much harder. To have a tractable optimization problem for TAN it is necessary to slightly modify the definition of the credal set w.r.t. to NCC; this is the subject of the next section.

Global IDM, local IDM and extreme IDM

Three kinds of IDM can be used with credal networks: the *global*, the *local* and the *extreme* (EDM) [98]. The differences between these approaches are explained in the following by considering the simple network

$$C \to A.$$

Let us focus on the class node. The constraints which define the set of Dirichlet distributions for the IDM (both local and global) are:

$$\begin{cases} \sum_c t(c) = 1 \\ 0 < t(c) < 1 \qquad \forall c \in C. \end{cases} \tag{10.16}$$

The credal set $K(C)$ is a set of mass functions $P(C)$, within which $P(c)$ is estimated as:

$$P(c) \in \left[\frac{n(c)}{s+n}, \frac{s+n(c)}{s+n} \right]. \tag{10.17}$$

The EDM restricts the set of priors defined by Equation (10.16) to the most extreme mass functions, i.e., each $t(c)$ can be only zero or one. Consequently, the probability of class c can only assume two values, namely the upper or the lower extreme of interval (10.17).

Let us now consider the feature node. The local IDM, similarly to Equation (10.16), adopts the constraints:

$$\begin{cases} \sum_a t(a,c) = 1 & \forall c \in C \\ t(a,c) > 0 & \forall a \in \mathcal{A}, \forall c \in C. \end{cases} \tag{10.18}$$

Yet, there is no relation between the $t(a,c)$ of Equation (10.18) and the $t(c)$ of Equation (10.16). For each c, the credal set $K(A|c)$ is a set of mass functions $P(A|c)$, where $P(a|c)$ is estimated as:

$$P(a|c) \in \left[\frac{n(a,c)}{s+n(c)}, \frac{s+n(a,c)}{s+n(c)} \right]. \tag{10.19}$$

The credal sets $\{K(A|c)\}_{c\in C}$ and $K(C)$ are thus independent of each other; the credal network is *separately specified*.

Instead, the *global* IDM is based on a set of *joint* Dirichlet distributions, defined by the constraints (already given in Section 10.2):

$$\begin{cases} \sum_{c \in \Omega_C} t(c) = 1 & \\ t(c) > 0 & \forall c \in C \\ \sum_a t(a,c) = t(c) & \forall c \in C \\ t(a,c) > 0 & \forall a \in \mathcal{A}, \forall c \in C. \end{cases} \tag{10.20}$$

In particular, the third constraint links $t(a, c)$ and $t(c)$, unlike in the local IDM; therefore, the model is *not* separately specified. For a specific value of $t(c)$, the credal set $K(A|c)$ contains the mass functions $P(A|c)$ such that:

$$P(a|c) \in \left[\frac{n(c,a)}{st(c) + n(c)}, \frac{st(c) + n(c,a)}{st(c) + n(c)} \right]. \tag{10.21}$$

The global IDM estimates narrower intervals than the local, as can be seen by comparing Equation (10.21) and Equation (10.19):[10] this implies less indeterminate classifications. Yet, the global IDM is more difficult to compute; so far, exact computation with the global IDM has been possible only with NCC. Instead, the local IDM can be computed for any network topology and is in fact often adopted with general credal networks; yet, it returns wider intervals, which are sometimes too little informative in real applications.

The EDM restricts the global IDM to its extreme distributions; it therefore allows $t(a, c)$ to be either 0 or $t(c)$, keeping the constraint $\forall c \in \Omega_C : \sum_{a \in \Omega_A} t(a, c) = t(c)$ inherited from the global IDM. The extreme points of the EDM corresponds in this case to the bounds of the interval in Equation (10.21); but in general, they are an inner approximation of the extremes of the global IDM [98]. The EDM can be interpreted as treating the s hidden instances as s rows of missing data; the rows are assumed *identical*; the ignorance is due to the fact that it is unknown which values the hidden instances contain. Therefore, the replacement considered by the EDM for the hidden instances are finite, while the global IDM consider infinite possible replacements, by letting the parameters $t(c)$ vary in a continuous fashion, as in the constraints of Equation (10.9).

The reliability of the EDM has been validated [149] by showing that NCC return almost identical classification when induced with the global IDM and the EDM.

Credal TAN

A credal TAN was firstly proposed in [720]; the classifier was reliable and very accurate when returning a single class, but raised a problem of excessive indeterminacy, due to the local IDM.

More recently [149] the credal TAN has been developed using the EDM. Experimental results [149] show that the novel credal TAN provides overall a better performance than the previous one, returning less indeterminate classifications without decreasing reliability. Yet, in some cases it is still much more indeterminate than NCC; this is due to the TAN structures (learned using Bayesian procedures), which sometimes assign the second parent to a feature node even though it induces a contingency table full of 0s; this generates

[10] Recall that $\sum_c t(c) = 1$ and that $t(c) > 0 \; \forall c \in C$.

large indeterminacy when used under imprecise probability. Future research could be about learning parsimonious TAN structures, more suited to usage with imprecise probability.

10.5 Tree-based credal classifiers

A decision tree[11] (also called a classification tree) is a simple structure, a tree, that can be used as a classifier. Within a decision tree, each node represents an attribute variable and each branch represents one of the states of this variable. A tree leaf specifies the expected value of the class variable conditioned to the path from the root to this leaf and depending on the information contained in the training data set. Quinlan's ID3 algorithm [523] for building decision trees is based on uncertainty-based information theory taking probability theory as the model to represent uncertainty. Building a tree from data implies the selection of the attributes for the nodes (branching attributes), stopping and pruning rules. Usually, these rules are based on measures of information and entropy.

In the last years, studies about properties and behaviour of measures of uncertainty on more general theories than probability, have been presented, principally in the Dempster-Shafer theory of evidence (DST) (Dempster [222] and Shafer [576]), and in the theory of general credal sets (Walley [672]). In DST, Yager [711] extends the concept of uncertainty in probability theory, considering two main origins of uncertainty: conflict and nonspecificity. In Abellán and Moral [4] and Abellán *et al.* [2], functions are presented to quantify these types of uncertainty on credal sets (Klir [399]).

It is possible to apply these measures of uncertainty on credal sets to build classification trees taking imprecise probability as the basic model. In Abellán and Moral [5], a method to build decision trees based on credal sets and measures of uncertainty is presented. This method starts with an empty tree and, for branching at each node, it selects the variable with the greatest degree of total uncertainty reduction in relation to the class variable. In the theory of probability, branching always implies a reduction in entropy. It is, therefore, necessary to include an additional criterion in order to not create excessively complex models. For this aim, a total uncertainty criterion it is used, i.e., a measure of total uncertainty which quantifies conflict and nonspecificity. In credal sets, although the conflict part of the uncertainty produced by branching is smaller, the nonspecificity part is greater. So, in this case a very simple stopping criterion can be proposed: when branching produces an increase in uncertainty (a decrease in conflict is not offset by an increase in nonspecificity). Finally, it is used a dominance criterion to the set values of the variable to be classified in the corresponding leaf.

In the following sections, we will analyze the measure of information used and the procedure to build decision trees.

10.5.1 Uncertainty measures on credal sets: The maximum entropy function

The study of uncertainty measures in the Dempster-Shafer theory of evidence [222, 576] has been the starting point for the development of these measures in more general theories

[11] The term 'decision tree' is used in different meaning in different contexts, namely as a way to describe sequential decision making (see Section 8.2) or as a synonym to classification tree.

(a study of the most important measures proposed in the literature can be seen in [399]). As a reference for the definition of an uncertainty measure on credal sets, Shannon's entropy [586] has been used due to its operation on probabilities. In any theory which is more general than probability theory, it is essential that a measure be able to quantify the uncertainty that a credal set represents: the parts of *conflict* and *nonspecificity* [399].

In recent years, Klir and Smith [400] and Abellán and Moral [8] justified the use of the maximum of entropy on credal sets as a good measure of total uncertainty that verifies a set of basic properties [401]. The problem lies in separating this function into others which really do measure the parts of conflict and nonspecificity, respectively. More recently, Abellán *et al.* [2] presented a separation of the maximum of entropy into functions which are capable of coherently measuring the conflict and nonspecificity of a credal set $\mathcal{P}$ on a finite variable X, as well as algorithms for facilitating its calculation in capacities of order 2 [7, 9] and this may be expressed in the following way:

$$S^*(\mathcal{P}) = S_*(\mathcal{P}) + (S^* - S_*)(\mathcal{P}),$$

where S^* represents the maximum of entropy and S_* represents the entropy minimum on the credal set $\mathcal{P}$:

$$S^*(\mathcal{P}) = \max_{P \in \mathcal{P}} \sum_x P(x) \log(P(x)), \quad S_*(\mathcal{P}) = \min_{P \in \mathcal{P}} \sum_x P(x) \log(P(x)),$$

where $S_*(\mathcal{P})$ coherently quantifies the conflict part of the credal set $\mathcal{P}$ and $(S^* - S_*)(\mathcal{P})$ represents the nonspecificity part of $\mathcal{P}$ [2].

In the particular case of belief functions, Harmanec and Klir [348] have already considered that upper entropy is a measure of total uncertainty. They justify it by using an axiomatic approach. However, uniqueness is not proved. But, perhaps the most compelling reason to use this function is given in Walley [672]. We start by explaining the case of a single probability distribution, P, on a finite set $\mathcal{X}$. It is based on the logarithmic scoring rule . To be subject to this rule means that we are forced to select a probability distribution Q on $\mathcal{X}$, and if the true value is x then we must pay $-\ln(Q(x))$. For example, if we say that $Q(x)$ is very small and x is found to be the true value, we must pay a lot. If $Q(x)$ is close to one, then we must pay a small amount. If our information about X is represented by a subjective probability P, then we should choose Q so that $P(-\ln(Q(X)))$ is minimum, where P is the mathematical expectation (prevision) with respect to P. This minimum is obtained when $Q = P$ and the value of $P(-\ln(P(x)))$ is the entropy of P: the expected loss or the minimum amount that we would require to be subject to the logarithmic scoring rule. This rule is widely used in statistics. The entropy is the negative of the expected logarithm of the likelihood under distribution P. The reason for taking logarithms is that if we do the prediction in two independent experiments at the same time, then the payment should be the addition of the payments in the two experiments.

For the case of a credal set, $\mathcal{P}$, we can also apply the logarithmic scoring rule, but now we choose Q in such a way that the upper expected loss $\overline{P}(-\ln(Q(x)))$ (the supremum of the expectations with respect to the probabilities in $\mathcal{P}$) is minimum. Under fixed Q, $\overline{P}(-\ln(Q(x)))$ is the maximum loss we can have (the minimum we should be given to accept this gamble). As we have freedom to choose Q, we should select it, so that this amount $\overline{P}(-\ln(Q(x)))$ is minimized.

Walley shows that this minimum is obtained for the distribution $\widehat{P} \in \mathcal{P}$ with maximum entropy.[12] Furthermore, $\overline{P}(-\ln(\widehat{P}(x)))$ is equal to $S^*(\mathcal{P})$, the upper entropy in $\mathcal{P}$. This is the minimum payment that we should require before being subject to the logarithmic scoring rule. This argument is completely analogous with the probabilistic one, except that we change expectation to upper expectation. This is really a measure of uncertainty, as the better we know the true value of X, then the less we should need to be paid to accept the logarithmic scoring rule (lower value of $S^*(\mathcal{P})$).

The approach used here is different of what it is called *principle of maximum entropy* [381]. This principle always considers a unique probability distribution: the one with maximum entropy compatible with available restrictions. Here we are not saying that $\mathcal{P}$ can be replaced by the probability distribution of maximum entropy. We continue using the credal set to represent uncertainty. We only say that the uncertainty of the credal set can be measured by its upper entropy.

In order to obtain the maximum of entropy on a set of probability intervals in the application of the IDM from a sample, we can use the algorithm presented in Abellán and Moral [6] for probability intervals. When using values between 1 and 2 for the parameter s, we can use a simpler procedure of Abellán [1], which is a simplification of the one presented in Abellán and Moral [6].

Assume that we have a variable X taking values on a set $\mathcal{X}$. If we have a sequence of independent and identically distributed sequence of observations of this variable. Then, applying the IDM, the estimated probability intervals for any value $x_i \in \mathcal{X}$ (see Equation (10.17)) are:

$$\left[\frac{n(x_i)}{N+s}, \frac{n(x_i)+s}{N+s}\right], \quad \forall x_i \in \mathcal{X},$$

where $n(x_i)$ is the frequency of observations in which $X = x_i$ in the sequence and N the total sample size.

To compute the maximum entropy of the credal set associated to these intervals, we must first determine the set $W = \{x_j | \ n(x_j) = \min_i\{n(x_i)\}\}$. Let $|W|$ be the cardinality of the set W. If we use $\widehat{P}$ to denote the distribution where the maximum of entropy will be reached, the procedure of Abellán [1] can be expressed in the following way:

Case 1. $|W| > 1$ or $s = 1$

$$\widehat{P}(x_i) = \begin{cases} \dfrac{n(x_i)}{N+s} & x_i \notin W \\[2ex] \dfrac{n(x_i) + s/|W|}{N+s} & x_i \in W. \end{cases}$$

Case 2. $|W| = 1$ and $s > 1$.

— Assign:

$n(x_j) \leftarrow n(x_j) + 1$ (where $W = \{x_j\}$),

$s \leftarrow s - 1$.

[12] The proof is based on the Minimax theorem which can be found in Appendix E of Walley's book [672].

— Obtain new W.

— Obtain $\widehat{P}$ as in Case 1.

10.5.2 Obtaining conditional probability intervals with the imprecise Dirichlet model

We adopt the same notation of Section 10.2; we recall in particular that the class variable is denoted as C (taking values in $\mathcal{C}$) while the *discrete* features are denoted by $\mathbf{A} = A_1, \ldots, A_k$ (the k feature variables taking values from the sets $\mathcal{A}_1, \ldots, \mathcal{A}_k$). Again, we assume the data to be complete. If $\mathbf{Y}$ is a subset of all the variables $\mathbf{A}$, then $\mathbf{y}$ will denote a generic value of it (a value for each one of the variables in the subset).

We work with the local IDM, described in Section 10.4.4; thus, the values of the attribute variables select a subset of data, which is used to estimate a credal set only for variable C. When classifying a new case, the values of attribute variables will be used to select the appropriate credal set for C.

Definition 10.1 A configuration, σ, about $\mathbf{A}$ is an assignment of values for a subset of variables: $\mathbf{Y} = \mathbf{y}$, where $\mathbf{Y} \subseteq \mathbf{A}$.

If $\mathcal{D}$ is a data set and σ is a configuration, then $\mathcal{D}[\sigma]$ will denote the subset of $\mathcal{D}$ given by the cases which are compatible with configuration σ (cases in which the variables in σ have the same values as the ones assigned in the configuration).

Definition 10.2 Given a data set and a configuration σ, we consider the credal set $\mathcal{P}^\sigma$ for variable C with respect to σ defined by the set of probability distributions, P, such that

$$P(c) \in \left[\frac{n(c)^\sigma}{N^\sigma + s}, \frac{n(n)^\sigma + s}{N^\sigma + s}\right], \quad \forall c \in \mathcal{C},$$

where $n(c)^\sigma$ is the number of occurrences of $(C = c)$ in $\mathcal{D}[\sigma]$, N^σ is the number of cases in $\mathcal{D}[\sigma]$, and $s > 0$ is a parameter.

We denote this interval as

$$\left[\underline{P}^\sigma(c), \overline{P}^\sigma(c)\right].$$

This credal set is the one obtained on the basis of the local IDM [673] applied to the subsample $\mathcal{D}[\sigma]$. This set is also associated with a reachable set of probability intervals and with a belief function (Abellán [1])

Example 10.1 Assume that we have a class variable with three possible values: $\mathcal{C} = \{c_1, c_2, c_3\}$.

Suppose that we have a database and a configuration σ such that:

$$n(c_1)^\sigma = 4, \quad n(c_2)^\sigma = 0, \quad n(c_3)^\sigma = 0.$$

With $s = 1$, we have the following vector of probability intervals, using the IDM:

$$([\frac{4}{5}, 1]; [0, \frac{1}{5}]; [0, \frac{1}{5}]).$$

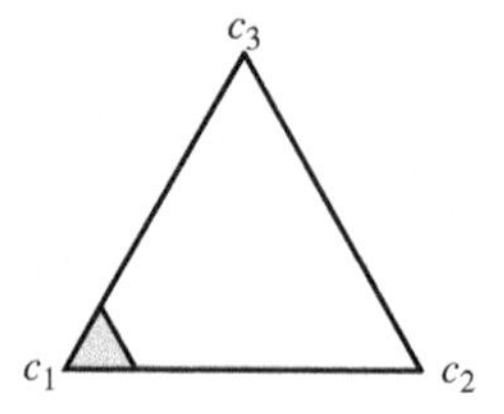

Figure 10.2 Simplex representation of the credal set in Example 10.1.

The credal set $\mathcal{P}^\sigma$ has three vertices:

$$\{(1, 0, 0); (\frac{4}{5}, \frac{1}{5}, 0); (\frac{4}{5}, 0, \frac{1}{5})\}.$$

This credal set is represented in Figure 10.2. Each point of the triangle represents a probability distribution in which the probability of c_i is the distance to the edge opposite to vertex c_i. The credal set is represented by the shadowed triangle. A general algorithm to find all the extreme points of a credal set that is defined by a system of intervals is given by Walley [672] and Campos, Huete and Moral [194].

It is simple to compute the upper entropy of this credal set applying the above algorithm. We obtain: $S^*(\mathcal{P}^\sigma) = H(\frac{4}{5}, \frac{1}{10}, \frac{1}{10}) = 0.639$, where H is the classical Shannon entropy. ♦

If we have a different database with the same relative frequencies for the values of C, but different sample size, then the credal set changes. If the sample size is smaller then the intervals are wider and if the sample size is larger then the intervals are more precise.

10.5.3 Classification procedure

In a classification tree each interior node is labelled with a variable of the data set $A_i \in \mathbf{A}$. Each leaf will have a decision rule to assign a value of the class variable C. In traditional classification trees, the decision rule assigns a single value of C (see Figure 10.3).

In a classification tree there is a correspondence between nodes and configurations. Each node defines a configuration: the set of variables that can be found in the path to that node from the root, with the values associated with the children that lie in this path. A complete configuration (a value for each one of the variables in $\mathbf{A}$) defines a leaf: we start at the root, and at each inner node with label A_i, we select the child corresponding to the value of A_i in the configuration.

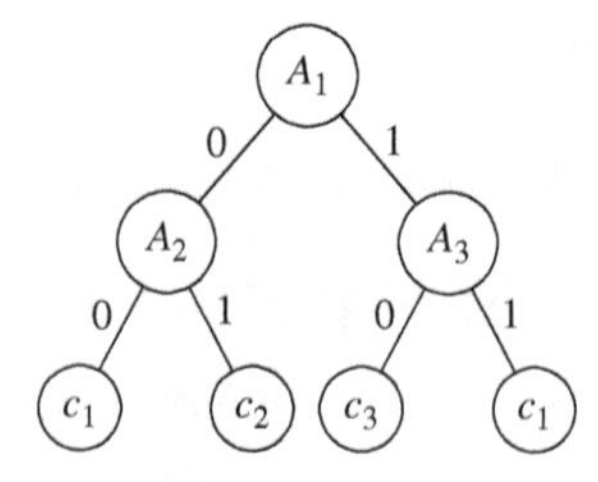

Figure 10.3 Example of a classification tree.

Given a data set $\mathcal{D}$, each node of the tree defines a credal set for C in the following way: we first consider the configuration σ associated to it, and then the credal set, $\mathcal{P}^\sigma$, as in Definition 10.2. For example, we have seen in Figure 10.3 that the node with label c_3 determines a configuration $\sigma = (A_1 = 1, A_3 = 0)$. This configuration has an associated data set, $\mathcal{D}[\sigma]$, which is the subset of the original $\mathcal{D}$ given by those cases for which $A_1 = 1$ and $A_3 = 0$. $\mathcal{P}^\sigma$ is the credal set built from this restricted data set, using the local IDM.

The method for building classification trees is based on measuring the total uncertainty of the credal set associated with each leaf. In the following we shall describe how to build the structure of the tree. The decision rules will be considered later.

The method starts with a tree with a single node. We shall describe it as a recursive algorithm, which is started with the empty node (the root node) with no label associated to it. Each node will have a list $\mathcal{L}^*$ of possible labels of variables which can be associated to it. The procedure will initially be started with the complete list of variables.

We will consider the following function implemented: $\texttt{Inf}(\sigma, A_i)$:

$$\texttt{Inf}(\sigma, A_i) = \left(\sum_{a_i \in \mathcal{A}_i} r(a_i)^\sigma TU(\mathcal{P}^{\sigma \cup (A_i = a_i)}) \right),$$

where $r(a_i)^\sigma$ is the relative frequency with which A_i ($r(a_i)^\sigma = n(a_i)^\sigma / N^\sigma$) takes value a_i in $\mathcal{D}[\sigma]$, $\sigma \cup (A_i = a_i)$ is the result of adding the value $A_i = a_i$ to configuration σ and TU is a total uncertainty measure. In the procedure, we will use the maximum entropy function as total uncertainty measure.[13]

If No is a node and σ a configuration associated with it, $\texttt{Inf}$ tries to measure the weighted average total uncertainty of the credal sets associated with the children of this node if variable A_i is added to it (and there is a child for each one of the possible values of this node). The average is weighted by the relative frequency of each one of the children in the data set.

In the following we describe the method. The basic idea is very simple and it is applied recursively to each one of the nodes we obtain. For each one of these nodes, we consider whether the total uncertainty of the credal set at this node can be decreased by adding one of the nodes. If this is the case, then we add a node with a maximum decrease of uncertainty. If the uncertainty cannot be decreased, then this node is not expanded and it is transformed into a leaf of the resulting tree. This procedure is explained in the algorithm of Figure 10.4.

In this algorithm, A_l is the branching variable of node No. The intuitive idea is that when we assign this variable to No, we divide the database associated with this node among its different children. In each one of the children, we can have more precise average knowledge about C but based on a smaller sample. We consider that the total uncertainty of the associated credal sets can be a good measure of the appropriate trade-off between the precision gained by dividing the database according to the different values of A_l and the precision lost by estimating the probability distribution of C from a smaller database.

Traditional probabilistic classification trees, are built in a similar way, but with the difference that we have precise estimations of probability values and the uncertainty measure is Shannon entropy. The quantity that is used to decide what variable to use to add a branch to a node is called information gain and it is similar to $TU(\mathcal{P}^\sigma) - \texttt{Inf}(\sigma, A_i)$, which is what

[13] As we see, the procedure is opened to any total uncertainty measure.

Procedure `BuilTree` (No, $\mathcal{L}^*$)

1. If $\mathcal{L}^* = \emptyset$, then `Exit`.
2. Let σ be the configuration associated with node No
3. Compute the credal set associated with σ and compute its total uncertainty $TU(\mathcal{P}^\sigma)$.
4. Compute the value
   $\alpha = \min_{A_{i} \in \mathcal{L}^*} \texttt{Inf}(\sigma, A_i)$
5. If α is greater than or equal to $TU(\mathcal{P}^\sigma)$ then
   6. `Exit`
7. If α is smaller than $TU(\mathcal{P}^\sigma)$, then
   8. Let A_l be the variable for which the minimum α is attained
9. Remove A_l from $\mathcal{L}^*$
10. Assign A_l to node No
11. For each possible value a_l of A_l
    12. Add a node No_l
    13. Make No_l a child of No
    14. Call `BuilTree` (No_l, $\mathcal{L}^*$)

Figure 10.4 Procedure to build credal decision trees.

it is computed to decide the branching variable. The only difference is that information gain is applied to precise probabilities. If P^σ is a precise probability estimation of probabilities about C in $\mathcal{D}[\sigma]$ (maximum likelihood is the usual estimation method), then the information gain is given by

$$\texttt{InfGain}(\sigma, A_i) = H(P^\sigma) - \left(\sum_{a_i \in A_i} r(a_i)^\sigma H(P^{\sigma \cup (A_i = a_i)}) . \right)$$

The information gain is also called the mutual information between A_i and C in sample $\mathcal{D}[\sigma]$ and it is always a nonnegative number. It is important to remark that the quantity we use, $TU(\mathcal{P}^\sigma) - \texttt{Inf}(\sigma, A_i)$, is not the supremum, nor the infimum of the mutual information, as in fact we are computing a difference of two upper values. If we were computing the supremum or the infimum of mutual information, we would always obtain a nonnegative value. On the other hand, $TU(\mathcal{P}^\sigma) - \texttt{Inf}(\sigma, X_i)$ can be negative and this is important as it is the criterion to stop branching. So, the procedure is not based on a sensitivity analysis of mutual information under imprecision. It can better understood as a method to choose between models: by comparing the information they give about the class variable.

Complexity

We are going to estimate the complexity as a function of the sample size, N, and the number of variables. As we never add a branch to a node which is compatible with no cases of the data, then the number of leaves is of order $O(N)$. The total number of nodes of the tree is of the same order. So we call procedure `BuilTree` a maximum of $O(N)$ times. Each time we call the procedure, we have to evaluate the weighted average of total uncertainty. To compute frequencies, we revise all the cases of the data set (N cases). In the method, for each call, we have to compute `Inf` for each variable. So, taking into account that m is the number of variables, we have a complexity for a call (without the recursive part) of $O(N \cdot m)$. Finally, the total complexity of building the complete tree in the method is $O(N^2 \cdot m)$.

We have considered that the number of possible values of a variable is constant. In fact, this is not an important factor, as the upper entropy for our interval probabilities can be found in linear time as a function of the number of possible classes by using algorithm in [6].

Decision at the leaves

In order to classify a new example, with observations of all the variables except the class variable C, we obtain the leaf corresponding to the observed configuration, i.e. we start at the root of the tree and follow the path corresponding to the observed values of the variables in the interior nodes of the tree, i.e., if we are at a node with variable A_i which takes the value a_i, then we choose the child corresponding to this value. We then use the associated credal set for C, $\mathcal{P}^\sigma$, to classify the new example.

To do that, we first follow *dominance under maximality*,[14] as discussed in Section 10.3. We recall that class c_1 is dominated under maximality if and only if for every P in $\mathcal{P}^\sigma$, there is a class value c_2 such that $P(c_2) > P(c_1)$. In this particular application of the IDM, dominance under the strict preference ordering is equivalent to *interval dominance*: c_1 is dominated if and only if there is a class value c_2 such that $\overline{P}^\sigma(c_1) < \underline{P}^\sigma(c_2)$.

The decision rule is to assign to every leaf with credal set $\mathcal{P}^\sigma$, the set of non-dominated class values corresponding to $\mathcal{P}^\sigma$. In this way, we obtain a credal classifier, in which we obtain a set of predicted values for class variable, non-dominated cases, instead of a unique prediction.

Growing forests

The method of [5] splits on the attribute with the greatest total uncertainty reduction. In [168] it has been proposed a different approach in order to decide on which variable to split: this implies considering a credal set for each candidate attribute, computing the uncertainty reduction for each distribution in the credal set and thus obtaining, for each attribute, an interval of uncertainty reduction. The variable to be used for splitting is decided by comparing such intervals; but the comparison can well identify more non-dominated attributes for splitting. In this case, it is proposed to grow a forest, each tree of the forest being obtained by splitting on one of the non-dominated attributes. Experiments with this method are however not yet available.

10.6 Metrics, experiments and software

The overall performance of a credal classifier is fully characterized by four indicators [152]:

- *determinacy*, i.e., the percentage of instances determinately classified;
- *single-accuracy*, i.e., the accuracy on the *determinately* classified instances;
- *set-accuracy*, i.e., the accuracy on the *indeterminately* classified instances;
- *indeterminate output size*: the average number of classes returned on the indeterminately classified instances.

[14] See also Section 8.1.

Note that set-accuracy and indeterminate output size are meaningful only if the data set has more than two classes.

These metrics completely characterize the performance of a credal classifier. As for comparing credal and traditional classifiers, this has been often done by assessing the drop of accuracy of the traditional classifier on the instances indeterminately classified by the credal; see for instance the empirical results in [152].

As for the problem of choosing among alternative credal classifiers, two metrics have been proposed in [153]. The first metric, borrowed from multi-label classification,[15] is the *discounted-accuracy*:

$$\text{d-acc} = \frac{1}{N} \sum_{i=1}^{N} \frac{(accurate)_i}{|Z_i|},$$

where $(accurate)_i$ is a 0-1 variable, showing whether the classifier is accurate or not on the i-th instance; $|Z_i|$ is the number of classes returned on the i-th instance and N is the number of instances of the test set. Although discounting *linearly* the accuracy on the output size might look arbitrary, a principled justification for it has been given in [718].

The non-parametric *rank test* overcomes this problem. On each instance, it ranks two classifiers CL_1 and CL_2 as follows:

- if CL_1 is accurate and CL_2 inaccurate: CL_1 wins;
- if both classifiers are accurate but CL_1 returns less classes: CL_1 wins;
- if both classifiers are wrong: tie;
- if both classifiers are accurate with the same output size: tie.

The wins, ties and losses are mapped into ranks and then analyzed via the Friedman test [224]. The rank test is more robust than d-acc, as it does not encode an arbitrary function for the discounting; yet, it uses less pieces of information and can therefore be less sensitive. Overall, a cross-check of both indicators is recommended.

However, both the discounted-accuracy and the rank test are subject to the drawback shown in the following example. Let us consider: a classification problem with two classes; a *random* classifier, which returns the class at random from a uniform distribution; a *vacuous* classifier, which returns both classes, being thus non-informative. Both discounted-accuracy and rank test would judge as equivalent the performance of the random and the vacuous classifier; in particular, discounted-accuracy would assign to both classifiers a score of 1/2.

In fact, both classifiers do not know how to predict the class, but only the vacuous classifier declares it. From this viewpoint, one might argue that the vacuous should be rewarded with more than 1/2: for instance, common sense suggests that a doctor who admits to be unable to diagnose is preferable to another one who issues random diagnoses. However, the reliability of a classifier is tightly related to the variability of its accuracy [718]; the aversion to this variability is what makes some people prefer credal classifier to precise ones. In the previous example, the vacuous classifier obtains 1/2 on each instance, while the score of the random continuously changes between 0 and 1. Therefore both classifiers have the same *expected* discounted-accuracy, but the discounted-accuracy of the vacuous is subject to much less variance. Let us consider the discounted-accuracy score as a reward. A risk-averse decision maker would prefer the vacuous classifier over the random one, since

[15] The metric is referred to as *precision* in [636].

they have the same expected reward of the random, but the former is subject to less variance. In [718, 719], this point is explored representing the subjective preferences of the decision maker via a utility function applied to the discounted-accuracy scores. It is formally shown that, under any concave utility function, the vacuous classifier has higher expected utility than the random one: this provides a way out of the indecision between random and vacuous classifier. Empirical experiments show that if a risk-averse utility is adopted, NCC is generally preferable to NBC. Utility-based accuracy measures constitute a promising approach to soundly compare traditional classifiers with classifiers based on imprecise probability.

It remains however open the problem of how to deal with cost-sensitive settings, in which different types of misclassification incur in different costs: so far, the analysis with the utility has been developed for settings in which all errors are regarded as equal.

10.7 Scoring the conditional probability of the class

If the final aim is to approximate the conditional information of the class C given the attributes instead of optimizing the accuracy, then a different measure should be used. If conditioned to the attributes, what we obtain is a precise probability for the probability of the class P, then the usual score for each case is the logarithm of the likelihood score: $\log(P(c))$, where c is the true value of the class. This scored is added for the different cases in the test set. If instead of a single probability, P, what we obtain is a credal set $\mathcal{P}$, then there is not an obvious way of generalizing this score. Here we propose to use the following one: Let $\widehat{P}$ the probability having maximum entropy in $\mathcal{P}$, then we use the logarithm of the likelihood score computed with this maximum entropy probability. This should not be misinterpreted as a substitution of $\mathcal{P}$ by $\widehat{P}$. In fact, it is only a method of scoring an imprecise probability through that distribution with maximum entropy. It is a situation similar to d-CCU metric. This metric can be seen as the result of distributing in an uniform way the set Z_i assigned by a credal classifier (if we obtain a precise classifier by this procedure, then in average it will obtain the same value of the metric than the imprecise one). This does not mean that the credal classifier is going to be transformed in a precise one. Analogously, using the maximum entropy distribution, $\widehat{P}$, is only a procedure to scoring the imprecise classifier given by $\mathcal{P}$.

10.7.1 Software

JNCC2 [151] is the Java implementation of NCC; it is available from www.idsia.ch/ giorgio/ jncc2.html and runs through a command-line interface.

On occasion of this book, we also realized a plug-in for WEKA [707] which includes a number of credal classifiers, namely NCC, LNCC, CMA, and the credal classification trees. The plug-in implements the metrics of Section 10.6 and allows to run and compare all such classifier through the graphical user interface of WEKA. The integration with WEKA makes also available many tools for data analysis (e.g., feature selection), which can be thus used with credal classifiers. The WEKA plug-in is available under the GNU GPL license from the webpage http://decsai.ugr.es/ andrew/weka-ip.html.

10.7.2 Experiments

This section gives an idea of how the different algorithms compare and show that the integration with WEKA makes accessible several useful tools (e.g., feature selection) which can be used to improve the performance of credal classifiers. We show results obtained running the credal classifiers available within the WEKA plug-in (NCC, LNCC, CMA, IPtree) on 39 public data sets from the UCI repository. Firstly, we have analyzed the performance of the classifiers without feature selection; later, we have exploited the integration with WEKA to analyze the effect of feature selection on credal classifiers. In a couple of cases, NCC turns out to be more determinate than LNCC; this is due to the fact that NCC implements the restriction of the prior credal set of [148] ($\epsilon = 0.01$), while this feature is currently not yet implemented for LNCC. We do not compare credal classifiers with their Bayesian counterparts (e.g., NCC vs NBC), as this kind of comparison is already extensively covered in the provided references.

The experiments have been performed using 10 runs of 10 folds cross-validation; numerical features have been discretized using the supervised discretization by [270]. For the sake of simplicity, we focus our analysis only on determinacy and discounted-accuracy; however, it would have been necessary to comprehensively consider the metrics of Section 10.6, if our goal was a deep analysis of the behaviours of the classifiers. The determinacy is important as it characterize the main practical difference between credal and traditional classifiers; the d-acc gives an overall evaluation of the performance of a credal classifier, which however tends to favour the more determinate classifiers. We can therefore expect some correlation between determinacy and d-acc. We have tested the significance of the differences through the t-test corrected for resampling ($\alpha = 0.01$), available within WEKA.

Without using feature selection, we find some known results: for instance, out of 39 data sets, LNCC is significantly more determinate than NCC in 18 cases and CMA in 23 cases (although in 5 cases NCC is significantly more determinate than CMA). In remaining cases, the differences are not significant. The IPtree is however the most determinate classifier: for instance, compared to CMA, it is 10 times significantly more determinate and only 3 times significantly less determinate (in the remaining 26 cases, the differences are not significant). The average determinacy of the classifiers across all data sets are respectively 0.81%, 0.85%, 0.93% and 0.95% for NCC, LNCC, CMA and IPtree.

As for the discounted-accuracy, LNCC has generally higher discounted-accuracy than NCC (12 wins, 25 ties, 2 losses) and so does CMA (12 wins, 25 ties, 2 losses). However, on the selected data sets, IPtree achieves better discounted-accuracy than both CMA (10 wins, 24 ties, 5 losses) and LNCC (8 wins, 27 ties, 4 losses). The average discounted-accuracy of the classifiers are respectively 0.72, 0.72., 0.75 and 0.76 for NCC, LNCC, CMA and IPtree.

We now analyze the effect of feature selection; we choose the Correlation-Based Feature Selection [334], which has been shown to be quite effective both with NBC and the C4.5 tree; it therefore appears as a reasonable choice for the set of credal classifiers which we are considering. We have designed the experiments to perform feature selection from scratch on each single experiment of cross-validation, using the data from the current training set; in this way, we avoid the optimistic bias due to feature selection, when feature selection is performed offline on the whole amount of available data [532].

A first effect of feature selection is that often it significantly increases determinacy: this happens in 10, 11, 7 and 3 times for respectively NCC, LNCC, CMA and IPtree; in no case feature selection implies a significant decrease of determinacy. Under feature selection, the average determinacy is respectively 86%, 89%, 95% and 95% for NCC, LNCC, CMA and IPtree.

Under feature selection, the discounted-accuracy of NCC significantly increases on 10 data sets, which are roughly the same over which it also significant increases its determinacy.

A possible explanation is that feature selection mitigates, as side effect, both the class and the feature problem.

Feature selection is beneficial also for LNCC, leading in 6 cases to a significant improvement of discounted-accuracy (although in one case there is a significant degradation); it is instead less effective for CMA (4 significant improvements, but also 1 significant deterioration); it should be however considered that the model averaging algorithms of CMA automatically remove irrelevant features. For the IPtree the effect of feature selection is more controversial: there is one significant improvement, but also three significant deteriorations. This can be due to the fact that trees automatically perform a kind of feature selection, when deciding on which feature to split, thus taking less advantage from an external step of feature selection. Under feature selection, the average discounted-accuracy is respectively 0.73, 0.74, 0.75, 0.75 for NCC, LNCC, CMA and IPtree. Results data by data set are available in Tables 10.1 and 10.2 of the Appendix.

10.7.3 Experiments comparing conditional probabilities of the class

A final experiment has been carried out to compare the estimation of probabilities of the IDM versus a Bayesian procedure. For this aim for each database, we have built a classification with the procedure used for the imprecise classification tree (IPtree). In the leaves, we have estimated the probabilities of the class variable, instead of assigning a set of values. We have considered different procedures the IDM and Bayesian methods based on a prior Dirichlet distribution with different values of s parameter.

Finally, we have scored the procedures with log $\widehat{P}(c_i|\mathbf{n})$ where c_i is the true value of the class, $\mathbf{n}$ the vector of observations of the attributes, and $\widehat{P}$ is the maximum entropy distribution in the case of IDM model, or the estimated probability in the case of a precise Bayesian classifier.

The results can be seen in Table 10.3. The results for the IDM are in IDM-s columns (s = 0.5,1,2) and the results for the Bayesian procedures are in B-s columns. For a value s the Bayesian procedures estimates the probabilities with a Dirichlet prior distribution with parameters uniform vector $\mathbf{t}$ and global sample size s. In these columns k is the number of different classes.

In the results we can see that $s = 1$ in the IDM provides the best results (in accordance with recommendations of this value of the parameter). Furthermore, this IDM provides better results than any of the Bayesian procedures, being the differences meaningful with 5 of these procedures using the Wilkonson signed-ranks test (see Table 10.4).

These results support the fact that there is no justification in using too precise procedures as the Bayesian ones. An imprecise method with a procedure to select one of the possible probabilities which is not based on information obtained from the data, can outperform the precise Bayesian procedures with a standard score as the log likelihood. The method to select the probability is maximum entropy, which can be justified as the distribution $\widehat{P}$ in the credal set with greatest value $\underline{P}(\ln(\widehat{P}(c)))$ (see the justification of maximum entropy as a global uncertainty measure in Subsection 10.5.1). That is, a procedure to select the distribution among a set of possible ones taking the log likelihood score as basis can be better than any of the precise Bayesian procedures. So, the imprecise credal set can be seen as a more faithful representation of the information content of the data, without adding any additional information necessary to make the final probabilities precise.

Table 10.1 Determinacy of the classifiers with and without feature selection. NCC represents the baseline, to which refer the increases and decreases.

Dataset	NCC	NCC-fsel	LNCC	LNCC-Fsel	CMA	CMA-fsel	IPtree	IPtree-fsel
audiology	0.07	0.38 ∘	0.06	0.43 ∘	0.96 ∘	1.00 ∘	0.95 ∘	0.94 ∘
cmc	0.97	0.98	1.00 ∘	1.00 ∘	0.93 •	1.00 ∘	0.93 •	0.98
contact-lenses	0.65	0.65	0.64	0.63	0.96 ∘	0.94 ∘	1.00 ∘	0.96 ∘
credit	0.98	0.98	0.99 ∘	1.00 ∘	0.98	0.99	0.96	0.96
german-credit	0.96	0.98 ∘	1.00 ∘	1.00 ∘	0.87 •	0.97	0.89 •	0.97
pima-diabetes	0.99	0.99	1.00	1.00	1.00	1.00	0.97 •	0.99
ecoli	0.90	0.90	0.94 ∘	0.94 ∘	1.00 ∘	1.00 ∘	0.96 ∘	0.96 ∘
eucalyptus	0.80	0.99 ∘	0.95 ∘	0.98 ∘	0.99 ∘	0.99 ∘	0.95 ∘	0.99 ∘
glass	0.68	0.71	0.77 ∘	0.81 ∘	1.00 ∘	1.00 ∘	0.97 ∘	0.96 ∘
grub-damage	0.61	0.57	0.74 ∘	0.68	0.56	0.69	0.75 ∘	0.76 ∘
haberman	0.95	0.95	1.00 ∘	1.00 ∘	1.00 ∘	1.00 ∘	0.96	0.96
hayes-roth	0.52	0.52	0.52	0.52	0.59	0.59	0.59	0.59
hepatitis	0.95	0.96	0.98	0.99 ∘	0.94	0.98	0.96	0.95
hypothyroid	0.87	1.00 ∘	0.96 ∘	1.00 ∘	1.00 ∘	1.00 ∘	1.00 ∘	1.00 ∘
ionosphere	0.97	0.96	0.97	0.99	1.00 ∘	1.00 ∘	0.97	0.97
iris	0.98	0.98	0.98	0.97	1.00	0.99	0.99	0.98
kr-vs-kp	0.99	1.00 ∘	1.00 ∘	1.00 ∘	0.96 •	1.00 ∘	1.00 ∘	1.00 ∘
labor	0.86	0.87	0.89	0.93	0.93	0.97 ∘	0.96	0.96
liver-disorders	1.00	1.00	1.00	1.00	1.00	1.00	1.00	1.00
lymphography	0.58	0.81 ∘	0.54	0.82 ∘	0.90 ∘	0.96 ∘	0.97 ∘	0.95 ∘
nursery	1.00	1.00	1.00 ∘	1.00 ∘	1.00	1.00	1.00	1.00
optdigits	0.98	0.98	0.99 ∘	0.99 ∘	1.00 ∘	1.00 ∘	0.90 •	0.90 •
page-blocks	0.97	0.99 ∘	0.99 ∘	1.00 ∘	1.00 ∘	1.00 ∘	1.00 ∘	1.00 ∘
pasture-prod.	0.63	0.76	0.61	0.76	0.99 ∘	0.99 ∘	0.96 ∘	0.96 ∘
primary-tumor	0.10	0.23 ∘	0.17 ∘	0.38 ∘	0.88 ∘	0.88 ∘	0.80 ∘	0.80 ∘
segment	0.96	0.97 ∘	0.96	0.98 ∘	1.00 ∘	1.00 ∘	0.98 ∘	0.98 ∘
solar-flare-C	0.85	0.96 ∘	0.98 ∘	1.00 ∘	0.98 ∘	0.98 ∘	1.00 ∘	1.00 ∘
sonar	0.96	0.96	0.99 ∘	0.99 ∘	0.99 ∘	0.99 ∘	0.95	0.95
soybean	0.93	0.94	0.92 •	0.93	1.00 ∘	1.00 ∘	0.98 ∘	0.99 ∘
spambase	1.00	1.00	1.00	1.00 ∘	1.00 ∘	1.00 ∘	0.98 •	0.99 •
spect-reordered	0.95	0.95	1.00 ∘	1.00 ∘	0.85 •	0.96	0.95	0.97
splice	0.99	0.99 ∘	1.00 ∘	1.00 ∘	1.00 ∘	1.00 ∘	0.98 •	0.98
squash	0.42	0.43	0.36	0.48	0.84 ∘	0.97 ∘	0.95 ∘	0.93 ∘
tae	0.97	0.97	1.00	1.00	0.43 •	0.43 •	1.00	1.00
vehicle	0.93	0.92	0.99 ∘	0.99 ∘	1.00 ∘	1.00 ∘	0.97 ∘	0.97 ∘
vowel	0.75	0.87 ∘	0.94 ∘	0.98 ∘	1.00 ∘	1.00 ∘	0.94 ∘	0.93 ∘
white-clover	0.10	0.49 ∘	0.25 ∘	0.66 ∘	0.95 ∘	0.95 ∘	0.99 ∘	0.99 ∘
wine	0.97	0.97	0.91 •	0.91 •	1.00 ∘	1.00 ∘	0.99	0.99
yeast	0.97	0.97	0.99 ∘	0.99 ∘	1.00 ∘	1.00 ∘	0.97	0.97
Average	0.81	0.86	0.85	0.89	0.93	0.95	0.95	0.95

∘, • statistically significant increase or decrease

Table 10.2 Discounted accuracy of the classifiers with and without feature selection. NCC represents the baseline, to which refer the increases and decreases.

Dataset	NCC	NCC-fsel	LNCC	LNCC-Fsel	CMA	CMA-fsel	IPtree	IPtree-fsel
audiology	0.21	0.47 ∘	0.20 •	0.48 ∘	0.73 ∘	0.74 ∘	0.79 ∘	0.76 ∘
cmc	0.50	0.51	0.50	0.51	0.50	0.51	0.49	0.52
contact-lenses	0.76	0.73	0.76	0.72	0.85	0.81	0.84	0.77
credit	0.86	0.86	0.84	0.85	0.86	0.86	0.84 •	0.84
german-credit	0.75	0.73	0.74	0.73	0.73	0.74	0.68 •	0.72
pima-diabetes	0.75	0.76	0.75	0.76	0.75	0.76	0.74	0.75
ecoli	0.80	0.80	0.81	0.81	0.81	0.81	0.80	0.80
eucalyptus	0.53	0.58 ∘	0.59 ∘	0.58 ∘	0.56 ∘	0.58 ∘	0.62 ∘	0.58 ∘
Glass	0.63	0.63	0.66 ∘	0.67 ∘	0.72 ∘	0.70 ∘	0.69 ∘	0.67
grub-damage	0.46	0.44	0.46	0.45	0.34 •	0.31 •	0.36 •	0.38 •
haberman	0.72	0.72	0.73	0.73	0.72	0.72	0.73	0.73
hayes-roth	0.58	0.58	0.58	0.58	0.58	0.58	0.58	0.58
hepatitis	0.84	0.83	0.84	0.83	0.84	0.82	0.80	0.81
hypothyroid	0.93	0.98 ∘	0.97 ∘	0.97 ∘	0.99 ∘	0.98 ∘	0.99 ∘	0.98 ∘
ionosphere	0.89	0.90	0.88	0.90	0.90	0.91	0.90	0.91
iris	0.93	0.94	0.93	0.94	0.93	0.94	0.93	0.94
kr-vs-kp	0.88	0.92 ∘	0.95 ∘	0.93 ∘	0.88	0.92 ∘	0.99 ∘	0.94 ∘
labor	0.89	0.85	0.89	0.85	0.88	0.86	0.84	0.84
liver-disorders	0.57	0.57	0.57	0.57	0.57	0.57	0.57	0.57
lymphography	0.68	0.75 ∘	0.66 •	0.75 ∘	0.81 ∘	0.78 ∘	0.73	0.75
nursery	0.90	0.90	0.96 ∘	0.95 ∘	0.90 •	0.90 •	0.96 ∘	0.95 ∘
optdigits	0.92	0.92	0.94 ∘	0.94 ∘	0.92	0.92	0.77 •	0.77 •
page-blocks	0.93	0.95 ∘	0.96 ∘	0.97 ∘	0.94 ∘	0.96 ∘	0.96 ∘	0.96 ∘
pasture-prod.	0.75	0.73	0.72	0.73	0.80	0.76	0.73	0.73
primary-tumor	0.19	0.26 ∘	0.22 ∘	0.29 ∘	0.36 ∘	0.36 ∘	0.38 ∘	0.37 ∘
segment	0.92	0.93 ∘	0.93 ∘	0.94 ∘	0.93 ∘	0.94 ∘	0.94 ∘	0.94 ∘
solar-flare-C	0.81	0.86 ∘	0.87 ∘	0.88 ∘	0.89 ∘	0.89 ∘	0.90 ∘	0.89 ∘
sonar	0.76	0.77	0.76	0.75	0.77	0.77	0.73	0.72
soybean	0.92	0.91	0.91 •	0.90 •	0.92	0.90	0.92	0.92
spambase	0.90	0.92 ∘	0.94 ∘	0.93 ∘	0.90	0.92 ∘	0.92 ∘	0.92 ∘
spect	0.79	0.79	0.83	0.84	0.77 •	0.80	0.79	0.81
splice	0.95	0.96	0.65 •	0.80 •	0.96 ∘	0.96	0.93 •	0.93 •
squash-stored	0.57	0.49 •	0.50 •	0.46 •	0.59	0.61	0.65	0.61
tae	0.46	0.46	0.47	0.47	0.38 •	0.38 •	0.47	0.47
vehicle	0.61	0.60	0.69 ∘	0.67 ∘	0.61	0.60	0.69 ∘	0.69 ∘
vowel	0.58	0.59	0.66 ∘	0.65 ∘	0.63 ∘	0.60	0.76 ∘	0.68 ∘
white-clover	0.39	0.50 ∘	0.46 ∘	0.55 ∘	0.54 ∘	0.54 ∘	0.62 ∘	0.61 ∘
wine	0.98	0.98	0.93 •	0.94 •	0.98	0.98	0.92 •	0.92 •
yeast	0.57	0.57	0.58	0.58	0.58	0.58	0.57	0.57
Average	0.72	0.73	0.72	0.74	0.75	0.75	0.76	0.75

∘, • statistically significant improvement or degradation

Table 10.3 Maximum entropy log-likehood of IDM versus Bayesian scores.

Dataset	IDM-0.5	IDM-1.0	IDM-2.0	B-0.5*k	B-1.0*k	B-2.0*k	B-0.5	B-1.0	B-2.0
anneal	−0.051	−0.070	−0.107	−0.117	−0.185	−0.295	−0.047	−0.062	−0.091
audiology	−1.388	−1.354	−1.374	−1.748	−2.063	−2.459	−1.395	−1.355	−1.365
autos	−1.132	−1.138	−1.228	−1.261	−1.436	−1.671	−1.146	−1.128	−1.172
breast-c.	−0.901	−0.859	−0.859	−0.917	−0.875	−0.847	−0.973	−0.917	−0.875
cmc	−1.505	−1.449	−1.434	−1.491	−1.446	−1.416	−1.591	−1.524	−1.471
horse-c.	−0.668	−0.626	−0.626	−0.685	−0.643	−0.618	−0.742	−0.685	−0.643
german-c.	−0.940	−0.855	−0.855	−0.971	−0.890	−0.835	−1.076	−0.971	−0.890
pima-dia.	−0.775	−0.761	−0.761	−0.779	−0.766	−0.755	−0.794	−0.779	−0.766
glass2	−0.735	−0.720	−0.720	−0.739	−0.722	−0.718	−0.766	−0.739	−0.722
hepatitis	−0.755	−0.676	−0.676	−0.771	−0.694	−0.640	−0.864	−0.771	−0.694
hypoth.	−0.045	−0.047	−0.054	−0.048	−0.055	−0.067	−0.045	−0.045	−0.048
ionosphere	−0.445	−0.423	−0.423	−0.451	−0.425	−0.418	−0.489	−0.451	−0.425
kr-vs-kp	−0.033	−0.039	−0.039	−0.033	−0.038	−0.048	−0.031	−0.033	−0.038
labor	−0.663	−0.616	−0.616	−0.672	−0.613	−0.586	−0.755	−0.672	−0.613
lymph.	−1.216	−1.119	−1.075	−1.079	−1.047	−1.071	−1.277	−1.160	−1.079
mushroom	−0.002	−0.004	−0.004	−0.002	−0.004	−0.008	−0.001	−0.002	−0.004
segment	−0.284	−0.297	−0.339	−0.366	−0.460	−0.603	−0.285	−0.291	−0.320
sick	−0.114	−0.113	−0.113	−0.115	−0.113	−0.112	−0.116	−0.115	−0.113
solar-flare	−0.172	−0.168	−0.168	−0.174	−0.169	−0.167	−0.180	−0.174	−0.169
sonar	−0.940	−0.833	−0.833	−0.965	−0.862	−0.789	−1.091	−0.965	−0.862
soybean	−0.451	−0.482	−0.560	−0.962	−1.306	−1.746	−0.449	−0.476	−0.546
sponge	−0.419	−0.415	−0.426	−0.423	−0.422	−0.435	−0.441	−0.428	−0.421
vote	−0.229	−0.219	−0.219	−0.231	−0.220	−0.215	−0.246	−0.231	−0.220
vowel	−1.306	−1.305	−1.378	−1.549	−1.792	−2.095	−1.314	−1.299	−1.343
zoo	−0.346	−0.398	−0.515	−0.582	−0.798	−1.088	−0.339	−0.376	−0.463
Average	−0.621	−0.600	−0.616	−0.685	−0.722	−0.788	−0.658	−0.626	−0.614

Table 10.4 Test results: z is the value of the normal two tailed distribution and p the significance level.

Wilcoxon Signed-Ranks Test					
B-0.5	B-1.0*k	B-2.0*k	B-0.5	B-1.0	B-2.0
$z = -3.86$ $p < 0.01$	$z = -3.21$ $p < 0.01$	$z = -0.79$ $p > 0.1$	$z = -2.89$ $p < 0.01$	$z = -2.38$ $p < 0.05$	$z = -3.26$ $p < 0.01$

Acknowledgements

Work partially supported by the Swiss NSF grant n. 200021-118071/1, 200020_137680 / 1, 200020_134759 / 1 and by the Hasler Foundation grant n. 10030.

11

Stochastic processes

Filip Hermans[1] **and Damjan Škulj**[2]

[1]*SYSTeMS Research Group, Ghent University, Belgium*
[2]*Faculty of Social Sciences, University of Ljubljana, Slovenia*

11.1 The classical characterization of stochastic processes

11.1.1 Basic definitions

In many applications we are interested in the evolution of an uncertain value. For example, we may want to study the evolution of a stock on the stock market over a certain period or we may be interested in the inter arrival time between packages on a router or it might be interesting to know how the radio-active decay of an isotope evolves over time. This however means that we have to incorporate dynamics in our modelling.

The evolution will be addressed using an index. Often this index is the time, but this is not necessarily so. Think for example about a sequence of nucleotides in a chromosome which can be considered as a random sequence. Throughout the majority of this chapter we will assume the index to be discrete. This does not necessarily mean that the actual dynamics behind the studied uncertain value have to be of discrete nature as well. The discrete index is usually more like an observation index than the representation of the real dynamics behind the process which is just sampled at regular times.

In the rest of this section we give the basic definitions of random processes with the emphasis on Markov chains. In the next section the so-called event driven approach to random processes is presented, which allows a natural generalization to imprecise probability models. In Section 11.3 imprecise Markov chains are described. Particularly, we emphasize on the comparison of their possible interpretations using different independence notions. Finally, in Section 11.4 we describe the limit behaviour of imprecise Markov chains.

Introduction to Imprecise Probabilities, First Edition.
Edited by Thomas Augustin, Frank P. A. Coolen, Gert de Cooman and Matthias C. M. Troffaes.

Definition 11.1 A *random process* (*stochastic process*) $\{X_t\}_{t\in\mathbb{H}}$ is a family of random variables X_t indexed by a (partially) ordered set $\mathbb{H}$. The domain of all random variables X_t is a set Ω, called the sample space.

Note that X_t corresponds to a random variable for each fixed value of $t \in \mathbb{H}$. For each fixed value $\omega \in \Omega$ we see however that $X_.(\omega)$ is a *random function* on $\mathbb{H}$. Using this we can look at a random process as a mapping from the sample space Ω into an ensemble of random functions $X_.(\omega)$ meaning that with every $\omega \in \Omega$ there corresponds a function of the index $X_t(\omega)$. The term random function is sometimes interchanged with *sample function*. Often the random variable X that describes the uncertain value is also called the *process variable*. A particular instance x_t of a random process is called a *realization*

Example 11.1 A single random variable X is a random process with $\mathbb{H}$ a singleton. 

Example 11.2 Let Ω be a finite set then $\{X_n\}_{n\in\mathbb{N}}$ is a one-sided discrete random process. Markov chains are an example of this kind of random process. They will be investigated in more detail in the following sections. ♦

Example 11.3 The random process $\{X_z\}_{z\in\mathbb{Z}^2}$ is also called a 2-dimensional *random field*. For example the 2d Ising model, or the pixels in a picture. ♦

Random processes provide models for various fields of applications. They play a prominent role in finance, providing models for the analysis of stock market prices and exchange rate fluctuations. In natural sciences, such as physics or biology, examples of random processes are population growth models, models in thermodynamics and statistical mechanics. Other areas of applications include social sciences, information sciences and also music and sports.

11.1.2 Precise Markov chains

In this chapter we will focus mainly on generalizations of particular stochastic processes known as *(imprecise) Markov chains*. The classical Markov process is a stochastic process $\{X_t\}$ where X_t can take values in a set $\mathcal{X}$, the *state space*. We consider a *finite state* and *discrete time* Markov processes, which means that we assume that $\mathcal{X}$ is a finite set and that the process changes values in consecutive intervals called *steps*. The elements of $\mathcal{X}$ are called *states*, and are denoted with $x, x_i, y \ldots$. We also introduce the notation $X_{1:k}$ which stands for the k-dimensional tuple of random variables X_1 till X_k on the product space $\mathcal{X}_{1:k} := \mathcal{X}_1 \times \mathcal{X}_2 \times \cdots \times \mathcal{X}_k$ and $x_{1:k}$ a particular element on this space. Of course, in our particular case of Markov chains we have that $\mathcal{X}_i = \mathcal{X}$ for every index i. The index is merely used to denote the (time) step we are considering. If we speak of a gamble on $f \in \mathcal{L}(\mathcal{X}_n)$ then we mean that the f assumes values depending on the outcome of the random variable X_n.

Usually, Markov chains are described in terms of *transition probabilities* which are conditional probabilities $P(X_{n+1} = \cdot|X_n = x)$ describing the agent's belief about transitions between states on consecutive time points. Another, equivalent way to describe beliefs is the expectation approach advocated by Whittle [700] and stands closer to the event-driven random process approach published by De Cooman and Hermans [203]. We will follow the latter approach, because it allows smoother transition to the imprecise model.

Beliefs are put down numerically, concretised as lower previsions defined for all gambles $f \in \mathcal{L}(\mathcal{X}_{n+1})$ conditional on its history; the states the system has been through. So, $P(f|X_{1:n} = x_{1:n})$ will denote the linear prevision of the gamble f denoting the expectation with respect to the transition probability model $P(X_{n+1} = \cdot|X_{1:n} = x_{1:n})$. A Markov process is assumed to be *memoryless*, which means that previsions about future states, conditional on the current state, do not depend on the history of the process. Specifically this means that at every time n we have that the Markov property has to be fulfilled.

Definition 11.2 A one parameter discrete random process $\{X_n\}_{n\in\mathbb{N}_0}$ satisfies the *Markov property* if the following condition holds for any $n \in \mathbb{N}_0$ and for any gamble f on $\mathcal{X}_{n+1}$:

$$P(f|X_{1:n} = x_{1:n}) = P(f|X_n = x_n).$$

A one parameter discrete random process that satisfies the Markov property is called a *Markov chain.*

In system theory, a state summarises the history of a system. It is the Markov property that makes every instantiation of a random variable a true state. When modelling a random process by means of a Markov chain it satisfies therefore to specify all possible one step previsions. To ease the notation we will introduce the following map.

Definition 11.3 The *transition operator* of a Markov chain is a map $T_n : \mathcal{L}(\mathcal{X}_{n+1}) \to \mathcal{L}(\mathcal{X}_n)$ that is defined by

$$T_n f(x) := P(f|X_n = x)$$

where $f \in \mathcal{L}(\mathcal{X}_{n+1})$.

In the classical description of Markov chains, beliefs about transitions between states are moulded into a transition matrix. This assumes that the state space is indexed and that we can without confusion speak of the i-th state where $i \in \{1, \ldots, |\mathcal{X}|\}$. The element on the i-th row and j-th column of the transition matrix at time step n is defined as $P(X_{n+1} = x_j|X_n = x_i)$ which can be written in terms of the transition operator as $T_n I_{\{x_j\}}(x_i)$. So in a way, the transition matrix is nothing more than the matrix representation of the transition operator in a canonical basis. Because of this correspondence, we will use the notation T both for a transition operator and a transition matrix.

Since any Markov chain will surely go from a specific state $X_n = x$ to a state z in X_{n+1}, it holds that $1 = \sum_{z\in\mathcal{X}} P(X_{n+1} = z|X_n = x)$ whence

$$\sum_{x\in\mathcal{X}} T_n I_{\{x\}} = I_{\mathcal{X}} \tag{11.1}$$

for every relevant $n \in \mathbb{N}$. This means that $I_{\mathcal{X}}$ is always an eigenfunction with eigenvalue 1 of the transition operator T_n. In general, nonnegative matrices with rows that sum up to one are called *stochastic matrices*. If the transition operators do not depend on the index n, then we are dealing with a *stationary* or *time homogeneous Markov chain*, in which case the subscript n is dropped from the notation. This means that the dynamics of such a Markov chain are completely described by the transition operator T and the initial prevision Q_0 on X_1.

Using the law of iterated expectation, the expectation for a gamble f that *depends only on* X_n ($f \in \mathcal{L}(\mathcal{X}_n)$) can be calculated as shown in [700, 203].

Proposition 11.1 *Given a Markov chain* $\{X_n : n \in \mathbb{N}_0\}$ *with initial prevision* Q_0 *and transition operators* $T_1, T_2, \ldots, T_n$, *the prevision of gambles* $f \in \mathcal{L}(\mathcal{X}_{n+1})$ *is given by*

$$P(f) = Q_0(T_1 T_2 \ldots T_n f). \tag{11.2}$$

If the Markov chain is stationary, this is equal to[1]

$$P(f) = Q_0(T^n f). \tag{11.3}$$

Sometimes previsions for more general gambles have to be calculated. Such gambles depend on the whole history of the chain instead of just the last term. The space of all such gambles is denoted with $\mathcal{L}(\mathcal{X}_{1:n+1})$. In order to describe the prevision of a gamble on $\mathcal{L}(\mathcal{X}_{1:n+1})$ in a similar fashion, we introduce another operator $\mathbb{T}_n$ on gambles in $\mathcal{L}(\mathcal{X}_{1:n+1})$ such that

$$\mathbb{T}_n f(x_{1:n}, \cdot) := P(f(x_{1:n}, \cdot)|X_n = x_n) = T_n f(x_{1:n}, \cdot)(x_n) \tag{11.4}$$

for every $f \in \mathcal{L}(\mathcal{X}_{1:n+1})$ and for all $x_{1:n} \in \mathcal{X}_{1:n}$. From their definition, we see immediately that $\mathbb{T}_1 = T_1$, but for indices i higher than 1, the equality $T_i = \mathbb{T}_i$ will surely not hold as T_i is an upper prevision conditional on the state X_i whereas $\mathbb{T}_i$ assumes that all the previous states are known. We can therefore interpret $\mathbb{T}$ as a transition operator that depends on the complete history. In [205] the following proposition is shown.

Proposition 11.2 *Given a Markov chain* $\{X_n\}_{n \in \mathbb{N}_0}$ *with initial prevision* Q_0 *and transition operators* $T_1, T_2, \ldots, T_n$, *then*

$$P(f) = Q_0(\mathbb{T}_1 \mathbb{T}_2 \ldots \mathbb{T}_n f) \tag{11.5}$$

for every gamble $f \in \mathcal{L}(\mathcal{X}_{1:n})$.

When restricted to the space of gambles $\mathcal{L}(\mathcal{X}_n)$ the operator $\mathbb{T}_n$ coincides with T_n and can therefore be regarded as its generalization to the extended space of gambles.

11.2 Event-driven random processes

In the classical definition of stochastic processes, the dynamics of one particular random variable is modelled. The random variable evolves over time and usually probability distributions are given to describe the dynamics. In a way, the definition for stochastic processes is very restrictive. At every timepoint the state the described systems resides in is picked from one and the same state-space. In their game-theoretic approach of probability [582] (see also Chapter 6 of this book), Shafer and Vovk describe an appealing alternative to model uncertain processes which is event rather than time-driven. The description of Markov chains in the previous section is a particular case of this approach.

[1] Here we define the power of the transition operators to be $T^{k+1} = T^k \circ T$ with $T^1 := T$.

The driving events in Shafer and Vovk's game-theoretic account (see Chapter 6) for random processes are the moves that a player called Reality makes. At every moment in time Reality is in a particular well defined situation and from this situation reality can go only to a specific set of new situations. Here a situation should be seen as a complete summary of the evolution of the system so far. Consequently, Reality can never return to the same situation. This means that a partial order $\sqsubseteq$ on the set of all possible situations $\Omega^{\Diamond}$ is assumed. Given two situations $s_1, s_5 \in \Omega^{\Diamond}$ we say that s_1 *precedes* s_5 if $s_1 \sqsubseteq s_5$. If moreover $s_1 \neq s_5$, then we say that s_1 strictly precedes s_5 and denote this $s_1 \sqsubset s_5$. Assuming that the experiment starts in a well known *initial situation* $\square$ (If this is not the case, then such a state can always be added) then the map of all possible realizations of the system can be graphically represented as a tree: the *event tree* $(\Omega^{\Diamond}, \sqsubseteq)$.

The minimal elements of all the situations that the node is strictly preceding are called the *children* of the node. In Figure 11.1, the children of node s_7 are $\text{ch}(s_7) = \{s_8, s_{12}\}$ and s_7 is said to be the mother of both s_8 and s_{12}. Situations $\{s_3, s_4, s_5, s_6, s_9, s_{10}, s_{11}, s_{12}\}$ are not preceding any situation and we call them *terminal situations* and denote them Ω. A *path* is a maximal chain and can be identified with a terminal situation: the path $\{\square, s_1, s_5\}$ can for example be identified with the terminal situation s_5. A *cut* is a maximal antichain. For instance, $\{s_2, s_5, s_6, s_7\}$ is a cut. We will always assume that the number of situations in a path is finite which leads to the property that every path and every cut intersect in precisely one situation.

When dealing with nondeterministic processes, we do not exactly know what situation Reality will go to. It is assumed however that there is another player, an agent which we call Subject that has beliefs about transitions to possible next states. If Subject has in every nonterminal node a conditional probability on the situation's children, then the event tree combined with the local probabilities is called a *probability tree*. In an imprecise probability tree, the local probability models are replaced with imprecise probabilistic models. In [203] these local models are assumed to be sets of desirable gambles. Here, we will assume that they are lower or upper previsions.

Definition 11.4 An *imprecise probability tree* $(\Omega^{\Diamond}, \sqsubseteq, \{\underline{Q}_s\}_{s\in\Omega^{\Diamond}\setminus\Omega})$ is an event tree $(\Omega^{\Diamond}, \sqsubseteq)$ with paths of finite length and equipped with a coherent lower previsions $\underline{Q}_s$ on $\mathcal{L}(\text{ch}(s))$ for any nonterminal situation $s \in \Omega^{\Diamond}\setminus\Omega$.

The local lower prevision $\underline{Q}_s$ should be interpreted as the supremal price Subject is willing to pay for a gamble $f \in \mathcal{L}(\text{ch}(s))$ contingent on the occurrence of the mother

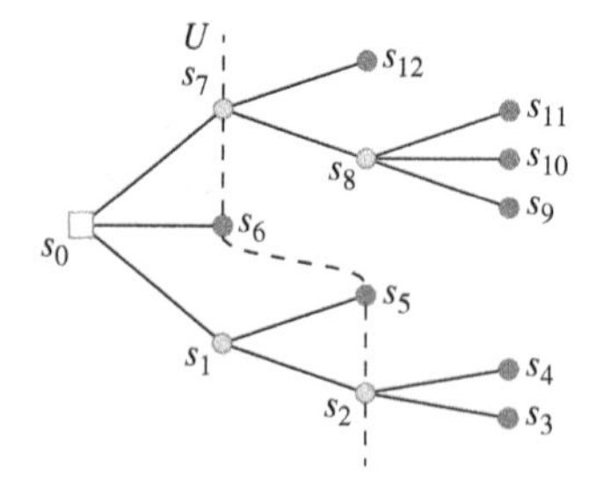

Figure 11.1 Event tree with initial situation s_0 and terminal situations $\{s_3, s_4, s_5, s_6, s_9, s_{10}, s_{11}, s_{12}\}$.

situation s. Here we have assumed Walley's updating principle [672, § 6.1.6] (see also Chapter 2 of this book) which moreover requires that Subject specifies all of his local models beforehand (in situation $\square$), which is in strong contrast with the *dynamical interpretation*.

How to combine the local models to answer questions about global gambles? For example, how to calculate the lower prevision of a general gamble? This question is very similar to the one posed in Section 2.3.3. In their attempt to glue the different conditional previsions together into a joint coherent set, the authors showed that an extra property was required. This property was $\mathcal{B}$-conglomerability and here we will make a very similar rationality assumption. As in [203], we assume *cut-conglomerability* which says that whenever a gamble is desirable contingent on every situation of a cut, then the complete gamble has to be desirable. Observe that the set $\mathcal{B}_U := \{\{s \in \Omega : u \sqsubseteq s\} : u \in U\}$ is a partition of Ω when U is a cut which shows correspondence between cut-conglomerability and $\mathcal{B}_U$-conglomerability.

By letting the contingent and updating interpretation (see Section 2.3) coincide and assuming cut-conglomerability, the following proposition has been shown in [203].

Proposition 11.3 (Concatenation formula). *For any gamble $f \in \mathcal{L}(\Omega)$ and any two cuts $V \sqsubseteq U$ of an imprecise probability tree, the following equality holds*

$$\underline{P}(f|V) = \underline{P}(\underline{P}(f|U)|V)$$

Note the similarity with the marginal extension formulae in Section 2.3.3. In [203], a recursive algorithm is given to calculate lower prevision in an imprecise probability tree.

Theorem 11.1 *The smallest coherent joint lower prevision $\underline{P}(f)$ for a gamble $f \in \mathcal{L}(\Omega)$ given an imprecise probability tree $(\Omega^{\diamond}, \sqsubseteq, \{\underline{Q}_s\}_{s \in \Omega^{\diamond} \setminus \Omega})$, can be recursively solved from*

$$\underline{P}(f|s) = \begin{cases} \underline{Q}_s(\underline{P}(f|\mathrm{ch}(s))) & \text{if } s \in \Omega^{\diamond} \setminus \Omega, \\ f(s) & \text{if } f \in \Omega, \end{cases}$$

where $\underline{P}(f) = \underline{P}(f|\square)$.

Using this recursion demands every local prediction model $\underline{Q}_s$ to be evaluated only once. Whenever such a local prevision is used in the calculation of $\underline{P}(f)$, the gamble gets measurable on an increasingly coarser partition of Ω or in other words, the gamble gets resolved on a cut that gets closer and closer to $\square$. A proof of this theorem can be found in [203, § 4.2.4].

The model where decision nodes are added to a probability tree corresponds to the sequential decision making which is studied in Section 8.2.

11.3 Imprecise Markov chains

The first systematic treatment of Markov chains with parameters not precisely known has been given by Hartfiel in [350], where he collects the results on this topic that have been published before in various journals. His theory of so-called *Markov set-chains* does not formally rely on the theory of imprecise probabilities; however, the techniques and interpretation of the results are closely related with those used in the theory of imprecise probabilities. The uncertainty in parameters in Hartfiel's work is represented in terms of intervals, which in the common language of imprecise probabilities corresponds to *probability intervals* (see Section 4.4), for the contrast with the more general *interval probabilities* (for a systematic treatment see [692]). The imprecise knowledge about the *probability vectors* or *probability mass functions* appearing in the theory of Markov set-chains is expressed in terms of *intervals* which denote the sets of probability mass functions of the form

$$[p, q] = \{x : x \text{ is a } 1 \times n \text{ stochastic vector with } p \leq x \leq q\}. \tag{11.6}$$

Similarly, the imprecision in transition matrix is represented by an *interval transition matrix*

$$[P, Q] = \{A : A \text{ is a } n \times n \text{ stochastic matrix with } P \leq A \leq Q\}. \tag{11.7}$$

The problem that Hartfiel recognizes is that the set of all possible products of matrices or between vectors and matrices in an interval are not intervals any more. Therefore, in a large part of the theory more general compact sets of probability vectors and transition matrices are allowed.

Throughout his book Hartfiel assumes that all possible behaviours of the process allow any transition matrix from the given interval at every step. In [407] Kozine and Utkin study the related problem where they assume the transition matrix to be constant in time, but its value is unknown. The analysis of the set of possible classical Markov chains that follows from this assumption is in many ways quite different from the analysis that follows form Hartfiel's assumptions. Results such as [94] suggest that finding exact bounds for corresponding interval probabilities is even more complex with the assumption of constant transition probabilities. From now on, though, we will always assume that transition matrices may be different at every step.

The formal connection between Hartfiel's model and the standard models of imprecise probabilities has been established by De Cooman et al. [205], who study imprecise Markov chains in the wider context of imprecise random processes by means of imprecise probability trees; and Škulj [595] who involves imprecision by means of interval probabilities (see e.g. [691, 692]).

In the related field of Markov decision processes modelling imprecision has for quite some time been an important topic too, starting with Satia and Lave [554] and followed by many other authors (see, e.g., [347, 376, 495, 699]).

11.3.1 From precise to imprecise Markov chains

The classical theory of Markov chains makes the assumption that the probability distribution over the states corresponding to the initial state and transition probabilities are all precisely known. Moreover, time homogeneity is often assumed, which means that the transition operator is constant in time. The models of *imprecise Markov chains* that we present here do not in general assume either of those. Thus, we will assume that the initial distribution as well as the transition operators are not known precisely and, as argued above, that the transition operators are not necessarily constant in time.

A Markov chain has to adopt the Markov property (see Definition 11.2), but what does this Markov property mean when imprecise belief models are assumed? In the precise case, the Markov property states that beliefs about the next state are (stochastically) independent of previous states given the current state. We know however that stochastic independence is a degenerate version of the imprecise independence concepts, which are described in depth in Chapter 3. Two of the main independence concepts that are currently assumed when using imprecise belief models are epistemic irrelevance – usually expressed in terms of lower previsions – and strong independence which is usually expressed in terms of credal sets, i.e. closed convex sets of probabilities. Both independence concepts lead to different types of imprecise Markov chains. When assuming epistemic irrelevance we get the Markov chains as defined by De Cooman, Hermans and Quaeghebeur [205], if we assume strong independence in our definition of the Markov property, then we get imprecise Markov chains as defined by Škulj [595].

Example 11.4 In Figure 11.2 we unroll the possible situations of an imprecise Markov chain of length three. At each step a transition of an element of the state-space $\mathcal{X} = \{a, b\}$ to the same state-space is possible with an exception from the initial step where an element from $\mathcal{X}$ is chosen but not conditional on being in a state. After two steps, a possible path is given for example by (a, b, b), meaning that initially the Markov chain was in state a, afterwards it went from a to b and in the next step it stayed in state b. ♦

The *epistemic irrelevance Markov property* says that the belief model for the next state given the current state is epistemically irrelevant on the previous states. The *strong independence Markov property* interpretation has a more mechanistic nature and states that at every step, there exists a precise transition probability which is taken and assumed to be stochastically independent of previous states. We will first discuss both notions in some detail and then demonstrate why and when the difference matters.

11.3.2 Imprecise Markov models under epistemic irrelevance

As in the classical definition of Markov chains, we assume that we are given a prior belief model and single step transition belief models, however now they will be given as upper (or lower) previsions instead of linear previsions. In this interpretation of Markov chains, epistemic irrelevance is the cornerstone when it comes to the interpretation of independence. In concrete, this means that whenever the agent (Subject) knows the current situation, then his beliefs about future states will not be altered upon learning earlier states were visited.

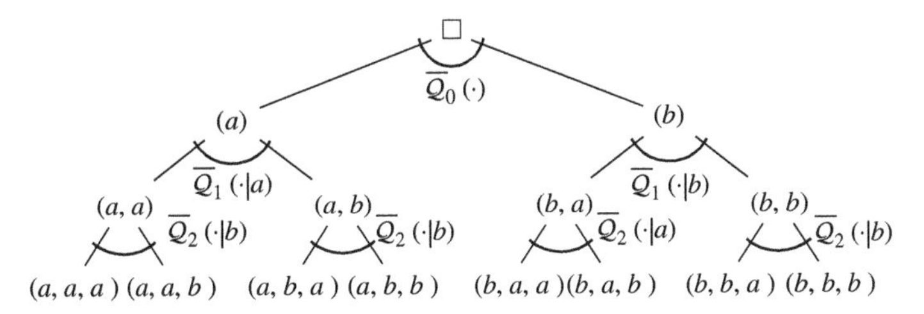

Figure 11.2 Imprecise probability tree unfolding of an imprecise Markov chain.

Definition 11.5 The (finite) random process $\{X_i\}_{i \in \{1,\dots,N\}}$ with $N \in \mathbb{N}_0$ satisfies the *epistemic irrelevance Markov condition* if

$$\underline{P}(f|X_{1:k}) = \underline{P}(f|X_k)$$

for any $f \in \mathcal{L}(X_{k+1:N})$.

If we try to interpret the imprecise Markov chain as an imprecise event tree, then we define the set of all situations as the set of all partial sample paths $\Omega^\Diamond = \{x \in \mathcal{X}_{1:k} : k \in \{1, \dots, n\}\} \cup \{\square\}$. Clearly a situation $x_{1:k}$ precedes $y_{1:l}$ if $x_{1:k} = y_{1:k}$. If the local models $\underline{Q}_s$ are such that $\underline{Q}_{x_{1:k+1}} = \underline{Q}_{y_{1:k+1}}$ whenever $x_{k+1} = y_{k+1}$, then it follows immediately from Theorem 11.1 that the defined imprecise event tree indeed fulfils the epistemic Markov property and the modelled process is an imprecise Markov model under epistemic irrelevance.

The epistemic irrelevance interpretation of imprecise Markov models can be dealt with easily as it is a special case of the imprecise probability trees (Section 11.2). If interested in the upper (or lower) prevision of a gamble, then the backwards recursion (See figures 11.3–11.5) formula given in Theorem 11.1 can be used. Let us assume that we have an imprecise Markov chain of length 2 where the epistemic irrelevance Markov property is satisfied. In this particular example we know furthermore that the state space $\mathcal{X} = \{a, b\}$ is binary, and we know that all the local models – which are immediate prediction models – are summarised by $\overline{Q}_0, \overline{T}_1, \overline{T}_2$ where the upper transition operators are defined in analogy to their precise counterparts (Definition 11.3):

$$\overline{T}_n f(x) := \overline{Q}_n(f|x) \tag{11.8}$$

for each $f \in \mathcal{L}(\mathcal{X}_{n+1})$ and each $x \in \mathcal{X}_n$. As in the linear case, the upper transition operator $\overline{T}$ is a map from $\mathcal{L}(\mathcal{X}_{n+1})$ to $\mathcal{L}(\mathcal{X}_n)$. If we are interested in the upper prevision of a gamble $f \in \mathcal{L}(\mathcal{X}_{1:N+1})$, then we can start the computation of $\overline{P}(f)$ by 'unrolling' the imprecise Markov chain into its corresponding imprecise probabilistic tree.

Now, the only thing to do when trying to get $\overline{P}(f)$ is iterate back as explained in Section 11.2. Using $\overline{P}(f) = \overline{P}(\overline{P}(f|U_{\mathcal{X}_2}))$ where $U_{\mathcal{X}_2} := \{(a,a), (a,b), (b,a), (b,b)\}$

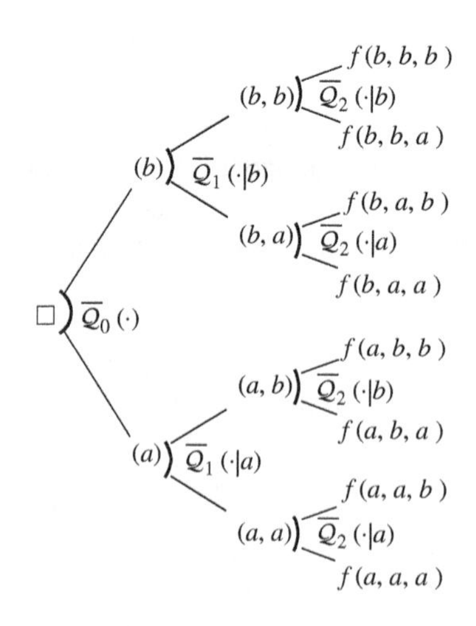

Figure 11.3 Backwards recursion in an imprecise probability tree: before first iteration.

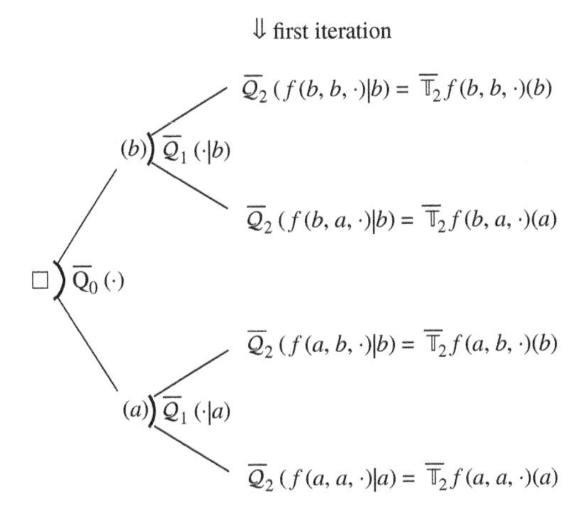

Figure 11.4 Backwards recursion in an imprecise probability tree: the first iteration.

corresponds to the cut that resolves gambles that are only depending on $X_{1:2}$, i.e. $X_{1:2}$-measurable gambles, we can start the calculation. Notationally, everything becomes clearer if we introduce the *history dependent upper transition operator* as in the precise case (11.4):

$$\overline{\mathbb{T}}_n f(x_1, x_2, \ldots, x_n, \cdot)(x_n) := \overline{Q}_n(f(x_1, x_2, \ldots, x_n, \cdot)|x_n) = \overline{T}_n f(x_1, x_2, \ldots, x_n, \cdot)(x_n)$$

which has to hold for every gamble $f \in \mathcal{L}(\mathcal{X}_{1:n+1})$ and every $(x_1, x_2, \ldots, x_n) \in \mathcal{X}_{1:\underline{n}}$. After applying the first iteration we have gone from a gamble f on $\mathcal{X}_{1:3}$ to a gamble $\overline{\mathbb{T}}_2 f$ on $\mathcal{X}_{1:2}$ and we get effectively a new, reduced imprecise probability tree. We can apply the same procedure again and we get a new gamble on $\mathcal{X}_1$: $\overline{\mathbb{T}}_1 \overline{\mathbb{T}}_2 f$. The more compact notation comes from the fact that $I_a \overline{\mathbb{T}}_2 f(b, a, \cdot)(a) + I_b \overline{\mathbb{T}}_2 f(b, b, \cdot)(b)$ is equal to $\overline{\mathbb{T}}_2 f(b, \cdot, \cdot)$.

As we keep iterating until the root node, we finally get the expression for $\underline{P}(f)$ which is equal to $\overline{Q}_0(\overline{\mathbb{T}}_1 \overline{\mathbb{T}}_2 f)$ and is remarkably similar to the result in the precise case (Theorem 11.1). The reasoning followed here in no way depends on the dimension of the state-space, nor does it depend on the specific form of the local prediction models. As long as the length of the Markov chain is finite, so will be the depth of the corresponding

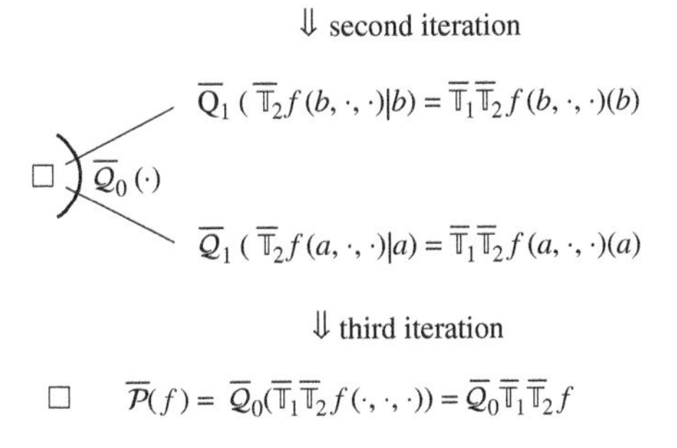

Figure 11.5 Backwards recursion in an imprecise probability tree: the second and the third iteration.

imprecise probability tree. From this we can conclude that there indeed is an imprecise probabilistic version of Theorem 11.1. An explicit proof is given in [205].

Theorem 11.2 (Concatenation Formula). *For an imprecise Markov chain with (history dependent) upper transition operators* $\overline{\mathbb{T}}_1, \overline{\mathbb{T}}_2, \overline{\mathbb{T}}_3, \ldots$, *and initial upper prevision* $\overline{Q}_0$ *and any real-valued map* f *on* $\mathcal{X}_{1:N}$ *we have that*

$$\overline{P}(f) = \overline{Q}_0(\overline{\mathbb{T}}_1\overline{\mathbb{T}}_2 \ldots \overline{\mathbb{T}}_{N-1}f).$$

Sometimes we are interested in a gamble that depends only on the last state X_N, independent on the path the process took to get there. In this case, Theorem 11.2 simplifies into the following form.

Theorem 11.3 *For an imprecise Markov chain with upper transition operators* $\overline{T}_1, \overline{T}_2, \overline{T}_3, \ldots$ *and initial upper prevision* $\overline{Q}_0$ *and any real-valued map* f *on* $\mathcal{X}_N$ *we have that*

$$\overline{P}(f) = \overline{Q}_0(\overline{T}_1\overline{T}_2 \ldots \overline{T}_{N-1}f).$$

Notice that even if we are interested in calculating upper or lower probabilities only, the formulation in terms of general gambles turns out to be advantageous, as already after one application of the upper transition operator the indicator function corresponding to the queried event is converted into a gamble which is most likely not an indicator function any more.

11.3.3 Imprecise Markov models under strong independence

Strong independence in the context of imprecise Markov chains means that convex envelopes of sets of precise Markov chains are considered that satisfy independence in the usual sense for precise probabilities. With the strong independence interpretation an imprecise Markov chain is interpreted as a set of nonstationary chains where the transition model at each step is chosen independently from the transitions at previous steps. The set of possible transition models is assumed to be constant over time. If the sets of probability distributions and transition matrices are convex, the one step transitions can be equivalently modelled with the corresponding upper or lower expectation and transition operators. Under the strong independence assumption, which is also implicitly assumed in Hartfiel's model [350] discussed at the beginning of this section, the same approach works for general sets of initial distributions and transition operators that are not necessarily convex. It is then only the matter of computational convenience to assume credal sets that allow the calculations with the backward recursion, as described earlier.

Let C_0 be an initial set of probabilities corresponding to a Markov chain $\{X_n : n \in \mathbb{N}_0\}$. Since sets of probabilities are equivalent to the sets of the corresponding expectations or linear previsions, and therefore represent belief models equivalent to upper previsions, we will do calculations in terms of expectations to be consistent with the rest of the chapter. We will denote an arbitrary element of C_0 by Q_0 and consider it as a possible candidate for the linear prevision corresponding to X_0. Further let $\mathcal{T}$ be a set of transition matrices. Again, in order to stay in accordance with the notation in the rest of this chapter, we will, instead of sets of transition matrices, use the equivalent notation with the corresponding sets of transition operators. At every time n one of the elements T from $\mathcal{T}$ is the actual transition operator.

Given $\mathcal{C}_0$ and $\mathcal{T}$ we would like to calculate the sets of linear previsions that model the distributions at future steps. We will denote the set of linear previsions corresponding to X_n with $\mathcal{C}_n$. Let Q_n be a possible linear prevision corresponding to X_n. The set $\mathcal{C}_n$ can be constructed in the following way. At every time i the transition probabilities are given by a transition operator T_i. This corresponds to a nonstationary Markov chain and, by Equation (11.2) of Proposition 11.1, the linear prevision at time n is $Q_n = Q_0 T_1 \dots T_n$. The set of linear previsions at time n is therefore exactly the set

$$\mathcal{C}_n = \{Q_0 T1 \dots T_n : Q_0 \in \mathcal{C}_0, T_i \in \mathcal{T}, 1 \leq i \leq n\}, \tag{11.9}$$

or, equivalently,

$$\mathcal{C}_n = \{Q_{n-1} T : Q_{n-1} \in \mathcal{C}_{n-1}, T \in \mathcal{T}\} =: \mathcal{C}_{n-1}\mathcal{T}. \tag{11.10}$$

There is yet another difficulty that comes quite unexpectedly. That is, even in the case where the set of initial distributions and the set of transition operators are convex and closed, the sets $\mathcal{C}_n$ may not be convex anymore. We demonstrate this by a simple example.

Example 11.5 Let $\mathcal{X} = \{x_1, x_2, x_3\}$ and $q_1 = (0.5, 0.5, 0)$ and $q_2 = (0, 0.5, 0.5)$ be probability mass functions. Further let

$$T = \begin{bmatrix} 0 & 1 & 0 \\ 0 & 0 & 1 \\ 1 & 0 & 0 \end{bmatrix}$$

and I the identity matrix. Now let

$$\mathcal{C}_0 = \{\alpha q_1 + (1-\alpha) q_2 : 0 \leq \alpha \leq 1\} \text{ and}$$
$$\mathcal{T} = \{\beta I + (1-\beta) T : 0 \leq \beta \leq 1\}.$$

It is an elementary exercise to verify that there is no $q \in \mathcal{C}_0$ and $T' \in \mathcal{T}$ such that $qT' = 0.5 q_1 I + 0.5 q_2 T$, which implies that the set $\mathcal{C}_0 \mathcal{T}$ is not convex. ◆

However, in the most important cases, convexity of all sets $\mathcal{C}_n$ is ensured by the following property.

Definition 11.6 A set of transition operators $\mathcal{T}$ has *separately specified rows* if for every two transition operators T and $T' \in \mathcal{T}$ and every $x \in \mathcal{X}$ there exists a transition operator $\tilde{T}$ such that $\tilde{T}f(y) = Tf(y)$ for every $y \in \mathcal{X}$ such that $y \neq x$, and $\tilde{T}f(x) = T'f(x)$.

In other words, a set of transition operators $\mathcal{T}$ has separately specified rows if the selection of a transition probability conditional on a state does not affect the possible selection of the transition probability conditional on another state.

The following lemma can be found in [595] or, in a slightly less general form, in [350].

Lemma 11.1 *Let $\mathcal{T}$ be a set of transition operators with separately specified rows and $\mathcal{C}$ a convex set of linear previsions. Then the set of linear previsions $\mathcal{C}\mathcal{T}$ (see (11.10)) is convex.*

Lemma 11.1 has the following important corollary.

Corollary 11.1 *Let $\mathcal{T}$ be a convex set of transition operators with separately specified rows and let C_0 be a credal set. Then C_n, as defined with (11.9), is a credal set for every $n \in \mathbb{N}$.*

From now on we will assume that rows of transition operators are separately specified, which together with the assumption of convexity of the initial set of linear previsions implies that the sets of distributions for every random variable X_n are credal sets, which allows equivalent representation in terms of upper previsions.

It is not yet clear how to link the definition of an upper transition operator from definition (11.8) with sets of transition operators adopted in the strong independence settings. The following proposition will allow this.

Proposition 11.4 *Let $\mathcal{T}$ be a set of transition operators with separately specified rows and $f \in \mathcal{L}(\mathcal{X})$ an arbitrary gamble. Then the set $\{Tf : T \in \mathcal{T}\}$ has a unique maximal element $\overline{T}f$ such that*

$$Tf \leq \overline{T}f$$

for every $T \in \mathcal{T}$.

The above proposition induces the canonical correspondence between sets of transition operators and the upper transition operators defined by (11.8). Without the requirement that rows are separately specified such a relation would not be possible, because the set $\{Tf : T \in \mathcal{T}\}$ can in general have multiple maximal elements, which makes the upper transition operator undefined.

11.3.4 When does the interpretation of independence (not) matter?

The question arises whether the interpretation of the Markov condition really matters. The next example shows that it actually does.

Example 11.6 Let $\mathcal{X} = \mathcal{X}_1 = \mathcal{X}_2 = \mathcal{X}_3 = \{a, b\}$ and $C_0 = \{(\frac{1}{2}, \frac{1}{2})\}$. The upper transition operator is given by a vacuous model, independent of the state: $\overline{T} = I_{\mathcal{X}} \max$. The gamble $f \in \mathcal{L}(\mathcal{X}_{1:3})$ of interest is $f = I_{(a,a,a)} + I_{(b,a,b)}$.

Relying on the strong independent interpretation we get that

$$\begin{aligned}
\overline{P}(f) &= \max\{\sum_{x\in\mathcal{X}_{1:3}} q(x)f(x) \, : \, q \in C_2\}, \\
&= \max\{q(a,a,a) + q(b,a,b) \, : \, q \in C_2\}, \\
&= \max\{m(a)r(a|a)s(a|a) + m(b)r(a|b)s(b|a) \, : \, m \in C_0 \text{ and } r, s \in \mathcal{T}\} \\
&= \frac{1}{2}\max\{\alpha\gamma + \beta(1-\gamma) \, : \, 0 < \alpha, \beta \leq 1, 0 \leq \gamma \leq 1\} = \frac{1}{2},
\end{aligned}$$

while under the epistemic irrelevance interpretation we infer from Theorem 11.2 that

$$\overline{P}(f) = \frac{1}{2}\max\{\max\{1,0\},\max\{0,0\}\} + \frac{1}{2}\max\{\max\{0,1\},\max\{0,0\}\},$$
$$= \frac{1}{2} + \frac{1}{2} = 1 \neq \frac{1}{2}.$$

Another way to write the epistemically irrelevant case gives more insight where the difference between the two interpretations comes from

$$\overline{P}(f) = \max\{m(a)r(a|a)s(a|a) + m(b)r(a|b)t(b|a)\ :\ m \in \mathcal{C}_0 \text{ and } r,s,t \in \mathcal{T}\},$$
$$= \frac{1}{2}\max\{\alpha\gamma + \beta\delta\ :\ 0 < \alpha,\beta \leq 1, 0 \leq \gamma,\delta \leq 1\} = \frac{1+1}{2} = 1.$$

The inequality still holds if the transition operator is assumed to be a linear-vacuous mixture (not purely linear). ♦

The example shows that the interpretation of independence does matter. It also demonstrates that the strong independence interpretation is the epistemic case with extra constraints. The extra constraint is the demand that at a given time point it is only allowed to choose one transition matrix. Hence, the imprecise Markov chain is a set of (nonstationary) precise Markov models in the strong independence interpretation. This is not the case in the epistemic irrelevance case where the transition matrices may depend on the complete history.

The discrepancy between the two interpretations is most unfortunate, however, if the gamble depends only on one state, then it can be shown that the interpretation of independence is irrelevant. This is so, because the local optimization done in the epistemic irrelevance case will take the same transition matrix which will of course be the one that was used in the strong independent case.

Theorem 11.4 *Let $\{X_n : n \in \mathbb{N}_0\}$ be a stationary imprecise Markov chain and let the upper prevision $\overline{Q}_0$ represent the beliefs about the initial state X_0 and $\overline{T}$ be the upper transition operator. Then the upper prevision of a gamble f depending only on one state X_n is given by*

$$\overline{P}(f) = \overline{Q}_0(\overline{T}^n f), \tag{11.11}$$

independent of the interpretation of the Markov property.

Proof. In the case where the epistemic irrelevance approach is adopted, the Equation (11.11) clearly coincides with the corresponding equation in Theorem 11.3.

Let us show that the same equation is valid with the strong independence approach. For an arbitrary $f \in \mathcal{L}(\mathcal{X})$, the upper prevision $\overline{Q}_n(f)$ is the maximum of all linear previsions $Q_n(f) = Q_0(T_1 \ldots T_n f)$, where $Q_0 \in \mathcal{C}_0$ and $T_i \in \mathcal{T}$ for every $1 \leq i \leq n$, as follows from (11.9). Equivalently, by (11.10), it is the maximum over all linear previsions $Q_n(f) = Q_{n-1}(TF)$ where $Q_{n-1} \in \mathcal{C}_{n-1}$ and $T \in \mathcal{T}$.

We can use mathematical induction to n to prove this case where the case $n = 0$ follows from the definition. Suppose that Equation (11.11) holds for $n - 1$, and take an arbitrary gamble $f \in \mathcal{L}(\mathcal{X})$. By Proposition 11.4 there is some $T \in \mathcal{T}$ such that $Tf = \overline{T}f$, which together with the induction assumption implies that

$$\begin{aligned}\overline{Q}_n(f) &= \max_{Q_{n-1} \in C_n-1} \max_{T \in \mathcal{T}} Q_{n-1}(Tf) = \max_{Q_{n-1} \in C_n-1} Q_{n-1}(\overline{T}f) = \overline{Q}_{n-1}(\overline{T}f) \\ &= \overline{Q}_0(\overline{T}^{n-1}(\overline{T}f)) = \overline{Q}_0(\overline{T}^n f).\end{aligned}$$

□

As a result of the theorem, the study of the limit behaviour in the following section will not depend on the Markov property that is assumed.

11.4 Limit behaviour of imprecise Markov chains

An important question in the theory of Markov chains is how the initial state or the distribution over the set of states influences the behaviour of the process after a long time. Often the process eventually reaches an equilibrium that is independent of the initial conditions. Such limit behaviour is usually described in terms of a limit probability distribution or prevision.

In the precise case we therefore consider the existence of the limit

$$Q_\infty := \lim_{n \to \infty} Q_n, \tag{11.12}$$

while in the imprecise case the linear previsions are replaced with upper previsions, and the limit in question is

$$\overline{Q}_\infty := \lim_{n \to \infty} \overline{Q}_n. \tag{11.13}$$

In the case where more general sets of distributions are possible, the convergence for the sequence of sets of linear previsions $\{C_n\}_{n \in \mathbb{N}_0}$ may be considered instead.

The limit behaviour of imprecise Markov chains has been explored by several authors. In [350] sufficient conditions for convergence of Markov set-chains are given, whose imprecision is described by general closed sets of probabilities, not necessarily convex. Most other approaches focus on Markov chains with imprecise probability models based on credal sets or equivalently, expectation functionals. In [205] the unique convergence is explored by defining an accessibility relation using an imprecise transition operator. In [595] the analysis of invariant sets of probabilities and their uniqueness is studied, and in [597] metric properties of imprecise probabilities are explored and used to define a coefficient of ergodicity suitable for models using upper previsions, which besides providing necessary and sufficient conditions for unique convergence allows measuring the rate of convergence. Another coefficient of ergodicity, more suitable for models using general compact sets of probability distributions, has also previously been defined in [350].

11.4.1 Metric properties of imprecise probability models

Convergence of probability distributions is studied using appropriate metrics. Usually in the set of probability measures the *total variation* metric is used:

$$d_p(P, P') := d_p(p, p') = \frac{1}{2} \sum_{x \in \mathcal{X}} |p(x) - p'(x)| = \max_{A \subset \mathcal{X}} |P(A) - P'(A)|, \tag{11.14}$$

where p and p' are the probability mass function corresponding to the probability measures P and P' respectively.

For a pair of nonempty compact subsets of probabilities C and C' the *Hausdorff metric* is defined by

$$d_H(C, C') = \max\left\{ \sup_{P\in C} \inf_{P'\in C'} d(P, P'), \sup_{P'\in C'} \inf_{P\in C} d(P, P') \right\}, \quad (11.15)$$

where d is some metric. Here we will assume that $d = d_p$. This metric makes the set of nonempty compact sets of probabilities a compact metric space. Note also that every compact metric space is complete, which means that every Cauchy sequence[2] converges.

The set of all (upper) previsions can also be endowed with the structure of a metric space. One of the possible metrics is

$$d_o(\overline{Q}, \overline{Q}') = \max_{f\in\mathcal{F}_1} \left|\overline{Q}(f) - \overline{Q}'(f)\right|, \quad (11.16)$$

where $\mathcal{F}_1 = \{f \in \mathcal{L}(\mathcal{X}) : 0 \leq f(x) \leq 1 \text{ for all } x \in \mathcal{X}\}$.

The above three metrics d_p, d_H and d_o all induce metrics on the space of all probabilities on the finite set $\mathcal{X}$, and there they all coincide:

$$d_p(P, P') = d_H(\{P\}, \{P'\}) = d_o(\mathbb{E}_P, \mathbb{E}_{P'}), \quad (11.17)$$

where $\mathbb{E}_P$ and $\mathbb{E}_{P'}$ are the expectation functionals corresponding to P and P'. Moreover, when the Hausdorff metric is restricted to the space of all credal sets, it induces a metric on the set of the upper expectation operators with respect to the credal sets (note that upper expectation operators on finite spaces are equivalent to upper previsions). The induced metric coincides with d_o:

$$d_H(\mathcal{M}, \mathcal{M}') = d_o(\overline{\mathbb{E}}_{\mathcal{M}}, \overline{\mathbb{E}}_{\mathcal{M}'}). \quad (11.18)$$

The space of all upper previsions on $\mathcal{L}(\mathcal{X})$ is a complete metric space, meaning that every Cauchy sequence $\{\overline{Q}_n\}_{n\in\mathbb{N}}$ of upper previsions converges to some upper prevision $\overline{Q}_\infty$.

11.4.2 The Perron-Frobenius theorem

For classical Markov chains with positive probability that after a certain finite number of steps the chain will move from any state x to any other state y the conditions for convergence follow from the Perron-Frobenius theorem, which is one of the classical results in the theory of Markov chains. It turns out that the same conclusions are also valid if upper previsions and upper transition operators are considered instead; therefore, we give here this more general form.

Definition 11.7 An upper transition operator $\overline{T}$ is *regular* if there exists a positive integer r such that $\min_{x\in\mathcal{X}} \overline{T}^r I_y(x) > 0$[3] for every pair of elements x and $y \in \mathcal{X}$. A Markov chain whose transition operator is regular is called a *regular Markov chain.*

---

[2] Cauchy sequence means that it has the property that for every $\varepsilon > 0$ there is some $N \in \mathbb{N}$ such that $d_H(C_m, C_n) < \varepsilon$ for every $m, n \geq N$.

[3] In terms of transition matrices this means that the rth power of the corresponding transition matrix P contains only strictly positive elements. Here, this means that the upper probability of making a transition from x to y in r steps is positive.

Theorem 11.5 (Perron-Frobenius theorem, the upper expectations form). *Let $\overline{T}$ be a regular transition operator. Then for every $f \in \mathcal{L}(\mathcal{X})$ the limit*

$$\overline{T}^{\infty} f := \lim_{n\to\infty} \overline{T}^{n} f$$

exists and is a constant gamble. Moreover, whenever $f \geq 0$ and $f(x) > 0$ for at least some $x \in \mathcal{X}$, then $\overline{T}^{\infty} f > 0$.

Let $\overline{Q}_0$ be an arbitrary prevision corresponding to the initial distribution and $\overline{Q}_n = \overline{Q}_0\overline{T}^{n}$. Then it follows from the above theorem that

$$\overline{Q}_{\infty}(f) := \lim_{n\to\infty} \overline{Q}_n(f) = \lim_{n\to\infty} \overline{Q}_0(\overline{T}^{n} f) = \lim_{n\to\infty} \overline{T}^{n} f(x) \tag{11.19}$$

for every $x \in \mathcal{X}$, which means that $\overline{Q}_{\infty}$ is independent of the initial distribution given by $\overline{Q}_0$.

The Perron-Frobenius gives a simple sufficient condition for convergence, which however is not necessary. In the next subsection we will derive necessary and sufficient conditions using coefficients of ergodicity.

11.4.3 Invariant distributions

Consider a convergent precise Markov chain with the initial probability mass function m_0 and the transition matrix P. Then the sequence of probability mass functions $\{m_0P^n\}_{n\in\mathbb{N}_0}$ converges to

$$m_{\infty} = \lim_{n\to\infty} m_0P^n. \tag{11.20}$$

Clearly,

$$m_{\infty}P = \left(\lim_{n\to\infty} m_0P^n\right)P = \lim_{n\to\infty} m_0P^{n+1} = m_{\infty}. \tag{11.21}$$

Any probability mass function m satisfying $mP = m$ is called *invariant* for P.

Similarly we say that a set C is an *invariant set of probabilities* for $\mathcal{T}$ if $C\mathcal{T} = C$. Let $\mathcal{P}$ be the set of all probability distributions on $\mathcal{X}$. Then the sequence $\{\mathcal{P}\mathcal{T}^n\}_{n\in\mathbb{N}_0}$ forms a nested chain with respect to set inclusion, and the intersection

$$C_{\infty} := \bigcap_{n\in\mathbb{N}_0} \mathcal{P}\mathcal{T}^n \tag{11.22}$$

exists and is an *invariant set of probabilities*. This set is the *limit set of probabilities* for the initial set $\mathcal{P}$; however, it is not necessarily the unique limit set.

In terms of upper previsions we will say that an upper prevision $\overline{Q}$ is an *invariant upper prevision* for an upper transition operator $\overline{T}$ if $\overline{Q}\,\overline{T} = \overline{Q}$ holds. Taking $\overline{V}$ to be the vacuous upper prevision, i.e. $\overline{V}(f) = \max_{x\in\mathcal{X}} f(x)$, the limit upper prevision

$$\overline{Q}_{\infty} = \lim_{n\to\infty} \overline{V}\,\overline{T}^{n} \tag{11.23}$$

corresponds to the credal set (11.22), and is an invariant upper prevision.

11.4.4 Coefficients of ergodicity

Coefficients of ergodicity or *contraction coefficients* are an efficient tool for the analysis of convergence in the theory of (imprecise) Markov chains. Besides providing necessary and sufficient conditions for convergence they also measure the rate of convergence, that is, they allow estimating how close the probability distribution at time n is to the limit distribution.

In the classical theory, a coefficient of ergodicity is a function $\tau(P)$ that assigns a value on the interval $[0, 1]$ to every stochastic matrix P, such that

- $\tau(PP') \le \tau(P)\tau(P')$;
- $\tau(P) = 0$ if and only if rank of P is 1, i.e. all rows of P are the same and equal to some probability mass vector m.

In general, a coefficient of ergodicity for a stochastic matrix is calculated as

$$\tau(P) = \sup_{m,m'} \frac{d(mP, m'P)}{d(m, m')}, \tag{11.24}$$

over all pairs of different probability mass functions m and m' on $\mathcal{X}$, and where d is some metric on the space of probability mass functions. In the case where d_p (see (11.14)) is used, $\tau(P)$ can be calculated directly as

$$\tau(P) = \frac{1}{2} \max_{i,j} \sum_{x_s \in \mathcal{X}} |P_i(x_s) - P_j(x_s)| = \max_{i,j} d_p(P_i, P_j), \tag{11.25}$$

where P_i and P_j denote the ith and jth row's probabilities with mass function p_i and p_j.

In terms of a transition operator $\overline{T}$, the coefficient of ergodicity (11.25) is equivalently calculated as

$$\tau(T) = \max_{x,y \in \mathcal{X}} d_o(\overline{T}(\cdot|x), \overline{T}(\cdot|y)). \tag{11.26}$$

11.4.5 Coefficients of ergodicity for imprecise Markov chains

Extending coefficients of ergodicity to imprecise Markov chains depends on the type of convergence we are considering. Suppose a very general imprecise Markov chain with an initial closed set of probabilities C_0 and a closed set of transition matrices $\mathcal{T}$. If no further assumptions are made, all that we can say about the sets of distributions at following times is that they are closed, which in the case of a finite state space, implies that they are compact. The convergence of the sequence of sets $\{C_n\}_{n \in \mathbb{N}_0}$ is then studied using the Hausdorff metric (11.15).

In [350] the *uniform coefficient of ergodicity* is defined for a set of transition matrices $\mathcal{T}$ with

$$\tau(\mathcal{T}) = \max_{P \in \mathcal{T}} \tau(P), \tag{11.27}$$

where $\tau(P)$ is the coefficient of ergodicity (11.25). The uniform coefficient of ergodicity can be interpreted as a contraction coefficient of the mapping $C \mapsto C\mathcal{T}$ because

$$d_H(C\mathcal{T}, C'\mathcal{T}) \le \tau(\mathcal{T}) d_H(C, C') \tag{11.28}$$

for every pair of compact sets C, C' and every compact set of transition matrices $\mathcal{T}$. Therefore, if $\tau(\mathcal{T}) < 1$ the mapping $C \mapsto C\mathcal{T}$ is a contraction mapping in the complete space of all compact sets of probabilities and therefore by Banach fixed point theorem it contains a unique invariant set (11.22), which is also the limit set, independent of the initial set of probabilities.

It is often the case that $\tau(\mathcal{T}) = 1$ but $\tau(\mathcal{T}^r) < 1$ for some positive integer r. The sets of transition matrices with this property are called *product scrambling*, and since the limit set C of $\mathcal{T}^r$ is also the limit set for $\mathcal{T}$, the unique convergence for a Markov chain with the transition set $\mathcal{T}^r$ implies the unique convergence of the one with the set $\mathcal{T}$, and to the same limit set.

The uniform coefficient of ergodicity depends on the contraction property of all linear transition operators that are possible candidates for the 'true' transition operator, and if they are all contractions, then the set-transition operator is a contraction too in the Hausdorff metric. Therefore, the product scrambling property is a sufficient condition for unique convergence; though, not also necessary. Consider for instance an upper transition operator with equal rows, i.e. $\overline{T}(\cdot|x) = \overline{T}(\cdot|y)$ for every pair of x and $y \in \mathcal{X}$. Clearly then $\overline{Q}_n(f) = \overline{Q}_1(f) = \overline{T}f(x)$ for any $x \in \mathcal{X}$, which is also the unique limit upper prevision. But the corresponding set $\mathcal{T}$ of transition operators does not necessarily only contain transition operators that are contractions. In the most extreme case, for instance, where $\overline{T}(\cdot|x)$ are vacuous upper previsions for every $x \in \mathcal{X}$, the credal set contains all possible transition operators.

Let us now consider a uniquely convergent imprecise Markov chain with the corresponding sequence of upper previsions $\{\overline{Q}_n\}_{n\in\mathbb{N}_0} = \{\overline{Q}_0\overline{T}^n\}_{n\in\mathbb{N}_0}$. If the limit

$$\lim_{n\to\infty} \overline{Q}_n(f) = \lim_{n\to\infty} \overline{Q}_0(\overline{T}^n f) \tag{11.29}$$

does not depend on $\overline{Q}_0$, this clearly means that the limit upper transition operator

$$\overline{T}^\infty f = \lim_{n\to\infty} \overline{T}^n f \tag{11.30}$$

is constant for every $f \in \mathcal{L}(\mathcal{X})$, or equivalently, that the row upper previsions $\overline{T}_i^n$ are converging to the same upper prevision. Thus the row upper previsions $\overline{T}_i^n$ are becoming more and more similar when n increases beyond all limits, or in other words, that the distance between them tends to zero.

We therefore extend the coefficient of ergodicity (11.26) to the *weak coefficient of ergodicity* defined with

$$\rho(\overline{T}) = \max_{x,y\in\mathcal{X}} d_o(\overline{T}(\cdot|x), \overline{T}(\cdot|y)). \tag{11.31}$$

It satisfies the conditions

- $\rho(\overline{T}\,\overline{T}') \leq \rho(\overline{T})\rho(\overline{T}')$ and
- $\rho(\overline{T}) = 0$ if and only if $\overline{T}$ is a constant operator: $\overline{T}(\cdot|x) = \overline{T}(\cdot|y)$ for every x and $y \in \mathcal{X}$.

The weak coefficient of ergodicity is a contraction coefficient of the mapping $\overline{Q} \mapsto \overline{Q}\,\overline{T}$ because

$$d_o(\overline{Q}\,\overline{T}, \overline{Q}'\,\overline{T}) \leq \rho(\overline{T}) d_o(\overline{Q}, \overline{Q}'). \tag{11.32}$$

Consequently, if for some $r > 0$, $\rho(\overline{T}^r) < 1$ then the limit

$$\overline{Q}_\infty = \lim_{n\to\infty} \overline{Q}_n = \lim_{n\to\infty} \overline{Q}_0 \overline{T}^n \tag{11.33}$$

always exists and is independent of $\overline{E}_0$. Moreover, we have the following upper bound for the distance between the upper prevision at time n and the limit upper prevision:

$$d(\overline{Q}_n, \overline{Q}_\infty) \leq \rho(\overline{T}^r)^{\frac{n-r}{r}} d(\overline{Q}_0, \overline{Q}_\infty) \leq \rho(\overline{T}^r)^{\frac{n-r}{r}}. \tag{11.34}$$

We say that an upper transition operator that satisfies $\rho(\overline{T}^r) < 1$ is *weakly product scrambling*. It follows from the above considerations that an imprecise Markov chain whose upper transition operator is weakly product scrambling uniquely converges. But the converse also holds, if an imprecise Markov chain converges uniquely then its upper transition operator is weakly product scrambling.

11.5 Further reading

Here we give a brief overview of the related literature (see also Section 11.3). The approach followed when studying the limit behaviour, was directed towards strong independence as was done in [349, 597]. Another approach to the analysis of the long term behaviour of imprecise Markov chains has been taken in [205]. They define an accessibility relation between states based on the upper transition operators, which allows them to classify states in a way similar to the one used in the precise Markov chains. Moreover, using the accessibility relation, they give necessary conditions for unique convergence of imprecise Markov chains. In [359], the necessary and suffient conditions are given and called ergodicity. Moreover, an algorithm to decide upon ergodicity is represented. This algorithm is essentially linear in the cardinality of state space where the calculation of the transition operator is considered an oracle. Almost everything that is explained in this chapter can also be found in [358] in a more extensive form.

In [169] the convergence of imprecise Markov chains with an absorbing state is analyzed. They study the problem of convergence of the probability distributions over the set of states conditional on nonabsorption. They find that under conditions analogous to those found in [171], stated for precise Markov chains with an absorbing state, the conditional imprecise distributions converge to a unique imprecise limit distribution.

Another random process that has been reported in the imprecise probabilistic literature is the Hidden Markov model [182]. This model can also be seen as a special type of a Markov tree [204], which on its turns can be seen as a special type of random process. The last two papers are situated more in the graphical models field. Given the graphical background of imprecise probability trees, it is also possible to consider random processes as a special type of graphical models and a lot of inspiration can be found in the papers from that field too (see also Chapter 9).

This chapter, and most of the published papers, assume discrete time random processes. The field of continuous time random processes where the uncertainty is described in terms of lower previsions, sets of desirable gambles, or other generalized uncertainty models is practically unexplored. An exception is [349], where some results about the solutions of a set-valued matrix stochastic differential equations are given. Recently, in [559] a contribution has been made to the analysis of continuous time stochastic processes with uncertain parameters with a study of the solutions of stochastic differential equations with parameters that are random sets (see also Section 1.3.5).

In [203] where imprecise probability trees are introduced, the match between these imprecise probability trees and Shafer and Vovk's [582] game-theoretical account to probability theory is also described. Shafer and Vovk's theory finds its foundation in betting strategies and martingale theory (see Chapter 6) and lends itself excellently to – for example – the study of laws of large numbers and concentration inequalities. One advantage of this theory is that also continuous time random processes are described, although it seems that in this case – as in the case with discrete event trees with infinitely long paths – one should be comfortable with a (weakened) σ-additivity assumption that is implicitly made.

12

Financial risk measurement

Paolo Vicig
Department DEAMS 'B. de Finetti', University of Trieste, Italy

12.1 Introduction

The statement that the theory of imprecise probabilities offers potentially many applications in the realms of finance and economics should seem quite obvious to a reader with some general knowledge of the theory, for instance to someone acquainted with just the introductory chapters of this book. In fact, a strong motivation for introducing imprecise probabilities is that they avoid from the very beginning the difficulty of assessing precisely beliefs which might be more or less vague. This motivation applies to these areas of investigation too, perhaps to an even larger extent than in physics, engineering or other fields of knowledge, since measurement of the relevant quantities may be more imprecise, or rely more heavily upon subjective evaluations, feelings, etc. Any model in finance making use of precise probabilities (think for instance of option pricing models) could be generalized by choosing instead imprecise probabilities as uncertainty evaluations.

Further, recall from Chapter 2 that a lower prevision $\underline{P}(f)$ (alternatively, an upper prevision $\overline{P}(f)$) is itself interpreted as a supremum buying price (an infimum selling price) for the gamble f. This provides evidence for the importance of imprecise probability theory in betting and insurance, whose core activity may be synthesized just through the operation of buying or selling a random quantity for a nonrandom price. Yet, and perhaps surprisingly, contributions in finance, economics or insurance making explicit use of imprecise probabilities are not numerous, and rather sparse. These areas offer nowadays wide, and still largely unexplored, spaces for applying imprecise probabilities.

The focus in this chapter is on a few applications. Precisely, a closer look at the betting scheme in the definition of coherence is given in Section 12.2, partly complementing ideas from previous chapters, partly to discuss how the betting scheme could be applied

Introduction to Imprecise Probabilities, First Edition.
Edited by Thomas Augustin, Frank P. A. Coolen, Gert de Cooman and Matthias C. M. Troffaes.

in real-world bets. Section 12.3 illustrates the role of imprecise previsions in risk measurement, showing that risk measures can be interpreted as upper previsions. This key fact makes it possible to present reinterpretations and extensions of results from classical risk measurement literature as well as recent new ideas, developed by means of the theoretical tools of imprecise probabilities.

There are various reasons for selecting just risk measures among the many potential financial applications of imprecise probabilities:

- Unlike other applications, risk measurement actually needs imprecise previsions as uncertainty measures, rather than the less general imprecise probabilities.
- Current risk measurement practice does not require much statistical reasoning, at least in its basic problems. What often matters is just to *'give a number'* to evaluate some risk, rather than apply statistical procedures, generally more complex with imprecise probabilities than with standard probability theory (cf. Chapter 7).
- Specific risk measurement problems are helpful in developing useful notions in the theory of imprecise probabilities. For instance, (centered) convex risk measures correspond, but were introduced prior to, (centered) convex previsions, a class of imprecise previsions with interesting properties, intermediate between those which avoid sure loss and the coherent ones.

After supplying basic ideas in Section 12.3.1, two important families of risk measures are considered: *coherent* risk measures (Section 12.3.2) and *convex* risk measures (Section 12.3.3). To keep the presentation simpler, conditional risk measures are not discussed. Something about this and other topics is mentioned in Section 12.4.

12.2 Imprecise previsions and betting

As shown in Chapter 2, both linear previsions and upper/lower previsions which are coherent or avoid sure loss may be defined resorting to suitable *betting schemes*. To give a closer look at these schemes, take the condition in Equation (2.2) for coherence of a lower prevision $\underline{P}$ defined on $\mathcal{K}$. Defining $\underline{g}_0 = -m[f_0 - \underline{P}(f_0)] = m[\underline{P}(f_0) - f_0]$, the condition requires that if the agent *buys* any (not necessarily distinct) $f_1, \ldots, f_n \in \mathcal{K}$ at her supremum buying prices $\underline{P}(f_1), \ldots, \underline{P}(f_n)$ and (when $m > 0$) *sells* any $f_0 \in \mathcal{K}$ at her supremum *buying* price for it, the supremum of the ensuing overall gain is nonnegative. Note that coherence reduces to the condition of avoiding sure loss both when $m = 0$ and also if the agent were to sell f_0 at her infimum selling price $\overline{P}(f_0)$: the latter case, using $\overline{P}(f_0) = -\underline{P}(-f_0)$, is equivalent to checking the condition of avoiding sure loss, when the gambles selected are $f_1, \ldots, f_n, -f_0$ ($-f_0$ is selected m times). Hence a distinguishing feature of coherence is that the agent might be forced to sell just one gamble at her (supremum) buying price for it. Intuitively, this prevents the agent from adopting unrealistic prices, as shown in the following simple example.

Example 12.1 If $\mathcal{K} = \{A\}$, where A is (the indicator of) any event, the assessment $\underline{P}(A) = c < 0$ does avoid sure loss, but is not coherent (taking $n = 0, m = 1$ in Equation (2.2), it holds that $\sup\{-A + \underline{P}(A)\} \leq c < 0$). An agent having to buy $A \geq 0$ would well be willing to 'pay' $c < 0$ for it, which means receiving a positive amount $-c$ for buying A. The point is that no vendor can be expected to accept such a transaction. To come to a realistic price, the agent has to consider whether she would accept c being at the other side of the transaction (being the vendor). This is precisely the role of the term $\underline{g}_0$. ♦

The same arguments apply to the definition of linear prevision in Equation (2.3), which displays anyway a relevant difference: there may be finitely many distinct gambles ($g_1, \ldots, g_n$ in Equation (2.3)) having the role of f_0. This ensures that a linear prevision on $\mathcal{K}$ is a *fair* price for either buying or selling any gamble in $\mathcal{K}$, as well known [218].

Recall that there are equivalent definitions for the consistency notions we mentioned that use not necessarily integer *stakes* λ_i.

For an instance that will be useful in the sequel, $\overline{P}$ is a *coherent upper prevision* on $\mathcal{K}$ if and only if $\forall n \geq 0, \forall f_o, f_1, \ldots, f_n \in \mathcal{K}$ and for all real and nonnegative $\lambda_0, \lambda_1, \ldots, \lambda_n$, it holds that

$$\sup \overline{G} \geq 0, \qquad \overline{G} = \sum_{i=1}^{n} \lambda_i(\overline{P}(f_i) - f_i) - \lambda_0(\overline{P}(f_0) - f_0). \tag{12.1}$$

The term $\overline{g}_0 = \lambda_0(f_0 - \overline{P}(f_0))$ has a normative function symmetric to that of $\underline{g}_0$: it obliges the agent to consider (at least hypothetically) *buying* f_0 at its infimum *selling* price $\overline{P}(f_0)$.

How do the betting schemes discussed so far and real–world betting relate? Often, a bookie or a bet organizer in a broader sense (for instance, an insurer), to be called *House* in the sequel, fixes most of the rules of the bet, including prices, being confronted with a multitude of bettors (insured, in the insurance case), none of which individually has the power of modifying the rules, but only the faculty of accepting or rejecting the bet. In deciding which bets to engage and their prices, House aims at achieving a *positive expected gain*.

Given this, it is clear that House's prices are not linear previsions: in fact, the linear prevision of the gain from betting on $f_1, \ldots, f_m, g_1, \ldots, g_n$ in Equation (2.3) is zero, as follows easily by linearity and homogeneity of linear previsions (Properties P1., P2. in Section 2.2.2). This means that House expects to neither gain nor lose anything when choosing linear previsions as betting prices, but in practice the expectation turns to negative when the costs House has to bear are considered. Hence, House's selling prices must be higher than her linear previsions.

Thus the betting scheme for linear previsions appears appropriate for eliciting beliefs, rather than describing real prices for betting. These can be interpreted instead as upper previsions, and coherence is a reasonable requirement for them.

However, House does not usually take up all bets which are (at least marginally) acceptable in Equation (12.1) in the definition of coherent upper prevision. Defining the gain $\overline{G}_{ASL} = \sum_{i=1}^{n} \lambda_i(\overline{P}(f_i) - f_i)$, we obtain $\overline{G} = \overline{G}_{ASL} + \overline{g}_0$ in (12.1), and it can be shown [658] that

$$\underline{P}(\overline{G}_{ASL}) \geq \underline{P}(\overline{G}), \quad \overline{P}(\overline{g}_0) \leq 0. \tag{12.2}$$

Here $\overline{P}(\overline{g}_0)$ is the natural extension of $\overline{P}$ (on the gamble $\overline{g}_0$), and $\underline{P}$ its conjugate (on the set of gambles $\{\overline{G}_{ASL}, \overline{G}\}$). In words, House has no convenience in buying anything at her (infimum) selling prices. Modifying the bet in (12.1) by not betting on f_0 ensures, *ceteris paribus*, better gain prospects.

Thus, unlike linear previsions, coherent upper previsions may be real prices in betting situations. Even though House can not impose prices that avoid sure loss only, since they might be too openly at her exclusive advantage and hence unacceptable to the counterparts (cf. Example 12.1), the bets she proposes are included (by (12.2)) into those of the scheme for avoiding sure loss. Apart from other practical reasons, the inclusion is strict because of the following motivation: House is interested in confining the *discounting* effect of the subadditivity (or sublinearity) property $\overline{P}(f+g) \leq \overline{P}(f) + \overline{P}(g)$. This can be most easily achieved accepting bets on f and g, but not on $f+g$. For an event example, betting *separately* on $A =$ *X wins the match against Y* and $C =$ *neither X nor Y wins* ensures a greater gain for House than betting on $A \vee C =$ *Y does not lose*, unless her $\overline{P}$ is additive on $\{A, C, A \vee C\}$.[1]

In the case that House is an insurer, her basic activity is roughly described by exchanges where she buys a gamble f from the insured for a price $\underline{P}(f)$. Since typically $f \leq 0$, as f is a claim the insured might be liable for, the price $\underline{P}(f)$ is negative, i.e. the insurer actually receives a sum $-\underline{P}(f)$, called *premium*, to 'buy' f. Often, for instance in motor insurance, the insurer has many available data, sufficient to assess a linear prevision $P(f)$ rather accurately and reliably. Yet her buying price is $\underline{P}(f) < P(f)$, because of the need to earn a profit, and also to cover costs. The difference $P(f) - \underline{P}(f) > 0$ is termed *loading*.

Remark. It is found sometimes convenient in insurance (and finance) to reason in terms of losses, for instance in terms of $-f$ instead of f. This makes the use of upper, rather than lower, previsions more natural. In particular, the premium is then $\overline{P}(-f)$. ■

12.3 Imprecise previsions and risk measurement

A basic risk measurement problem is that of evaluating the riskiness of the gambles in some set $\mathcal{K}$, gambles which typically are financial assets or also portfolios (i.e., sets of financial assets) held by an individual, society or institution.[2] The problem is practically very important, and different solutions are available. Assessing a *risk measure* is a very straightforward and popular one, because of its simplicity. In fact, it requires just 'to give a number' $\rho(f)$ to any $f \in \mathcal{K}$; $\rho(f)$ measures how risky f is, and operationally identifies the amount of money *to be reserved* for facing potential future losses arising from f. Apart from asset owners, further subjects interested in such evaluations include governmental or supervising authorities, called *regulators*.

Formally, a risk measure $\rho(\cdot)$ is thus a mapping from $\mathcal{K}$ into the real line. The question of what properties ρ should satisfy to be a 'good' risk measure is controversial, and originated a great deal of different proposals among academicians and practitioners. A good number of these assume that a probability distribution P_f is known or estimated for each $f \in \mathcal{K}$ and define $\rho(f)$ by means of some indexes, moments, or other quantities depending on P_f.

Value–at–Risk or *VaR*, a very common if not even the most popular risk measure, is also defined in this way. A general warning: like other risk measures, there exist more definitions of *VaR*, not all equivalent, even though the unifying idea is that of defining a quantile–based

[1] The *Pari–mutuel model* (cf. Section 4.7.1) with upper probability $\overline{P}_{pm}(A) = \min\{(1+\epsilon)P_0(A), 1\}, \forall A$, where $\epsilon \in [0, 1]$ is a (loading) parameter chosen by House, is a simple model of a coherent upper probability which is additive as far as the evaluation of A given by $\overline{P}_{pm}(A) = (1+\epsilon)P_0(A)$ does not exceed 1. Therefore, because of its precluding any discounting effect whenever this does not clash against coherence, the Pari–mutuel model is quite suited for this kind of problems, and was in fact named after the so-called Pari–mutuel betting systems.

[2] Unbounded random numbers could more generally be considered, which we shall not do here. This is not overly restrictive, given that financial assets are generally bounded. Even a debtor's *unlimited reliability* is practically bounded by the amount of her overall wealth.

measure. Following [30], say that, given $\alpha \in]0,1[$, q is an *α-quantile* for f if $P(f < q) \leq \alpha \leq P(f \leq q)$. Define then

$$q_\alpha^+ = \inf\{x : P(f \leq x) > \alpha\}, \quad VaR_\alpha(f) = -q_\alpha^+(f). \tag{12.3}$$

The idea underlying the definition of *VaR* consists of determining a lower threshold (q_α^+) which, with a confidence level α fixed *a priori* – usually $\alpha = 0.05$ or 0.01 – will not be exceeded. While summarizing f, *VaR* obviously neglects other potentially important aspects. Among these, it is quite uninformative about the maximum loss f may cause, which might instead be evaluated, for instance, by the conditional expectation $E(f|f < q_\alpha^+)$. Further, it does not necessarily ensure the subadditivity property $\rho(f+g) \leq \rho(f) + \rho(g)$. Subadditivity is often regarded as a desirable property for risk measures, since the risk of the sum is generally less than the sum of the risks, because of diversification.

As an alternative to *VaR*, Artzner, Delbaen, Eber and Heath introduced a new family of risk measures in a series of papers, including [29, 30]. While not referring to the theory of imprecise probabilities, they called them *coherent risk measures*, and this is the name reserved to them in most of the literature. To better emphasize their generalization, presented in Section 12.3.2, we shall call them *ADEH–coherent* risk measures. They are defined by means of axioms. Precisely,

Definition 12.1 Let $\mathcal{L}$ be a linear space of gambles which contains real constants. A mapping $\rho : \mathcal{L} \to \mathbb{R}$ is an *ADEH–coherent risk measure* if and only if it satisfies the following axioms:

(T) $\forall f \in \mathcal{L}, \forall \alpha \in \mathbb{R}, \rho(f + \alpha) = \rho(f) - \alpha$ (translation invariance)

(PH) $\forall f \in \mathcal{L}, \forall \lambda \geq 0, \rho(\lambda f) = \lambda \rho(f)$ (positive homogeneity)

(M) $\forall f, g \in \mathcal{L}$, if $f \leq g$ then $\rho(g) \leq \rho(f)$ (monotonicity)

(S) $\forall f, g \in \mathcal{L}, \rho(f + g) \leq \rho(f) + \rho(g)$ (subadditivity).

Many other risk measures were proposed in the literature. The topic is generally treated independently of imprecise probability theory, with few exceptions [46, 442, 508, 507, 509, 511]. We shall now see that the relationship between these research areas is indeed very close.

12.3.1 Risk measures as imprecise previsions

The interpretation we are about to discuss, suggested in [508], points out the nature of imprecise prevision of a risk measure. Precisely, how could an agent elicit $\rho(f)$? She might identify $\rho(f)$ with the infimum of the amounts of money she would ask in order to shoulder f. Since getting a certain amount for receiving f is the same as selling $-f$ for the same sum, $\rho(f)$ can be viewed as the agent's infimum selling price for $-f$, which is indeed the behavioural interpretation given in [672] for an upper prevision $\overline{P}$ of $-f$. Recalling also the conjugacy property $\overline{P}(-f) = -\underline{P}(f)$, we come to the *basic identity*

$$\rho(f) = \overline{P}(-f) = -\underline{P}(f). \tag{12.4}$$

The fundamental Equation (12.4) ensures that a risk measure $\rho(f)$ can be equivalently interpreted as either an upper prevision (for $-f$) or as the opposite of a lower prevision (for f). This makes it possible to exploit results from imprecise probability theory in risk measurement, and vice versa. This process will be exemplified in the next sections. The main improvements it allows on the risk measurement side are:

- A general formal framework becomes available for studying the consistency of a risk measure. For an instance, see the case of *VaR* in the next section.
- Several existing risk measures can be easily generalized. This is for instance the case of ADEH–coherent risk measures, see the next section.
- New risk measures can be introduced by means of more general uncertainty measures, for instance replacing expectations with upper/lower previsions. For an example, see the Dutch risk measures in Example 12.3. Other instances are discussed in [511, § 4].

The approach outlined here applies more generally to *conditional* risk measures, i.e. risk measures for evaluating conditional gambles $f|B$, where B is a non-impossible event. The starting point is the generalization of Equation (12.4) to $\rho(f|B) = \overline{P}(-f|B) = -\underline{P}(f|B)$.

12.3.2 Coherent risk measures

Recalling the discussion of the previous section and Equation (12.4), it is straightforward to define coherent risk measures on an arbitrary set $\mathcal{K}$ of gambles:

The mapping $\rho: \mathcal{K} \to \mathbb{R}$ is a *coherent risk measure* on $\mathcal{K}$ if and only if there exists a coherent upper prevision $\overline{P}$ on $\mathcal{K}^- = \{-f : f \in \mathcal{K}\}$ such that $\rho(f) = \overline{P}(-f), \forall f \in \mathcal{K}$.

We may use Eqs. (12.1) and (12.4) to reword equivalently the definition as follows:

Definition 12.2 $\rho: \mathcal{K} \to \mathbb{R}$ is a *coherent risk measure* on $\mathcal{K}$ if and only if $\forall n \geq 0, \forall f_o, f_1, \ldots, f_n \in \mathcal{K}$ and for all real and nonnegative $\lambda_0, \lambda_1, \ldots, \lambda_n$, it holds that

$$\sup \overline{G} \geq 0, \qquad \overline{G} = \sum_{i=1}^{n} \lambda_i(\rho(f_i) + f_i) - \lambda_0(\rho(f_0) + f_0). \tag{12.5}$$

The definition of risk measure that avoids sure loss on $\mathcal{K}$ is introduced analogously.

Given this, a general methodology to investigate consistency of known risk measures is available: we may use tools from imprecise probability theory to determine if and when they are coherent, or avoid at least sure loss.

This investigation was performed in the case of *VaR* in [508, § 3] showing that *VaR* may or may not be coherent, and when not it may even incur sure loss.[3] Sufficient conditions for VaR_α to either be coherent or avoid sure loss are available, but their actual implementation is limited by their requiring some constraints on the *VaR* parameter α, which is instead fixed *a priori* in practice.

ADEH–coherent risk measures are on the contrary *a special case* of coherent risk measures. In fact, it was shown in [508] that:

Theorem 12.1 *When $\mathcal{K}$ is a linear space of gambles that contains real constants, a mapping $\rho: \mathcal{K} \to \mathbb{R}$ is ADEH–coherent (on $\mathcal{K}$) if and only if it is a coherent risk measure (on $\mathcal{K}$).*

[3] The consistency of some other known risk measures is discussed in [657, § 5.4].

In general, that is when $\mathcal{K}$ is arbitrary, the axioms for ADEH–coherent risk measures are (necessary, but) not sufficient to ensure that a risk measure is coherent.

A practical way of obtaining coherent risk measures requires first each of n experts to assess a linear prevision P on the gambles in $\mathcal{K}$ (or equivalently on the gambles of $\mathcal{K}^-$: one of the two assessments implies the other, by linearity of linear previsions). It secondly defines $\rho(f)$ as the maximum over all linear previsions $P(-f) = -P(f)$, for each $f \in \mathcal{K}$. This method exploits the characterization of coherent upper previsions (hence also of coherent risk measures) as *upper envelopes* of linear previsions, dual to the characterization of coherent lower previsions as lower envelopes of linear previsions, cf. Proposition 2.3. It was already considered for ADEH–coherent risk measures in [30] and elsewhere in the literature under the name of method of *scenarios*. A scenario may be simply a single expert's linear prevision on $\mathcal{K}$, even though it is rather a collection of expectations (one for each $f \in \mathcal{K}$) in the standard approach, where further $\mathcal{K}$ is a linear space. To compute such expectations, it is required preliminarily that each expert elicits a cumulative distribution function for every gamble in $\mathcal{K}$, an often non-trivial task which as we have seen is actually unnecessary to get ρ as an upper envelope of linear previsions.

If a risk measure ρ_{ASL} is not coherent, but avoids sure loss, it is possible to correct it to a coherent risk measure, for instance computing its natural extension ρ_E. However, the peculiarity of risk measurement makes it not so obvious that ρ_E should always be the favoured correction. In fact, from the properties of the natural extension (cf. (12.4), Proposition 2.1 and [672]) we know that $\rho_E(f) \leq \rho_{ASL}(f), \forall f$. This means that the natural extension correction is *less prudential* than ρ_{ASL}, i.e. it requires reserving a smaller amount of money to face the same risks. Even though ρ_E is the most prudential among the coherent corrections of ρ_{ASL} that are less prudential than ρ_{ASL}, that is if ρ^* is coherent and $\rho^* \leq \rho_{ASL}$ it is guaranteed that $\rho^* \leq \rho_E$, often a *more prudential* correction $\rho_c \geq \rho_{ASL}$ may be advisable, or even imposed by some regulator. In this case the natural extension is inappropriate, while under mild constraints it is possible to resort to other coherent corrections, termed *upper extensions* and discussed in [508], originally introduced in [692]. An upper extension ρ_U ensures in fact that the inequality $\rho_U \geq \rho_{ASL}$ holds.

12.3.3 Convex risk measures (and previsions)

Convex risk measures were introduced in [296, 297] with the purpose of weakening the positive homogeneity axiom (PH), $\rho(\lambda f) = \lambda \rho(f), \forall \lambda > 0$, of ADEH–coherent risk measures. This axiom may be too strict in several circumstances. For instance, $\rho(\lambda f)$ might be greater than $\lambda \rho(f)$ for a large λ because of *liquidity risk*, that is due to the difficulty in selling a large amount of some asset quickly and without lowering prices. In [296, 297], Föllmer and Schied defined convex risk measures on linear spaces of random variables by replacing axioms (S) and (PH) in Definition 12.1 of ADEH–coherent risk measures (while keeping the domain $\mathcal{L}$ and the remaining axioms (T), (M)) with the *convexity axiom*

(C) $\rho(cf + (1 - c)g) \leq c\rho(f) + (1 - c)\rho(g)\ \forall f, g \in \mathcal{L}, c \in [0, 1]$.

We shall term these measures *FS–convex risk measures*. In fact, similarly to coherent risk measures, it is possible to define convex risk measures on arbitrary sets of gambles, resorting to a betting interpretation similar to that of coherent previsions:

Definition 12.3 The map $\rho : \mathcal{K} \to \mathbb{R}$ is a *convex risk measure* on $\mathcal{K}$ if and only if $\forall n \geq 1$, $\forall f_o, f_1, \ldots, f_n \in \mathcal{K}$ and for all real and nonnegative $\lambda_1, \ldots, \lambda_n$ such that $\sum_{i=1}^{n} \lambda_i = 1$

(convexity condition), it holds that

$$\sup \overline{G} \geq 0, \qquad \overline{G} = \sum_{i=1}^{n} \lambda_i(\rho(f_i) + f_i) - (\rho(f_0) + f_0). \tag{12.6}$$

Further, ρ is a *centered convex* or *C−convex* risk measure if it is convex and ($0 \in \mathcal{K}$ and) $\rho(0) = 0$.

Hence, the convexity constraint $\sum_{i=1}^{n} \lambda_i = 1$ on the stakes is formally the distinguishing feature of convex risk measures, with respect to the coherent ones. At an interpretation level, this constraint, which is equivalent to $\sum_{i=1}^{n} \lambda_i = \lambda_0 > 0$, appears to be a sort of *bounded rationality* condition. This can be understood recalling from Section 12.2 that coherence for, say, lower previsions corrects the possibly unbalanced evaluations of the condition of avoiding sure loss by allowing selling at most one gamble (f_0) at one's buying price for it, with an *arbitrary* stake $\lambda_0 \geq 0$. Convexity asks for the same kind of correction, but puts a bound on it by including the constraint $\lambda_0 = 1$ (or $\lambda_0 = \sum_{i=1}^{n} \lambda_i$).

Unlike coherent risk measures, the corresponding concepts in imprecise probability theory, i.e. *convex* and *C−convex previsions*, had not been already introduced, prior to Föllmer and Schied's proposal. They were studied in [507, 509].

It can be shown that FS−convex risk measures and convex risk measures are equivalent concepts *when* $\mathcal{K}$ is a linear space of gambles.

Note that FS−convex risk measures are not necessarily C−convex, even though the additional condition for C−convexity $\rho(0) = 0$ is quite natural: it means that 0 is what we would reserve to cover the risk of owning the gamble identically equal to 0.

Not surprisingly then, C−convex risk measures have more satisfactory properties than purely convex measures. They always avoid sure loss (see some more detail on this below, in terms of C−convex lower previsions), and there exists also an analogue of the natural extension, termed *convex natural extension* , which allows extending a C−convex risk measure on any superset of $\mathcal{K}$ while preserving C−convexity. The convex natural extension characterizes C−convexity: a risk measure is C−convex on $\mathcal{K}$ if and only if it coincides on $\mathcal{K}$ with its convex natural extension.

To illustrate the greater flexibility of C−convex compared to coherent risk measures, we report the following example, developed at length in [507].

Example 12.2 An agent is interested in buying an amount λf of asset f, and has not settled yet a value for $\lambda > 0$. Thus $\mathcal{K} = \{\lambda f : \lambda > 0\}$. It is easy to check that any coherent risk measure on such a $\mathcal{K}$ reduces to a *linear* prevision, $\rho(\lambda f) = \overline{P}(-\lambda f) = P(-\lambda f) = -\lambda P(f)$. Hence there is a limited degree of freedom in using coherent risk measures, in particular it is not possible to incorporate the agent's possibly increasing aversion or difficulty in buying λf when λ is raised considerably. This can be done with C−convex risk measures, for instance defining $\rho(\lambda f)$ as a piecewise linear function of λ, getting steeper as λ grows higher than some fixed threshold. Given $\rho(f)$, and conventionally setting the threshold at $\lambda = 1$ (which is not restrictive, we could redefine f and λ by a convenient scaling factor to reduce any different threshold to one), we may thus define

$$\rho(\lambda f) = \begin{cases} \lambda\rho(f) & \text{if} \quad \lambda \in [0, 1] \\ k(\lambda - 1) + \rho(f) & \text{if} \quad \lambda > 1 \end{cases} \qquad \text{with} \quad k \in [\rho(f), -\inf f]$$

which is a C−convex risk measure. See [507, Example 4.1] for details. ◆

Given the correspondence (12.4), results for C−convex or convex risk measures are easily reworded in terms of either upper or lower previsions. Some properties of convex lower previsions are studied in [507]. Among these, an analogue of Proposition 2.1 holds, with an identical statement but for the substitution of 'coherent' with 'convex'. In words, lower envelopes, pointwise limits and convex linear combinations of convex lower previsions are convex lower previsions. In general, however, convex lower previsions do not always avoid sure loss: when $0 \in \mathcal{K}$, they avoid sure loss if and only if $\underline{P}(0) \leq 0$. C−convex lower previsions always satisfy this condition, hence they are a subclass of the imprecise previsions that avoid sure loss. As such, they have special properties, close to those of coherent lower previsions. We mention here a relevant one, which extends the characterization of coherent lower previsions by means of dominating linear previsions discussed in Section 2.2.2., cf. Proposition 2.3. It is based on the following theorem [507].

Theorem 12.2 (Generalized Envelope Theorem). *$\underline{P}$ is C−convex on $\mathcal{K}$ if and only if there exist a set $\mathcal{P}$ of linear previsions on $\mathcal{K}$ and a function $\alpha : \mathcal{P} \to \mathbb{R}$ such that:*

(a) $\underline{P}(f) = \min_{P \in \mathcal{P}}\{P(f) + \alpha(P)\}, \forall f \in \mathcal{K}$

(b) $\min_{P \in \mathcal{P}} \alpha(P) = 0$.

Clearly, we obtain the familiar lower envelope theorem (Proposition 2.3) when the function α is identically equal to 0, and $\mathcal{P}$ coincides with the set $\mathcal{M}$ of all linear previsions that dominate $\underline{P}$ on $\mathcal{K}$. In risk measurement versions of this theorem [296, 297], function α is termed *penalty function*. The generalized envelope theorem allows for a sensitivity analysis interpretation of C−convex lower previsions: if an agent is given a set $\mathcal{P}$ of linear previsions, but believes that some of them are systematically biased because they assign lower values than reasonable to relevant gambles, she may correct each such prevision P by raising its value by $\alpha(P)$. For instance, if $\mathcal{P}$ is formed by the previsions of n experts, the agent might in this way diminish or even (for a large enough $\alpha(P)$) rule out the influence of an expert reputed to be little reliable, prior to forming the lower envelope in condition (a). The correction is of a special type, being a translation from $P(f)$ to $P(f) + \alpha$ for every $f \in \mathcal{K}$. There is a bound to this practice, set by condition (b): at least one $P \in \mathcal{P}$ must not be corrected.

We already mentioned the need of weakening the positive homogeneity axiom (PH) as a basic motivation for introducing FS−convex risk measures. The way C−convex risk measures relate to positive homogeneity is not immediate from Definition 12.3, but it can be shown [507] that any C−convex risk measure ρ must satisfy the following, radically weaker forms of positive homogeneity:

$$\rho(\lambda f) \leq \lambda \rho(f), \forall \lambda \leq 1, \quad \rho(\lambda f) \geq \lambda \rho(f), \forall \lambda \geq 1. \tag{12.7}$$

To better illustrate the usefulness of imprecise probability theory in risk measurement, as well as how ideas originally introduced in risk measurement may suggest useful imprecise probability counterparts, we present the following example concerning Dutch risk measures. The topic is discussed in greater detail in [46].

Example 12.3 (Dutch risk measure). In [653], van Heerwarden and Kaas discuss the following risk measure, known as Dutch risk measure:

$$\rho_D^*(f) = E(-f) + cE[\max(E(f) - f, 0)], c \in [0, 1], \tag{12.8}$$

where E means expectation (given an underlying probability distribution) and c is a parameter. They derive ρ_D^* imposing suitable properties to a more general measure, which refers to certain insurance contracts historically termed Dutch treats.

The measure ρ_D^* can be given the interesting interpretation to be outlined now. Suppose an agent's risk measure for f is $E(-f)$, which is intuitively inadequate: a risk assessor tends to overweight lower values of f with respect to higher ones, which $E(-f)$ does not do. We might anyway think of correcting the first approach risk measure $E(-f)$ by taking account of its potential inadequacy, which is actually done in the second term in (12.8). In fact, the gamble $\max(E(f) - f, 0) = \max(-f - E(-f), 0)$ is the *shortfall*, or the *residual loss* in absolute value induced by reserving only $E(-f)$ to cover losses from f. For instance, if f turns out to be -4, while $E(-f) = 1$, we have $\max(E(f) - f, 0) = 3$, and 3 are the units of the loss of 4 not covered by $E(-f)$.

Hence ρ_D^* is a sort of *second–order* risk measure, increasing the naive risk measure $E(-f)$ by $cE[\max(E(f) - f, 0)]$.

This approach can be generalized. As a first step, consider

$$\rho_D(f) = P_0(-f) + cP_1[\max(P_0(f) - f, 0)], c \in [0, 1]. \tag{12.9}$$

In (12.9), the expectation has been replaced by two linear previsions, possibly distinct. This latter fact is interesting, because it ensures that the correction could be decided by an agent (for instance, a regulatory authority) other than that making the first attempt evaluation $P_0(-f)$. Further, ρ_D is known to be a coherent risk measure [657]. ♦

A second generalizing step replaces the linear previsions P_0, P_1 with generic risk measures (hence, by (12.4), with imprecise previsions). This is quite natural a choice, given that employing E as a risk measure seems to rely more on the well–established habit of using precise uncertainty measures than on a real need in the measurement problem.

This extension preserves the consistency properties of the original measures, as stated more precisely in the following theorem.

Theorem 12.3 *Define*

$$\rho_c(f) = \rho(f) + c \cdot \rho_1(-\max(-\rho(f) - f, 0)), c \in [0, 1], \tag{12.10}$$

where ρ, ρ_1 are coherent (alternatively: C–convex, convex) risk measures on $\mathcal{K}, \mathcal{K}_1$ respectively ($\mathcal{K}_1$ is such that (12.10) is well defined). Then ρ_c is a coherent (alternatively: C–convex, convex) risk measure on $\mathcal{K}$. □

Thus (12.10) suggests a way of 'correcting' a risk measure, such that the new risk measure inherits its consistency properties. ■

Of course, using (12.4) the argument can be presented directly, and is relevant also, in terms of imprecise previsions. We may modify a lower prevision $\underline{P}$, which we feel inadequate for some reason, maybe because it is just a first approach evaluation, by replacing it with $\underline{P}_c(f) = \underline{P}(f) + c \cdot \underline{P}_1(\min(f - \underline{P}(f), 0))$. The new lower prevision $\underline{P}_c$ is guaranteed to be coherent (alternatively, C–convex or convex) if $\underline{P}, \underline{P}_1$ are so.

12.4 Further reading

The approach followed in this chapter is generally based on the theory of imprecise probabilities as introduced in [672] and on some of its developments. More specifically, the material of Section 12.2 is treated more extensively in [658], that of Section 12.3 in [46, 508, 507, 509, 657]. A different unifying approach, not presented here, to imprecise previsions, risk measures and exact cooperative games by means of exact functionals is developed in [442].

A major known issue we did not discuss regards the extension of the relationship between imprecise previsions and risk measures to conditioning. In particular, coherent and C–convex conditional risk measures may be introduced [509]. It has been shown that they retain several important properties of their correspondent unconditional measures. For coherent conditional risk measures, this depends also on the definition of coherence, which corresponds to that of Williams, a very general one in the conditional framework (cf. [703] and Section 2.4.1), while conditional C–convexity is an even more general consistency concept. For instance, there always exists a coherent or, respectively, C–convex natural extension on any superset of the domain of a coherent, or C–convex, conditional risk measure. The envelope theorem is also extended to coherent conditional risk measures and applies in all cases, while it is not clear yet whether there exists a manageable version of the generalized envelope theorem for C–convex conditional risk measures, outside some special cases. The generalized Bayes rule (GBR) is instead a necessary consistency requirement for both coherent and C–convex risk measures.[4] The GBR and other conditions are also jointly sufficient for C–convexity if the domain of the risk measure has a special structure (which reduces to a linear space in the unconditional case), see [509, Theorem 8].

There are various essentially open further questions about C–convex risk measures or previsions. One concerns modelling C–convexity in terms of desirable gambles: intuitively, and recalling inequalities (12.7), at least the rationality criterion of positive scaling for desirable gambles (see Section 1.2.1) should be softened. Another issue is exploring weaker conditions than C–convexity. Some work in this direction has already been done, starting from FS–convexity and introducing quasiconvex risk measures and comonotone convex risk measures, see among others [304]. The betting scheme or desirability interpretation of these models is still to come.

Certainly, several further issues in risk measurement deserve a reinterpretation, and possibly an extension, through the methods of imprecise probabilities. Risk measurement is in fact a wide research area having important connections with, for instance, utility theory or stochastic dominance concepts. For a general textbook on (traditional) risk measurement, presenting also these topics, see e.g. [228].

Game–theoretic probability, discussed in Chapter 6, offers another interesting methodology for developing consistency notions for imprecise probabilities which may be quite naturally be applied to financial and economic problems, see the basic reference [582]. We point out the application in [666], where the well known Capital Asset Pricing Model

[4] The GBR we refer to was introduced by Williams [703], see also Section 2.4.1, and is more general than that presented in Section 2.3.3. In the lower prevision version, it requires that $\underline{P}(I_B(f - \underline{P}(f|A \wedge B))|A) = 0$, which reduces to the GBR equation in Section 2.3.3 if $A = \Omega$.

(CAPM) is derived under just a single, general but simple assumption instead of the many ones in standard financial theory.

Finally, to give an idea of the variety of financial or economic themes where imprecise probabilities can be applied, we mention some topics among those where they have already been employed: decision making under imprecise risk [380], evaluation of the net present value of an investment with interval–valued data [641], incomplete preferences [490], risk and variability measures [443], connections between coherence and arbitrage with unbounded random quantities [557].

13

Engineering

Michael Oberguggenberger

Unit of Engineering Mathematics, University of Innsbruck, Austria

13.1 Introduction

From small gadgets of daily use, houses, cars, ships, aerospace transport to large scale structures such as nuclear power plants, the public requires usability, safety and reliability. The designing engineer is expected to guarantee these functions or at least make every reasonable effort to achieve these goals (we leave aside the trade-off between affordable performance and available resources). In the design process, models of the prospective structures are set up, at the hand of which their expected behaviour can be investigated, understood and predicted. It is here that the major issue of uncertainty enters the scene, and with it *uncertainty analysis*.

Scientific modelling in engineering has to deal with three facets. First, there is reality (with materials, soils etc.). Second, there is the model of reality (formulated in mathematical terms and containing physical laws and constitutive equations). Third, correspondence rules (prescribing how to translate one into the other) are needed. The physical model establishes what are the state variables and what are the material constants, the *parameters* to be observed. Once this has been decided, the values of the parameters have to be determined from information extracted from the real world and will serve as input in the physical model. This plus the design of the structure enters in numerical computations that result in the proper dimensions, construction procedures, and assessment of safety measures. The actual implementation usually undergoes several transitional phases of construction – each requiring different modelling – until the final structure is completed. Ideally, the model output should provide a design that works, reliable guidelines for action, and aids for decision making (as emphasized e.g. in Einstein [258]).

Introduction to Imprecise Probabilities, First Edition.
Edited by Thomas Augustin, Frank P. A. Coolen, Gert de Cooman and Matthias C. M. Troffaes.

Here the question of accuracy, or at least robustness (against variability of operating conditions), becomes decisive. Predictions are valuable only to the extent that their uncertainty is taken into account and assessed. Mathematical methods of capturing the uncertainty play an essential role by now.

To narrow down the vast scope of engineering tasks, the focus in this chapter will be on civil engineering. As an illustrative example, we will use a geotechnical design problem. Geotechnical problems (foundations, slope stability, excavations of tunnels, buried pipes and the like) exhibit the issues of uncertainty in a paradigmatic way.

There are two major categories of uncertainty: model uncertainties and parameter uncertainties, the latter encompassing the uncertainty of the data used to determine the parameter values.

Model uncertainties. The choice of the structural model is one of the central engineering decisions to be made. For example, for the description of the soil, there are continuum models and granular models, there are two- and three-dimensional models, there are multi-phase models (solids, liquids, gases), and so on. The next point is the adequate selection of state variables and parameters (constant or not). Failure in engineering is described by the so called *limit state function* that separates the safe states from the unsafe states. The choice of the limit state function is again an engineering decision. For example in geotechnics, is failure due to bounds exceeded by average values of the total loads or due to localized disturbances? Another issue is the proper incorporation of transitional states during construction.

Parameter uncertainties. Parameter variability can be attributed to a large number of causes. There are random fluctuations, lack of information, random measurement errors, but also systematic measurement errors (deriving e.g. from uncontrollable changes of the properties of the soil material caused by its extraction in bore holes before being analysed in the laboratory). There are fluctuations due to spatial inhomogeneity, and errors made by assigning parameter status to state variables. This is one of the essential and often unavoidable errors in engineering, because models typically are valid in certain ranges only. If the state of the structure exits the intended range, constants may turn into variables depending on external forces (for example, the friction coefficient of most materials is approximately constant for small loads, but starts becoming a function of the internal stresses for larger loads). Finally, there is variability arising from the fact that parameters have to carry the burden of model insufficiency. The available information on data uncertainty may range from frequency distributions obtained from large samples, values from small samples or single measurements, interval bounds, to experts' point estimates and educated guesses from experience.

Traditionally, engineers have dealt with uncertainty by employing *safety factors*. That is, the traditional codes would require that the load carrying capacity of the structure exceeds the design loads by a certain factor > 1, typically 1.35 for permanent loads (such as dead weight) and 1.5–2.0 for temporary loads. These factors have been negotiated in the committees of standards. This state of affairs has been considered as unsatisfactory: no information about the *actual* distance to failure can be extracted from such a procedure. Based on the desire for a more analytical description of the uncertainties, engineering codes have been put on a probabilistic foundation, starting with the pioneering work of Freudenthal [301], Bolotin [79] and others in the 1950ies. Under this point of view, every relevant parameter of the engineering model is a random variable. There is no absolute safety, but rather a probability of failure.

To make it more precise, let the vector R comprise all random variables describing the resistance of a structure, S the loads and denote by $g(R, S)$ the limit state function (that is, $g(R, S) < 0$ means failure, $g(R, S) > 0$ signifies a safe state). Then $p_f = P(g(R, S) < 0)$ is the *failure probability*; $R = 1 - p_f$ is called the *reliability* of the structure. To determine the probability p_f, the types and parameters of the probability distributions of R and S are needed. This multiplies the number of parameter values that have to be provided by the designing engineer: each model parameter comes with a distribution type and set of (uncertain) distribution parameters. Actually, the current codes employ critical values R_k and S_k (certain percentiles of R and S) and partial safety factors γ_R and γ_S, so that the designing engineer has to verify a relation of the type

$$R_k/\gamma_R \geq \gamma_S S_k.$$

In theory, the critical values and the partial safety factors are computed in such a way that this inequality holds if and only if p_f attains a certain required value p_{fr}. In practice, γ_R and γ_S are not computed but rather prescribed in the codes. Starting with the 1980ies and 1990ies, the European codes have been changed into probability based codes. By now, this is the standard in civil engineering (see e.g. EN 1990:2002 [266]); other branches in engineering partially lag behind, but there is no question that the probabilistic safety concept is the current paradigm in engineering. In addition, risk analysis and risk management have become a major ingredient in construction and project management. Especially calls for tender by public organizations ask the contractor to supply a risk analysis together with the proposed design.

The civil engineering codes require that the designed structure obtains an instantaneous probability of failure of $p_f = 10^{-6}$ and a long-term failure probability of $p_f = 10^{-5}$. To credibly estimate tail probabilities of such a small magnitude, *a lot* of information is needed. In addition, if time dependent reliability is to be assessed, failure rates and the additional parameters of the reliability function are required. It is no surprise that simultaneously with the establishment of the probabilistic codes, criticism started in the engineering community, especially from the geotechnical side. The uneasiness about the concept has focused on three points:

- the meaning of the failure probability;
- the need for information which as a rule is not available in engineering practice;
- the artificial separation into resistance R and loads S.

Already the pioneers warned that the failure probability cannot have the meaning of a frequency of failure (e.g., Bolotin [79, § 26] stressing the comparative role of probability assessments). This is quite contrary to public perception and has been widely disputed from many aspects in the probabilistic and engineering literature. A selection of papers on this question is in Section 13.7. It is true that the probabilistic foundations play a fruitful role in putting arguments and negotiation of the codes on a rational basis. Accordingly, some of the authors of the codes refer to the failure probability as an *operational* or *normative probability* [266, Annex C4(3); 398, § 1.3]. The way the probabilistic concept is applied in practice quite clearly puts it into the realm of *subjective probability*.

As described above, there are many types of uncertainties in an engineering model. It has been questioned whether a purely probabilistic approach is capable of catching all aspects, for example, ignorance or fluctuations due to systematic model errors. In addition, large samples allowing a frequentist assessment of data are rarely available. Due to high costs, sample sizes in laboratory experiments are in the range of 5–10. The number of soundings in soil investigations depend on the situation, but typically may be in the range of 0–3 on small construction sites. However, what is generally known about a model parameter is a central value and a coefficient or range of fluctuation. In geotechnics, geologists can provide interval estimates of soil parameters; a geological report in tunnelling may deliver rock classes and interval probabilities for their occurrence. Thus, in practice, the determination of distribution parameters takes recourse to normative assumptions (types of distributions extracted from the engineering literature) and to expert estimates (sometimes put in a Bayesian framework).

Finally, the requirement of separating resistance and loads brings the geotechnician into a delicate situation: in the stability of a foundation, say, the soil acts both as resistance and as load (this specialized question will not be pursued here).

The uneasiness about the probabilistic safety concept and the desire for models of the data uncertainty that reflect and incorporate the level of available information led to the search for alternative concepts in the engineering community. On the one hand, probabilistic models and probabilistic reasoning were considered as too tight a concept. On the other hand, engineering practice quite blatantly showed that interval estimates should be incorporated in the framework of uncertainty analysis.

Starting in the 1990s, the usability of fuzzy sets has been explored in the engineering literature. One of the earliest ventures is in Haldar and Reddy [328], who modelled the material properties of existing buildings by fuzzy sets (acknowledging ignorance about the exact values which would not be available in a nondestructive way). In the subsequent years, a vast literature on fuzzy methods in civil engineering appeared, starting from fuzzy finite element methods, fuzzy geotechnical stability, fuzzy rule based decision models, project planning with fuzzy durations and costs, to the proposal of a fuzzy safety concept as a complete replacement of the probabilistic approach (Möller *et al.* [481]). Later, many special imprecise probability models, as generally discussed in Chapter 4 have been investigated as well, starting with Ferson and Oberkampf *et al.* [277, 499] as well as Blockley and Hall *et al.* [76, 329], who explored capabilities of evidence theory and interval probability. In the decade after the year 2000, the usefulness of random sets was increasingly acknowledged, starting with the pioneering work of Bernardini *et al.* [622, 623] and Hall *et al.* [333]. In a sense, random sets were seen as a framework bridging the gap between probability and interval analysis and admitting easily accessible visualization tools such as probability boxes (p-boxes) (cf. Section 4.6.4) (Ferson *et al.* [277]). Lower and upper previsions have found applications in systems reliability (Chapter 14, see also [648, 650]). In addition to these approaches, a number of further alternative frameworks have been proposed in the engineering literature: for example, ellipsoidal models (cf. [53]), clouds (Neumaier [492]) or info-gap-analysis (Ben-Haim [52]), see also Chapter 4.

As pointed out in the beginning of this section, the probabilistic safety concept has become the wide-spread standard in engineering uncertainty analysis. It appears mandatory that any imprecise probability approach must be formulated in a way that can be communicated to an engineering community trained in probability theory. For this reason, fuzzy set theory seems to have lost some of its popularity in civil engineering in the past years. It is true that fuzzy sets can be interpreted as describing sets of probability measures, and profound work has been done on the relations of fuzzy sets and possibility measures, e.g., [249, 318]. However, the probabilistic interpretation does not properly capture the semantics inherent to fuzzy set theory. For example, it seems there is no probabilistic counterpart of the idea of a fuzzy controller.

Among the vast number of approaches listed above, the framework of choice in the subsequent sections will be random set theory as a prototype for imprecise probability methods. As mentioned, random sets naturally encompass intervals and probability, and come with the concepts of upper and lower probabilities and probability boxes, both of which can be easily communicated. There is a frequentist and a subjectivist interpretation of random sets, and the random set language can be seen as complementing the established probabilistic language.

An adequate understanding of the uncertainties in an engineering task cannot be reduced to the calculation of a single number – the failure probability. It requires a multitude of steps, some of which may be isolated as follows:

- reflection about the choice of model and the failure mechanisms;
- assessing the variability of input and output variables and model parameters;
- sensitivity analysis;
- assessing the reliability of the structure;
- decision about acceptance or nonacceptance of the design.

Some of the tasks can be aided by quantitative methods, some have to be dealt with in a qualitative way. Thinking in scenarios is essential (possibly aided by fault trees and decision trees), as is the continuous revision of the model in a feedback loop. For example, the results of the sensitivity analysis may lead to changes in design towards increased robustness (against variations of input parameters and side conditions). Validation of parts of the model and calibration using experimental data is important as well. Care should be taken that the framework used for describing the data uncertainty reflects and incorporates the level of information available. Finally, one should keep in mind that the role of practical engineering is not so much the production of factual knowledge but rather decision making with the help of scientific tools.

The next sections will demonstrate how some of the tasks can be tackled using imprecise probability methods. In order to keep matters accessible, we shall use a single (and simple) example from geotechnical engineering – the design of a buried pipeline. We shall first present the standard probabilistic approach (and put it into question) in Section 13.2. Next, in Section 13.3, random sets will be employed to describe the overall output variability. Section 13.4 addresses random set methods in sensitivity analysis. In Section 13.5 we show how a hybrid model can be set up, combining stochastic processes and intervals. Section 13.6 is devoted to reliability and decision making in engineering under imprecise information. Finally, Section 13.7 contains a small subsample of references to each of the subjects touched upon.

13.2 Probabilistic dimensioning in a simple example

This section sets the stage with highlighting a few aspects from classical probabilistic modelling in engineering. We demonstrate this at the hand of an infinite beam on a linear elastic bedding. This will also serve as the accompanying practical example in the following sections. The simplest, one-dimensional model is the so-called Winkler beam, extending along the real line with coordinate $x \in \mathbb{R}$. The displacement $u(x)$ is described by the bending equation

$$EI\, u^{IV}(x) + bc\, u(x) = q(x), \quad -\infty < x < \infty\ ,$$

see e. g. [79, § 61]. Here EI is the flexural rigidity of the beam, b its effective width, c the bearing coefficient of the foundation and $q(x)$ the loading. One may imagine that the beam describes a buried pipeline, the loading $q(x)$ resulting from the covering soil. The parameters EI and b of the beam may be considered as precisely known, whereas the soil properties c and q vary in an imprecisely known fashion. We will study the singular boundary value problem for the standardized equation

$$u^{IV}(x) + 4k^4 u(x) = p(x), \quad -\infty < x < \infty$$

with $bc/EI = 4k^4$, $p(x) = q(x)/EI$, requiring that the solution should remain bounded at $\pm\infty$. In case k is a constant and $p(x)$ is an integrable function, both deterministic, its unique deterministic solution is given by

$$u(x) = \int_{-\infty}^{\infty} G(x, y)\, p(y)\, dy$$

in terms of its Green function

$$G(x, y) = \frac{1}{8k^3} e^{-k|x-y|}(\sin k|x - y| + \cos k|x - y|)\ .$$

In case the loading $q(x) \equiv q$ (and hence $p(x) \equiv p$) is constant, the displacement is constant as well and simply given by

$$u(x) \equiv \frac{p}{4k^4} = \frac{q}{bc}.$$

For the computational examples to follow we let the parameters vary around central moduli of $k = 10^{-2}$, $p = 10^{-8}$. Approximately, this corresponds to the case of a buried cast-iron pipeline with an effective diameter of 6 [cm], covered by about 100 [cm] of top soil ($q =$ 10 [N/cm]) and bedded in loosely packed sand ($c \approx 6.7$ [N/cm^3]). The resulting overall displacement would amount to $u(x) \equiv 0.25$ [cm] in the deterministic case.

In a probabilistic design, one would assume that the input parameters are random variables. In a standard engineering approach one would argue that their mean values are given by the deterministic design values; further, a coefficient of variation is assumed (for material properties, usually around 5%, for soil parameters up to 15%, cf. [524, 565]). In the simple example, the most uncertain parameters are q (mean $\mu_q = 10$) and bc (mean $\mu_{bc} = 40$); we take here a coefficient of variation of 10%. Next, a type of distribution for the parameters has to be assumed. Both q and c are soil- and bedding related parameters, whose values strongly depend on what actually happens at construction site. Thus little evidence about the type of distribution is available. We make the assumption that both parameters are normally distributed, that is, $q \sim \mathcal{N}(10, 1)$, $bc \sim \mathcal{N}(40, 16)$. Under this assumption, we can compute the probability density of the displacement $u = q/bc$, see Figures 13.1, and read off the quantiles. For example, the probability that the displacement is larger than 0.5 [cm] is $\approx 3.9 \cdot 10^{-6}$.

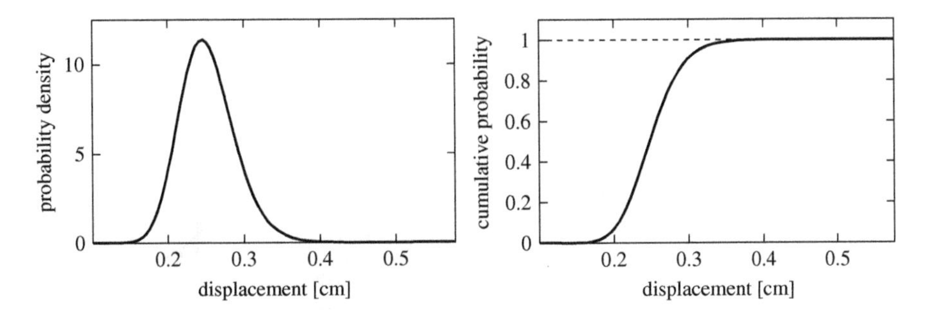

Figure 13.1 Probability density (left) and distribution function (right) of displacement under constant, but random load.

It is quite clear that the model is too simple to be credible. In particular, the load will certainly not be given by a single, albeit random, value along the whole beam, but will rather vary from point to point.

This suggests describing the load as a random process $q(x), x \in \mathbb{R}$ (cf. Chapter 11 – we will follow the engineering convention, whereby random processes indexed by points in space are termed random fields), and thus brings us to a second important aspect of probabilistic modelling in engineering. The standard assumption in soil engineering is that the soil load is given by a stationary Gaussian field. Thus we need to provide the mean value, the variance and the autocorrelation function. As above, the mean value is assumed to be $\mu_q = 10$; for the field variance we take $\sigma_q^2 = 4$. A typical autocorrelation function is of the form

$$C(\rho) = \exp\left(-|\rho|/\ell\right)$$

at spatial lag ρ, where ℓ is the so-called correlation length (available in the literature, e.g. [524]). In the following, we assumed a moderate correlation length of $\ell = 100$ [cm] and took bc fixed at its deterministic design value 40 [N/cm^2]. A realization of the load and the corresponding displacement is shown in Figure 13.2; Figure 13.3 (left) shows the corresponding realization of the bending moment.

A critical quantity for assessing the safety against failure is the maximal bending moment $M_{\max}$ in the beam, which is given by $M_{\max} = \max(EIu''(x))$. A typical failure criterion would require that the maximal stress $M_{\max}/W$ (with the section modulus W)

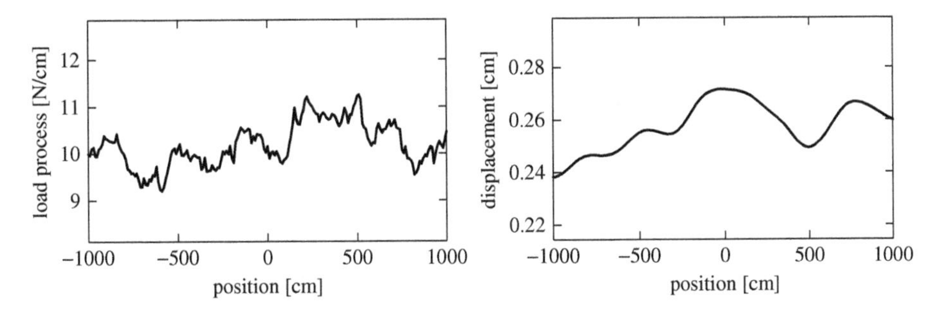

Figure 13.2 Random field model: trajectories of load process (left) and corresponding displacement (right).

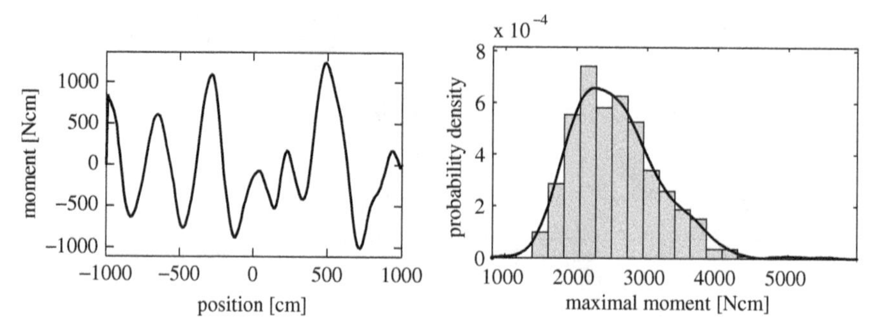

Figure 13.3 Random field model: trajectory of bending moment (left) and simulation of maximal bending moment, based on $N = 500$ trajectories (right).

does not exceed the admissible stress (corresponding to the 0.1% yield strength, that is, the stress after which 0.1% plastic deformations remain). We drop the lengthy details and just show how $M_{\max}$ would be assessed probabilistically. To this end, a Monte Carlo simulation of $N = 500$ trajectories has been undertaken, yielding an estimate for the distribution of $M_{\max}$. Figure 13.3 (right) shows the result; the histogram has been extrapolated with the aid of a kernel smoother. In this way, we get the estimate $P(M_{\max} > 6000) \approx 6.52 \cdot 10^{-5}$, for example. We record this value for reasons of comparison with Section 13.5.

The possibilities of more refined statistical methods, like confidence intervals for quantiles or robust estimates, are left aside here. This concludes our brief survey of classical probability in engineering.

13.3 Random set modelling of the output variability

As mentioned in the previous section, assuming that q and bc are normally distributed random variables is rather artificial. The available information consists of a nominal value and a coefficient of variation. One way of organizing this information is by means of a Tchebycheff random set. If the loading q, say, is preliminarily viewed as a random variable with unknown probability distribution, but with expectation value μ_q and variance σ_q^2, Tchebycheff's inequality asserts that the probability of the event

$$\left\{ |q - \mu_q| > \frac{\sigma_q}{\sqrt{\omega}} \right\}$$

is less or equal to ω, where $\omega \in (0, 1]$. Let

$$Q(\omega) = \left[\mu_q - \sigma_q / \sqrt{\omega}, \mu_q + \sigma_q / \sqrt{\omega} \right].$$

By Tchebycheff's inequality, the probability of $Q(\omega)$ is greater or equal to $1 - \omega$, while the probability of its complement $Q(\omega)^c$ is less or equal to ω. Thus $Q(\omega)$ may be viewed as an approximate two-sided $(1 - \omega)$-fractile range for the parameter q. This is a conservative, nonparametric estimate valid for whatever distribution of the random variable q. It encodes the minimal information that can be extracted from the expectation and the variance of a random variable without further parametric assumptions. We formalize this as a random set (actually a random interval)

$$\omega \to Q(\omega)$$

on the space $\Omega = (0, 1]$, equipped with the uniform probability distribution P. In engineering, events of the form $(-\infty, \mu_q - x] \cup [\mu_q + x, \infty)$ are often of interest. The upper probability of such an event is exactly ω if and only if $x = \sigma_q/\sqrt{\omega}$, confirming the intuition about the contents of Tchebycheff's inequality. For further explanations of the semantics of Tchebycheff random sets, see [281, 285, 497].

We take up the example of the previous section and apply this construction to the loading q with $\mu_q = 10$ and $\sigma_q = 1$ (from a coefficient of variation of 10%). This results in a random set Q, whose covering function is depicted in Figure 13.4 (left).

In a similar way, we construct a Tchebycheff random set BC for the variable bc, using $\mu_{bc} = 40$ and $\sigma_{bc} = 4$. To form the joint random set (Q, BC), the dependence of Q and BC has to be modelled. Various notions are available, from unknown interaction, random set independence, strong independence to fuzzy set independence, each of which comes with a certain interpretation (see e.g. [284] as well as Chapters 3 and 5). It has also been suggested to deform the focal intervals ([625]) or to use copulas on the underlying probability spaces ([13, 15]). To make computations easy, we settle for fuzzy set independence; that is, the joint random set is also defined on $\Omega = (0, 1]$ with focal elements $Q(\omega) \times BC(\omega)$, $\omega \in \Omega$. The random set data can be propagated through the mapping that gives the displacement $u(q, bc) = q/bc$, resulting in a random set U with focal sets $U(\omega) = u(Q(\omega) \times BC(\omega))$, the set of values attained when (q, bc) range in $Q(\omega) \times BC(\omega)$. This is consistent with the extension principle for fuzzy sets and the transport of a probability distribution through a function.

The evaluation of the interval bounds for $U(\omega)$ requires a global optimization and thus the computational effort may be high in large models. It is useful to describe the output random set as a probability box, which is bounded by the lower and the upper distribution functions

$$\underline{F}(x) = \underline{P}(-\infty, x], \qquad \overline{F}(x) = \overline{P}(-\infty, x], \tag{13.1}$$

see [277] and Section 4.6.4, where $\underline{P}(A) = P(\omega \in \Omega : U(\omega) \subset A)$ is the lower probability of a Borel event A and $\overline{P}(A) = P(\omega \in \Omega : U(\omega) \cap A \neq \emptyset)$ its upper probability. The resulting probability box for the displacement is shown in Figure 13.4 (right). The probability

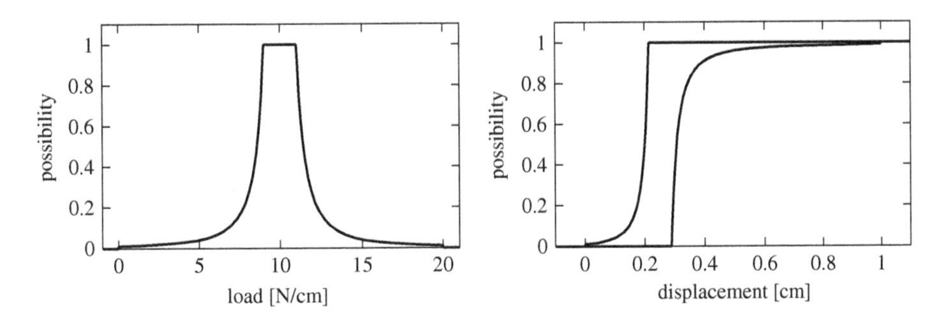

Figure 13.4 Tchebycheff random set for load (left) and probability box for resulting displacement (right).

box immediately gives information on quantile ranges. For the event $A = \{U \geq 0.5\}$, for example, we get the probability interval $[\underline{P}(A), \overline{P}(A)] = [0, 0.04]$, which is more credible than the point estimate $P(A) \approx 3.9 \cdot 10^{-6}$ from the previous section (obtained under the stringent assumption that q and bc were normally distributed).

The even more realistic combination of the random field model with random set parameters will be described in Section 13.5.

13.4 Sensitivity analysis

An important issue in assessing the behaviour of a structural model is sensitivity analysis. It aims at determining the input parameters that have the largest influence on critical output. Sensitivity analysis does not necessarily require knowledge of the probabilistic properties of the input and thus is a nonparametric method. If the input-output function is explicitly given, one may use partial derivatives to assess the sensitivity. Otherwise, sampling based methods can be used, among them variance decomposition [608] or various input-output correlation measures. We show here how pinching of variables can be used in conjunction with a random set model.

We take up the simple random set model with $u = q/bc$ from the previous section. The probability box in Figure 13.4 (right) shows the total variability of the output. We successively freeze the variables q and bc at their nominal value, compute the displacement u and observe whether the probability box gets narrower or not. The narrower it gets, the more influence the pinched variable has on the variability of the output. Alternatively, one may use the contour functions for the assessment and even quantify the reduction through the use of Hartley-like measures [401, 498]. The result of the pinching analysis can be seen in Figure 13.5. In the left picture, bc was pinched at its nominal value; the pinching of q in the right picture shows a more pronounced reduction of the variability, thus q is seen to be more influential on the displacement than bc.

The computational effort for the pinching of random sets is low, once the computation with the overall random set input has been performed: the values needed for the calculation with pinched variables are already among the computed values. In contrast to this, the pinching strategy is expensive when applied to Monte Carlo based sensitivity analyses: each pinching requires a complete Monte Carlo simulation of the output.

To round off the picture, we take a look at Monte Carlo based sensitivity analysis in the random field model. In this approach, a random sample of the parameters of interest is

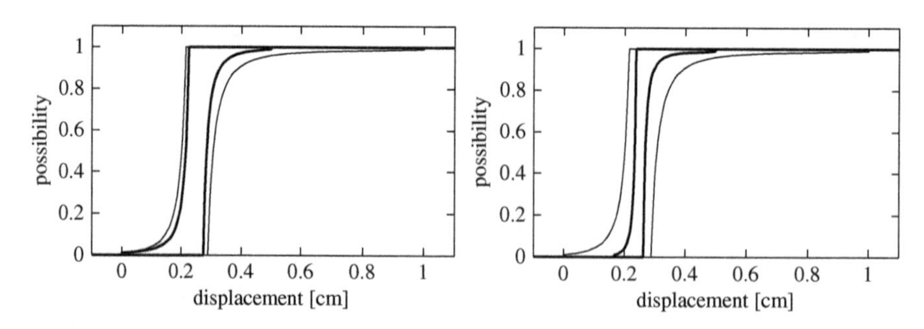

Figure 13.5 Sensitivity analysis: probability box with bc pinched (left) and with q pinched (right). The outer thin lines indicate the p-box without pinching.

drawn, and the corresponding sample of output values is computed. Then various correlation measures can be used to assess the input-output relationship. A survey of these and other methods of sampling based sensitivity analysis can be found in [357].

The input parameters under scrutiny are q, bc, the flexural rigidity EI of the beam and the field parameters σ and ℓ. To assess the sensitivities, we started with uniformly distributed input, each in an interval around its nominal value with spread $\pm 15\%$. Then an independent random sample of size $N = 500$ was generated and put through the input-output function. The correlation coefficients of the five input parameters with the displacement u turned out to be

$$R(u, bc) \approx -0.67, \qquad R(u, q) \approx 0.71$$

with $R(u, EI)$, $R(u, \sigma_q)$, $R(u, \ell)$ all around 0.035. Thus the simple probabilistic sensitivity analysis confirms that q and bc are the most influential parameters.

13.5 Hybrid models

In engineering practice, information on the uncertainties of the structural model and its parameters is usually diverse. It may range from single point estimates and intervals to statistical data and subjective assessments. In order to properly incorporate all available knowledge, a hybrid approach is desirable that combines different uncertainty models. Here random set theory comes in as a useful tool, because random sets encompass, among others, interval-valued random variables, single valued distribution functions, intervals, and fuzzy sets.

A combination of stochastic differential equations with random set parameters has been recently worked out in [559]. This can be used in the dynamics of structures. Earthquake induced vibrations can be modelled by stochastic processes, like coloured noise, whereas uncertain material parameters can be modelled by random intervals. We shall demonstrate a more modest hybrid model, returning to the elastically bedded beam. The load q will be modelled as a random field as in Section 13.2, while the bedding parameter bc will be modelled as a random set. Of course, the model can be generalized to higher levels by also taking the field parameters σ_q and ℓ as random sets, etc. For the sake of simplicity, we shall take bc as an interval and the field parameters as in Section 13.2, i.e., $\mu_q = 10$, $\sigma_q = 2$, $\ell = 100$. For bc we choose the interval $[20, 40]$, which has the previously assumed mean value for bc as its upper boundary.

The resulting output u will be a set-valued stochastic process; more precisely, each trajectory will be interval-valued. Figure 13.6 (left) shows a single trajectory of the bending moment. In order to assess the statistical properties of the output, a sample of trajectories has to be generated. From there, one can compute, e.g., the upper and lower distribution functions of the maximal bending moments in the beam. This is a critical quantity on which the failure criterion from Section 13.2 is based. A probability box of the maximal bending moment is shown in Figure 13.6 (right), based on $N = 500$ trajectories of the field. From the list of computed values (interpolated using a kernel smoother) one may obtain upper and lower probabilities that given limits are being exceeded, e.g.,

$$\underline{P}(M_{\max} > 6000) \approx 6.51 \cdot 10^{-5}, \qquad \overline{P}(M_{\max} > 6000) \approx 2.05 \cdot 10^{-2},$$

$$\underline{P}(M_{\max} > 8000) \approx 0, \qquad \overline{P}(M_{\max} > 8000) \approx 1.21 \cdot 10^{-3}.$$

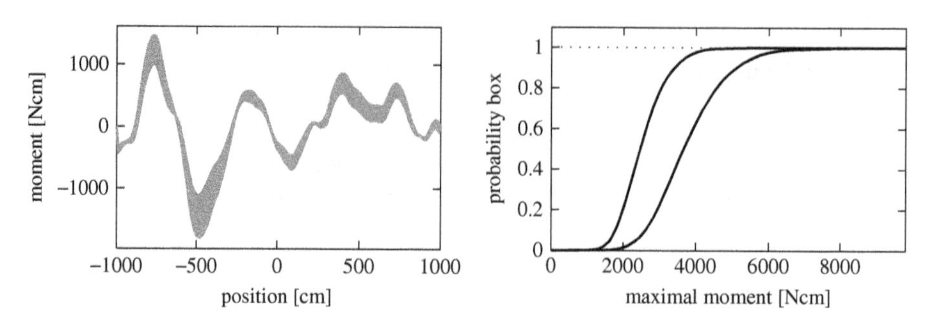

Figure 13.6 Hybrid model: Single interval trajectory of bending moment (left) and p-box for maximal bending moment, based on $N = 500$ interval trajectories (right).

Due to the fact that the parameters of the standard probabilistic model lie in the range of the parameters of the hybrid model, the value $P(M_{\max} > 6000) \approx 6.52 \cdot 10^{-5}$ obtained in Section 13.2 is within the probability bounds of the hybrid model. However, the probability intervals obtained by taking interval uncertainty into account are typically large and possibly of quite a different magnitude than the sharp probabilistic estimates. If the decision maker is not satisfied with the relatively high probabilities, more information about the distribution of the parameters has to be gathered. If the imprecise probability model properly reflects the available data, the results have to be accepted and should not be distorted by artificial probabilistic assumptions.

13.6 Reliability analysis and decision making in engineering

The dream of a decision maker in engineering companies is to get a single number stating that the design of the structure to be built is safe. This is also reflected in the Eurocodes that purport and require a reliability $R = 1 - p_f$ of 0.999999. At best, such a number can be viewed as an *operational probability* (as explicitly stated in Eurocode EN 1990:2002 [266]). It is hardly credible (nor verifiable or falsifiable) that such a number says anything about the frequency of failure.

Thus the designing engineer is left with the question how big the distance to failure *actually* is. There is no answer to this question – there are many examples of structural failures due to reasons *not* taken into account in the model and the reliability analysis, or due to false reactions to unexpected behaviour during construction. The key to a well-founded decision about the safety of a structure is an interplay between model building, numerical calculations, laboratory experiments, in-situ investigations, and uncertainty analysis in a feed-back loop. (A showcase of such a procedure is described in [427].) This loop extends to monitoring during construction and possible re-design.

As mentioned, the available knowledge should be properly captured by the uncertainty model – thus the need for a framework going beyond probability and using various methods from imprecise probability. For example, if knowledge about a parameter consists in a lower and an upper bound, intervals should be used. If different plausibilities can be assigned to different intervals, random sets may be helpful, etc. Generally, the basis for reliability analysis will be a sensitivity analysis of the model, identifying the most influential parameters.

This may already result in a re-design in order to eliminate one or the other influence. A complete probabilistic or nonprobabilistic reliability analysis is usually beyond available computing power and time. Instead, the analysis will usually be done with the aid of a reduced model, possibly given by interpolating or approximating the input-output function at a finite number of points (meta-model or response surface).

When proceeding with an imprecise probability model, the output parameters measuring the distance to failure come with upper and lower probabilities. In view of the lack of a total ordering, the designing engineer will have to base the decision about acceptance or rejection of the design on considerations as in multiple criteria decision making (cf. Chapter 8). In the end, a decision based on interval-valued, but accurate estimates will be more reliable than a decision based on sharp, but false numbers. As the methods become more widespread, this is more and more appreciated by practising engineers who are aware of the need of a rational analysis of uncertainty.

13.7 Further reading

Here is a short extraction of the literature on the aspects touched upon in the previous sections. Only references to engineering topics are included. Even so, it would be impossible to give a complete account of the hundreds of papers that have appeared so far.

The general concepts of probabilistic structural design are laid out in the Eurocode EN1990:2002 [266]; for a concise survey of computational methods, see [566]. As introductory literature to the probabilistic safety concept, we recommend [42, 87]; for risk analysis concepts, see [475]. For models with random fields and stochastic processes, especially finite elements, see [309, 556]. A probabilistic approach to the uncertainty of the structural model has been developed by [459, 609, 610]; for an imprecise probability approach, see [280, 282, 331].

Various authors have critically appraised the probabilistic approach, notably the concept of failure probabilities [259, 273, 274].

The topic of sensitivity analysis has been treated with sampling based methods [357] as well as with methods from imprecise probability theory [278, 332, 356, 498]. A monograph on random set models in engineering is [69]. These authors have also applied random set models in rock engineering, the assessment of seismic damage, and in reliability [67, 621, 622, 623]. See also the work [333, 545] on hydrological problems and on random sets for assessing the reliability [284, 286, 497]. Random sets in finite element models have been developed in [513, 567, 568]; finite element analysis with probability boxes has been undertaken in [724] (for fuzzy methods and interval methods with finite elements, see [283, 483, 484, 501]; a recent survey is given in [476]). A hybrid approach combining random sets and stochastic processes has been developed in [559, 560] and applied to structures under earthquake excitation [561].

Evidence theory and p-boxes have been introduced into engineering models in [277, 499], see also [574]. For upper and lower previsions and other imprecise probability models in reliability, see [197, 646, 648, 650]; a comparative survey has been recently given in [305], also covering the concept of clouds [492].

Interval probabilities for aiding engineering decisions have been proposed in [76, 330].

We should not fail to mention the literature using hybrid stochastic/fuzzy approaches [480], in particular, fuzzy reliability [479, 481]. The rather extensive engineering literature using fuzzy sets only is beyond the scope of this presentation; we just refer to a monograph comparing the fuzzy and the probabilistic approach in engineering applications [541] and to the textbooks [346, 540]; see also the hybrid fuzzy set/random set modelling of seismic damage in [68].

Various alternative approaches to uncertainty quantification in engineering have been proposed: ellipsoidal modelling [53], info-gap theory [52], anti-optimization [261], robust statistics [37].

Finally, we mention five compilations with various aspects of uncertainty modelling in engineering. The volume [41] emphasizes fuzzy approaches, the volume [260] compares and discusses worst case scenarios, probability and random sets, while [494] contains a wealth of information on imprecise probability methods in engineering, including generalized information theory, evidence theory, interval methods, info-gap decision theory, but also classical probabilistic ideas. The volume [272] critically appraises diverse aspects of uncertainty in civil engineering. A synopsis of different approaches to uncertainty modelling in nondeterministic mechanics is presented in [262].

14

Reliability and risk

Frank P. A. Coolen[1] and Lev V. Utkin[2]
[1]*Department of Mathematical Sciences, Durham University, UK*
[2]*Department of Control, Automation and System Analysis, Saint-Petersburg State Forest Technical University, Russia*

14.1 Introduction

Reliability and risk analysis are important application areas of statistics and probability theory in engineering and related fields, with several specific features which often complicate application of standard methods. Such features include data censoring, lack of knowledge and information on dependence between random quantities, and required use of expert judgements. These may, for example, result from the use of new or upgraded systems, multiple failure modes and maintenance activities. While mathematical approaches for dealing with such issues have been presented within the framework of statistics using precise probabilities, the use of imprecise probabilities provides exciting new ways for dealing with such challenges in reliability and risk. Attractive features of imprecise probability theory for this application area include the ability to model and reflect partial information about random quantities or the dependence of these and the ability to take expert judgement into account in a realistic way when only limited elicitation is possible due to time constraints. All kinds of partial information that might be available in practice can be formulated as constraints which have to be satisfied by sets of probabilities, with the inferential problem formulated as an optimisation problem. This approach, which uses the imprecise probability concept of natural extension, is illustrated with an example from structural ('stress-strength') reliability in Section 14.2.

Imprecise probabilistic approaches to statistics are of great value to reliability and risk problems. Chapter 7 provides an introduction to popular statistical approaches that

Introduction to Imprecise Probabilities, First Edition.
Edited by Thomas Augustin, Frank P. A. Coolen, Gert de Cooman and Matthias C. M. Troffaes.

can be applied in this field. For example, Walley's imprecise Dirichlet model [673] (cf. Sections 7.4.3 and 10.3) can be used to model system reliability without detailed assumptions on dependence of components [631]. Nonparametric predictive inference (NPI) (cf. Section 7.6) has also been presented for a variety of problems in reliability and risk analysis, and provides the opportunity to deal with unobserved or unknown risks. Section 14.3 gives a flavour of the opportunities for imprecise statistical methods in risk and reliability, some examples of NPI applications to several basic problems are presented in Section 14.4. This chapter ends with a brief guide to further reading and a discussion of main research challenges in Section 14.5.

14.2 Stress-strength reliability

An important problem in structural reliability is determination of strength of a structure under stress. Frequently, there is only little information available about the random strength and stress involved. This scenario provides an interesting example of the use of natural extension (see Section 2.2.4) in imprecise reliability as it shows the mathematical reasoning and methods typically involved in such problems. Let Y denote the random strength of a structure and X the random stress that it is exposed to, throughout this section these are assumed to be continuous nonnegative random quantities. The event of interest is $Y \geq X$, so the structure is strong enough to successfully withstand the stress, this is a crucial measure of reliability. In this section, we suppose that, due to limited available information, one does not feel able to quantify precise probability distributions for X and Y but meaningful bounds for their expected values and standard deviations can be assigned. To deal with such information while interest is in the event $Y \geq X$, we need some preliminary results which are presented first.

The variance of a random quantity Z is $\sigma_Z^2 = \mathbb{E}Z^2 - (\mathbb{E}Z)^2$, which cannot be represented as a prevision because it contains the non-linear term $(\mathbb{E}Z)^2$. Suppose that lower and upper bounds for σ_Z^2 have been specified, with values $\underline{\sigma}_Z^2$ and $\overline{\sigma}_Z^2$, respectively, while a precise value for the expectation of Z is known, say $h = \mathbb{E}Z$. Suppose further that interest is in the expected value (prevision) of a function of Z, say $g(Z)$. For example, $g(Z) = I_{[0,\infty)}(Z)$ with $I_A(Z)$ the indicator function which is equal to 1 if $Z \in A$ and 0 else, corresponds to the probability for $Z \geq 0$. Given only the information h, $\underline{\sigma}_Z^2$ and $\overline{\sigma}_Z^2$, we can derive optimal bounds for $\mathbb{E}g$, which are lower and upper previsions, by solving the following optimization problems (with p_Z as the density function of the probability law P_Z of the random variable Z):

$$\underline{\mathbb{E}}g = \inf_{\underline{\mathbb{E}}Z \leq h \leq \overline{\mathbb{E}}Z} \inf_{P_Z} \mathbb{E}_{P_Z} g = \inf_{\underline{\mathbb{E}}Z \leq h \leq \overline{\mathbb{E}}Z} \inf_{p_Z} \int_{\mathbb{R}_+} g(z) p_Z(z) \mathrm{d}z,$$

$$\overline{\mathbb{E}}g = \sup_{\underline{\mathbb{E}}Z \leq h \leq \overline{\mathbb{E}}Z} \sup_{P_Z} \mathbb{E}_{P_Z} g = \sup_{\underline{\mathbb{E}}Z \leq h \leq \overline{\mathbb{E}}Z} \sup_{p_Z} \int_{\mathbb{R}_+} g(z) p_Z(z) \mathrm{d}z,$$

subject to

$$\mathbb{E}_{P_Z} Z = \int_{\mathbb{R}_+} z p_Z(z) \mathrm{d}z = h,$$

$$\underline{\sigma}_Z^2 + h^2 \leq \mathbb{E}_{P_Z} Z^2 = \int_{\mathbb{R}_+} z^2 p_Z(z) \mathrm{d}z \leq \overline{\sigma}_Z^2 + h^2.$$

The dual forms of these problems are

$$\underline{\mathbb{E}}g = \inf_{\underline{\mathbb{E}}Z \leq h \leq \overline{\mathbb{E}}Z} \sup_{c_i, d_i, c} \left\{ c + (c_1 - d_1)h + c_2(\underline{\sigma}_Z^2 + h^2) - d_2(\overline{\sigma}_Z^2 + h^2) \right\},$$

subject to $c_i, d_i \in \mathbb{R}^+$, $c \in \mathbb{R}$, $i = 1, 2$, and $\forall z \in \mathbb{R}_+$,

$$c_0 + (c_1 - d_1)z + (c_2 - d_2)z^2 \leq g(z),$$

and

$$\overline{\mathbb{E}}g = \sup_{\underline{\mathbb{E}}Z \leq h \leq \overline{\mathbb{E}}Z} \inf_{c_i, d_i, c} \left\{ c + (c_1 - d_1)h + c_2(\overline{\sigma}_Z^2 + h^2) - d_2(\underline{\sigma}_Z^2 + h^2) \right\},$$

subject to $c_i, d_i \in \mathbb{R}^+$, $c \in \mathbb{R}$, $i = 1, 2$, and $\forall z \in \mathbb{R}_+$,

$$c_0 + (c_1 - d_1)z + (c_2 - d_2)z^2 \geq g(z).$$

Here c, c_i, d_i are optimization variables such that c corresponds to the constraint $\int_{\mathbb{R}_+} \rho_Z(z)\mathrm{d}z = 1$, $c_1 - d_1$ corresponds to the constraint $\int_{\mathbb{R}_+} \rho_Z(z)\mathrm{d}z = h$, c_2 corresponds to the constraint $\int_{\mathbb{R}_+} z^2 \rho_Z(z)\mathrm{d}z \leq \overline{\sigma}_Z^2 + h^2$, and d_2 corresponds to the constraint $\int_{\mathbb{R}_+} z^2 \rho_Z(z)\mathrm{d}z \geq \underline{\sigma}_Z^2 + h^2$.

This problem formulation enables us to derive upper and lower probabilities for the event $Z \leq t$ when in addition to $\underline{\sigma}_Z^2$ and $\overline{\sigma}_Z^2$ only lower and upper expectations for Z, say $\underline{m} = \underline{\mathbb{E}}Z$ and upper $\overline{m} = \overline{\mathbb{E}}Z$, are available.

Proposition 14.1 *Suppose that lower $\underline{m} = \underline{\mathbb{E}}Z$ and upper $\overline{m} = \overline{\mathbb{E}}Z$ expectations and lower and upper variances $\underline{\sigma}_Z^2$ and $\overline{\sigma}_Z^2$ of a continuous random variable Z defined on the sample space $\mathbb{R}_+$ are known. Then the upper probability is*

$$\overline{P}(Z \leq t) = \overline{\mathbb{E}}I_{[0,t]}(Z) = \begin{cases} \dfrac{\overline{\sigma}_Z^2}{(\underline{m} - t)^2 + \overline{\sigma}_Z^2}, & t < \underline{m} \\ 1, & t \geq \underline{m} \end{cases}. \quad (14.1)$$

The corresponding lower probability is

$$\underline{P}(Z \leq t) = \underline{\mathbb{E}}I_{[0,t]}(Z) = \begin{cases} 0, & t < \overline{m} \\ 1 - \dfrac{\overline{m}}{t}, & \overline{m} \leq t \leq \overline{m} + \dfrac{\overline{\sigma}_Z^2}{\overline{m}} \\ \dfrac{(\overline{m} - t)^2}{(\overline{m} - t)^2 + \overline{\sigma}_Z^2}, & t > \overline{m} + \dfrac{\overline{\sigma}_Z^2}{\overline{m}} \end{cases}. \quad (14.2)$$

It should be noted that these upper and lower probabilities (14.1) and (14.2) do not depend on the lower variance $\underline{\sigma}_Z^2$. The middle term on the right-hand side of (14.2) results from the constraint that $Z > 0$.

The proposed bounds (14.1) and (14.2) can be derived from the well known one-sided Chebyshev-Cantelli inequality taking into account the fact that the considered random variables are nonnegative. At the same time, they can be proven by using the dual optimization problems (natural extension) and analyzing relationships between the parabola $c_0 + (c_1 - d_1)x + (c_2 - d_2)x^2$ and the indicator function $g(x) = I_{[0,t]}(x)$ (see constraints to the dual form above).

We can now consider the stress-strength reliability under the assumption that information about the random stress X and strength Y is available consisting of lower and upper expectations, $\underline{m}_X$, $\overline{m}_X$, $\underline{m}_Y$, $\overline{m}_Y$, and lower and upper standard deviations $\underline{\sigma}_X$, $\overline{\sigma}_X$, $\underline{\sigma}_Y$, $\overline{\sigma}_Y$ (these are the square roots of lower and upper variances).

We start again with the assumption that there is actually information in the form of precise expectations $\mathbb{E}X = h_X$ and $\mathbb{E}Y = h_Y$. Then the lower and upper variances can be represented through the lower and upper second moments $\underline{\mathbb{E}}X^2 = \underline{\sigma}_X^2 + h_X^2$, $\overline{\mathbb{E}}X^2 = \overline{\sigma}_X^2 + h_X^2$, $\underline{\mathbb{E}}Y^2 = \underline{\sigma}_Y^2 + h_Y^2$, $\overline{\mathbb{E}}Y^2 = \overline{\sigma}_Y^2 + h_Y^2$. The lower and upper bounds for stress-strength reliability are the solutions to the following optimization problems:

$$\underline{R} = \underline{P}(Y \geq X) = \inf_{h_X, h_Y, p_X, p_Y} \int_{\mathbb{R}_+^2} I_{[0,\infty)}(y - x) p_X(x) p_Y(y) \mathrm{d}x\mathrm{d}y, \tag{14.3}$$

$$\overline{R} = \overline{P}(Y \geq X) = \sup_{h_X, h_Y, p_X, p_Y} \int_{\mathbb{R}_+^2} I_{[0,\infty)}(y - x) p_X(x) p_Y(y) \mathrm{d}x\mathrm{d}y, \tag{14.4}$$

subject to

$$\int_{\mathbb{R}_+} x p_X(x) \mathrm{d}x = h_X, \ \underline{\sigma}_X^2 + h_X^2 \leq \int_{\mathbb{R}_+} x^2 p_X(x) \mathrm{d}x \leq \overline{\sigma}_X^2 + h_X^2, \tag{14.5}$$

$$\int_{\mathbb{R}_+} y p_Y(y) \mathrm{d}y = h_Y, \ \underline{\sigma}_Y^2 + h_Y^2 \leq \int_{\mathbb{R}_+} y^2 p_Y(y) \mathrm{d}y \leq \overline{\sigma}_Y^2 + h_Y^2. \tag{14.6}$$

Here the infimum and supremum are taken over the set of all densities p_X and p_Y satisfying constraints (14.5) and (14.6) and over the sets $h_X \in [\underline{m}_X, \overline{m}_X]$ and $h_Y \in [\underline{m}_Y, \overline{m}_Y]$. Let us investigate the first optimization problem (14.3). The objective function can be rewritten as follows:

$$\begin{aligned}
\underline{R} &= \inf_{h_X, h_Y, p_X, p_Y} \int_{\mathbb{R}_+} \left(\int_{\mathbb{R}_+} I_{[0,\infty)}(y - x) p_X(x) \mathrm{d}x \right) p_Y(y) \mathrm{d}y \\
&= \inf_{h_Y, p_Y} \int_{\mathbb{R}_+} \left(\inf_{h_X, p_X} \int_{\mathbb{R}_+} I_{[0,\infty)}(y - x) p_X(x) \mathrm{d}x \right) p_Y(y) \mathrm{d}y \\
&= \inf_{h_Y, p_Y} \int_{\mathbb{R}_+} \underline{R}(y) p_Y(y) \mathrm{d}y,
\end{aligned}$$

where

$$\underline{R}(y) = \inf_{h_X, p_X} \int_{\mathbb{R}_+} I_{[0,\infty)}(y - x) p_X(x) \mathrm{d}x,$$

subject to

$$\int_{\mathbb{R}_+} x p_X(x) \mathrm{d}x = h_X, \quad \underline{\sigma}_X^2 + h_X^2 \leq \int_{\mathbb{R}_+} x^2 p_X(x) \mathrm{d}x \leq \overline{\sigma}_X^2 + h_X^2. \tag{14.7}$$

Note that $I_{[0,\infty)}(y - x) = I_{[0,y]}(x)$ and

$$\underline{R}(y) = \inf_{h_X, p_X} \int_{\mathbb{R}_+} I_{[0,y]}(x) p_X(x) \mathrm{d}x. \tag{14.8}$$

The objective function (14.8) and constraints (14.7) are none other than the optimization problem for computing the lower probability of failure before time y whose solution is (14.2), i.e. $\underline{R}(y) = \underline{P}_X(y)$. The distribution function $\underline{P}_X(y)$ can be represented as

$$\underline{P}_X(y) = \int_{\mathbb{R}_+} I_{[0,y]}(z)\underline{\rho}_X(z)\mathrm{d}z,$$

where $\underline{\rho}_X$ is a density function corresponding to $\underline{P}_X(y)$. It should be noted that lower and upper probability distributions do not necessarily have the corresponding parameters $\underline{m}_X$, $\overline{m}_X$ and $\underline{\sigma}_X$, $\overline{\sigma}_X$. They are bounds (like a convex hull) for arbitrary probability distributions having these parameters.

Hence

$$\begin{aligned}\underline{R} &= \inf_{h_Y, p_Y} \int_{\mathbb{R}_+} \left(\int_{\mathbb{R}_+} I_{[0,y]}(z)\underline{\rho}(z)\mathrm{d}z \right) p_Y(y)\mathrm{d}y \\ &= \int_{\mathbb{R}_+} \left(\inf_{h_Y, p_Y} \int_{\mathbb{R}_+} I_{[z,\infty)}(y) p_Y(y)\mathrm{d}y \right) \underline{\rho}_X(z)\mathrm{d}z \\ &= \int_{\mathbb{R}_+} (1 - \overline{P}_Y(z))\underline{\rho}_X(z)\mathrm{d}z.\end{aligned}$$

Here $\overline{P}_Y(z)$ is the upper probability of failure before time z, which is equal to (14.1). The density $\underline{\rho}_X$ is computed from (14.2) as

$$\underline{\rho}_X(t) = \begin{cases} 0, & t < \overline{m}_X \\ \dfrac{1}{t^2}\overline{m}_X, & \overline{m}_X \leq t \leq \overline{m}_X + \dfrac{\overline{\sigma}_X^2}{\overline{m}_X} \\ \dfrac{2\overline{\sigma}_X^2\left(t - \overline{m}_X\right)}{\left(\overline{\sigma}_X^2 + \left(t - \overline{m}_X\right)^2\right)^2}, & t > \overline{m}_X + \dfrac{\overline{\sigma}_X^2}{\overline{m}_X} \end{cases}.$$

Three cases should be considered.

Case 1. $\underline{m}_Y \leq \overline{m}_X$. Then $\underline{R} = 0$.

Case 2. $\overline{m}_X \leq \underline{m}_Y \leq \overline{m}_X + \overline{\sigma}_X^2/\overline{m}_X$. Then

$$\underline{R} = \overline{m}_X \int_{\overline{m}_X}^{\underline{m}_Y} \frac{(\underline{m}_Y - t)^2}{(\underline{m}_Y - t)^2 + \overline{\sigma}_Y^2} \frac{1}{t^2}\mathrm{d}t.$$

Case 3. $\overline{m}_X + \overline{\sigma}_X^2/\overline{m}_X \leq \underline{m}_Y$. Then

$$\begin{aligned}\underline{R} &= \overline{m}_X \int_{\overline{m}_X}^{\overline{m}_X + \overline{\sigma}_X^2/\overline{m}_X} \frac{(\underline{m}_Y - t)^2}{(\underline{m}_Y - t)^2 + \overline{\sigma}_Y^2} \frac{1}{t^2}\mathrm{d}t \\ &+ 2\overline{\sigma}_X^2 \int_{\overline{m}_X + \overline{\sigma}_X^2/\overline{m}_X}^{\underline{m}_Y} \frac{(\underline{m}_Y - t)^2}{(\underline{m}_Y - t)^2 + \overline{\sigma}_Y^2} \frac{(t - \overline{m}_X)}{\left(\overline{\sigma}_X^2 + (t - \overline{m}_X)^2\right)^2}\mathrm{d}t.\end{aligned}$$

The upper bound $\overline{R}$ can be derived similarly

$$\overline{R} = \int_{\mathbb{R}_+} (1 - \underline{P}_Y(z))\overline{\rho}_X(z)\mathrm{d}z = 1 - \int_{\mathbb{R}_+} \underline{P}_Y(z)\overline{\rho}_X(z)\mathrm{d}z.$$

Again three cases should be considered.

Case 1. $\underline{m}_X \leq \overline{m}_Y$. Then $\overline{R} = 1$.

Case 2. $\overline{m}_Y \leq \underline{m}_X \leq \overline{m}_Y + \overline{\sigma}_Y^2/\overline{m}_Y$. Then

$$\overline{R} = 1 - 2\overline{\sigma}_Y^2 \int_{\overline{m}_Y}^{\underline{m}_X} \frac{(\overline{m}_Y - t)}{\left((\overline{m}_Y - t)^2 + \overline{\sigma}_Y^2\right)^2} \left(1 - \frac{\underline{m}_X}{t}\right) \mathrm{d}t.$$

Case 3. $\overline{m}_Y + \overline{\sigma}_Y^2/\overline{m}_Y \leq \underline{m}_X$. Then

$$\overline{R} = 1 - 2\overline{\sigma}_Y^2 \int_{\overline{m}_Y}^{\overline{m}_Y + \overline{\sigma}_X^2/\overline{m}_Y} \frac{(\overline{m}_Y - t)}{\left((\overline{m}_Y - t)^2 + \overline{\sigma}_Y^2\right)^2} \left(1 - \frac{\underline{m}_X}{t}\right) \mathrm{d}t$$

$$- 2\overline{\sigma}_Y^2 \int_{\overline{m}_Y + \overline{\sigma}_X^2/\overline{m}_Y}^{\underline{m}_X} \frac{(\overline{m}_Y - t)}{\left((\overline{m}_Y - t)^2 + \overline{\sigma}_Y^2\right)^2} \frac{(\underline{m}_X - t)^2}{(\underline{m}_X - t)^2 + \overline{\sigma}_X^2} \mathrm{d}t.$$

Example 14.1 (Stress-strength reliability). Let us consider a system with the following parameters of the stress and the strength: $m_X \in [90, 140]$, $\sigma_X \leq 40$, $m_Y \in [300, 350]$, $\sigma_Y \leq 10$. It follows from the parameters that $\overline{m}_X + \overline{\sigma}_X^2/\overline{m}_X = 151.43 \leq \underline{m}_Y = 300$. This corresponds to the third case for computing the lower bound and the first case for computing the upper bound. Hence, the reliability bounds are $\underline{R} = 0.846$ and $\overline{R} = 1$. It is interesting to note that the lower reliability bound depends on the upper mean value of the stress but not on the corresponding lower mean value, and on the lower mean value of the strength but not on the corresponding upper mean value. Similarly, the upper reliability depends on the lower mean value of the stress but not on the corresponding upper mean value, and on the upper mean value of the strength but not on the corresponding lower mean value. Both the reliability bounds depend on the upper variance of the stress and the strength but not on the corresponding lower variances. ♦

14.3 Statistical inference in reliability and risk

Standard models for inference usually require a large number of observations of events, e.g. failures, or assume that an appropriate precise prior probability distribution is available (for Bayesian models). A possible way to avoiding these assumptions is by use of imprecise probability models. One popular statistical model for use with multinomial data (see Page 177) is Walley's imprecise Dirichlet model (IDM) [673], as discussed in Sections 7.4.3 and 10.3. The IDM has also been applied to reliability problems. Coolen [128] presented a generalization of the IDM, suitable for lifetime data including right-censored observations, which are common in reliability theory and survival analysis. The NPI has also been applied to basic reliability systems with partial information on dependencies in failure data [631].

For statistical inference on interval data,[1] as frequently occur in reliability applications with interest in time to failure in case failures are not directly observed, Utkin [639, 640] considered a set of IDMs produced by these data. The set of IDMs in this case does not require partitioning of the time-axis into a number of intervals for constructing the multinomial model. These intervals are produced by bounds of interval-valued data. The following example illustrates the above.

Example 14.2 Suppose that we observe $N = 5$ intervals of Total Time to Failure (TTF), $A_1 = [10, 14], A_2 = [12, 16], A_3 = [9, 11], A_4 = [12, 14], A_5 = [13, \infty)$. Then the lower and upper probabilities of an arbitrary interval A are of the form:

$$\underline{P}(A|s) = \frac{\sum_{i:A_i \subseteq A} 1}{N+s}, \qquad \mathrm{P}(A|s) = \frac{\sum_{i:A_i \cap A \neq \emptyset} 1 + s}{N+s}.$$

Let $A = [0, 14]$ and $s = 1$. Then $\underline{P}([0, 14]|1) = 3/6$, $\overline{P}([0, 14]|1) = 1$. It can be seen from the above expressions that the lower and upper probabilities do not depend on the division of the time-axis into intervals and right-censored observations (A_5) can be analyzed by the set of IDMs. ♦

For restricting to a set of possible distribution functions of TTF, and for formalizing judgements about the *ageing* aspects of lifetime distributions, various nonparametric or semi-parametric classes of probability distributions can be used. In particular, the classes of all IFRA (increasing failure rate average) and DFRA (decreasing failure rate average) distributions have been studied by Barlow and Proschan [44]. In order to formalize judgements about the ageing aspects of lifetime distributions, new flexible classes of distributions, denoted as $\mathcal{H}(r, s)$ *classes* [647], have been proposed and investigated. The probability distribution of the component (or system) lifetime X can be written as $H(t) = \Pr(X \geq t) = \exp(-\Lambda(t))$, where $\Lambda(t) = \int_0^t \lambda(x)\mathrm{d}x$ and $\lambda(t)$ is the time-dependent failure rate, also known as the hazard rate. Let r and s be numbers such that $0 \leq r \leq s \leq +\infty$. A probability distribution belongs to a class $\mathcal{H}(r, s)$ with parameters r and s if $\Lambda(t)/t^r$ increases and $\Lambda(t)/t^s$ decreases as t increases. In particular, $\mathcal{H}(1, +\infty)$ is the class of all IFRA distributions; $\mathcal{H}(r, s)$ with $1 \leq r < s$ is the class of all IFRA distributions whose failure rate increases with rate bounded by r and s; $\mathcal{H}(0, 1)$ is the class of all DFRA distributions; $\mathcal{H}(r, s)$ with $r < s \leq 1$ is the class of all DFRA distributions whose failure rate decreases with rate bounded by r and s; and $\mathcal{H}(r, s)$, $r < 1 < s$ is a class containing distributions whose failure rate is nonmonotone. Inferences for such classes, and solutions to corresponding computational problems, were presented by Utkin and others in the papers referred to above. To make these promising distributional classes available for imprecise reliability analysis in practice, a number of interesting research problems are still open, including the important question of how to fit such classes to available data.

As a further example of an imprecise reliability application, Fetz and Tonon [286] consider bounds for the probability of failure of a series system when no information is available about dependence between failure probabilities of different modes. They consider several models, including random sets and p-boxes, and they provide a detailed list of references to

[1] For a general discussion of interval data, see Section 7.8.

the literature on such topics. They also discuss some computational methods, which is an important aspect of application of imprecise reliability to medium or larger size practical problems. One of the possible ways in which output from imprecise probability methods can be useful is in the study of sensitivity of model outputs with respect to variations in input parameters. An interesting recent study [498] presents such an approach to a large-scale modelling problem to assess reliability in an aerospace engineering application, comparing the use of classical probabilities and a variety of imprecise probability methods.

14.4 Nonparametric predictive inference in reliability and risk

To illustrate and discuss attractive features of imprecise probabilistic methods, we will next present several applications of nonparametric predictive inference (NPI, see Section 7.6) to basic problems in reliability and risk. A first introduction to NPI in reliability was presented in [137], and theory for dealing with right-censored observations in NPI in [140]. Several aspects of data as typically occur in reliability and risk applications lead to interesting inferences when lower and upper probabilities are used, where NPI particularly shows such aspects due to the limited influence of modelling assumptions. These aspects include data sets consisting of relatively few observations and often including right-censored observations, success-failure data which actually contain few or even zero failures, and the wish to draw inferences on failure modes that have not yet been observed or have not even been specified. Some of these topics are discussed in this section via four illustrative examples. For details on the theoretical results and their justifications, together with further examples and more discussion of properties, we refer to the original papers in which these approaches have been presented.

Example 14.3 (System reliability). NPI for Bernoulli data has been implemented for system reliability, with test information on components of several types that are exchangeable with those in the system. This is illustrated in the following example, see [10] for further references, discussion and presentation of general formulae and their derivation.

Consider a system consisting of two subsystems in series configuration with components of two types, A and B. Subsystem 1 is a 2-out-of-4 system (so this functions if at least two of its components function), with two type A components and two type B components; subsystem 2 is a 1-out-of-4 system, also with two components of each type. Assume that two components of type A were tested, both of which functioned successfully, and also two components of type B were tested of which only one functioned successfully. The NPI lower probability for the event that this system functions successfully is equal to 0.664. Suppose that, to increase the system's reliability by increasing redundancy, extra components can be added to the system, keeping the requirements that at least two components function in subsystem 1 and at least one in subsystem 2, but adding extra components to subsystems. It is assumed that there are no cost considerations, only the number of extra components that can be added is restricted, and these extra components can be of any type and added to any of the two subsystems. Table 14.1 presents the optimal allocation of 1 to 11 extra components ('Extra' in the first column; m_a^1 denotes the total number of components of type A in subsystem 1, *et cetera*), in the sense of maximum resulting NPI lower probability for the event that the system functions (denoted by $\underline{P}$).

Table 14.1 Example system reliability.

Extra	$(m_a^1, m_b^1, m_a^2, m_b^2)$	$\underline{P}$
1	(3,2,2,2)	0.775
2	(4,2,2,2)	0.827
3	(5,2,2,2)	0.856
4	(5,2,3,2)	0.882
5	(6,2,3,2)	0.901
6	(6,2,4,2)	0.914
7	(7,2,4,2)	0.972
8	(8,2,4,2)	0.936
9	(8,2,5,2)	0.944
10	(9,2,5,2)	0.950
11	(9,3,5,2)	0.956

If one extra component is allowed, it is optimal to add a component of type A to subsystem 1. This is fully as expected, since type A components seem to be more reliable than type B components based on the test results, and subsystem 1 has less redundancy in the original system than subsystem 2. With two further extra components allowed, both would be chosen and assigned similarly. However, if four extra components are allowed, they would still all be of type A but now one of these would be added to subsystem 2. For up to 10 extra components, it is optimal to choose them all to be of type A and added to the subsystems as presented in Table 14.1. However, if an eleventh extra component is allowed, a component of type B would be added to subsystem 1. This illustrates an important aspect of NPI for system reliability, namely that it takes explicitly into account that the reliabilities of components of one type in the system are statistically dependent, as a result of the limited information from the test data. Effectively, if one has the system with already 10 extra components added in the optimal manner, it has become quite a reliable system. If, however, this system does not function, it implies that the components of type A are far less reliable than had been expected following the test results. Hence, at this point it is better to add a component of type B than a further one of type A. This illustrates that diversity in redundancy allocation can result directly from maximization of reliability, and is due to the limited knowledge about the reliability of the components of different types. ♦

Example 14.4 (Survival functions with right-censored observations). Statistical methods for right-censored observations are crucial in many reliability applications. Coolen and Yan [140] presented NPI for such data, leading to NPI lower and upper survival functions which are an attractive alternative to the well-known Kaplan-Meier (KM) estimator, which is the nonparametric maximum likelihood estimator. Throughout, independent censoring is assumed, which effectively means that at the time of censoring no further information is available about the residual lifetime of the right-censored unit. As a small example, consider data $2^c, 4, 7, 9^c, 10, 12^c, 12^c, 12^c, 12^c$ where e.g. 2^c indicates that one unit was withdrawn from the study at time 2, leading to a right-censored observation. Figure 14.1 presents the NPI lower and upper survival functions together with the Kaplan-Meier estimator.

Let us remember that the lower probability reflects the evidence in support of the event while the upper probability reflects the evidence against the event. This is particularly illustrated by the fact that, at a right-censoring time, the lower survival function decreases, as indeed from then on there is less evidence in support of survival, while the upper survival

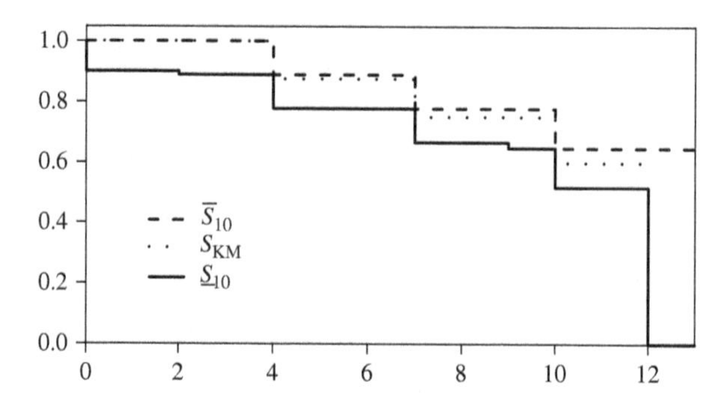

Figure 14.1 NPI lower and upper survival functions and KM estimator.

Table 14.2 Failure data for electrical appliances.

# cycles	FM	# cycles	FM	# cycles	FM
381	6	1594	2	2471	9
708	6	1925	9	2702	10
958	10	2327	6	2831	2
1167	9	2451	5	3112	9

function does not change, as indeed such an observation does not provide actual evidence against survival. The KM estimate only decreases at the actual event observation times $4, 7, 10$. ♦

Example 14.5 (Competing risks). The NPI approach for competing risks has been presented recently [454], it uses the NPI lower and upper survival functions [140]. In a situation with competing risks, which are assumed to be independent, NPI lower and upper survival functions are derived conditional on each failure mode, where failures due to other failure modes are right-censored observations. The assumed independence of the competing risks allow the overall NPI lower and upper survival functions for the units considered, affected by all competing risks, to be calculated by taking the products of all the corresponding NPI lower and upper survival functions conditioned on the failure modes. A question particularly considered in [454] is which failure mode will cause the next unit to fail, which is illustrated in the following example [141]. This also shows an important advantage of the NPI approach, namely it can include unobserved and even (any number of) unknown failure modes in the analysis.

The data in Table 14.2 are a subset of a well-known data set from the literature that was also used in [454]. Twelve units of a new model of a small electrical appliance were tested, the lifetime observation per unit consists of the number of completed cycles of use until the unit failed. There were 18 possible failure modes identified, the specific failure mode (FM) that caused the unit to fail is given in the table.

The two most frequently occurring failure modes in these data are FM9 and FM6, which caused 4 and 3 units to fail, respectively. The event of interest is that the next unit, say unit 13, will fail due to a specific failure mode, assuming it would undergo the same test and its number of completed cycles would be exchangeable with these numbers for the 12 units reported. Six cases are presented, (A) to (F), each involving a different grouping of the

Table 14.3 Results for Example 14.5.

	$\underline{P}$	$\overline{P}$		$\underline{P}$	$\overline{P}$
(A)			(B)		
9	0.237	0.481	6	0.208	0.452
OFM	0.519	0.764	OFM	0.548	0.793
	0.755	1.245		0.755	1.245
(C)			(D)		
9	0.190	0.481	9	0.190	0.481
6	0.187	0.452	3	0	0.245
OFM	0.255	0.556	OFM	0.442	0.764
	0.632	1.489		0.632	1.489
(E)			(F)		
9	0.157	0.481	3	0	0.245
6	0.168	0.452	4	0	0.245
3	0	0.245	7	0	0.245
OFM	0.212	0.556	OFM	0.536	1
	0.536	1.734		0.536	1.734

failure modes in 2, 3 or 4 re-defined failure modes, the corresponding NPI lower and upper probabilities are presented in Table 14.3. The term OFM is introduced for 'other failure modes' grouped into one single re-defined failure mode. For example in case (A), OFM is considered to be a single new failure mode containing all originally defined failure modes except failure mode 9. For each case the sums of the lower probabilities and of the upper probabilities are also given, these illustrate that they are constant for a fixed number of re-defined failure modes. For further discussion of properties of these inferences we refer to [141], where also the corresponding NPI lower and upper survival functions are investigated and illustrated.

In case (A) all failure modes except FM9 are together re-defined as one failure mode (OFM). Suppose that there may be an additional unknown failure mode that may cause the units to fail, but this has not happened for the tested units. To illustrate the effect of such a possible further failure mode, consider case (D) where FM3 has been defined although it had not led to a failure for the tested units. Including such an unobserved failure mode has led to decrease of the lower probabilities for the event that unit 13 will fail due to FM9 or due to OFM, which follows by comparison of cases (A) and (D) and can also be seen in cases (C) and (E), while the corresponding upper probabilities have remained the same. This reflects that the unknown failure mode could possibly lead to the next failure, which would make the other failure modes less likely causes for it, but it cannot be excluded that this unknown failure mode may not have any effect at all, as reflected by those unchanged upper probabilities. Case (F) shows that multiple unobserved failure modes can be considered, each of these has lower probability 0 and the same positive upper probability for the event that it causes unit 13 to fail. Any number of such unobserved failure modes could be included, they would all have the same lower and upper probabilities. This uses the sub-additivity of upper probabilities to great advantage. The upper probability for OFM in Case (F) is one, reflecting that all units observed failed due to one of the failure modes in OFM. ♦

Example 14.6 (Unobserved failure modes). While NPI for competing risks enables inference on unobserved or even unknown failure modes, it does not quantify how likely such failure modes are to occur. The NPI approach has also been presented for multinomial data [133, 134] (cf. also Page 177), which explicitly enables inference on the event that the next observation will fall into either a defined but not yet observed category or even into an undefined category, which can be interpreted as an 'unknown unknown' event. This approach was presented explicitly for prediction of unobserved failure modes in [132]. Typically, if outcome categories have not occurred yet, the NPI lower probability of the next observation falling in such a category is zero, but the corresponding NPI upper probability is positive and depends on whether or not the category is explicitly defined, on the total number of categories or whether this number is unknown, and on the number of categories observed so far. Such NPI upper probabilities can be used to support cautious decisions, which are often deemed attractive in reliability and risk analysis. The following example [132] illustrates this approach.

Suppose that a database contains detailed information on failures experienced during warranty periods of a particular product. Currently 200 failures have been recorded, with five different failure modes specified. The producer is interested in the event that the next recorded failure of such a product during its warranty period is caused by another failure mode than these five already recorded. First assume that there is no clear assumption or knowledge about the number of possible failure modes. Suppose that interest is in the event that the next reported failure is caused by any as yet unseen failure mode, represented by $X_{201} \in UN$, then the NPI-based upper probability for this event is equal to $5/200$. If, however, the producer has actually specified two further possible failure modes, which have not yet been recorded so far, and interest is in the event that the next failure mode will be one of these two, then this method gives a different answer. In this case, let these two failure modes be denoted by DN_1 and DN_2, then the NPI-based upper probability for the event $X_{201} \in DN_1 \cup DN_2$ is equal to $2/200$. The NPI-based lower probabilities for both these events are 0, reflecting that there is no actual evidence in the data for these events to occur.

Now suppose that these 200 failures were instead caused by 25 different failure modes, not including the defined DN_1 and DN_2. Then the upper probability for $X_{201} \in UN$ changes to $25/200$, but the upper probability for $X_{201} \in DN_1 \cup DN_2$ remains $2/200$. It is in line with intuition that the changed data affect this first upper probability, as the fact that more failure modes have been recorded suggests that there may be more different failure modes that can cause this product to fail (no knowledge is assumed on the number of possible failure modes). For the second event considered here, the reasoning is somewhat different, as effectively interest is in two specific, as yet unseen, failure modes, and there is no actual difference in the data available that is naturally suggesting that either of these two failure modes has become more likely, as in both cases there is actually no real evidence that they can lead to failures during the warranty period.

Suppose now that the producer is absolutely certain that at most 40 different failure modes exist for this product. This would not affect the above upper probabilities in case so far only five failure modes had been observed, but in the case of 25 observed failure modes, it reduces the upper probability for the event $X_{201} \in UN$ to $15/200$, reflecting that there are only 15 possible failure modes not yet recorded. Assuming that DN_1 and DN_2 are among those 15 failure modes, the upper probability for $X_{201} \in DN_1 \cup DN_2$ remains $2/200$. Clearly, if there had been 40 well defined failure modes, of which 25 had already caused failures and with no other failure modes possible, then the 15 which had not yet been recorded could be denoted by DN_i for $i = 1, \dots, 15$, in which case

$$P(X_{201} \in UN) = P(X_{201} \in \underline{P} \bigcup_{i=1}^{15} DN_i) = \frac{15}{200}$$

♦

In addition to these four examples, there are several further examples of applications of NPI in reliability and risk that are of interest in themselves, for the further applications that can be considered using similar approaches, for the clear advantages of the use of imprecise probability in these fields that they clearly illustrate, and for the related research opportunities. Powerful methods for replacement decisions of technical units have been presented which are fully adaptive to process data and in simulation studies have shown to perform very well [142, 143, 145]. As NPI is particularly attractive in situations with zero failures, its applications for reliability demonstration [135] and probabilistic safety assessment [131] provide interesting alternatives to classical approaches. It is worth noting as well that NPI can also deal with grouped data [139] as often occur in real-world applications, for example if lifetime and failure data are collected only on a monthly basis for noncritical hardware.

14.5 Discussion and research challenges

Imprecise reliability is a relatively new area of research, with methods presented that are inspired by practical problems but that are not yet suitable for applications of substantial size. The main challenges are in upscaling the methods to become useful for practical problems, together with aspects of implementation which include consideration of elicitation and model choice. The combination of substantial optimization problems and statistical modelling and updating may also lead to a level of complexity that requires attention to methods for computation, for example it is not clear how modern simulation-based methods, that are for example popular in Bayesian statistics, can best be used or adapted for imprecise approaches.

The models for imprecise reliability that have been presented so far are still pretty basic, and for example inclusion of covariates requires further research, see also Section 7.10. Generally, imprecise approaches can be found that generalize the established methods in varying ways, so in addition to developing new methods one must find ways to decide on how useful they are, which requires careful consideration of fundamental aspects of uncertainty and information. Hybrid methods, which combine imprecise models where useful to model indeterminacy with precise models where possible due to sufficient data or information, provide exciting opportunities for research, with issues that must be addressed including interpretation of results, choice of models and methods, and computation.

15

Elicitation

Michael Smithson
Research School of Psychology, The Australian National University, Canberra, Australia

15.1 Methods and issues

The elicitation of imprecise probabilities is a relatively unexplored topic, and so some parts of this review will draw from work on precise probability judgements. Most methods of eliciting imprecise probabilities fall into three groups: Direct numerical estimates, numerical translations of verbal probability expressions (e.g., 'somewhat likely'), and evaluations of candidate intervals or bounds. We shall begin by surveying these in turn, elaborating the issues and methods in evaluating judged imprecise probabilities. The section thereafter examines methods for evaluating judged probabilities that can be adapted and applied to imprecise probability judgements. The third section discusses factors that influence probability judgements and ways in which elicitation methods may be matched with the purposes intended to be served by elicitation. The final section provides a guide to further reading on this topic.

Numerical elicitation methods range from simply asking for numerical estimates to inferring such estimates from other indicators. Until recently relatively little work had been done comparing these methods with one another or with other alternatives. The most obvious direct numerical estimation task is to ask for a lower and upper bound on the probability of an event. This is rather problematic because no criterion is given for the bounds. The most commonly employed criterion is a confidence level (e.g., within what bounds are you 95% confident that $P(A)$ lies?). There is a sizeable empirical literature on how people construct predictive 'confidence intervals' of this sort, although it is not based on estimations of probabilities ([436, 550]). We shall review this literature in the third subsection.

Introduction to Imprecise Probabilities, First Edition.
Edited by Thomas Augustin, Frank P. A. Coolen, Gert de Cooman and Matthias C. M. Troffaes.

A second criterion relies on the interpretation of lower and upper probabilities as buying and selling prices or bid-ask spreads. The judge is asked to nominate a maximum buying price, X_b, to purchase a gamble paying 1 unit if event A occurs and a minimum selling price, X_s, for the same gamble. The judge also is asked for a maximum buying and a minimum selling price, Y_b and Y_s, for a gamble paying 1 unit if the complement A^c occurs. Assuming $\underline{P}(A) = 1 - \overline{P}(A^c)$, we may infer

$$\underline{P}(A) = \frac{1 - \dfrac{Y_b}{Y_s}}{\dfrac{X_s}{X_b} - \dfrac{Y_b}{Y_s}}. \tag{15.1}$$

A third possible criterion involves the application of a 'scoring rule', such that the estimator is rewarded in accordance with how much weight she gives the event that occurs relative to others. The scoring rule can be used as a motivating device. For precise probabilities assigned to a finite exhaustive collection of events $E_1, E_2, \ldots, E_n$, the score $f(P_i)$ is the payoff if E_i occurs. Popular examples include the quadratic rule $f(P_i) = 2P_i - P_i^2$ and the logarithmic rule $f(P_i) = ln(P_i)$. A simple adaptation to lower and upper probabilities would be to use $f(\underline{P}_i)$.

A fourth approach, related to the buying-selling price interpretation, is inferring utilities from imprecise probability assignments and vice-versa [603]. We begin with the (non)occurrence of event A consideration and an ordered collection of options R_0, R_1, and R_2, one of which is to be selected depending on the probability $P(A)$ and the decision-maker's utilities regarding the outcomes produced by the R_j given A versus outcomes given A^c.

Example 15.1 R_0 = bet on A^c, R_1 = don't bet, and R_2 = bet on A. Table 15.1 shows the utilities given A and A^c, associated with acts R_j. We assume $H_j > H_{j-1}$ and $G_{j-1} > G_j$ for any j. ♦

A straightforward argument shows that if the odds of A, w_A, exceed the threshold $w_{j-1j} = (G_{j-1} - G_j)/(H_j - H_{j-1})$ then the decision-maker should prefer act R_j over R_{j-1}. The threshold odds w_{j-1j} therefore are determined by the ratio of utility differences. Thus, elicitation of the H_j and G_j yield a lower odds w_{01} and an upper odds w_{12}. Of course the converse also can be applied. Thus, given a judge's lower and upper probabilities, the resulting utilities of being right or wrong about the occurrence of the event given the decision to be taken can be fed back to the judge to ascertain whether they accord with her preferences.

Table 15.1 Utility Setup for Three Options Acts.

State:	R_2	R_1	R_0
A	H_2	H_1	H_0
A^c	G_2	G_1	G_0

Example 15.2 An illustration in medical diagnostic classification is an actuarial violence risk assessment tool presented in [612], using discharged hospital patients' follow-up data. Commission of a violent act during the first 20 weeks following hospital discharge was coded as a dichotomous outcome. Of the sample of 939 discharged patients included for study, 18.7% committed a violent act within 20 weeks after discharge. Without any apparent justification, Steadman et al. classify any case with a predicted probability of violence greater than twice the base prevalence rate ($P(A) = .37$) in the 'high-risk' category, and any case whose predicted probability of violence is less than half the base prevalence rate ($P(A) = .09$) in the 'low-risk' category. The middle group is 'unclassified.' The corresponding threshold odds are $w_{12} = .37/(1 - .37) = .587$ and $w_{01} = .09/(1 - .09) = .099$.

These thresholds correspond to the preferences of a rational agent for whom the differences between consequences of correctly and wrongly classifying a violent patient are weighted much more heavily than the consequences of classifying a nonviolent patient, because both threshold odds fall below 1. Suppose the agent says the best outcomes are classifying a patient who ends up being violent as high-risk or classifying a nonviolent patient as low-risk, and the worst is misclassifying a violent patient as low-risk. Without loss of generality we may then assign $H_2 = 1$, $G_0 = 1$, and $H_0 = 0$. G_2 and H_1 then have the simple relationship

$$G_2 = 1 - w_{12} + (w_{12} - w_{01})H_1, \tag{15.2}$$

which along with the restriction $H_0 < H_1 < H_2$ implies $1 - w_{12} < G_2 < 1 - w_{01}$. For example, if the agent claims that $G_2 = 0$ then this utility assignment is incompatible with the threshold odds $w_{12} = .587$ and $w_{01} = .099$. ♦

Verbal probability expressions may be regarded as imprecise probabilities. If such expressions have an agreed-upon numerical translation then the elicitation task reduces to the judge simply choosing from a set of these expressions. There is a long history of attempts to translate verbal probability expressions into numerical form and debates over whether the results are sufficiently useful.

Early researchers asked people to nominate precise probabilities corresponding to verbal phrases. On the positive side, these investigations yielded reports of reasonably high intra-subjective reliability (e.g., [437, 73, 89, 681]) and reliable aggregate means ([591, 592, 528]). Some of these findings highlighted relevant differences between meanings or usages in natural vs. formal language. For instance, negation was found to be asymmetric in its effects, so that 'unlikely' is not subjectively equivalent to 1 - 'likely'.

However, the same research also revealed considerable inter-subjective variability and overlap among phrases ([614, 437, 73, 77]). [89] argued that verbal probability phrases may lead to ordinal confusion in usual communication, and [90] provided evidence that people vary widely in the probability phrases they regularly use.

One reasonable interpretation of these findings is that they are symptomatic of fuzziness instead of unreliability. Wallsten *et al.* [682] established an experimental paradigm in which subjects constructed fuzzy membership functions over the unit interval to translate probabilistic quantifiers into numerical terms (see the material on interval evaluation below). Kent [392] anticipated this idea by proposing probability intervals as translations of a set of verbal expressions he hoped would be adopted by American intelligence operatives. However, although the British intelligence community eventually adopted this approach, the American intelligence community has not [393]. Translations of verbal probabilities into numerical imprecise probabilities seem likely to succeed only in small communities of experts who can agree on nomenclature and, of course, the translation.

Example 15.3 Finally, turning to interval-evaluation methods, these are broadly compatible with a random-set interpretation of imprecise probabilities. A simple interval-evaluation method is the 'staircase' method [638]. Participants are instructed:

> For the following questions, please assign probability estimates for the event described by marking one of the five categories (from Clearly too low to Clearly too high) for each probability value (0 to 100 per cent). Mark each column until each has one X in it (as illustrated in the following example). For example, you may represent the probability of rain tomorrow by checking each column in the manner below.

The example table is displayed in Table 15.2 (extremal columns have been omitted to fit the table onto the page). Lower and upper probabilities were determined by the upper bound on the columns checked in the 'Slightly too low' row and the lower bound on the columns checked in the 'Slightly too high' row, respectively (in Table 15.1, this would yield a lower probability .55 and an upper probability .80). In perhaps the only empirical comparison between the staircase method and direct numerical estimation, [606] reported that the staircase method tended to yield wider intervals but otherwise behaved identically to direct numerical estimates.

A related evaluation method treats the 'stairs' in the staircase method as degrees of membership in a fuzzy set (e.g., [682]). [528] compared three methods of number-word translations for eighteen probabilistic concepts: (a) frequency distributions of what single number best represents each expression; (b) word-to-number acceptability functions from what range of numbers from 0% to 100% best represents each expression; and (c) number-to-word acceptability functions from which expressions are appropriate for multiples of 5% from 5% to 95%. The latter two are types of fuzzy membership functions. ♦

The fuzzy membership approach generalizes the staircase method, but brings with it an additional issue, namely the meaning of a 'degree of membership' (see [607, § 3]). When restricted to judgements of one fuzzy set, this does not pose any more difficulties for measurement than many other self-reported numerical estimates. Nevertheless, degree of membership lacks an agreed-upon behavioural definition of the kind exemplified in the betting interpretation of probabilities.

A natural generalization of interval evaluation that incorporates probability judgements is to present an interval to respondents and ask them for their subjective probability that the

Table 15.2 Instructional example for staircase estimation.

Probability	40	45	50	55	60	65	70	75	80	85	90	95
Clearly too high											X	X
Slightly too high									X	X		
About right					X	X	X	X				
Slightly too low			X	X								
Clearly too low	X	X										

interval contains the correct answer. This approach has become popular in recent years because of evidence that it yields reasonably well-calibrated estimates, and there is an empirical literature comparing this method with direct interval elicitation (e.g., [706]).

15.2 Evaluating imprecise probability judgements

Standardized methods for evaluating imprecise probability judgements have yet to be developed. This subsection describes methods for evaluating judged probabilities that can be adapted and applied to imprecise probability judgements. We address four issues:

1. Factors affecting responses;
2. Calibration;
3. Response style heterogeneity; and
4. Coherence.

Factors affecting the response given may include characteristics of the respondent, the task, and the context or setting in which the elicitation took place. Several well-known influences on probability judgments are reviewed in the next subsection. An appropriate GLM for modelling such influences assumes that judged probabilities follow a Beta(ω, τ) distribution. We reparameterize these component distributions in terms of a location (mean), $\mu = \omega/(\omega + \tau)$ and precision parameter $\phi = \omega + \tau$ (for further details see [607]). The GLM uses a logit link for the location submodel and a log link for the dispersion submodel. The *location submodel* is

$$g(\mu_i) = \sum_k \beta_k X_{ki}, \tag{15.3}$$

For $i = 1, \ldots, N$ and $k = 0, 1, \ldots, K$ where the link function is $g(v) = log(v/(1 - v))$, the X_{ki} are predictors and the β_k are coefficients. The *dispersion submodel* is

$$h(\phi_i) = \sum_m -\delta_m W_{mi}, \tag{15.4}$$

for $m = 0, 1, \ldots, M$ where $h(v) = \log(v)$.

Generalizing this to judged lower and upper probabilities entails modelling both simultaneously via a beta GLM with a random intercept in the location submodel to model the respondent effect. A ‘null’ model with no predictors is

$$g(\mu_{ij}) = \beta_j + b_i, \tag{15.5}$$

$$h(\phi_{ij}) = -\delta_j, \tag{15.6}$$

for $i = 1, \ldots, N$ and $j = 1, 2$, where $b_i \sim N(\mu_b, \sigma_b)$. This model is quite similar to a number commonly employed in the behavioural sciences, including unidimensional item-response theory models or those for repeated measures experiments (for a detailed treatment of this kind of model, see [655]).

Calibration refers to the extent to which judged probabilities track true probabilities of predicted events, versus systematically over-estimating them (over-confidence) or under-estimating them (under-confidence, see next subsection). For precise probabilities, beta GLMs can be used in a straightforward way to assess calibration, i.e., by testing whether the coefficient for a regression model with the logit of the true probability as predictor differs from 1. For imprecise probabilities, standard methods have yet to be developed. An obvious approach suited to lower and upper probabilities is comparing the number of times the true probability lies above or below the lower-upper probability assignments.

A somewhat more sophisticated approach is to use the random-effects model in Equations (15.7) and (15.8) with $g(P_i^{\lambda_j})$ substituted for the β_j term in equation (15.7), where P_i denotes the true probability for the ith case. Let $j = 1$ correspond to the lower probability assignments and $j = 2$ correspond to the upper probabilities. If $\lambda_1 > 1$ and $\lambda_2 < 1$ then the lower and upper probabilities appear to be well-calibrated. On the other hand, $\lambda_1 < 1$ indicates over-estimation and $\lambda_2 > 1$ indicates under-estimation.

Response style heterogeneity refers to the existence of subpopulations whose probability assignments are distinguishable from one another. The study of partition dependence in [606] used a beta-mixture GLM to identify two subgroups, one anchoring their probabilities around 1/2 and the other anchoring around 1/7. Smithson, Merkle and Verkuilen [605] describe these mixture models and present examples of their application. One of these involves modelling adherence to and deviations from coherence in lower-upper probability assignments, and so we reproduce that example here.

Example 15.4 [606] investigated two judgement tasks (the Boeing Stock and Sunday Weather tasks), each from two independent samples, and in one experimental condition elicited lower and upper probability estimates from participants. For both tasks half of the participants were asked to provide lower and upper probability estimates of how likely each event was to occur and how likely to not occur. A normative requirement for coherency in lower-upper probability judgements is conjugacy, i.e., $\underline{P}(A) = 1 - \overline{P}(A^c)$. A simple test of conjugacy therefore is $\underline{P}(A) + \overline{P}(A^c) = 1$, which provides a specified anchor. A beta-mixture model can estimate the proportion of respondents displaying conjugacy and, for those whose $\underline{P}(A) + \overline{P}(A^c) \neq 1$, investigate the behaviour of this sum.

The [605] model uses $Y = \left(\underline{P}(No) + \overline{P}(Yes)\right)/2$ as the dependent variable (dividing by 2 to map the sum into the unit interval). The mixture model has two components, one whose location and dispersion parameters are left free for estimation, and the second whose location is fixed at 1/2 (the conjugate respondents' distribution) with a small pre-set dispersion parameter. Thus, we have three submodels: location, dispersion and composition. The first component's location submodel has two predictors, task and $\overline{P}(Yes) - \underline{P}(Yes)$ (see [605] for the rationale behind this second predictor). Task also is a predictor in the composition submodel. The resultant model is:

$$Y \sim \gamma \text{Beta}(\mu_1, \phi_1) + (1 - \gamma)\text{Beta}(\mu_2, \phi_2), \tag{15.7}$$

for $0 < \gamma < 1$, with

$$\begin{aligned} g(\mu_{1i}) &= \beta_{10} + \beta_{11}X_{1i} + \beta_{12}Z_{1i} + \beta_{13}Z_{1i}X_{1i} \\ g(\mu_{2i}) &= 1/2 \\ h(\phi_{1i}) &= -\delta_{10} \\ h(\phi_{2i}) &= -\delta_{20} \\ g(\gamma_i) &= \theta_0 + \theta_1 Z_{1i}, \end{aligned} \tag{15.8}$$

where $\delta_{20} = \{-8, -10, -12, -14, -16, -18\}$, $X_{1i} = \overline{P}(Yes) - \underline{P}(Yes)$ and $Z_{1i} = -1$ for the Boeing stock task and +1 for the Sunday task. The δ_{20} values were varied to effect a sensitivity analysis, which indicated that results stabilized for values of −14 or less. Briefly, the findings indicated that about 42.1% of the respondents adhered to conjugacy for the Sunday Weather task whereas only about 22.5% did so for the Boeing Stock task, and the relationship between Y and X_1 was positive and stronger for the Sunday Weather task (see [605] for further details). ♦

15.3 Factors affecting elicitation

Elicitations of numerical estimates, particularly those involving probability judgements, are notorious for their vulnerability to extraneous factors as well as human heuristics and biases. We shall limit this review to the three influences on subjective interval estimates and probability judgements that have generated the greatest concerns about their validity: Over- and under-weighting, overconfidence, and partition dependence.

There is a large body of empirical and theoretical work on subjective probability weighting. A widely accepted account may be summarized by saying that people over-weight small and under-weight large probabilities [386]. Note that this claim does not mean that people are under- or over-estimating the probabilities, but instead treating them in a distorted fashion when making decisions based on them. Rank-dependent expected utility theory (e.g., [522]) reconfigures the notion of a probability weighting function by applying it to a cumulative distribution whose ordering is determined by outcome preferences. Cumulative prospect theory [637] posits separate weighting functions for gains and losses.

Two psychological influences have been offered to explain the properties of probability weighting functions. The hypothesis proposed in [637] is 'diminishing sensitivity' to changes that occur further away from the reference-points of 0 and 1. This would account for the inverse-S shape, or curvature, of the weighting function typically found in empirical studies (e.g., [93]). [317] add the notion of 'attractiveness' of a gamble to account for the elevation of the weighting curve. The magnitude of consequences affects both the location of the inflection-point of the curve and its elevation, in the expected ways. Large gains tend to move the inflection point to the left and large losses move it to the right (e.g., [265]).

Diminishing sensitivity has an implication for judgements and decisions based on imprecise probabilities that goes beyond treating their location in a distorted fashion. [317] presented respondents with the following question:

> You have two lotteries to win \$250. One offers a 5% chance to win the prize and the other offers a 30% chance to win the prize.
>
> A: You can improve the chances of winning the first lottery from 5 to 10%.
>
> B: You can improve the chances of winning the second lottery from 30 to 35%.
>
> Which of these two improvements, or increases, seems like a more significant change?

The majority of respondents (75%) viewed option A as the more significant improvement. In this way they demonstrated that a change from .05 to .10 is seen as more significant than a change from .30 to .35, but a change from .65 to .70 is viewed as less significant than a change from .90 to .95. An implication is that for decisional purposes people might treat a probability *interval* [.05, .10] as *less precise* than [.30, .35], and so on.

We now turn to overconfidence. A large empirical literature on subjective confidence-interval estimation tasks suggests that people tend to be overconfident in the sense that they construct intervals that are too narrow for their confidence criteria (e.g., [12, 396]). If directly elicited subjective probability intervals are psychologically similar to other kinds of confidence-intervals, then overconfidence will occur there too. Nor is this confined to laypeople or novices. In a study of experts' judgemental estimates ([550]), in which business managers estimated 90% confidence-intervals for uncertain quantities in their areas of expertise (e.g. petroleum, banking, etc), the hit rates obtained in various samples of managers ranged from 38% to 58%. These are performance levels similar to those typically found in studies of lay people.

Yaniv and Foster [712, 713] suggest that the precision of subjective interval estimates is a product of two competing objectives: accuracy and informativeness. The patterns of preference ranking for judgements supported a simple trade-off model between precision (width) of interval estimates and absolute error which they characterized by the error-to-precision ratio. Both papers present arguments and evidence that people tend to prefer narrow but inaccurate interval estimates over wide but accurate ones, i.e., they prioritize informativeness over accuracy. For instance, in study 3 in [712] participants were asked to choose between two estimates, (A) [140, 150] and (B) [50, 300]. They were told that the correct answer was 159. Most (90%) of the respondents preferred estimate A over B, although only the latter includes the correct answer.

Some researchers found that the format for eliciting interval estimates influences overconfidence. [611] compared overconfidence in interval estimates using three elicitation methods: Range, two-point and three-point. The range method simply asks for, say, a 90% subjective confidence interval. The two-point method asks for a lower limit with a 95% chance of falling below the true value and then an upper limit with a 95% chance of falling above it. The three-point method adds a 'midpoint' estimate with a 50% criterion. They found the least overconfidence for the three-point method, and the greatest for the range method. Their explanation is that the two-and three-point methods encourage people to sample their knowledge at least twice whereas the range method is treated by most people as a single assessment of their knowledge base. Other aspect of the task that have been investigated include the extremity of the confidence criterion (e.g., [307] found that tertiles yielded better calibration than more extreme confidence levels), and the nature of the scale used for elicitation (e.g., the study of graininess effects in [713]).

For some time researchers puzzled over the repeated finding that while people are overconfident when they construct intervals, they are reasonably well-calibrated when asked to assign probabilities to two-alternative questions with the same estimation targets [396]. An example of such a task is asking the respondent whether the population of Thailand exceeds 25 million (yes or no) and then asking for the subjective probability that her answer is correct.

A breakthrough came when [706] extended the two-alternative question format to probability judgements about interval estimates provided to the respondent (e.g., estimating the probability that Thailand's population is between 25 million and 35 million). Comparing these judgements with the intervals elicited from respondents with a fixed confidence criterion (e.g., a 90% subjective confidence interval for Thailand's population), they found that overconfidence was nearly absent in the former but, as always, quite high in the latter.

These findings have since been replicated in most, but not all, comparison experiments of this kind [500].

Juslin, Winman and Hansson [384] accounted for these and related findings, in part, by noting that while a sample proportion is an unbiased estimate, the sample interval coverage-rate is upwardly biased. Their explanation is that people are 'native intuitive statisticians' who are relatively accurate in sampling their own knowledge but treat all sample estimates as if they are unbiased. The clear implication is that subjective probability judgements of intervals provided to respondents are better-calibrated than intervals produced by respondents to match a coverage-rate.

Finally, let us consider partition dependence. On grounds of insufficient reason, a probability of $1/K$ is assigned to K mutually exclusive possible events when nothing is known about the likelihood of those events. Fox and Rottenstreich [299] present evidence that subjective probability judgements are typically biased towards this ignorance prior. Thus, probability judgements influenced by the ignorance prior are partition dependent. Fox and Clemen [298] find evidence that this dependence decreases as domain knowledge increases, and that experts in decision analysis are susceptible to it.

Partition dependence poses problems for probability assignments in two respects. First, it may be unjustified because there is a normatively correct partition. For instance, [299] pose the question of how likely Sunday is to be the hottest day of the week. The principle of insufficient reason would suggest that 1/7 is the correct answer, so their demonstration that people can be induced to partition the events into just two possibilities (Sunday is or is not the hottest day) and therefore assign a probability of 1/2 indicates that those people are anchoring onto an incorrect partition.

A more profound difficulty is that partitions may be indeterminate or the sample space may be ambiguous. Consider a bag containing 10,000 marbles whose colours are completely unknown to us. How should we use the principle of insufficient reason to judge the probability of drawing a red marble from this bag? Also, consider a scenario in which we are told that Chris, James and Pat are applying to an engineering firm and are then asked to estimate the probability that each of them is hired by the firm. It is unclear whether there is only one position available or multiple positions. Thus, equally defensible partitions could yield ignorance priors of 1/2 each or 1/3 each. Smithson, Hatori and Gurr [604] present experiments demonstrating that a simple cue asking people to nominate which candidate has the highest probability of being hired induces more people to constrain their probabilities to sum to 1. See also the discussion of the principle of insufficient reason in Footnote 49 of Chapter 7.

In many real-world situations the sample space partition is only incompletely known (e.g., in complex technical systems where not all failure modes have been observed). A seminal study by [291] concerning people's assignments of probabilities to possible causes of a given outcome (e.g., an automobile that will not start) revealed that possible causes that were explicitly listed received higher probabilities than when the same causes were implicitly incorporated into a 'Catch-All' category of additional causes. The effect has since been referred to as the 'catch-all underestimation bias' and also sometimes the 'pruning bias' [549]. In a more general vein, [638] show that unpacking a compound event (e.g., cancer) into disjoint components (e.g., breast cancer, lung cancer, etc.) or repacking its complement tends to increase the perceived likelihood of that event. Partition dependence accounts for this effect as well as the catch-all underestimation bias.

Walley [672] argues that while partition dependence poses a problem for likelihood judgements that yield a single probability, imprecise probability judgements should not depend on the state-space partition. Whether or when imprecise probability judgements are or should be partition-dependent is an open question. Smithson and Segale [606] compared elicited probability intervals with precise probabilities and found that the location of the intervals were as influenced by partition priming as precise probabilities. However, they also found that some respondents constructed wider intervals when they were primed with an incorrect partition, indicating that they still bore the correct partition in mind.

No research appears to have been conducted on ways of dealing with partition dependence in the face of sample space incompleteness or ambiguity. There are several unresolved issues regarding partition dependence, especially where imprecise probabilities are involved. Nevertheless, the literature on partition dependence should remind elicitors to take care in specifying a partition clearly when a preferred partition is available, and to find out which partition(s) respondents have in mind when providing their estimates.

15.4 Matching methods with purposes

What methods can be recommended for eliciting subjective imprecise probabilities? The choice of method will depend in good part on the goals of elicitation and the availability of effective tools for assessing the results. Owing to the many gaps in knowledge about the effectiveness of various elicitation methods, this subsection can only provide some incomplete guidelines.

Beginning with numerical estimation methods, we have seen that three key concerns are diminishing sensitivity to probabilities further away from the 0 and 1 bounds, overconfidence, and partition dependence. There are tradeoffs between dealing with these issues and obtaining informative estimates. Evaluations of intervals provided by an elicitor may be better calibrated than intervals produced by respondents, but two obvious shortcomings arise from the fixed interval approach. First, the elicitor cannot specify an a priori confidence-level, and second, the respondent cannot direct the elicitor towards an interval that 'best' represents her or his assessment. If a specific confidence-level is desired, [706] recommend an iterative procedure involving the evaluation of a series of intervals that homes in on the appropriate confidence-level. However, this approach does not yield a unique 'best' interval.

[698] propose an elicitation method (MOLE, or 'more or less') that arguably relies less on absolute judgements than either interval production or interval evaluation methods. Their rationale begins with the arguments in [706] and adds the claim that people are better at relative judgements than absolute ones [310]. Respondents are asked to choose which of two randomly generated numbers is closer to the true value of the quantity to be estimated, and then to judge the subjective probability that their choice is correct. A subjective probability density function is then built up from ten of these choices and confidence judgements, and it is then fitted by a beta pdf. Welsh et al. compared this procedure with three interval production techniques, and found that it nearly eliminated overconfidence. If further investigations of the MOLE technique find that its calibration is comparable to that of the interval evaluation approach then the MOLE would arguably be superior because it yields more information about the respondent's state of knowledge or belief.

Regardless of the numerical estimation method, elicitors will need to be careful about specifying sample space partitions, especially where the partition is incompletely known or ambiguous. The chief dangers here lie in the catch-all underestimation bias and in respondents having unwanted alternative or multiple partitions in mind when they make probability judgements. Elicitors also will need to take into account the diminishing sensitivity issue, especially if some probability intervals include 0 or 1 or bounds close to either of these reference-points while others' bounds are toward the middle of the unit interval.

Verbal probability expressions may not have much to recommend them even when users can agree on a numerical translation, mainly because of the prospect of overconfidence arising from the probability intervals nominated by users. However, the elicitation of such translations could be improved by using interval evaluation instead of interval construction. The most justifiable use of verbal expressions would be among users with an agreed-upon translation in settings characterized by deep uncertainty and a need for rapid communication or decision making.

Finally, utility-based methods (e.g., buying and selling prices or decisional threshold probabilities) could be useful for consistency checks on numerical elicitation results. However, their primary applications would be in settings where such utilities are naturally evaluated. Examples are bid-ask spreads in pricing negotiations, establishment of burden-of-proof or safety standards, and three-option decisional tasks (e.g., go, no-go, wait).

15.5 Further reading

The answer to the question what are the best starting-points in the literature on probability judgements depends on the reader's purpose. Thus, the recommendations provided here will refer to particular topics covered in this review. Beginning with direct numerical elicitation, [500] provides a book-length overview. The classic work on eliciting subjective confidence intervals is [436], with a more recent treatment in [550]. Good reviews and perspectives regarding over- and under-weighting in subjective probabilities are [637] and [317]. Likewise, [713] describes and reviews research on the accuracy-informativeness tradeoff that can influence the width of elicited intervals.

Turning now to psychological influences on probability judgements, the literature on the phenomenon of over-confidence manifested in subjective confidence intervals is exemplified by [396], and recent developments in overcoming this bias may be found in [706] and [384]. The other major influence reviewed here, namely partition priming, is first described in [299] and applied to imprecise probability judgements in [606].

Finally, key works on the topic of words-to-numbers translations of verbal probability phrases are [682, 90, 528]. Issues in the communication of such phrases have been raised by authors such as [89], and a recent paper by [88] describes attempts to improve public understanding of verbal uncertainty expressions employed in the Intergovernmental Panel on Climate Change reports, by interpreting them in terms of imprecise probabilities.

16

Computation

Matthias C. M. Troffaes[1] **and Robert Hable**[2]
[1]*Department of Mathematical Sciences, Durham University, UK*
[2]*Department of Mathematics, University of Bayreuth, Germany*

16.1 Introduction

In this chapter, we will very briefly discuss the main aspects of implementing the practical calculations for inference and decision making. Specialized computational techniques for graphical models were discussed in Section 9.5.

We start with discussing computational aspects of the natural extension of a conditional lower prevision. Of course, when dealing with problems that involve imprecise probabilities, finding the natural extension is usually only part of the solution. Nevertheless, because the idea of natural extension is one of the core aspects of imprecise probability theory, it seems worth to spend at least a few pages on its computational aspects.

Next, we discuss the computational aspects of solving static decision problems.

What is discussed here draws mostly from the result presented in earlier chapters, particularly Chapters 2, 4, and 8.

For an implementation of the algorithms presented here, see for instance the Python library `improb` [627].

16.2 Natural extension

As we must be able to deal with finite domains for computations, we start this section with a discussion of conditional lower previsions on arbitrary domains. A general purpose algorithm for natural extension is then presented, followed by simpler and faster implementations for important special cases.

Introduction to Imprecise Probabilities, First Edition.
Edited by Thomas Augustin, Frank P. A. Coolen, Gert de Cooman and Matthias C. M. Troffaes.

16.2.1 Conditional lower previsions with arbitrary domains

Let $\underline{P}(\cdot|\cdot)$ be an arbitrary conditional lower prevision defined on a set

$$\mathcal{C} \subseteq \mathcal{L} \times (\wp(\mathcal{X}) \setminus \{\emptyset\}).$$

Its interpretation was discussed in Section 2.3.

In order to simplify the expressions of avoiding partial loss, coherence and natural extension, it was assumed in Section 2.3.1 that

$$\mathcal{C} = \bigcup_{i=1}^{n} \mathcal{H}_i \times \mathcal{B}_i \tag{16.1}$$

where

$$\mathcal{H}_i = \left\{ \sum_{B \in \mathcal{B}_i} I_B f_B : f_B \in \mathcal{K}_B \right\} \tag{16.2}$$

for some partitions $\mathcal{B}_1, \dots, \mathcal{B}_n$ and linear spaces $\mathcal{K}_B$ such that $I_B \in \mathcal{K}_B$.

Obviously, any nontrivial linear space has infinite cardinality. Therefore, to study computation with conditional lower previsions, we drop the restriction of Section 2.3.1, and allow $\underline{P}(\cdot|\cdot)$ to be defined on *any* collection $\mathcal{C}$ of gambles and events. The resulting generalization of the theory has been studied for instance by Williams [703] (whose work was discussed in Section 2.4.1) and, for finite sets, by Walley, Pelessoni and Vicig [679]. We summarise Williams's work briefly here.

Note that, in addition, the conditioning events do not need to make up a collection of partitions. That condition was introduced for the sole purpose of dealing with conglomerability in infinite partitions (see Equation (1.15) and following discussion in Section 1.3.5; also see Section 2.3.1). Since we are only concerned with a finite possibility space, we do not need to concern ourselves with conglomerability in infinite partitions, and not having to work with collections of partitions simplifies notation considerably.

Williams [703] apparently does not define a concept of avoiding partial loss, but obviously the expression for avoiding partial loss follows logically from his expression for coherence [703, Equation (A*), p. 5], so we have:

Definition 16.1 We say that $\underline{P}(\cdot|\cdot)$ *avoids partial loss* when

$$\sup\left(\sum_{i=1}^{n} \lambda_i B_i \left(f_i - \underline{P}(f_i|B_i)\right) \middle| B_1 \cup \dots \cup B_n \right) \geq 0 \tag{16.3}$$

for all $n \in \mathbb{N}_0$, all nonnegative $\lambda_1, \dots, \lambda_n \in \mathbb{R}$, and all $(f_1, B_1), \dots, (f_n, B_n) \in \mathcal{C}$.

Definition 16.2 We say that $\underline{P}(\cdot|\cdot)$ is *coherent* when

$$\sup\left(\sum_{i=1}^{n} \lambda_i B_i \left(f_i - \underline{P}\left(f_i|B_i\right)\right) - \lambda_0 B_0 \left(f_0 - \underline{P}\left(f_0|B_0\right)\right) \middle| B_0 \cup \dots \cup B_n \right) \geq 0 \tag{16.4}$$

for all $n \in \mathbb{N}_0$, all nonnegative $\lambda_0, \dots, \lambda_n \in \mathbb{R}$, and all $(f_0, B_0), \dots, (f_n, B_n) \in \mathcal{C}$.

The following expression for natural extension can be shown to be equivalent to the expression given by Williams [703, § 2.1.0, p. 12]:

Definition 16.3 The natural extension $\underline{E}(\cdot|\cdot)$ of $\underline{P}(\cdot|\cdot)$ is defined as follows. For any $f \in \mathcal{L}$ and $B \subseteq \mathcal{X}$, $B \neq \emptyset$, $\underline{E}(f|B)$ is the supremum value of $\alpha \in \mathbb{R}$ for which there are $n \in \mathbb{N}_0$, nonnegative $\lambda_1, \dots, \lambda_n \in \mathbb{R}$, and $(f_0, B_0), \dots, (f_n, B_n) \in C$, such that

$$\sup\left(\sum_{i=1}^{n} \lambda_i B_i \left(f_i - \underline{P}(f_i|B_i)\right) - B(f-\alpha)\middle| B \cup B_1 \cup \cdots \cup B_n\right) < 0. \tag{16.5}$$

It can be shown that, under the additional domain assumptions stated in Eqs. (16.1) and (16.2), and under the additional condition that all partitions $\mathcal{B}_i$ are finite, these conditions are equivalent to those given in Chapter 2: Equation (2.14) is equivalent to (16.3), (2.13) is equivalent to Equation (16.4), and Equation (2.15) is equivalent to Equation (16.5).

In computer calculations, we cannot work with infinite sets, whence, from now onwards, we will assume that

- the possibility space $\mathcal{X}$ is finite, and
- the domain C of the conditional lower prevision $\underline{P}(\cdot|\cdot)$ is finite

In the following, we want to find $\underline{E}(f|B)$ for some gamble $f \in \mathcal{L}$ and some event $B \subseteq \mathcal{X}$, $B \neq \emptyset$. We also use the notation:

$$\mathcal{A} := \{A : (g, A) \in C \text{ for at least one gamble } g \in \mathcal{L}\}$$
$$\mathcal{K}_A := \{g : (g, A) \in C\}.$$

16.2.2 The Walley–Pelessoni–Vicig algorithm

Walley, Pelessoni and Vicig [679, p. 142, Algorithm 4] provide an elegant general purpose algorithm for finding the natural extension, by means of a sequence of linear programs. First, use Algorithm 16.1 to calculate a subset $\mathcal{A}_B$ of $\mathcal{A}$; this depends only on B.

If $\mathcal{A}_B$ is empty, then $\underline{P}(\cdot|\cdot)$ does not avoid partial loss – the converse does not hold generally, but it does hold for $B = \emptyset$. Whence, to determine whether $\underline{P}(\cdot|\cdot)$ avoids partial loss, simply check that $\mathcal{A}_\emptyset$ is empty (effectively, this is [679, p. 132, Algorithm 2]).

Note that Algorithm 16.1 is slightly more efficient than the one given in [679], as it only uses at most one τ_A variable per conditioning event $A \in \mathcal{A}$ in the domain C, whereas the original algorithm uses one variable $\tau_{g|A}$ for every $(g, A) \in C$. Obviously, in general, the domain C may have far higher cardinality than the set $\mathcal{A}$ of conditioning events in the domain.

Once $\mathcal{A}_B$ is determined, we can calculate the natural extension $\underline{E}(f|B)$ through the following linear program:

$$\begin{aligned}
&\text{maximize} && \alpha \\
&\text{subject to} && \forall A \in \mathcal{A}_B,\ \forall g \in \mathcal{K}_A : \lambda_{g|A} \geq 0 \\
& && \forall x \in \mathcal{X} : \sum_{A \in \mathcal{A}_B} \sum_{g \in \mathcal{K}_A} \lambda_{g|A} I_A(x)(g(x) - \underline{P}(g|A)) \leq I_B(x)(f(x) - \alpha).
\end{aligned}$$

Algorithm 16.1 Calculate $\mathcal{A}_B$.

$\mathcal{A}_B \leftarrow \mathcal{A}$
if $B = \mathcal{X}$ **then**
  return $\mathcal{A}_B$ /* *no need to check further in unconditional case* */
end if
loop

$$
\begin{aligned}
&\text{maximise} && \sum_{A \in \mathcal{A}_B} \tau_A \\
&\text{subject to} && \forall A \in \mathcal{A}_B : 0 \leq \tau_A \leq 1 \\
& && \forall A \in \mathcal{A}_B,\ \forall g \in \mathcal{K}_A : \lambda_{g|A} \geq 0 \\
& && \forall x \in B^c : \sum_{A \in \mathcal{A}_B} I_A(x) \left[\tau_A + \sum_{g \in \mathcal{K}_A} \lambda_{g|A}(g(x) - \underline{P}(g|A)) \right] \leq 0
\end{aligned}
$$

  if $\forall A \in \mathcal{A}_B : \tau_A = 1$ **then**
    return $\mathcal{A}_B$
  end if
  $\mathcal{A}_B \leftarrow \{A \in \mathcal{A}_B : \tau_A = 1\}$
end loop

In practice, it makes sense to cache the sets $\mathcal{A}_B$, so it only needs to be calculated once for each conditioning event B, particularly in applications where we need $\underline{E}(f|B)$ for many different gambles f, but for the same conditioning event B.

16.2.3 Choquet integration

General case

A simple approximation of the natural extension goes as follows. First, observe that we can always uniquely decompose a gamble in terms of its level sets:

$$I_B f = \sum_{i=1}^{n} \lambda_i I_{A_i} \tag{16.6}$$

where $\lambda_1 \in \mathbb{R}$, $\lambda_2, \ldots, \lambda_n > 0$, and $B = A_1 \supsetneq A_2 \supsetneq \cdots \supsetneq A_n \supsetneq \emptyset$. If all $I_{A_i} \in \mathcal{K}_B$, then this decomposition can be used to approximate – and in some cases even calculate exactly – the natural extension $\underline{E}(f|B)$. Indeed, the following inequality holds, by the properties of coherence (see Proposition 2.2):

$$\underline{E}(f|B) \geq \sum_{i=1}^{n} \lambda_i \underline{P}(A_i|B). \tag{16.7}$$

The right hand side of this inequality is the Choquet integral mentioned earlier in Section 4.3. From what is mentioned there, it follows easily that the inequality becomes an equality when $\mathcal{A} = \{B\}$ and $\underline{P}(\cdot|B)$ is a 2-monotone capacity on $\wp(B)$ (see Definition 4.1 and Definition 4.2). The Choquet integral relies on evaluation of $\underline{P}(\cdot|B)$ on level sets only, which is quite straightforward for many special cases (pari-mutuel models, possibility measures, p-boxes, etc.; see Chapter 4).

A fast algorithm for calculating the Choquet integral is given in Algorithm 16.2. Its complexity is $O(m + k \log k)$, where m is the cardinality of B, and k is the cardinality of $\{f(x) : x \in B\}$.

Possibility measures

Note that Algorithm 16.2 requires evaluation of $\underline{P}(A|B)$ for a decreasing sequence of sets A. As a result, we can implement a further optimization of the algorithm in case $\underline{P}(\cdot|B)$ is induced by a possibility measure (see Section 4.6.1). Indeed, suppose that $\pi(x|B)$ is the possibility distribution inducing $\underline{P}(\cdot|B)$, that is:

$$\underline{P}(A|B) = 1 - \sup_{x \in A^c} \pi(x|B) \tag{16.8}$$

for all $A \subseteq B$. Then, for any $A' \subseteq A$, such as in Algorithm 16.2,

$$\underline{P}(A'|B) = \min \left\{ \underline{P}(A|B), 1 - \max_{x \in A \setminus A'} \pi(x|B) \right\} \tag{16.9}$$

so for a decreasing sequence of sets A, we need $\pi(x|B)$ only once for every x. As it turns out, in Algorithm 16.2, $A \setminus A'$ was already calculated as part of the algorithm: it is A_v. Algorithm 16.3 summarizes the technique.

Algorithm 16.2 Choquet integral of f with respect to $\underline{P}(\cdot|B)$.

$V \leftarrow \emptyset$ /* *set of values of f on B* */
for $x \in B$ **do**
  $v \leftarrow f(x)$
  if $v \notin V$ **then**
    $V \leftarrow V \cup \{v\}$
    $A_v \leftarrow \{x\}$ /* A_v *stores* $f^{-1}(v)$ */
  else
    $A_v \leftarrow A_v \cup \{x\}$
  end if
end for
$y \leftarrow 0$ /* *stores result* */
$v' \leftarrow 0$ /* *previous value* */
$A \leftarrow B$ /* *level set* */
for $v \in V$ sorted from low to high **do**
  $y \leftarrow y + (v - v')\underline{P}(A|B)$
  $v' \leftarrow v$
  $A \leftarrow A \setminus A_v$
end for
return y

Algorithm 16.3 Choquet integral of f with respect to a possibility distribution $\pi(\cdot|B)$. V and A_v are calculated as in Algorithm 16.2.

$y \leftarrow 0$ /* *stores result* */
$v' \leftarrow 0$ /* *previous value* */
$A \leftarrow B$ /* *level set* */
$\ell \leftarrow 1$ /* *lower probability of A conditional on B* */
for $v \in V$ sorted from low to high **do**
  $y \leftarrow y + (v - v')\ell$
  $v' \leftarrow v$
  $A \leftarrow A \setminus A_v$
  $\ell \leftarrow \min\{\ell, 1 - \max_{x \in A_v} \pi(x|B)\}$
end for
return y

16.2.4 Möbius inverse

If $\{I_A : A \subseteq B\} \subseteq \mathcal{K}_B$, then an alternative way to calculate the Choquet integral with respect to $\underline{P}(\cdot|B)$ goes via the Möbius inverse of $\underline{P}(\cdot|B)$ (see Definition 4.3):

$$m(A|B) := \sum_{C \subseteq A} (-1)^{|A \setminus C|} \underline{P}(C|B) \tag{16.10}$$

for every $A \subseteq B$. The expression

$$\sum_{A \subseteq B} m(A|B) \inf(f|A) \tag{16.11}$$

coincides with the Choquet integral with respect to $\underline{P}(\cdot|B)$, when $\underline{P}(\emptyset|B) = 0$. For the 2-monotone case, this relates obviously to [110, p. 275, Corollary 4]. The general case follows from the fact that the Choquet integral commutes with linear combinations.

Again, this can be used to approximate – and under the right conditions, calculate exactly – the natural extension. Equation (16.11) leads to efficient algorithms when $m(A|B)$ is zero for most $A \subseteq B$. Generally, however, direct Choquet integration will be much faster.

16.2.5 Linear-vacuous mixture

Suppose that $\underline{P}(\cdot|\cdot)$ specifies at least an unconditional lower probability for every singleton, that is,

$$\{(I_x, \mathcal{X}) : x \in \mathcal{X}\} \subseteq \mathcal{C}, \tag{16.12}$$

and, additionally,

$$\forall x \in \mathcal{X} : \underline{P}(x|\mathcal{X}) \geq 0 \text{ and } 0 < \sum_{x \in \mathcal{X}} \underline{P}(x|\mathcal{X}) < 1. \tag{16.13}$$

In this case, it follows for instance from [672, p. 309, §6.6.2] that

$$\underline{E}(f|B) \geq \frac{(1-\epsilon)\sum_{x\in B} p(x)f(x) + \epsilon \min_{x\in B} f(x)}{(1-\epsilon)\sum_{x\in B} p(x) + \epsilon} \tag{16.14}$$

where $\epsilon = 1 - \sum_x \underline{P}(x|\mathcal{X})$ and $p(x) = \underline{P}(x|\mathcal{X})/(1-\epsilon)$. The right hand side of this inequality is a *linear-vacuous* lower prevision, because it is a mixture of a linear prevision (determined by the probability mass function p), and a vacuous prevision (i.e., a minimum operator). Equation (16.14) leads to an algorithm of complexity $O(m)$, where m is the cardinality of B; whence, it is faster than Choquet integration. The downside is that, in general, the approximation is very crude.

The inequality becomes an equality when $\{(I_x, \mathcal{X}) : x \in \mathcal{X}\} = C$.

16.3 Decision making

This subsection gives an idea of how decision problems can be computationally solved. We restrict ourselves to the solution of nonsequential decision problems discussed in Section 8.1, and are concerned with the unconditional case only.

Let $\mathcal{X}$ be finite, and let $\underline{P}$ be an (unconditional) lower prevision defined on a finite set of gambles. Consider a finite set D of acts. As explained in Section 8.1, each act $d \in D$ induces a gamble $f_d \in \mathcal{L}$, which is interpreted as an uncertain reward (expressed in utility).

16.3.1 Γ-maximin, Γ-maximax and Hurwicz

Evidently, any of the techniques of Section 16.2 can be used to find $\underline{E}(f_d)$ and $\overline{E}(f_d) = -\underline{E}(-f_d)$ for every possible act d. Note that, in the unconditional case, $\mathcal{A}_{\mathcal{X}} = \mathcal{A}$, so natural extension requires us to solve only a single linear program per gamble.

Γ-maximin simply picks the acts with the highest $\underline{E}(f_d)$ (see Definition 8.2), and Γ-minimax (see Definition 8.3) picks those acts with the highest $\overline{E}(f_d)$. Both algorithms require the evaluation of the natural extension of n gambles, if n is the cardinality of D. An algorithm to calculate randomized Γ-maximin acts by a single linear optimization problem is given in [644].

Hurwicz (see Definition 8.4) picks those acts with the highest mixture $\beta\underline{E}(f_d) + (1-\beta)\overline{E}(f_d)$, and so requires the natural extension of $2n$ gambles.

16.3.2 Maximality

For maximality (see Definition 8.6), we need to identify those acts d for which[1]

$$\forall d' \in D : \overline{E}(f_d - f_{d'}) \geq 0. \tag{16.15}$$

Naively, it seems that we need to evaluate $\overline{E}(f_d - f_{d'})$ for every pair (d, d') of acts. Whence, it seems that maximality requires the evaluation of the natural extension of $n(n-1)$ (which is $O(n^2)$) gambles, if n is the cardinality of D.

[1] For simplicity, we ignore the pointwise condition.

Algorithm 16.4 Find the set of maximal acts in D.

$R \leftarrow D$ /* *remaining decisions* */
$M \leftarrow \emptyset$ /* *maximal decisions* */
while $R \neq \emptyset$ **do**
  $d \leftarrow$ next element from R
  $R \leftarrow R \backslash \{d\}$
  if $\exists d' \in M : \underline{E}(f_{d'} - f_d) > 0$ **then**
    /* *d is dominated by* $d' \in M$ */
  else if $\exists d' \in R : \underline{E}(f_{d'} - f_d) > 0$ **then**
    /* *d is dominated by* $d' \in R$ */
  else
    $M \leftarrow M \cup \{d\}$ /* *d is undominated* */
  end if
end while
return M

Fortunately, in practice, we do not need to make as many comparisons (except when every decision is maximal, but if this is the case, you probably have more serious things to worry about), by exploiting the properties of maximality.

Indeed, firstly, as soon as it is determined that a decision d is not maximal, we can omit it from any further comparisons, because

$$C_{>_{\underline{P}}}(\mathcal{K}) = C_{>_{\underline{P}}}(\mathcal{K}') \tag{16.16}$$

whenever $C_{>_{\underline{P}}}(\mathcal{K}) \subseteq \mathcal{K}' \subseteq \mathcal{K}$. Secondly, it holds that every nonmaximal gamble is dominated by a maximal one. Whence, we can speed up the algorithm by comparing against gambles that are known to be maximal already. Algorithm 16.4 details how these two properties can be exploited algorithmically.

If the first decision considered in Algorithm 16.4 happens to be the only maximal decision, then the algorithm requires the evaluation of the natural extension of only $2(n-1)$ gambles – obviously, this is the best possible case. In general, the algorithm performs best if the gambles f_d are sorted in advance, for instance by lower or upper prevision (or a mixture thereof) using a quick approximate version of the natural extension, such as Choquet integration or linear-vacuous approximation. Even if the order is not perfect, the algorithm runs faster if it can identify maximal gambles early on.

16.3.3 E-admissibility

An act $d \in D$ is E-admissible (see Definition 8.7) if there is a function $p : \mathcal{X} \to \mathbb{R}$ such that

$$\sum_{x \in \mathcal{X}} p(x) = 1,$$

$$\forall x \in \mathcal{X}:\ p(x) \geq 0$$

$$\forall f \in \mathcal{K}:\ \sum_{x \in \mathcal{X}} f(x)p(x) \geq \underline{P}(f)$$

$$\forall d' \in D \backslash \{d\} : \sum_{x \in \mathcal{X}} (f_d(x) - f_{d'}(x))p(x) \geq 0.$$

This feasibility check can be done by standard solvers of linear programs.

The above set of linear constraints is of course quite large. More efficient algorithms for E-admissibility can be found in [644] (also dealing with randomized acts) and [395]. Evidently, because every E-admissible act is also maximal, one can run maximality first, to eliminate as many non-E-admissible acts as possible, before checking the feasibility of the above constraints.

16.3.4 Interval dominance

An act $d \in D$ fulfils interval dominance if

$$\overline{E}(f_d) \geq \max_{d' \in D} \underline{E}(f'_d). \tag{16.17}$$

So, in essence, we first find the Γ-maximin decision, giving us the right hand side of the above inequality, and then, for every decision $d \in D$, we compare $\overline{E}(f_d)$ against the Γ-maximin value.

All in all, we need to evaluate the natural extension of $2n - 1$ gambles, so this is really efficient. Of course, interval dominance comes with many issues, discussed in Section 8.1.2. Nevertheless, because every maximal act is also interval dominant, it can be used to quickly eliminate nonmaximal acts, before running Algorithm 16.4. Again, a very fast approximate (but conservative) version of the natural extension can be used for this purpose. This stage would also be useful to sort the acts in advance, to the benefit of Algorithm 16.4.

References

1 J. Abellán. Uncertainty measures on probability intervals from the imprecise Dirichlet model. *International Journal of General Systems*, 35:509–528, 2006. (Cited on pages 244 and 245)

2 J. Abellán, G. Klir, and S. Moral. Disaggregated total uncertainty measure for credal sets. *International Journal of General Systems*, 35:29–44, 2006. (Cited on pages 242 and 243)

3 J. Abellán and A. Masegosa. Combining decision trees based on imprecise probabilities and uncertainty measures. In K. Mellouli, editor, *Symbolic and Quantitative Approaches to Reasoning with Uncertainty*, volume 4724, pages 512–523. Springer, Berlin, 2007. (Cited on page 228)

4 J. Abellán and S. Moral. A non-specificity measure for convex sets of probability distributions. *International Journal of Uncertainty, Fuzziness and Knowledge-Based Systems*, 8:357–367, 2000. (Cited on page 242)

5 J. Abellán and S. Moral. Building classification trees using the total uncertainty criterion. *International Journal of Intelligent Systems*, 18:1215–1225, 2003. (Cited on pages 242 and 249)

6 J. Abellán and S. Moral. Maximum entropy for credal sets. *International Journal of Uncertainty, Fuzzines and Kowledge-Based Systems*, 11:58–597, 2003. (Cited on pages 244 and 249)

7 J. Abellán and S. Moral. Difference of entropies as a non-specificity function on credal sets. *International Journal of General Systems*, 34:201–214, 2005. (Cited on page 243)

8 J. Abellán and S. Moral. Upper entropy of credal sets. Applications to credal classification. *International Journal of Approximate Reasoning*, 39:235–255, 2005. (Cited on page 243)

9 J. Abellán and S. Moral. An algorithm to compute the upper entropy for order-2 capacities. *International Journal of Uncertainty, Fuzziness and Knowledge-Based Systems*, 14:141–154, 2006. (Cited on page 243)

10 A. Aboalkhair. *Nonparametric Predictive Inference for System Reliability*. PhD thesis, Durham University, Durham, UK (available from www.npi-statistics.com), 2012. (Cited on page 312)

11 C. Alchourròn, P. Gärdenfors, and D. Makinson. On the logic of theory change: Partial meet contraction and revision functions. *Journal of Symbolic Logic*, 50:510–530, 1985. (Cited on page 95)

12 M. Alpert and H. Raiffa. A progress report on the training of probability assessors. In D. Kahneman, P. Slovic, and A. Tversky, editors, *Judgment under Uncertainty: Heuristics and Biases*, pages 294–305. Cambridge University Press, Cambridge, UK, 1982. (Cited on page 325)

13 D. Alvarez. Nonspecificity for infinite random sets of indexable type. *Fuzzy Sets and Systems*, 159:289–306, 2008. (Cited on page 299)

Introduction to Imprecise Probabilities, First Edition.
Edited by Thomas Augustin, Frank P. A. Coolen, Gert de Cooman and Matthias C. M. Troffaes.

14 D. Alvarez. A Monte Carlo-based method for the estimation of lower and upper probabilities of events using infinite random sets of indexable type. *Fuzzy Sets and Systems*, 160:384–401, 2009. (Cited on page 84)

15 D. Alvarez. Reduction of uncertainty using sensitivity analysis methods for infinite random sets of indexable type. *International Journal of Approximate Reasoning*, 50:750–762, 2009. (Cited on page 299)

16 D. Andrews and G. Soares. Inference for parameters defined by moment inequalities using generalized moment selection. *Econometrica*, 78:119–157, 2010. (Cited on page 187)

17 F. Anscombe and I. Guttman. Rejection of outliers. *Technometrics*, 2:123–147, 1960. (Cited on page 166)

18 A. Antonucci. The imprecise noisy-OR gate. In *FUSION '11: Proceedings of the 14th International Conference on Information Fusion*, pages 709–715. IEEE, 2011. (Cited on page 223)

19 A. Antonucci, R. Brühlmann, A. Piatti, and M. Zaffalon. Credal networks for military identification problems. *International Journal of Approximate Reasoning*, 50:666–679, 2009. (Cited on page 229)

20 A. Antonucci, M. Cattaneo, and G. Corani. Likelihood-based naive credal classifier. In F. Coolen, G. de Cooman, T. Fetz, and M. Oberguggenberger, editors, *ISIPTA '11: Proceedings of the Seventh International Symposium on Imprecise Probability: Theories and Applications*, pages 21–30, Innsbruck, 2011. SIPTA. (Cited on pages 179 and 239)

21 A. Antonucci, R. de Rosa, and A. Giusti. Action recognition by imprecise hidden Markov models. In H. R. Arabnia, L. Deligiannidis, and G. Schaefer, editors, *IPCV '11: Proceedings of the 2011 International Conference on Image Processing, Computer Vision and Pattern Recognition*, pages 474–478. CSREA Press, 2011. (Cited on page 229)

22 A. Antonucci, A. Salvetti, and M. Zaffalon. Credal networks for hazard assessment of debris flows. In J. Kropp and J. Scheffran, editors, *Advanced Methods for Decision Making and Risk Management in Sustainability Science*, pages 125–132. Nova Science, New York, 2007. (Cited on pages 160 and 229)

23 A. Antonucci, S. Yi, C. de Campos, and M. Zaffalon. Generalized loopy 2U: A new algorithm for approximate inference in credal networks. *International Journal of Approximate Reasoning*, 51:474–484, 2010. (Cited on pages 223 and 224)

24 A. Antonucci and M. Zaffalon. Equivalence between Bayesian and credal nets on an updating problem. In J. Lawry, E. Miranda, A. Bugarin, S. Li, M. Gil, P. Grzegorzewski, and O. Hryniewicz, editors, *Soft Methods for Integrated Uncertainty Modelling*, pages 223–230, Springer, Berlin, 2006. (Cited on page 222)

25 A. Antonucci and M. Zaffalon. Decision-theoretic specification of credal networks: A unified language for uncertain modeling with sets of Bayesian networks. *International Journal of Approximate Reasoning*, 49:345–361, 2008. (Cited on pages 215, 219, and 223)

26 K. Arrow and L. Hurwicz. An optimality criterion for decision-making under ignorance. In D. Carter and F. Ford, editors, *Uncertainty and Expectations in Economics*, pages 461–472. Oxford University Press, Oxford, 1972. (Cited on page 193)

27 G. Arts and F. Coolen. Two nonparametric predictive control charts. *Journal of Statistical Theory and Practice*, 2:499–512, 2008. (Cited on page 176)

28 G. Arts, F. Coolen, and P. van der Laan. Nonparametric predictive inference in statistical process control. *Quality Technology & Quantitative Management*, 1:201–216, 2004. (Cited on page 176)

29 P. Artzner. Application of coherent risk measures to capital requirements in insurance. *North American Actuarial Journal*, 3:11–25, 1999. (Cited on page 283)

30 P. Artzner, F. Delbaen, J.-M. Eber, and D. Heath. Coherent measures of risk. *Mathematical Finance*, 9:203–228, 1999. (Cited on pages 283 and 285)

31 T. Augustin. Modeling weak information with generalized probability assignments. In H. Bock and W. Polasek, editors, *Data Analysis and Information Systems. Statistical and Conceptual Approaches*, pages 101–113. Springer, Heidelberg, 1996. (Cited on page 147)

32 T. Augustin. *Optimale Tests bei Intervallwahrscheinlichkeit*. Vandenhoeck und Ruprecht, Göttingen, 1998. (Cited on page 170)

33 T. Augustin. Neyman-Pearson testing under interval probability by globally least favorable pairs: Reviewing Huber-Strassen theory and extending it to general interval probability. *Journal of Statistical Planning and Inference*, 105:149–173, 2002. (Cited on pages 147 and 171)

34 T. Augustin. On the suboptimality of the generalized Bayes rule and robust Bayesian procedures from the decision theoretic point of view: A cautionary note on updating imprecise priors. In J. Bernard, T. Seidenfeld, and M. Zaffalon, editors, *ISIPTA '03: Proceedings of the Third International Symposium on Imprecise Probabilities and their Applications*, pages 31–45, Waterloo, 2003. Carleton Scientific. (Cited on pages 165 and 206)

35 T. Augustin. Optimal decisions under complex uncertainty – basic notions and a general algorithm for data-based decision making with partial prior knowledge described by interval probability. *ZAMM. Zeitschrift für Angewandte Mathematik und Mechanik. Journal of Applied Mathematics and Mechanics*, 84:678–687, 2004. (Cited on page 165)

36 T. Augustin and F. Coolen. Nonparametric predictive inference and interval probability. *Journal of Statistical Planning and Inference*, 124:251–272, 2004. (Cited on pages 55 and 175)

37 T. Augustin and R. Hable. On the impact of robust statistics on imprecise probability models: A review. *Structural Safety*, 32:358–365, 2010. (Cited on pages 147, 187, 206, and 304)

38 R. Aumann. Utility theory without the completeness axiom. *Econometrica*, 30:445–462, 1962. (Cited on page 26)

39 R. Aumann. Utility theory without the completeness axiom: A correction. *Econometrica*, 32:210–212, 1964. (Cited on page 26)

40 D. Avis and K. Fukuda. A pivoting algorithm for convex hulls and vertex enumeration of arrangements and polyhedra. *Discrete and Computational Geometry*, 8:295–313, 1992. (Cited on pages 208 and 211)

41 B. Ayyub, editor. *Uncertainty Modelling and Analysis in Civil Engineering*, Boca Raton, 1998. CRC. (Cited on page 304)

42 G. Baecher and J. Christian. *Reliability and Statistics in Geotechnical Engineering*. Wiley, Chichester, 2010. (Cited on page 303)

43 M. Banerjee and D. Dubois. A simple modal logic for reasoning about revealed beliefs. In C. Sossai and G. Chemello, editors, *ECSQARU '09: Proceedings of the 10th European Conference on Symbolic and Quantitative Approaches to Reasoning with Uncertainty*, volume 5590, pages 805–816, Berlin, 2009. Springer. (Cited on pages 96 and 98)

44 R. Barlow and F. Proschan. *Statistical Theory of Reliability and Life Testing: Probability Models*. Holt, Rinehart and Winston, New York, 1975. (Cited on page 311)

45 V. Barnett. *Comparative Statistical Inference*. Wiley, Chichester, 3rd edition, 1999. (Cited on pages 178 and 187)

46 P. Baroni, R. Pelessoni, and P. Vicig. Generalizing Dutch risk measures through imprecise previsions. *International Journal of Uncertainty, Fuzziness and Knowledge–Based Systems*, 17:153–177, 2009. (Cited on pages 283, 287 and 289)

47 P. Bártfai and P. Révész. On a zero-one law. *Zeitschrift für Wahrscheinlichkeitstheorie und Verwandte Gebiete*, 7:43–47, 1967. (Cited on page 128)

48 O. Bataineh. *Imprecise Probability Models for Logistic Regression*. PhD thesis, Department of Mathematics and Statistics, University of Saskatchewan, Saskatoon (available from http://ecommons.usask.ca/bitstream/handle/10388/ETD-2012-09-741/BATAINEH-DISSERTATION.pdf), 2012. (Cited on page 188)

49 C. Baudrit, I. Couso, and D. Dubois. Joint propagation of probability and possibility in risk analysis: Towards a formal framework. *International Journal of Approximate Reasoning*, 45:82–105, 2007. (Cited on page 109)

50 C. Baudrit and D. Dubois. Practical representations of incomplete probabilistic knowledge. *Computational Statistics and Data Analysis*, 51:86–108, 2006. (Cited on pages 86 and 89)

51 T. Bednarski. On solutions of minimax test problems for special capacities. *Zeitschrift für Wahrscheinlichkeitstheorie und Verwandte Gebiete*, 58:397–405, 1981. (Cited on page 171)

52 Y. Ben-Haim. *Information-Gap Decision Theory. Decisions Under Severe Uncertainty*. Academic Press, San Diego, CA, 2001. (Cited on pages 112, 294, and 304)

53 Y. Ben-Haim and I. Elishakoff. *Convex Models of Uncertainty in Applied Mechanics*. Elsevier, Amsterdam, 1990. (Cited on pages 294 and 304)

54 B. Ben Yaghlane, P. Smets, and K. Mellouli. Belief function independence: I. The marginal case. *International Journal of Approximate Reasoning*, 29:47–70, 2002. (Cited on page 108)

55 B. Ben Yaghlane, P. Smets, and K. Mellouli. Independence concepts for belief functions. In B. Bouchon-Meunier, L. Magdalena, J. Gutierrez-Rios, and R. R. Yager, editors, *Technologies for Constructing Intelligent Systems 2*, pages 45–53. Springer, Berlin, 2002. (Cited on page 77)

56 A. Benavoli and A. Antonucci. Aggregating imprecise probabilistic knowledge: Application to Zadeh's paradox and sensor networks. *International Journal of Approximate Reasoning*, 51:1014–1028, 2010. (Cited on page 228)

57 A. Benavoli and M. Zaffalon. A model of prior ignorance for inferences in the one-parameter exponential family. *Journal of Statistical Planning and Inference*, 142:1960–1979, 2012. (Cited on pages 156, 157, 160, and 161)

58 A. Benavoli, M. Zaffalon, and E. Miranda. Reliable hidden Markov model filtering through coherent lower previsions. In *FUSION '09: Proceedings of the 12th International Conference on Information Fusion*, pages 1743–1750, Seattle, WA, 2009. IEEE. (Cited on page 90)

59 A. Benavoli, M. Zaffalon, and E. Miranda. Robust filtering through coherent lower previsions. *Automatic Control, IEEE Transactions on*, 56:1567–1581, 2011. (Cited on page 227)

60 S. Benferhat, D. Dubois, and H. Prade. Non-monotonic reasoning, conditional objects and possibility theory. *Artificial Intelligence*, 92:259–276, 1997. (Cited on page 113)

61 A. Beresteanu, I. Molchanov, and F. Molinari. Partial identification using random set theory. *Journal of Econometrics*, 166:17–32, 2012. (Cited on pages 181 and 186)

62 J. Berger. *Statistical Decision Theory and Bayesian Analysis*. Springer, New York, 2nd edition, 1985. (Cited on pages 193 and 206)

63 J. Berger, D. Ríos Insua, and F. Ruggeri. Bayesian robustness. In D. Rios Insua and F. Ruggeri, editors, *Robust Bayesian Analysis*, pages 1–31. Springer, Berlin, 2000. (Cited on pages 164 and 187)

64 J. Berger, E. Moreno, L. Pericchi, M. Bayarri, J. Bernardo, J. Cano, J. De la Horra, J. Martín, D. Ríos Insua, B. Betrò, A. Dasgupta, P. Gustafson, L. Wasserman, J. Kadane, C. Srinivasan, M. Lavine, A. O'Hagan, W. Polasek, C. Robert, C. Goutis, F. Ruggeri, G. Salinetti, and S. Sivaganesan. An overview of robust Bayesian analysis. *Test*, 3:5–124, 1994. (Cited on pages 90 and 93)

65 J.-M. Bernard. An introduction to the imprecise Dirichlet model for multinomial data. *International Journal of Approximate Reasoning*, 39:123–150, 2005. (Cited on page 160)

66 J.-M. Bernard. Special issue on the imprecise Dirichlet model. *International Journal of Approximate Reasoning*, 50:201–268, 2009. (Cited on pages 155 and 160)

67 A. Bernardini. Qualitative and quantitative measures in seismic damage assessment and forecasting of masonry buildings. In A. Bernardini, editor, *Seismic Damage to Masonry Buildings*, pages 169–177. A. A. Balkema, Rotterdam, 1999. (Cited on page 303)

68 A. Bernardini, S. Lagomarsino, A. Mannella, A. Martinelli, L. Milano, and S. Parodi. Forecasting seismic damage scenarios of residential buildings from rough inventories: A case-study in the Abruzzo Region (Italy). *Journal of Risk and Reliability*, 224:279–296, 2010. (Cited on page 304)

69 A. Bernardini and F. Tonon. *Bounding Uncertainty in Civil Engineering – Theoretical Background*. Springer, Berlin, 2010. (Cited on page 303)

70 J. Bernardo and A. Smith. *Bayesian Theory*. Wiley, Chichester, 2000. (Cited on pages 139, 150, 152, 158, and 187)

71 S. Bernstein. Démonstration du théorème de Weierstrass, fondée sur le calcul des probabilités. *Communications of the Kharkov Mathematical Society*, 13:1–2, 1912. (Cited on page 76)

72 B. Betrò. Numerical treatment of Bayesian robustness problems. *International Journal of Approximate Reasoning*, 50:279–288, 2009. (Cited on page 165)

73 R. Beyth-Marom. How probable is probable? A numerical translation of verbal probability expressions. *Journal of Forecasting*, 1:257–269, 1982. (Cited on pages 320)

74 K. Bhaskara Rao and M. Bhaskara Rao. *Theory of Charges*. Academic Press, London, 1983. (Cited on pages 36, 37, and 71)

75 M. Bickis. The imprecise logit-normal model and its application to estimating hazard functions. *Journal of Statistical Theory and Practice*, 3:183–195, 2009. Reprinted in [146], pages 255–267. (Cited on page 164)

76 D. Blockley and I. Ferraris. Managing the quality of a project. In Academia Nacional de Ingeniería, editor, *Seguridad en Ingeniería*, pages 69–89. Editorial Dunken, Buenos Aires, 2000. (Cited on pages 294 and 303)

77 W. Boettcher. Context, methods, numbers, and words: Prospect theory in international relations. *Journal of Conflict Resolution*, 39:561–583, 1995. (Cited on page 320)

78 R. Bogdan, editor. *Henry E. Kyburg, Jr. & Isaac Levi*. Reidel, Dordrecht, 1982. (Cited on page 179)

79 V. Bolotin. *Statistical Method in Structural Mechanics*. Holden-Day, San Francisco, 1969. (Cited on pages 292, 293, and 296)

80 G. Bontempi, M. Birattari, and H. Bersini. Lazy learning for local modelling and control design. *International Journal of Control*, 72:643–658, 1999. (Cited on page 237)

81 G. Boole. *An Investigation of the Laws of Thought on Which are Founded the Mathematical Theories of Logic and Probabilities*. Walton and Maberly, London, 1854. (Cited on page xiv)

82 A. Boratyńska. Stability of Bayesian inference in exponential families. *Statistics & Probability Letters*, 36:173–178, 1997. (Cited on page 156)

83 B. Bouchon-Meunier, G. Coletti, and C. Marsala. Independence and possibilistic conditioning. *Annals of Mathematics and Artificial Intelligence*, 35:107–123, 2002. (Cited on page 108)

84 G. Box and N. Draper. *Empirical Model-building and Response Surface*. Wiley, New York, 1987. (Cited on page 144)

85 L. Breiman. *Probability*. Addison-Wesley, Reading, Mass, 1968. (Cited on page 142)

86 A. Bronevich and T. Augustin. Approximation of coherent lower probabilities by 2-monotone measures. In T. Augustin, F. Coolen, M. Troffaes, and S. Moral, editors, *ISIPTA'09: Proceedings of the Sixth International Symposium on Imprecise Probability: Theories and Applications*, pages 61–70, Durham, UK, 2009. SIPTA. (Cited on page 81)

87 C. Bucher. *Computational Analysis of Randomness in Structural Mechanics*. CRC/Balkema, Leiden, 2009. (Cited on page 303)

88 D. Budescu, S. Broomell, and H. Por. Improving communication of uncertainty in the reports of the intergovernmental panel on climate change. *Psychological Science*, 20:299–308, 2009. (Cited on page 328)

89 D. Budescu and T. Wallsten. Consistency in interpretation of probabilistic phrases. *Organizational Behavior and Human Decision Processes*, 36:391–405, 1985. (Cited on pages 320 and 328)

90 D. Budescu, S. Weinberg, and T. Wallsten. Decisions based on numerically and verbally expressed uncertainties. *Journal of Experimental Psychology: Human Perception and Performance*, 14:281–294, 1988. (Cited on pages 320 and 328)

91 F. Bugni. Bootstrap inference in partially identified models defined by moment inequalities: Coverage of the identified set. *Econometrica*, 78:735–753, 2010. (Cited on page 187)

92 A. Buja. On the Huber-Strassen theorem. *Probability Theory and Related Fields*, 73:149–152, 1986. (Cited on page 170)

93 C. Camerer and T. Ho. Violations of the betweenness axiom and nonlinearity in probability. *Journal of Risk and Uncertainty*, 8:167–196, 1994. (Cited on page 324)

94 M. Campos, G.P. Dimuro, A. da Rocha Costa, and V. Kreinovich. Computing 2-step predictions for interval-valued finite stationary Markov chains. Technical report utep-cs-03-20a, University of Texas at El Paso, 2003. (Cited on page 264)

95 I. Canay. EL inference for partially identified models: Large deviations optimality and bootstrap validity. *Journal of Econometrics*, 156:408–425, 2010. (Cited on page 187)

96 A. Cano, J. Cano, and S. Moral. Convex sets of probabilities propagation by simulated annealing on a tree of cliques. In R. Yager, B. Bouchon-Meunier, and L. Zadeh, editors, *IPMU '94: Proceedings of the Fifth International Conference on Processing and Management of Uncertainty in Knowledge-Based Systems*, pages 4–8, Paris, 1994. Springer. (Cited on pages 208, 223, and 224)

97 A. Cano, M. Gómez Olmedo, and S. Moral. Application of a hill-climbing algorithm to exact and approximate inference in credal networks. In F. Cozman, R. Nau, and T. Seidenfeld, editors, *ISIPTA '05: Proceedings of the Fourth International Symposium on Imprecise Probabilities and Their Applications*, pages 88–97, Pittsburgh, PA, 2005. SIPTA. (Cited on page 224)

98 A. Cano, M. Gómez Olmedo, and S. Moral. Credal nets with probabilities estimated with an extreme imprecise Dirichlet model. In G. de Cooman, J. Vejnarová, and M. Zaffalon, editors, *ISIPTA'07: Proceedings of the Fourth International Symposium on Imprecise Probabilities: Theories and Applications*, pages 57–66, Prague, 2007. Action M Agency for SIPTA. (Cited on pages 240 and 241)

99 A. Cano and S. Moral. A genetic algorithm to approximate convex sets of probabilities. In B. Bouchon-Meunier, editor, *IPMU '96: Proceeding of the Sixth International Conference on Information Processing and Management of Uncertainty in Knowledge-Based Systems*, volume 2, pages 847–852, Grenada, 1996, Proyecto Sur. (Cited on pages 223 and 224)

100 A. Cano and S. Moral. Using probability trees to compute marginals with imprecise probabilities. *International Journal of Approximate Reasoning*, 29:1–46, 2002. (Cited on pages 223 and 224)

101 A. Capotorti, G. Coletti, and B. Vantaggi. Non additive ordinal relations representable by lower or upper probabilities. *Kybernetika*, 34:79–90, 1998. (Cited on page 113)

102 C. Carathéodory. Über das lineare Mass von Punktmengen – eine Verallgemeinerung des Längenbegriffs. *Nachrichten der Akademie der Wissenschaften zu Göttingen. II. Mathematisch-Physikalische Klasse*, 4:404–426, 1914. (Cited on pages 123 and 124)

103 M. Cattaneo. *Statistical Decisions Based Directly on the Likelihood Function*. PhD thesis, ETH Zurich (available from http://dx.doi.org/10.3929/ethz-a-005463829), 2007. (Cited on page 179)

104 M. Cattaneo. Fuzzy probabilities based on the likelihood function. In D. Dubois, M. Lubiano, H. Prade, M. Gil, P. Grzegorzewski, and O. Hryniewicz, editors, *Soft Methods for Handling Variability and Imprecision*, volume 48, pages 43–50. Toulouse, 2008. Springer. (Cited on pages 165 and 179)

105 M. Cattaneo. Likelihood-based inference for probabilistic graphical models: Some preliminary results. In P. Myllymaki, T. Roos, and T. Jaakkola, editors, *PGM '10: Proceedings of The Fifth European Workshop on Probabilistic Graphical Models*, pages 57–65. HIIT Publications, Helsinki, 2010. (Cited on page 239)

106 M. Cattaneo. Likelihood decision functions. *Electronic Journal of Statistics*, 7:2924–2946, 2013. (Cited on page 179)

107 M. Cattaneo and A. Wiencierz. Likelihood-based imprecise regression. *International Journal of Approximate Reasoning*, 53:1137–1154, 2012. (Cited on page 187)

108 A. Chateauneuf. Combination of compatible belief functions and relation of specificity. In R. Yager, J. Kacprzyk, and M. Fedrizzi, editors, *Advances in the Dempster-Shafer Theory of Evidence*, pages 97–114. Wiley, New York, NY, 1994. (Cited on page 112)

109 A. Chateauneuf and M. Cohen. Cardinal extensions of the EU model based on the Choquet integral. In D. Bouyssou, D. Dubois, M. Pirlot, and H. Prade, editors, *Decision-Making Process: Concepts and Methods*, pages 401–434. Wiley, Hoboken, NJ, 2009. (Cited on page 94)

110 A. Chateauneuf and J.-Y. Jaffray. Some characterizations of lower probabilities and other monotone capacities through the use of Möbius inversion. *Mathematical Social Sciences*, 17:263–283, 1989. (Cited on pages 81 and 334)

111 B. Chellas. *Modal Logic*. Cambridge University Press, Cambridge, UK, 1978. (Cited on page 96)

112 A. Chernov and V. Vovk. Prediction with expert evaluators' advice. In R. Gavaldà, G. Lugosi, T. Zeugmann, and S. Zilles, editors, *ALT '09: Proceedings of the 20th International Conference on Algorithmic Learning Theory*, volume 5809, pages 8–22, Berlin, 2009. Springer. (Cited on page 133)

113 G. Choquet. Theory of capacities. *Annales de l'Institut Fourier*, 5:131–295, 1954. (Cited on page 80)

114 L. Chrisman. Propagation of 2-monotone lower probabilities on an undirected graph. In E. Horvitz and F. Jensen, editors, *UAI '96: Proceedings of the Twelfth Annual Conference on Uncertainty in Artificial Intelligence*, pages 178–185, San Francisco, CA, 1996. Morgan Kaufmann. (Cited on page 81)

115 D. Cifarelli and E. Regazzini. De Finetti's contributions to probability and statistics. *Statistical Science*, 11:253–282, 1996. (Cited on page 78)

116 M. Clyde and E. George. Model uncertainty. *Statistical Science*, 19:81–94, 2004. (Cited on page 238)

117 G. Coletti. Coherent numerical and ordinal probability assessments. *IEEE Transactions on Systems, Man and Cybernetics*, 24:1747–1754, 1994. (Cited on page 47)

118 G. Coletti and R. Scozzafava. Zero probabilities in stochastic independence. In B. Bouchon-Meunier, R. Yager, and L. Zadeh, editors, *Information, Uncertainty and Fusion*, volume 516, pages 185–196. Kluwer, Norwell, MA, 2000. (Cited on page 47)

119 G. Coletti and R. Scozzafava. *Probabilistic Logic in a Coherent Setting*. Kluwer, New York, 2002. (Cited on pages 47)

120 G. Coletti and R. Scozzafava. Coherent conditional probability as a measure of uncertainty of the relevant conditioning events. In T. Nielsen and N. Zhang, editors, *ECSQARU '03: Proceedings of the 7th European Conference on Symbolic and Quantitative Approaches to Reasoning with Uncertainty*, volume 2711, pages 407–418, Berlin, 2003. Springer. (Cited on page 101)

121 G. Coletti and B. Vantaggi. Possibility theory: Conditional independence. *Fuzzy Sets and Systems*, 157:1491–1513, 2006. (Cited on page 77)

122 G. Coletti and B. Vantaggi. A view on conditional measures through local representability of binary relations. *International Journal of Approximate Reasoning*, 47:268–283, 2008. (Cited on page 113)

123 R. Cooke. *Experts in Uncertainty*. Oxford University Press, Oxford, 1991. (Cited on page 110)

124 F. Coolen. Imprecise conjugate prior densities for the one-parameter exponential family of distributions. *Statistics & Probability Letters*, 16:337–342, 1993. (Cited on pages 164 and 165)

125 F. Coolen. *Statistical Modeling of Expert Opinions using Imprecise Probabilities*. PhD thesis, Eindhoven Technical University (available from http://alexandria.tue.nl/extra3/proefschrift/PRF9B/9305256.pdf), 1993. (Cited on page 165)

126 F. Coolen. On Bernoulli experiments with imprecise prior probabilities. *The Statistician*, 43:155–167, 1994. (Cited on page 165)

127 F. Coolen. Comparing two populations based on low stochastic structure assumptions. *Statistics & Probability Letters*, 29:297–305, 1996. (Cited on page 177)

128 F. Coolen. An imprecise Dirichlet model for Bayesian analysis of failure data including right-censored observations. *Reliability Engineering and System Safety*, 56:61–68, 1997. (Cited on pages 160 and 310)

129 F. Coolen. Low structure imprecise predictive inference for Bayes' problem. *Statistics & Probability Letters*, 36:349–357, 1998. (Cited on page 176)

130 F. Coolen. On nonparametric predictive inference and objective Bayesianism. *Journal of Logic, Language and Information*, 15:21–47, 2006. (Cited on page 175)

131 F. Coolen. On probabilistic safety assessment in case of zero failures. *Journal of Risk and Reliability*, 220:105–114, 2006. (Cited on page 317)

132 F. Coolen. Nonparametric prediction of unobserved failure modes. *Journal of Risk and Reliability*, 221:207–216, 2007. (Cited on pages 177 and 316)

133 F. Coolen and T. Augustin. Learning from multinomial data: A nonparametric predictive alternative to the imprecise Dirichlet model. In F. Cozman, B. Nau, and T. Seidenfeld, editors, *ISIPTA '05: Proceedings of the Fourth International Symposium on Imprecise Probabilities and Their Applications*, pages 125–134, Pittsburgh, PA, 2005. SIPTA. (Cited on pages 160 and 316)

134 F. Coolen and T. Augustin. A nonparametric predictive alternative to the imprecise Dirichlet model: The case of a known number of categories. *International Journal of Approximate Reasoning*, 50:217–230, 2009. (Cited on pages 160, 177, and 316)

135 F. Coolen and P. Coolen-Schrijner. Nonparametric predictive reliability demonstration for failure-free periods. *Journal of Management Mathematics*, 16:1–11, 2005. (Cited on page 317)

136 F. Coolen and P. Coolen-Schrijner. Nonparametric predictive comparison of proportions. *Journal of Statistical Planning and Inference*, 137:23–33, 2007. (Cited on page 178)

137 F. Coolen, P. Coolen-Schrijner, and K. Yan. Nonparametric predictive inference in reliability. *Reliability Engineering and System Safety*, 78:185–193, 2002. (Cited on page 312)

138 F. Coolen and P. van der Laan. Imprecise predictive selection based on low structure assumptions. *Journal of Statistical Planning and Inference*, 98:259–277, 2001. (Cited on page 177)

139 F. Coolen and K. Yan. Nonparametric predictive inference for grouped lifetime data. *Reliability Engineering and System Safety*, 80:243–252, 2003. (Cited on page 317)

140 F. Coolen and K. Yan. Nonparametric predictive inference with right-censored data. *Journal of Statistical Planning and Inference*, 126:25–54, 2004. (Cited on pages 176, 312, 313, and 314)

141 T. Coolen-Maturi and F. Coolen. Unobserved, re-defined, unknown or removed failure modes in competing risks. *Journal of Risk and Reliability*, 225:461–474, 2010. (Cited on pages 314 and 315)

142 P. Coolen-Schrijner and F. Coolen. Adaptive age replacement based on nonparametric predictive inference. *Journal of the Operational Research Society*, 55:1281–1297, 2004. (Cited on pages 178 and 317)

143 P. Coolen-Schrijner and F. Coolen. Nonparametric adaptive age replacement with a one-cycle criterion. *Reliability Engineering and System Safety*, 92:74–84, 2007. (Cited on pages 317)

144 P. Coolen-Schrijner, F. Coolen, and I. MacPhee. Nonparametric predictive inference for systems reliability with redundancy allocation. *Journal of Risk and Reliability*, 222:463–476, 2008. (Cited on pages 176 and 178)

145 P. Coolen-Schrijner, F. Coolen, and S. Shaw. Nonparametric adaptive opportunity-based age replacement strategies. *Journal of the Operational Research Society*, 57:63–81, 2006. (Cited on pages 317)

146 P. Coolen-Schrijner, F. Coolen, M. Troffaes, T. Augustin, and S. Gupta, editors. *Imprecision in Statistical Theory and Practice*. Grace, Greensboro, NC, 2009. (Cited on pages 342, 345, 348, 353, and 355)

147 P. Coolen-Schrijner, T. Maturi, and F. Coolen. Nonparametric predictive precedence testing for two groups. *Journal of Statistical Theory and Practice*, 3:273–287, 2009. Reprinted in [146], pages 91–105. (Cited on pages 178)

148 G. Corani and A. Benavoli. Restricting the IDM for classification. In Hüllermeier, E. and Kruse, R. and Hoffmann, F., E. Hüllermeier, R. Kruse, and F. Hoffmann, editors, *IPMU '10: International Conference on Information Processing and Management of Uncertainty in Knowledge-Based Systems*, volume 80, pages 328–337, New York, 2010. Springer. (Cited on pages 235 and 252)

149 G. Corani and C. de Campos. A tree-augmented classifier based on the extreme imprecise Dirichlet model. *International Journal of Approximate Reasoning*, 51:1053–1068, 2010. (Cited on page 241)

150 G. Corani and M. Zaffalon. Credal model averaging: An extension of Bayesian model averaging to imprecise probabilities. In W. Daelemans, B. Goethals, and K. Morik, editors, *ECML PKDD '08: Proceeding of the 2008 European Conference on Machine Learning and Principles and Practice of Knowledge Discovery in Databases*, volume 5211, pages 257–271. Berlin, 2008, Springer. (Cited on page 239)

151 G. Corani and M. Zaffalon. Jncc2: The Java implementation of naive credal classifier 2. *Journal of Machine Learning Research*, 9:2695–2698, 2008. (Cited on pages 237 and 251)

152 G. Corani and M. Zaffalon. Learning reliable classifiers from small or incomplete data sets: The naive credal classifier 2. *Journal of Machine Learning Research*, 9:581–621, 2008. (Cited on pages 234, 236, 237, 249, and 250)

153 G. Corani and M. Zaffalon. Lazy naive credal classifier. In J. Pei, L. Getoor, and A. de Keijzer, editors, *Proceedings of the 1st ACM SIGKDD Workshop on Knowledge Discovery from Uncertain Data*, pages 30–37. ACM, 2009. (Cited on pages 237, 238, and 250)

154 I. Couso and D. Dubois. On the variability of the concept of variance for fuzzy random variables. *IEEE Transactions on Fuzzy Systems*, 17:1070–1080, 2009. (Cited on page 100)

155 I. Couso, S. Montes, and P. Gil. The necessity of the strong alpha-cuts of a fuzzy set. *International Journal on Uncertainty, Fuzziness and Knowledge-Based Systems*, 9:249–262, 2001. (Cited on pages 85 and 92)

156 I. Couso and S. Moral. Sets of desirable gambles and credal sets. In T. Augustin, F. Coolen, S. Moral, and M. Troffaes, editors, *ISIPTA '09: Proceedings of the Sixth International Symposium on Imprecise Probability: Theories and Applications*, pages 99–108, Durham, UK, 2009. SIPTA. (Cited on pages 17 and 26)

157 I. Couso and S. Moral. Independence concepts in evidence theory. *International Journal of Approximate Reasoning*, 51:748–758, 2010. (Cited on pages 77 and 109)

158 I. Couso, S. Moral, and P. Walley. A survey of concepts of independence for imprecise probabilities. *Risk, Decision and Policy*, 5:165–181, 2000. (Cited on pages 26, 60, 63, 77, 149, and 169)

159 F. Cozman. Irrelevance and independence relations in quasi-Bayesian networks. In G. Coopen and S. Moral, editors, *UAI '98: Proceedings of the Fourteenth Conference on Uncertainty in Artificial Intelligence*, pages 89–96, San Francisco, CA, 1998. Morgan Kaufmann. (Cited on page 60)

160 F. Cozman. Credal networks. *Artificial Intelligence*, 120:199–233, 2000. (Cited on pages 208 and 223)

161 F. Cozman. Constructing sets of probability measures through Kuznetsov's independence condition. In G. de Cooman, T. Fine, and T. Seidenfeld, editors, *ISIPTA '01: Proceedings of the Second International Symposium on Imprecise Probabilities and Their Applications*, pages 104–111, Ithaca, NY, 2001. Shaker. (Cited on pages 54 and 66)

162 F. Cozman. Concentration inequalities and laws of large numbers under epistemic irrelevance. In T. Augustin, F. Coolen, S. Moral, and M. Troffaes, editors, *ISIPTA '09: Proceedings of the Sixth International Symposium on Imprecise Probability: Theories and Applications*, pages 109–118, Durham, UK, 2009. SIPTA. (Cited on pages 61 and 114)

163 F. Cozman and C. de Campos. Local computation in credal networks. In *Workshop on Local Computation for Logics and Uncertainty*, pages 5–11, Valencia, 2004. IOS Press. (Cited on page 224)

164 F. Cozman, C. de Campos, and J. Ferreira da Rocha. Probabilistic logic with independence. *International Journal of Approximate Reasoning*, 49:3–17, 2008. (Cited on page 229)

165 F. Cozman, C. de Campos, J. Ide, and J. Ferreira da Rocha. Propositional and relational Bayesian networks associated with imprecise and qualitative probabilistic assessments. In J. Halpern, editor, *UAI '04: Proceedings of the 20th Conference on Uncertainty in Artificial Intelligence*, pages 104–111, Arlington, VA, 2004. AUAI Press. (Cited on pages 223 and 224)

166 F. Cozman and R. Polastro. Complexity analysis and variational inference for interpretation-based probabilistic description logics. In J. Bilmes and A. Ng, editors, *UAI '09: Proceeding of the Twenty-Fifth Conference on Uncertainty in Artificial Intelligence*, pages 120–133, Corvallis, OR, 2009. AUAI Press. (Cited on page 229)

167 F. Cozman and P. Walley. Graphoid properties of epistemic irrelevance and independence. *Annals of Mathematics and Artificial Intelligence*, 45:173–195, 2005. (Cited on page 227)

168 R. Crossman, J. Abellán, T. Augustin, and F. Coolen. Building imprecise classification trees with entropy ranges. In F. Coolen, G. de Cooman, T. Fetz, and M. Oberguggenberger, editors, *ISIPTA '11: Proceedings of the Seventh International Symposium on Imprecise Probability: Theories and Applications*, pages 129–138, Innsbruck, 2011. SIPTA. (Cited on page 249)

169 R. Crossman and D. Škulj. Imprecise Markov chains with absorption. *International Journal of Approximate Reasoning*, 51:1085–1099, 2010. (Cited on page 277)

170 G. Dantzig. *Linear Programming and Extensions*. Princeton University Press, Princeton, NJ, 1963. (Cited on page 209)

171 J. Darroch and E. Seneta. On quasi-stationary distributions in absorbing discrete-time finite Markov chains. *Journal of Applied Probability*, 1:88–100, 1965. (Cited on page 277)

172 A. Darwiche. Recursive conditioning. *Artificial Intelligence*, 126:5–41, 2001. (Cited on page 223)

173 A. Darwiche. *Modeling and Reasoning with Bayesian networks*. Cambridge University Press, New York, NY, 2009. (Cited on page 220)

174 D. Dash and G. Cooper. Model averaging for prediction with discrete Bayesian networks. *Journal of Machine Learning Research*, 5:1177–1203, 2004. (Cited on pages 238 and 239)

175 A. Dawid. Conditional independence in statistical theory. *Journal of the Royal Statistical Society, Series B*, 41:1–31, 1979. (Cited on page 214)

176 A. Dawid. Statistical theory: The prequential approach (with discussion). *Journal of the Royal Statistical Society, Series A*, 147:278–292, 1984. (Cited on page 117)

177 A. Dawid. Probability, symmetry, and frequency. *British Journal for the Philosophy of Science*, 36:107–128, 1985. (Cited on page 71)

178 A. Dawid. Probability forecasting. In S. Kotz, N. Johnson, and C. Read, editors, *Encyclopedia of Statistical Sciences*, volume 7, pages 210–218. Wiley, New York, 1986. (Cited on page 117)

179 A. Dawid. Conditional independence. In S. Kotz, C. Read, and D. Banks, editors, *Encyclopedia of Statistical Sciences, Update Volume 2*, pages 146–153. Wiley, 1998. (Cited on page 77)

180 A. Dawid, S. de Rooij, G. Shafer, A. Shen, N. Vereshchagin, and V. Vovk. Insuring against loss of evidence in game-theoretic probability. *Statistics and Probability Letters*, 81:157–162, 2011. (Cited on page 133)

181 A. Dawid and V. Vovk. Prequential probability: Principles and properties. *Bernoulli*, 5:125–162, 1999. (Cited on page 116)

182 J. De Bock and G. de Cooman. State sequence prediction in imprecise hidden Markov models. In F. Coolen, G. de Cooman, T. Fetz, and M. Oberguggenberger, editors, *ISIPTA '11: Proceedings of the Seventh International Symposium on Imprecise Probability: Theories and Applications*, pages 159–168, Innsbruck, 2011. SIPTA. (Cited on pages 227 and 277)

183 D. D. Mauà, C. P. de Campos, A. Benavoli, A. Antonucci. On the complexity of strong and epistemic credal networks. In A. Nicholson and P. Smyth, editors, *Proceedings of the 29th Conference on Uncertainty in Artificial Intelligence*, pages 391–400. AUAI Press, 2013. (Cited on pages 65, 223, and 227)

184 C. de Campos and F. Cozman. Inference in credal networks using multilinear programming. In E. Onaindia and S. Staab, editors, *STAIRS '04: Proceedings of the Second Starting Artificial Intelligence Researcher Symposium*, volume 109, pages 50–61, Amsterdam, 2004. IOS Press. (Cited on page 223)

185 C. de Campos and F. Cozman. Belief updating and learning in semi-qualitative probabilistic networks. In F. Bacchus and T. Jaakkola, editors, *UAI '05: Proceedings of the 21st Conference in Uncertainty in Artificial Intelligence*, pages 153–160, Arlington, Virginia, 2005. AUAI Press. (Cited on page 228)

186 C. de Campos and F. Cozman. The inferential complexity of Bayesian and credal networks. In L. Kaelbling and A. Saffiotti, editors, *IJCAI '05: Proceedings of the International Joint Conference on Artificial Intelligence*, pages 1313–1318, Edinburgh, 2005. (Cited on pages 65, 220, 223, and 228)

187 C. de Campos and F. Cozman. Computing lower and upper expectations under epistemic independence. *International Journal of Approximate Reasoning*, 44:244–260, 2007. (Cited on pages 65 and 227)

188 C. de Campos and F. Cozman. Inference in credal networks through integer programming. In G. de Cooman, J. Vejnarová, and M. Zaffalon, editors, *ISIPTA '07: Proceedings of the Fifth International Symposium on Imprecise Probability: Theories and Applications*, pages 145–154, Prague, 2007. Action M Agency for SIPTA. (Cited on page 223)

189 C. de Campos, F. Cozman, and J. Luna. Assembling a consistent set of sentences in relational probabilistic logic with stochastic independence. *Journal of Applied Logic*, 7:137–154, 2009. (Cited on page 229)

190 C. de Campos and Q. Ji. Strategy selection in influence diagrams using imprecise probabilities. In D. McAllester and P. Myllymaki, editors, *UAI '08: Proceedings of the Twenty-Fourth Conference in Uncertainty in Artificial Intelligence*, pages 121–128, Corvallis, OR, 2008. AUAI Press. (Cited on page 228)

191 C. de Campos, L. Zhang, Y. Tong, and Q. Ji. Semi-qualitative probabilistic networks in computer vision problems. *Journal of Statistical Theory and Practice*, 3:197–210, 2009. Reprinted in [146] pages 207–220. (Cited on pages 228 and 229)

192 L. de Campos and J. Huete. Independence concepts in possibility theory: part I. *Fuzzy Sets and Systems*, 103:127–152, 1999. (Cited on page 108)

193 L. de Campos and J. Huete. Independence concepts in possibility theory: part II. *Fuzzy Sets and Systems*, 103:487–505, 1999. (Cited on pages 108 and 109)

194 L. de Campos, J. Huete, and S. Moral. Probability intervals: A tool for uncertain reasoning. *International Journal of Uncertainty, Fuzziness and Knowledge-Based Systems*, 2:167–196, 1994. (Cited on pages 82, 84, 223, and 246)

195 L. de Campos, M. Lamata, and S. Moral. The concept of conditional fuzzy measure. *International Journal of Intelligent Systems*, 5:237–246, 1990. (Cited on page 105)

196 L. de Campos and S. Moral. Independence concepts for convex sets of probabilities. In P. Besnard and S. Hanks, editors, *UAI '95: Proceedings of Eleventh Conference on Uncertainty in Artificial Intelligence*, pages 108–115, San Francisco, CA, 1995. Morgan Kaufmann. (Cited on pages 63, 65, and 77)

197 G. de Cooman. On modeling possibilistic uncertainty in two-state reliability theory. *Fuzzy Sets and Systems*, 83:215–238, 1996. (Cited on page 303)

198 G. de Cooman. Possibility theory I: Measure and integral theoretics groundwork. *International Journal of General Systems*, 25:291–323, 1997. (Cited on pages 96, 100, and 113)

199 G. de Cooman. Possibility theory II: Conditional possibility. *International Journal of General Systems*, 25:325–351, 1997. (Cited on pages 100 and 113)

200 G. de Cooman. Possibility theory III: Possibilistic independence. *International Journal of General Systems*, 25:353–371, 1997. (Cited on pages 77, 100, 108, and 113)

201 G. de Cooman. Belief models: An order-theoretic investigation. *Annals of Mathematics and Artificial Intelligence*, 45:5–34, 2005. (Cited on page 26)

202 G. de Cooman and D. Aeyels. Supremum preserving upper probabilities. *Information Sciences*, 118:173–212, 1999. (Cited on pages 85 and 103)

203 G. de Cooman and F. Hermans. Imprecise probability trees: Bridging two theories of imprecise probability. *Artificial Intelligence*, 172:1400–1427, 2008. (Cited on pages 115, 119, 259, 261, 262, 263, and 278)

204 G. de Cooman, F. Hermans, A. Antonucci, and M. Zaffalon. Epistemic irrelevance in credal networks: The case of imprecise Markov trees. *International Journal of Approximate Reasoning*, 51:1029–1052, 2010. (Cited on pages 52, 60, 66, 223, 227, and 277)

205 G. de Cooman, F. Hermans, and E. Quaeghebeur. Imprecise Markov chains and their limit behavior. *Probability in the Engineering and Informational Sciences*, 23:597–635, 2009. (Cited on pages 261, 264, 265, 268, 272, and 277)

206 G. de Cooman and E. Miranda. Symmetry of models versus models of symmetry. In W. Harper and G. Wheeler, editors, *Probability and Inference: Essays in Honor of Henry E. Kyburg, Jr.*, pages 67–149. King's College Publications, London, 2007. (Cited on pages 26, 69, 70, 71, 75, and 77)

207 G. de Cooman and E. Miranda. Weak and strong laws of large numbers for coherent lower previsions. *Journal of Statistical Planning and Inference*, 138:2409–2432, 2008. (Cited on pages 45, 49, 58, 61, 66, and 114)

208 G. de Cooman and E. Miranda. Forward irrelevance. *Journal of Statistical Planning and Inference*, 139:256–276, 2009. (Cited on pages 60 and 61)

209 G. de Cooman and E. Miranda. Irrelevant and independent natural extension for sets of desirable gambles. *Journal of Artificial Intelligence Research*, 45:601–640, 2012. (Cited on pages 60, 63, and 66)

210 G. de Cooman, E. Miranda, and M. Zaffalon. Independent natural extension. *Artificial Intelligence*, 174:1911–1950, 2011. (Cited on pages 54, 62, 63, 66, 67, 77, and 227)

211 G. de Cooman and E. Quaeghebeur. Exchangeability for sets of desirable gambles. In T. Augustin, F. Coolen, S. Moral, and M. Troffaes, editors, *ISIPTA '09: Proceedings of the Sixth International Symposium on Imprecise Probabilities: Theories and Applications*, pages 159–168, Durham, UK, 2009. SIPTA. (Cited on page 26)

212 G. de Cooman and E. Quaeghebeur. Infinite exchangeability for sets of desirable gambles. In E. Hüllermeier, R. Kruse, and F. Hoffmann, editors, *Communications in Computer and Information Science*, volume 80, pages 60–69, Berlin, 2010. Springer. (Cited on pages 11 and 26)

213 G. de Cooman and E. Quaeghebeur. Exchangeability and sets of desirable gambles. *International Journal of Approximate Reasoning*, 53:363–395, 2011. (Cited on pages 5, 6, 14, 16, 17, 26, 68, 71, and 78)

214 G. de Cooman, E. Quaeghebeur, and E. Miranda. Exchangeable lower previsions. *Bernoulli*, 15:721–735, 2009. (Cited on pages 11, 58, 71, 74, 76, and 78)

215 G. de Cooman and M. Zaffalon. Updating beliefs with incomplete observations. *Artificial Intelligence*, 159:75–125, 2004. (Cited on pages 52, 182, and 222)

216 B. de Finetti. La prévision: Ses lois logiques, ses sources subjectives. *Annales de l'Institut Henri Poincaré*, 7:1–68, 1937. English translation in [421]. (Cited on pages 28, 71, 72, 75, 76, 77, and 112)

217 B. de Finetti. *Teoria delle Probabilità*. Einaudi, Turin, 1970. (Cited on page 349)

218 B. de Finetti. *Theory of Probability: A Critical Introductory Treatment*. Wiley, Chichester, 1974–1975. English translation of [217]. (Cited on pages 28, 34, 71, 77, 93, and 281)

219 R. Dechter. Bucket elimination: A unifying framework for probabilistic inference. In E. Horvitz and F. Jensen, editors, *UAI '96: Proceedings of the Twelfth Conference on Uncertainty in Artificial Intelligence*, pages 211–219, San Francisco, 1996. Morgan Kaufmann. (Cited on pages 223, 224, and 225)

220 K. Delgado, L. de Barros, F. Cozman, and R. Shirota. Representing and solving factored Markov decision processes with imprecise probabilities. In T. Augustin, F. Coolen, S. Moral, and M. Troffaes, editors, *ISIPTA '09: Proceedings of the Sixth International Symposium on Imprecise Probability: Theories and Applications*, pages 169–178, Durham, UK, 2009. SIPTA. (Cited on page 228)

221 K. Delgado, S. Sanner, L. de Barros, and F. Cozman. Efficient solutions to factored MDPs with imprecise transition probabilities. In A. Gerevini, A. Howe, A. Cesta, and I. Refanidis, editors, *ICAPS '09: Proceedings of the Nineteenth International Conference on Automated Planning and Scheduling*, pages 98–105. AAAI Press, 2009. (Cited on page 229)

222 A. Dempster. Upper and lower probabilities induced by a multivalued mapping. *Annals of Mathematical Statistics*, 38:325–339, 1967. (Cited on pages 83, 84, 99, 106, 179, 242, and xiv)

223 A. Dempster. A generalization of Bayesian inference. *Journal of the Royal Statistical Society, Series B*, 30:205–247, 1968. (Cited on page 179)

224 Janez Demšar. Statistical comparisons of classifiers over multiple data sets. *Journal of Machine Learning Research*, 7:1–30, 2006. (Cited on page 250)

225 T. Denoeux. Inner and outer approximation of belief structures using a hierarchical clustering approach. *International Journal of Uncertainty, Fuzziness and Knowledge-Based Systems*, 9:437–460, 2001. (Cited on page 84)

226 T. Denoeux. Constructing belief functions from sample data using multinomial confidence regions. *International Journal of Approximate Reasoning*, 42:228–252, 2006. (Cited on page 84)

227 T. Denoeux. Conjunctive and disjunctive combination of belief functions induced by nondistinct bodies of evidence. *Artificial Intelligence*, 172:234–264, 2008. (Cited on page 111)

228 M. Denuit, J. Dhaene, M. Goovaerts, and R. Kaas. *Actuarial Theory for Dependent Risks: Measures, Orders and Models*. Wiley, Chichester, 2005. (Cited on page 289)

229 L. DeRobertis and J. Hartigan. Bayesian inference using intervals of measures. *The Annals of Statistics*, 9:235–244, 1981. (Cited on page 165)

230 S. Destercke and D. Dubois. Idempotent conjunctive combination of belief functions: Extending the minimum rule of possibility theory. *Information Sciences*, 19:4075–4100, 2011. (Cited on page 111)

231 S. Destercke, D. Dubois, and E. Chojnacki. Unifying practical uncertainty representations: I. Generalized p-boxes. *International Journal of Approximate Reasoning*, 49:649–663, 2008. (Cited on page 88)

232 S. Destercke, D. Dubois, and E. Chojnacki. Unifying practical uncertainty representations: II. Clouds. *International Journal of Approximate Reasoning*, 49:664–667, 2008. (Cited on pages 87 and 89)

233 S. Destercke and O. Strauss. Using cloudy kernels for imprecise linear filtering. In E. Hüllermeier, R. Kruse, and F. Hoffmann, editors, *IPMU '10: Proceedings of the 13th International Conference on Information Processing and Management of Uncertainty*, pages 198–207, New York, 2010, Springer. (Cited on page 88)

234 P. Diaconis and D. Freedman. Finite exchangeable sequences. *The Annals of Probability*, 8:745–764, 1980. (Cited on page 78)

235 A. Dobra, E. Erosheva, and S. Fienberg. Disclosure limitation methods based on bounds for large contingency tables with applications to disability. In H. Bozdogan, editor, *Statistical Data Mining and Knowledge Discovery*, pages 93–116, Boca Raton, FL, 2004. Chapman & Hall/CRC. (Cited on page 180)

236 A. Dobra and S. Fienberg. Bounds for cell entries in contingency tables given marginal totals and decomposable graphs. *Proceedings of the National Academy of Sciences of the United States of America*, 97:11885–11892, 2000. (Cited on page 180)

237 P. Domingos and M. Pazzani. On the optimality of the simple Bayesian classifier under zero-one loss. *Machine Learning*, 29:103–130, 1997. (Cited on page 231)

238 L. Dubins. Finitely additive conditional probabilities, conglomerability and disintegrations. *Annals of Probability*, 3:89–99, 1975. (Cited on pages 51 and 58)

239 D. Dubois. Belief structure, possibility theory and decomposable confidence measures on finite sets. *Computers and Artificial Intelligence*, 5:403–416, 1986. (Cited on pages 112, 113, and 180)

240 D. Dubois, H. Fargier, and H. Prade. Ordinal and probabilistic representations of acceptance. *Journal of Artificial Intelligence Research*, 22:23–56, 2004. (Cited on page 100)

241 D. Dubois, L. Fariñas Del Cerro, A. Herzig, and H. Prade. A roadmap of qualitative independence. In D. Dubois, H. Prade, and E. Klement, editors, *Fuzzy Sets, Logics and Reasoning about Knowledge*, pages 325–350. Kluwer, Dordrecht, 1999. (Cited on page 113)

242 D. Dubois, J. Fortin, and P. Zielinski. Interval PERT and its fuzzy extension. In C. Kahraman and M. Yavuz, editors, *Production Engineering and Management Under Fuzziness*, volume 252, pages 171–199. Springer, Berlin, 2010. (Cited on page 87)

243 D. Dubois, L. Foulloy, G. Mauris, and H. Prade. Probability-possibility transformations, triangular fuzzy sets, and probabilistic inequalities. *Reliable Computing*, 10:273–297, 2004. (Cited on page 85)

244 D. Dubois, S. Moral, and H. Prade. A semantics for possibility theory based on likelihoods. *Journal of Mathematical Analysis and Applications*, 205:359–380, 1997. (Cited on page 101)

245 D. Dubois, S. Moral, and H. Prade. Belief change rules in ordinal and numerical uncertainty theories. In D. Dubois and H. Prade, editors, *Handbook of Defeasible Reasoning and Uncertainty Management Systems. Volume 3: Belief Change*, pages 311–392. Kluwer, Dordrecht, 1998. (Cited on pages 104 and 108)

246 D. Dubois, H. Nguyen, and H. Prade. Possibility theory, probability and fuzzy sets: Misunderstandings, bridges and gaps. In D. Dubois and H. Prade, editors, *Fundamentals of Fuzzy Sets*, pages 343–438. Kluwer, Dordrecht, 2000. (Cited on pages 84 and 100)

247 D. Dubois and H. Prade. A set-theoretic view of belief functions: Logical operations and approximation by fuzzy sets. *International Journal of General Systems*, 12:193–226, 1986. (Cited on page 111)

248 D. Dubois and H. Prade. Properties of measures of information in evidence and possibility theory. *Fuzzy Sets and Systems*, 24:161–182, 1987. (Cited on pages 84 and 100)

249 D. Dubois and H. Prade. *Possibility Theory: An Approach to Computerized Processing of Uncertainty*. Plenum, New York, 1988. (Cited on pages 94, 96, 101, and 295)

250 D. Dubois and H. Prade. Representation and combination of uncertainty with belief functions and possibility measures. *Computational Intelligence*, 4:244–264, 1988. (Cited on page 110)

251 D. Dubois and H. Prade. Possibility theory and data fusion in poorly informed environments. *Control Engineering Practice*, 2:811–823, 1994. (Cited on page 110)

252 D. Dubois and H. Prade. When upper probabilities are possibility measures. *Fuzzy Sets and Systems*, 49:65–74, 1996. (Cited on pages 105 and 107)

253 D. Dubois and H. Prade. Possibility theory: Qualitative and quantitative aspects. In D. Gabbay and P. Smets, editors, *Handbook of Defeasible Reasoning and Uncertainty Management Systems*, volume 1, pages 169–226. Kluwer, Dordrecht, 1998. (Cited on pages 84 and 100)

254 D. Dubois and H. Prade. An introduction to bipolar representations of information and preference. *International Journal of General Systems*, 23:866–877, 2008. (Cited on page 95)

255 D. Dubois, H. Prade, and P. Smets. Representing partial ignorance. *IEEE Transactions on Systems, Man, and Cybernetics*, 26:361–377, 1996. (Cited on pages 95 and 104)

256 N. Dunford and J. Schwartz. *Linear Operators Part I*. Interscience, New York, 1958. (Cited on page 1)

257 A. Edwards. *Likelihood*. Johns Hopkins University Press, Baltimore, MD, 1992. (Cited on page 179)

258 H. Einstein. Quantifying uncertain engineering geologic information. *Felsbau*, 19:72–84, 2001. (Cited on page 291)

259 I. Elishakoff. Are probabilistic and anti-optimization approaches compatible? In I. Elishakoff, editor, *Whys and Hows in Uncertainty Modelling*, pages 263–355. Springer, Wien, 1999. (Cited on page 303)

260 I. Elishakoff, editor. *Whys and Hows in Uncertainty Modelling: Probability, Fuzziness and Anti-Optimization*, volume 388, Wien, 1999. Springer. (Cited on page 304)

261 I. Elishakoff and M. Ohsaki. *Optimization and Anti-Optimization of Structures Under Uncertainty*. Imperial College Press, London, 2010. (Cited on page 304)

262 I. Elishakoff and C. Soize, editors. *Nondeterministic Mechanics*, Wien, 2012. Springer. (Cited on page 304)

263 C. Elkan. Magical thinking in data mining: Lessons from CoIL challenge 2000. In *Proceedings of the Seventh ACM SIGKDD International Conference on Knowledge Discovery and Data Mining*, pages 426–431, New York, 2001. ACM. (Cited on page 231)

264 D. Ellsberg. Risk, ambiguity, and the Savage axioms. *Quarterly Journal of Economics*, 75:643–669, 1961. (Cited on pages 94, 204, and xiv)

265 N. Etchart-Vincent. Is probability weighting sensitive to the magnitude of consequences? An experimental investigation on losses. *Journal of Risk and Uncertainty*, 28:217–235, 2004. (Cited on page 324)

266 European Committee for Standardization. *EN 1990:2002. Eurocode – Basis of Structural Design*. CEN, Brussels, 2002. (Cited on pages 293, 302, and 303)

267 M. Evans and H. Moshonov. Checking for prior-data conflict. *Bayesian Analysis*, 1:893–914, 2006. (Cited on page 145)

268 R. Fagin and J. Halpern. A new approach to updating beliefs. In P. Bonissone, M. Henrion, L. Kanal, and J. Lemmer, editors, *Uncertainty in Artificial Intelligence*, volume 6, pages 347–374. North-Holland, Amsterdam, 1991. (Cited on pages 52, 105, and 108)

269 L. Fahrmeir, T. Kneib, S. Lang, and B. Marx. *Regression – Models, Methods and Applications*. Springer, Berlin, 2013. (Cited on pages 139 and 187)

270 U. Fayyad and K. Irani. Multi-interval discretization of continuous-valued attributes for classification learning. In R. Bajcsy, editor, *IJCAI '93: Proceedings of the 13th International Joint Conference on Artificial Intelligence*, pages 1022–1027, San Mateo, CA, 1993. Morgan Kaufmann. (Cited on page 252)

271 W. Feller. On the Kolmogorov-Smirnov limit theorems for empirical distributions. *Annals of Mathematical Statistics*, 19:177–189, 1948. (Cited on page 89)

272 W. Fellin, H. Lessmann, M. Oberguggenberger, and R. Vieider, editors, *Analyzing Uncertainty in Civil Engineering*. Springer, Berlin, 2005. (Cited on page 304)

273 W. Fellin and M. Oberguggenberger. From probability to fuzzy sets: The struggle for meaning in geotechnical risk assessment. In R. Pöttler, H. Klapperich, and H. Schweiger, editors, *Probabilistics in Geotechnics: Technical and Economic Risk Estimation*, pages 29–38. Glückauf, Essen, 2002. (Cited on page 303)

274 W. Fellin and M. Oberguggenberger. The fuzziness and sensitivity of failure probabilities. In W. Fellin, H. Lessmann, M. Oberguggenberger, and R. Vieider, editors, *Analyzing Uncertainty in Civil Engineering*, pages 33–49. Springer, Berlin, 2005. (Cited on page 303)

275 J. Ferreira da Rocha and F. Cozman. Inference in credal networks: Branch-and-bound methods and the a/r+ algorithm. *International Journal of Approximate Reasoning*, 39:279–296, 2005. (Cited on page 223)

276 J. Ferreira da Rocha, F. Cozman, and C. de Campos. Inference in polytrees with sets of probabilities. In C. Meek and U. Kjærulff, editors, *UAI '03 : Proceedings of the Nineteenth Conference on Uncertainty in Artificial Intelligence*, pages 217–224, San Francisco, CA, 2003. Morgan Kaufmann. (Cited on page 223)

277 S. Ferson, L. Ginzburg, V. Kreinovich, D. Myers, and K. Sentz. Constructing probability boxes and Dempster-Shafer structures. Technical Report SAND2002-4015 (available from http://ebookbrowse.com/sand2002-4015-pdf-d22644198), Sandia National Laboratories, 2003. (Cited on pages 88, 89, 294, 299, and 303)

278 S. Ferson and W. Tucker. Sensitivity analysis using probability bounding. *Reliability Engineering and System Safety*, 91:1435–1442, 2006. (Cited on page 303)

279 T. Fetz. Sets of joint probability measures generated by weighted marginal focal sets. In G. de Cooman, T. Fine, and T. Seidenfeld, editors, *ISIPTA '01: Proceedings of the Second International Symposium on Imprecise Probabilities and Their Applications*, pages 171–178, Maastricht, 2001. Shaker. (Cited on page 109)

280 T. Fetz. Modelling uncertainties in limit state functions. In F. Coolen, G. de Cooman, T. Fetz, and M. Oberguggenberger, editors, *ISIPTA'11: Proceedings of the Seventh International Symposium on Imprecise Probability: Theory and Applications*, pages 179–188, Innsbruck, 2011. SIPTA. (Cited on page 303)

281 T. Fetz. Multivariate models and variability intervals: A local random set approach. *International Journal of Uncertainty, Fuzziness and Knowledge-based Systems*, 19:799–823, 2011. (Cited on page 299)

282 T. Fetz. Modelling uncertainties in limit state functions. *International Journal of Approximate Reasoning*, 53:1–23, 2012. (Cited on page 303)

283 T. Fetz, J. Jäger, D. Köll, G. Krenn, M. Oberguggenberger, and H. Lessmann. Fuzzy models in geotechnical engineering and construction management. *Computer-Aided Civil & Infrastructure Engineering*, 14:93–106, 1999. (Cited on page 303)

284 T. Fetz and M. Oberguggenberger. Propagation of uncertainty through multivariate functions in the framework of sets of probability measures. *Reliability Engineering and System Safety*, 85:73–87, 2004. (Cited on pages 299 and 303)

285 T. Fetz and M. Oberguggenberger. Multivariate models of uncertainty: A local random set approach. *Structural Safety*, 32:417–424, 2010. (Cited on page 299)

286 T. Fetz and F. Tonon. Probability bounds for series systems with variables constrained by sets of probability measures. *International Journal of Reliability and Safety*, 2:309–339, 2008. (Cited on pages 303 and 311)

287 P. Fierens. An extension of chaotic probability models to real-valued variables. *International Journal of Approximate Reasoning*, 50:627–641, 2009. (Cited on pages 175 and 188)

288 P. Fierens and T. Fine. Towards a frequentist interpretation of sets of measures. In G. de Cooman, T. Fine, and T. Seidenfeld, editors, *ISIPTA'01: Proceedings of the Second International Symposium on Imprecise Probabilities and Their Applications*, pages 179–187, Maastricht, 2001. Shaker. (Cited on pages 175 and 188)

289 P. Fierens and T. Fine. Towards a chaotic probability model for frequentist probability: The univariate case. In J. Bernard, T. Seidenfeld, and M. Zaffalon, editors, *ISIPTA'03: Proceedings of the Third International Symposium on Imprecise Probabilities and Their Applications*, pages 245–259, Waterloo, 2003. Carleton Scientific. (Cited on pages 175 and 188)

290 T. Fine. *Theories of Probability: An Examination of Foundations*. Academic Press, New York, 1973. (Cited on pages 102, 113, and 187)

291 B. Fischhoff, P. Slovic, and S. Lichtenstein. Fault trees: Sensitivity of estimated failure probabilities to problem representation. *Journal of Experimental Psychology: Human Perception Performance*, 4:330–344, 1978. (Cited on page 326)

292 P. Fishburn. The axioms of subjective probability. *Statistical Science*, 1:335–358, 1986. (Cited on pages 26, and 113)

293 R. Fisher. Inverse probability. *Proceedings of the Cambridege Philosophical Society*, 26:528–535, 1930. (Cited on page 179)

294 R. Fisher. The concepts of inverse probability and fiducial probability referring to unknown parameters. *Proceedings of the Royal Society of London Series A*, 139:343–348, 1933. (Cited on page 179)

295 F. Fleuret. Fast binary feature selection with conditional mutual information. *Journal of Machine Learning Research*, 5:1531–1555, 2004. (Cited on page 231)

296 H. Föllmer and A. Schied. Convex measures of risk and trading constraints. *Finance and Stochastics*, 6:429–447, 2002. (Cited on pages 285 and 287)

297 H. Föllmer and A. Schied. Robust preferences and convex measures of risk. In K. Sandmann and P. Schönbucher, editors, *Advances in Finance and Stochastics*, pages 39–56. Springer, Berlin, 2002. (Cited on pages 285 and 287)

298 C. Fox and R. Clemen. Subjective probability assessment in decision analysis: Partition dependence and bias toward the ignorance prior. *Management Science*, 51:1417–1432, 2005. (Cited on page 326)

299 C. Fox and Y. Rottenstreich. Partition priming in judgment under uncertainty. *Psychological Science*, 13:195–200, 2003. (Cited on pages 326 and 328)

300 E. Frank, M. Hall, and B. Pfahringer. Locally weighted naive Bayes. In C. Meek and U. Kjærulff, editors, *UAI '03: Proceedings of the 19th Conference on Uncertainty in Artificial Intelligence*, pages 249–256, San Francisco, CA, 2003. Morgan Kaufmann. (Cited on page 237)

301 A. Freudenthal. Safety and the probability of structural failure. *Transactions of the American Society of Civil Engineers*, 121:1337–1397, 1956. (Cited on page 292)

302 J. Friedman. On bias, variance, 0/1-loss, and the curse-of-dimensionality. *Data Mining and Knowledge Discovery*, 1:55–77, 1997. (Cited on page 231)

303 N. Friedman, D. Geiger, and M. Goldszmidt. Bayesian network classifiers. *Machine learning*, 29:131–163, 1997. (Cited on page 240)

304 M. Frittelli and E. Rosazza Gianin. On the penalty function and on continuity properties of risk measures. *International Journal of Theoretical and Applied Finance*, 14:163–185, 2011. (Cited on page 289)

305 M. Fuchs. Clouds, p-boxes, fuzzy sets, and other uncertainty representations in higher dimensions. *Acta Cybernetica*, 19:61–92, 2009. (Cited on page 303)

306 M. Fuchs and A. Neumaier. Potential based clouds in robust design optimization. *Journal of Statistical Theory and Practice*, 3:225–238, 2008. Reprinted in [146], pages 285–298. (Cited on page 88)

307 P. Garthwaite and A. O'Hagan. Quantifying expert opinion in the UK water industry: An experimental study. *The Statistician*, 49:455–477, 2000. (Cited on page 325)

308 S. Geisser. *Predictive Inference*. Chapman & Hall/CRC, Boca Raton, FL, 1993. (Cited on page 139)

309 R. Ghanem and P. Spanos. *Stochastic Finite Elements: A Spectral Approach*. Springer, New York, 1991. (Cited on page 303)

310 G. Gigerenzer and R. Selten. *Bounded Rationality: The Adaptive Toolbox*. MIT Press, Cambridge, MA, 2001. (Cited on page 327)

311 I. Gilboa and D. Schmeidler. Updating ambigous beliefs. *Journal of Economic Theory*, 59:33–49, 1993. (Cited on page 107)

312 W. Gilks, S. Richardson and D. Spiegelhalter, editors. *Markov Chain Monte Carlo in Practice*. Chapman & Hall/CRC, Boca Raton, FL, 1998. (Cited on page 155)

313 D. Gillies. Was Bayes a Bayesian? *Historia Mathematica*, 14:325–346, 1987. (Cited on page 140)

314 D. Gillies. *Philosophical Theories of Probability*. Routledge, New York, 2000. (Cited on pages 140 and 187)

315 F. Girón and S. Rios. Quasi-Bayesian behaviour: A more realistic approach to decision making? *Trabajos de Estadística y de Investigación Operativa*, 31:17–38, 1980. (Cited on page 26)

316 R. Givan, S. Leach, and T. Dean. Bounded parameter Markov decision processes. In S. Steel and R. Alami, editors, *Recent Advances in Artificial Intelligence Planning*, volume 1348, pages 234–246. Springer, Berlin, 1997. (Cited on page 228)

317 R. Gonzales and G. Wu. On the shape of the probability weighting function. *Cognitive Psychology*, 38:129–166, 1999. (Cited on pages 324 and 328)

318 I. Goodman and H. Nguyen. Fuzziness and randomness. In C. Bertoluzza, M. Gil, and D. Ralescu, editors, *Statistical Modeling Analysis and Management of Fuzzy Data*, pages 3–21. Physica, Heidelberg, 2002. (Cited on page 295)

319 M. Grabisch. K-order additive discrete fuzzy measures and their representation. *Fuzzy Sets and Systems*, 92:167–189, 1997. (Cited on page 84)

320 S. Greenland. Identifiability. In M. Lovric, editor, *International Encyclopedia of Statistical Science*, page 645. Springer, Berlin, 2011. (Cited on page 183)

321 P. Gustafson. Sample size implications when biases are modelled rather than ignored. *Journal of the Royal Statistical Society, Series A*, 169:865–881, 2006. (Cited on page 187)

322 R. Hable. *imprProbEst: Minimum distance estimation in an imprecise probability model*, 2008. Contributed R-Package on CRAN, Version 1.0, 2008-10-23 (available from http://cran.r-project.org/web/packages/imprProbEst/index.html); maintainer Hable, R. (Cited on page 174)

323 R. Hable. *Data-based Decisions under Complex Uncertainty*. PhD thesis, LMU Munich (available from http://edoc.ub.uni-muenchen.de/9874/), 2009. (Cited on pages 135 and 148)

324 R. Hable. Minimum distance estimation in imprecise probability models. *Journal of Statistical Planning and Inference*, 140:461–479, 2010. (Cited on page 174)

325 R. Hable. A minimum distance estimator in an imprecise probability model: Computational aspects and applications. *International Journal of Approxmimate Reasoning*, 51:111–1128, 2010. (Cited on page 174)

326 I. Hacking. *The Emergence of Probability*. Cambridge University Press, Cambridge, UK, 1975. (Cited on page 103)

327 R. Hafner. Konstruktion robuster Teststrategien. In S. Schach and G. Trenkler, editors, *Data Analysis and Statistical Inference (Festschrift for Friedhelm Eicker)*, pages 145–160. Eul, Bergisch Gladbach, 1992. (Cited on page 171)

328 K. Haldar and R. Reddy. A random-fuzzy analysis of existing structures. *Fuzzy Sets and Systems*, 48:201–210, 1992. (Cited on page 294)

329 J. Hall. Uncertainty-based sensitivity indices for imprecise probability distributions. *Reliability Engineering & System Safety*, 91:1443–1451, 2006. (Cited on page 294)

330 J. Hall, D. Blockley, and J. Davis. Uncertain inference using interval probability theory. *International Journal of Approximate Reasoning*, 19:247–264, 1998. (Cited on page 303)

331 J. Hall and J. Lawry. Imprecise probabilities of engineering system failure from random and fuzzy set reliability analysis. In G. de Cooman, T. Fine, and T. Seidenfeld, editors, *ISIPTA '01: Proceedings of the Second International Symposium on Imprecise Probabilities and their Applications*, pages 195–204, Maastricht, 2001. Shaker. (Cited on page 303)

332 J. Hall and J. Lawry. Generation, combination and extension of random set approximations to coherent lower and upper probabilities. *Reliability Engineering & System Safety*, 85:89–101, 2004. (Cited on pages 84 and 303)

333 J. Hall, E. Rubio, and M. Anderson. Random sets of probability measures in slope hydrology and stability analysis. *Journal of Applied Mathematics and Mechanics*, 84:710–720, 2004. (Cited on pages 294 and 303)

334 M. Hall and G. Holmes. Benchmarking attribute selection techniques for discrete class data mining. *IEEE Transactions on Knowledge and Data Engineering*, 15:1437–1447, 2003. (Cited on page 252)

335 P. Halmos. *Measure Theory*. Van Nostrand, New York, 1950. (Cited on pages 123 and 124)

336 J. Halpern. Lexicographic probability, conditional probability, and nonstandard probability. *Games and Economic Behavior*, 68:155–179, 2010. (Cited on page 27)

337 C. Hamblin. The modal "probably". *Mind*, 60:234–240, 1959. (Cited on page 94)

338 P. Hammond. Consequentialist foundations for expected utility. *Theory and Decision*, 25:25–78, 1988. (Cited on page 199)

339 O. Hamouda and J. Rowley. *Paradoxes, Ambiguity and Rationality*. Edward Elgar, Cheltenham, 1997. (Cited on page 204)

340 F. Hampel. On the foundations of statistics: A frequentist approach. In *Estatistica: A Diversidade Na Unidate*, pages 77–97. Salamandra, Lisbon, (available also from: ftp://stat.ehtz.ch/Research-Reports/85.pdf) 1998. (Cited on page 179)

341 F. Hampel. An outline of a unifying statistical theory. In G. de Cooman, T. Fine, and T. Seidenfeld, editors, *ISIPTA '01: Proceedings of the Second International Symposium on Imprecise Probabilities and their Applications*, pages 205–212, Ithaca, NY, 2001. Shaker. (Cited on page 179)

342 F. Hampel. The proper fiducial argument. In R. Ahlswede, L. Bäumer, N. Cai, H. Aydinian, V. Blinovsky, C. Deppe, and H. Mashurian, editors, *General Theory of Information Transfer and Combinatorics*, volume 4123, pages 512–526. Springer, Berlin, 2006. (Cited on page 179)

343 F. Hampel. Nonadditive probabilities in statistics. *Journal of Statistical Theory and Practice*, 3:11–23, 2009. Reprinted in [146], pages 13–25. (Cited on pages 187 and xiv)

344 F. Hampel, E. Ronchetti, P. Rousseeuw, and W. Stahel. *Robust Statistics: The Approach Based on Influence Functions*. Wiley, New York, 1986. (Cited on pages 144 and 187)

345 D. Hand and K. Yu. Idiot's Bayes - not so stupid after all? *International Statistical Review*, 69:385–398, 2001. (Cited on page 231)

346 M. Hanns. *Applied Fuzzy Arithmetic. An Introduction with Engineering Applications*. Springer, Berlin, 2005. (Cited on page 304)

347 D. Harmanec. Generalizing Markov decision processes to imprecise probabilities. *Journal of Statistical Planning and Inference*, 105:199–213, 2002. (Cited on page 264)

348 D. Harmanec and G. Klir. Measuring total uncertainty in Dempster-Shafer theory. *International Journal of General Systems*, 22:405–419, 1994. (Cited on page 243)

349 D. Hartfiel. On the solutions to $x'(t) = a(t)x(t)$ over all $a(t)$, where $p \leq a(t) \leq q$. *Journal of Mathematical Analysis and Applications*, 108:230–240, 1985. (Cited on page 277)

350 D. Hartfiel. *Markov Set-Chains*. Springer, Berlin, 1998. (Cited on pages 263, 268, 269, 272, and 275)

351 T. Hastie, R. Tibshirani, and J. Friedman. *The Elements of Statistical Learning. Data Mining, Inference, and Prediction*. Springer, New York, 2nd edition, 2009. (Cited on page 187)

352 D. Heath and W. Sudderth. De Finetti's theorem on exchangeable variables. *The American Statistician*, 30:188–189, 1976. (Cited on page 78)

353 D. Heckerman. A tutorial on learning Bayesian networks. Technical Report MSR-95-06, Microsoft Research, 1995. (Cited on page 238)

354 D. Heitjan and D. Rubin. Ignorability and coarse data. *The Annals of Statistics*, 19:2244–2253, 1991. (Cited on page 181)

355 H. Held, E. Kriegler, and T. Augustin. Bayesian learning for a class of priors with prescribed marginals. *International Journal of Approximate Reasoning*, 2008. (Cited on page 165)

356 J. Helton, J. Johnson, W. Oberkampf, and C. Sallaberry. Sensitivity analysis in conjunction with evidence theory representations of epistemic uncertainty. *Reliability Engineering and System Safety*, 91:1414–1434, 2006. (Cited on page 303)

357 J. Helton, J. Johnson, C. Sallaberry, and C. Storlie. Survey of sampling-based methods for uncertainty and sensitivity analysis. *Reliability Engineering & System Safety*, 91:1175–1209, 2006. (Cited on pages 301 and 303)

358 F. Hermans. *An operational approach to graphical uncertainty modelling*. PhD thesis, Ghent University (available from http://hdl.handle.net/1854/LU-2153963), 2012. (Cited on page 277)

359 F. Hermans and G. de Cooman. Characterisation of ergodic upper transition operators. *International Journal of Approximate Reasoning*, 53:573–583, 2012. (Cited on page 277)

360 E. Hewitt and L. Savage. Symmetric measures on Cartesian products. *Transactions of the American Mathematical Society*, 80:470–501, 1955. (Cited on page 78)

361 B. Hill. Posterior distribution of percentiles: Bayes' theorem for sampling from a population. *Journal of the American Statistical Association*, 63:677–691, 1968. (Cited on page 175)

362 B. Hill. De Finetti's theorem, induction, and $A_{(n)}$ or Bayesian nonparametric predictive inference (with discussion). In D. Lindley, J. Bernardo, M. DeGroot, and A. Smith, editors, *Bayesian Statistics 3*, pages 211–241. Oxford University Press, Oxford, 1988. (Cited on page 175)

363 Z. Hoare. Landscapes of naive Bayes classifiers. *Pattern Analysis & Applications*, 11:59–72, 2008. (Cited on page 231)

364 J. Hoeting, D. Madigan, A. Raftery, and C. Volinsky. Bayesian model averaging: A tutorial. *Statistical Science*, 14:382–401, 1999. (Cited on page 238)

365 P. Huber. A robust version of the probability ratio test. *Annals of Mathematical Statistics*, 36:1753–1758, 1965. (Cited on page 168)

366 P. Huber. Robust confidence limits. *Zeitschrift für Wahrscheinlichkeitstheorie und Verwandte Gebiete*, 10:269–278, 1968. (Cited on page 173)

367 P. Huber. The use of Choquet capacities in statistics. *Bulletin de l'Institut International de Statistique, Proceedings of the 39th Session of the International Statistical Institute (Vienna, 1973)*, 45:181–191, 1973. (Cited on pages 144 and xiv)

368 P. Huber. Kapazitäten statt Wahrscheinlichkeiten? Gedanken zur Grundlegung der Statistik. *Jahresbericht der Deutschen Mathematiker-Vereinigung*, 78:81–92, 1976/77. (Cited on page 144)

369 P. Huber. *Robust Statistics*. Wiley, New York, 1981. (Cited on pages 144 and 187)

370 P. Huber and V. Strassen. Minimax tests and the Neyman-Pearson lemma for capacities. *The Annals of Statistics*, 1:251–263, 1973. (Cited on page 170)

371 D. Hume. *A Treatise of Human Nature*. Everyman, London, UK, 1961. (Cited on pages 137 and 140)

372 N. Huntley and M. Troffaes. An efficient normal form solution to decision trees with lower previsions. In D. Dubois, M. Lubiano, H. Prade, M. Gil, P. Grzegorzewski, and O. Hryniewicz, editors, *Soft Methods for Handling Variability and Imprecision*, volume 48, pages 419–426. Springer, Berlin, 2008. (Cited on pages 199 and 228)

373 N. Huntley and M. Troffaes. Characterizing factuality in normal form sequential decision making. In T. Augustin, F. Coolen, S. Moral, and M. Troffaes, editors, *ISIPTA '09: Proceedings of the Sixth International Symposium on Imprecise Probability: Theories and Applications*, pages 239–248, Durham, UK, 2009. SIPTA. (Cited on page 199)

374 J. Ide and F. Cozman. IPE and L2U: Approximate algorithms for credal networks. In E. Onaindia and F. Cozman, editors, *STAIRS '04: Proceedings of the Second Starting Artificial Intelligence Researcher Symposium*, volume 109, pages 118–127, Amsterdam, 2004. IOS Press. (Cited on page 224)

375 J. Ide and F. Cozman. Approximate algorithms for credal networks with binary variables. *International Journal of Approximate Reasoning*, 48:275–296, 2008. (Cited on page 223)

376 H. Itoh and K. Nakamura. Partially observable Markov decision processes with imprecise parameters. *Artificial Intelligence*, 171:453–490, 2007. (Cited on pages 228 and 264)

377 J.-Y. Jaffray. Bayesian updating and belief functions. *IEEE Transactions on Systems, Man and Cybernetics*, 22:1144–1152, 1992. (Cited on pages 52, 105, and 106)

378 J.-Y. Jaffray. Rational decision making with imprecise probabilities. In G. de Cooman, F. Cozman, S. Moral, and P. Walley, editors, *ISIPTA '99: Proceedings of the First International Symposium on Imprecise Probabilities and their Applications*, pages 183–188, Zwijnaarde, 1999. Universiteit Gent. (Cited on page 198)

379 J.-Y. Jaffray and M. Jeleva. Information processing under imprecise risk with the Hurwicz criterion. In G. de Cooman, J. Vejnarová, and M. Zaffalon, editors, *ISIPTA '07: Proceedings of the Fifth International Symposium on Imprecise Probability: Theories and Applications*, pages 233–242, Prague, 2007. Action M Agency for SIPTA. (Cited on page 193)

380 J.-Y. Jaffray and M. Jeleva. Information processing under imprecise risk with an insurance demand illustration. *International Journal of Approximate Reasoning*, 49:117–129, 2008. (Cited on page 290)

381 E. Jaynes. Information theory and statistical mechanics. In K. Ford, editor, *Statistical Physics*, pages 182–218. Benjamin, New York, 1963. (Cited on page 244)

382 G. Jeantet and O. Spanjaard. Optimizing the Hurwicz criterion in decision trees with imprecise probabilities. In F. Rossi and A. Tsoukiàs, editors, *ADT '09: Proceedings of the 1st International Conference on Algorithmic Decision Theory*, volume 5783, pages 340–352, Berlin, 2009. Springer. (Cited on page 228)

383 N. Johnson, S. Kotz, and N. Balakrishnan. *Discrete Multivariate Distributions*. Wiley, New York, 1997. (Cited on page 73)

384 P. Juslin, A. Winman, and P. Hansson. The naïve intuitive statistician: A naïve sampling model of intuitive confidence intervals. *Psychological Review*, 114:678–703, 2000. (Cited on pages 326 and 328)

385 J. Kadane, M. Schervish, and T. Seidenfeld. *Rethinking the Foundations of Statistics*. Cambridge University Press, Cambridge, UK, 1999. (Cited on page 366)

386 D. Kahneman and A. Tversky. Prospect theory: An analysis of decision under risk. *Econometrica*, 47:263–291, 1979. (Cited on page 324)

387 O. Kallenberg. *Foundations of Modern Probability*. Springer, New York, 2nd edition, 2002. (Cited on page 78)

388 O. Kallenberg. *Probabilistic Symmetries and Invariance Principles*. Springer, New York, 2005. (Cited on page 78)

389 A. Karr. *Probability*. Springer, New York, 1993. (Cited on pages 138 and 142)

390 D. Kendall. Foundations of a theory of random sets. In E. Harding and D. Kendall, editors, *Stochastic Geometry*. Wiley, London, 1974. (Cited on page 99)

391 R. Kennes. Computational aspects of the Möbius transformation of graphs. *IEEE Transactions on Systems, Man, and Cybernetics*, 22:201–223, 1992. (Cited on page 84)

392 S. Kent. Words of estimative probability. *Studies in Intelligence*, 8:49–65, 1964. (Cited on page 320)

393 R. Kesselman. *Verbal Probability Expressions in National Intelligence Estimates: A Comprehensive Analysis of Trends from the Fifties through Post 9/11*. Masters Degree Thesis, Mercyhurst College, PA, 2008. (Cited on page 320)

394 J. Keynes. *A Treatise on Probability*. Macmillan, London, 1921. (Cited on page xiv)

395 D. Kikuti, F. Cozman, and C. de Campos. Partially ordered preferences in decision trees: Computing strategies with imprecision in probabilities. In R. Brafman and U. Junker, editors, *Multidisciplinary IJCAI - 05 Workshop on Advances in Preference Handling*, pages 118–123, 2005. (Cited on pages 199, 228, and 337)

396 J. Klayman, J. Soll, C. Gonzalez-Vallejo, and S. Barlas. Overconfidence: It depends on how, what, and whom you ask. *Organizational Behavior and Human Decision Processes*, 79:216–247, 1999. (Cited on pages 325 and 328)

397 E. Klement, R. Mesiar, and E. Pap. *Triangular Norms*. Kluwer, Dordrecht, 2000. (Cited on page 110)

398 O. Klingmüller and U. Bourgund. *Sicherheit und Risiko im Konstruktiven Ingenieurbau*. Vieweg, Braunschweig, 1992. (Cited on page 293)

399 G. Klir. *Uncertainty and Information. Foundations of Generalized Information Theory*. Wiley, Hoboken, NJ, 2006. (Cited on pages 242 and 243)

400 G. Klir and R. Smith. On measuring uncertainty and uncertainty-based information: Recent developments. *Annals of Mathematics and Artificial Intelligence*, 32:5–33, 2001. (Cited on pages 243)

401 G. Klir and M. Wierman. *Uncertainty-Based Information: Elements of Generalized Information Theory*. Physica, Heidelberg, 1998. (Cited on pages 243 and 300)

402 M. Kohl. *Numerical Contributions to the Asymptotic Theory of Robustness*. PhD thesis, Universität Bayreuth (available from http://opus.ub.uni-bayreuth.de/opus4-ubbayreuth/frontdoor/index/index/docId/178), 2005. (Cited on page 174)

403 M. Kohl and P. Ruckdeschel. *ROptEst: Optimally robust estimation*, 2008. Contributed R-Package on CRAN, Version 0.6.0, 2008-08-07 (available from http://cran.r-project.org/web/packages/ROptEst/index.html); maintainer Kohl, M. (Cited on page 174)

404 D. Koller and N. Friedman. *Probabilistic Graphical Models: Principles and Techniques*. MIT Press, Cambridge, MA, 2009. (Cited on pages 225 and 232)

405 A. Kolmogorov. *Grundbegriffe der Wahrscheinlichkeitsrechnung*. Springer, Berlin, 1933. English translation: *Foundations of Probability*, Chelsea, Providence, RI, 1950. (Cited on pages 58, 114, 122, and 128)

406 B. Koopman. The bases of probability. *Bulletin of American Mathematical Society*, 46:763–774, 1940. (Cited on pages 112 and 113)

407 I. Kozine and L. Utkin. Interval-valued finite Markov chains. *Reliable Computing*, 8:97–113, 2002. (Cited on page 264)

408 V. Krätschmer. When fuzzy measures are upper envelopes of probability measures. *Fuzzy Sets and Systems*, 138:455–468, 2003. (Cited on page 54)

409 P. Krauss. Representation of conditional probability measures on Boolean algebras. *Acta Mathematica Academiae Scientiarum Hungaricae*, 19:229–241, 1968. (Cited on page 47)

410 V. Kreinovich, G. Xiang, and S. Ferson. Computing mean and variance under Dempster-Shafer uncertainty: Towards faster algorithms. *International Journal of Approximate Reasoning*, 42:212–227, 2006. (Cited on page 183)

411 H. Küchenhoff, T. Augustin, and A. Kunz. Partially identified prevalence estimation under misclassification using the kappa coefficient. *International Journal of Approximate Reasoning*, 53:1168–1182, 2012. (Cited on page 186)

412 M. Kumon and A. Takemura. On a simple strategy weakly forcing the strong law of large numbers in the bounded forecasting game. *Annals of the Institute of Statistical Mathematics*, 60:801–812, 2008. (Cited on page 127)

413 M. Kumon, A. Takemura, and K. Takeuchi. Game-theoretic versions of strong law of large numbers for unbounded variables. *Stochastics*, 79:449–468, 2007. (Cited on page 128)

414 V. Kuznetsov. *Interval Statistical Models*. Radio i Svyaz Publ., Moscow, 1991. In Russian. (Cited on pages 52, 54, 66, and xiv)

415 V. Kuznetsov. Auxiliary problems of statistical data processing: Interval approach. In V. Kreinovich, editor, *Reliable Computing: Supplement with the Extended Abstracts of APIC '95*, pages 123–128, 1995. (Cited on page 54)

416 V. Kuznetsov. Interval methods for processing statistical characteristics. In V. Kreinovich, editor, *Reliable Computing: Supplement with the Extended Abstracts of APIC '95*, pages 116–122, 1995. (Cited on page 54)

417 S. Kwerel. Fréchet bounds. In S. Kotz and N. Johnson, editors, *Encyclopedia of Statistical Sciences*, volume 3, pages 203–209, Wiley, New York, 1983. (Cited on page 180)

418 H. Kyburg Jr. *The Logical Foundations of Statistical Inference*. Synthese Library. Reidel, 1974. (Cited on page 179)

419 H. Kyburg Jr. Rational belief. *Behavioral and Brain Sciences*, 6:231–245, 1983. (Cited on page 194)

420 H. Kyburg Jr. Bayesian and non-Bayesian evidential updating. *Artificial Intelligence*, 31:271–293, 1987. (Cited on page 106)

421 H. Kyburg Jr. and H. Smokler, editors. *Studies in Subjective Probability*. Wiley, New York, 1964. Second edition (with new material) 1980. (Cited on page 187)

422 H. Kyburg Jr. and C.-M. Teng. *Uncertain Inference*. Cambridge University Press, Cambridge, UK, 2001. (Cited on page 94)

423 S. Lauritzen and D. Spiegelhalter. Local computations with probabilities on graphical structures and their application to expert systems. *Journal of the Royal Statistical Society, Series B*, 50:157–224, 1988. (Cited on page 208)

424 I. LaValle and K. Wapman. Rolling back decision trees requires the independence axiom! *Management Science*, 32:382–385, 1986. (Cited on page 198)

425 J. Lawless and M. Fredette. Frequentist prediction intervals and predictive distributions. *Biometrika*, 92:529–542, 2005. (Cited on page 175)

426 S. Lee and R. Wilke. Reform of unemployment compensation in Germany: A nonparametric bounds analysis using register data. *Journal of Business & Economic Statistics*, 27:193–205, 2009. (Cited on page 187)

427 H. Lehar, G. Niederwanger, and G. Hofstetter. FE ultimate load analysis of pile-supported pipelines – Tackling uncertainty in a real design problem. In W. Fellin, H. Lessmann, M. Oberguggenberger, and R. Vieider, editors, *Analyzing Uncertainty in Civil Engineering*, pages 129–163. Springer, Berlin, 2005. (Cited on page 302)

428 E. Lehmann and G. Casella. *Theory of Point Estimation*. Springer, New York, 2nd edition, 1998. (Cited on page 187)

429 E. Lehmann and J. Romano. *Testing Statistical Hypothesis*. Springer, New York, 3rd edition, 2005. (Cited on pages 168, 169, 171, and 187)

430 I. Levi. *The Enterprise of Knowledge. An Essay on Knowledge, Credal Probability, and Chance*. MIT Press, Cambridge, 1980. (Cited on pages 179 and 195)

431 P. Lévy. *Théorie de l'Addition des Variables Aléatoires*. Gauthier-Villars, Paris, 1937. Second edition: 1954. (Cited on page 128)

432 D. Lewis. Counterfactuals and comparative possibility. *Journal of Philosophical Logic*, 2:418–446, 1973. (Cited on pages 94 and 100)

433 D. Lewis. *Counterfactuals*. Blackwell, Worcester, UK, 1986. (Cited on page 112)

434 C.-Y. Li, X. Chen, X.-S. Yi, and J.-Y. Tao. Interval-valued reliability analysis of multi-state systems. *IEEE Transactions on Reliability*, 60:323–330, 2011. (Cited on page 160)

435 Y. Liao and W. Jiang. Bayesian analysis in moment inequality models. *Annals of Statistics*, 3:275–316, 2010. (Cited on page 187)

436 S. Lichtenstein, B. Fischhoff, and L. Phillips. Calibration of subjective probabilities: The state of the art up to 1980. In D. Kahneman, P. Slovic, and A. Tversky, editors, *Judgment under Uncertainty: Heuristics and Biases*, pages 306–334. Cambridge University Press, Cambridge, UK, 1982. (Cited on pages 318 and 328)

437 S. Lichtenstein and J. Newman. Empirical scaling of common verbal phrases associated with numerical probabilities. *Psychonomic Sciences*, 9:563–564, 1967. (Cited on page 320)

438 R. Little and D. Rubin. *Statistical Analysis with Missing Data*. Wiley, New York, 1987. (Cited on pages 181 and 221)

439 K. Loquin and O. Strauss. Noise quantization via possibilistic filtering. In T. Augustin, F. Coolen, S. Moral, and M. Troffaes, editors, *ISIPTA '09: Proceedings of the Sixth International Symposium on Imprecise Probability: Theories and Applications*, pages 297–306, Durham, UK, 2009. SIPTA. (Cited on page 87)

440 G. Lorentz. *Bernstein Polynomials*. Chelsea Publishing, New York, 2nd edition, 1986. (Cited on page 76)

441 R. Luce and H. Raiffa. *Games and Decisions: Introduction and Critical Survery*. Wiley, 1957. (Cited on page 199)

442 S. Maaß. Exact functionals and their core. *Statistical Papers*, 43:75–93, 2002. (Cited on pages 283 and 289)

443 S. Maaß. Coherent and convex fair pricing and variability measures. *International Journal of Approximate Reasoning*, 49:130–139, 2008. (Cited on page 290)

444 M. Machina. Dynamic consistency and non-expected utility models of choice under uncertainty. *Journal of Economic Literature*, 27:1622–1668, 1989. (Cited on page 198)

445 I. MacPhee, F. Coolen, and A. Aboalkhair. Nonparametric predictive system reliability with redundancy allocation following component testing. *Journal of Risk and Reliability*, 223:181–188, 2009. (Cited on page 178)

446 M. Madden. On the classification performance of TAN and general Bayesian networks. *Knowledge-Based Systems*, 22:489–495, 2009. (Cited on page 240)

447 C. Manski. *Partial Identification of Probability Distributions*. Springer, New York, 2003. (Cited on pages 145, 182, 183, 186, and 187)

448 C. Manski. *Identification for Prediction and Decision*. Harvard University Press, Cambridge, MA, 2009. (Cited on page 187)

449 C. Manski and F. Molinari. Rounding probabilistic expectations in surveys. *Journal of Business & Economic Statistics*, 28:219–231, 2010. (Cited on page 180)

450 K. Mardia and S. El-Atoum. Bayesian inference for the Von Mises-Fisher distribution. *Biometrika*, 63:203–206, 1976. (Cited on page 161)

451 R. Maronna, D. Martin, and V. Yohai. *Robust Statistics: Theory and Methods*. Wiley, Chichester, 2006. (Cited on page 187)

452 G. Matheron. *Random Sets and Integral Geometry*. Wiley, New York, 1975. (Cited on pages 84 and 99)

453 T. Maturi, P. Coolen-Schrijner, and F. Coolen. Nonparametric predictive pairwise comparison for real-valued data with terminated tails. *International Journal of Approximate Reasoning*, 51:141–150, 2009. (Cited on page 178)

454 T. Maturi, P. Coolen-Schrijner, and F. Coolen. Nonparametric predictive inference for competing risks. *Journal of Risk and Reliability*, 224:11–26, 2010. (Cited on pages 176 and 314)

455 D. Mauá and C. de Campos. Solving decision problems with limited information. In J. Shawe-Taylor, R. Zemel, P. Bartlett, F. Pereira, and K. Weinberger, editors, *NIPS '11: Proceedings of the Twenty-Fifth Annual Conference on Neural Information Processing Systems*, pages 603–611, 2011. (Cited on page 228)

456 D. Mauá, C. de Campos, and M. Zaffalon. A fully polynomial time approximation scheme for updating credal networks of bounded treewidth and number of variable states. In F. Coolen, G. de Cooman, T. Fetz, and M. Oberguggenberger, editors, *ISIPTA '11: Proceedings of the Seventh International Symposium on Imprecise Probability: Theories and Applications*, pages 277–286, Innsbruck, 2011. SIPTA. (Cited on pages 219, 223, and 224)

457 G. Mauris. Inferring a possibility distribution from very few measurements. In D. Dubois, M. Lubiano, H. Prade, M. Gil, P. Grzegorzewski, and O. Hryniewicz, editors, *Soft Methods for Handling Variability and Imprecision*, volume 48, pages 92–99, Berlin, 2008. Springer. (Cited on page 87)

458 K. McConway. Marginalization and linear opinion pools. *Journal of the American Statistical Association*, 76:410–414, 1981. (Cited on page 110)

459 M. Mignolet and C. Soize. Stochastic reduced order models for uncertain geometrically nonlinear dynamical systems. *Computer Methods in Applied Mechanics and Engineering*, 197:3951–3963, 2008. (Cited on page 303)

460 E. Miranda. A survey of the theory of coherent lower previsions. *International Journal of Approximate Reasoning*, 48:628–658, 2008. (Cited on page 36)

461 E. Miranda. Updating coherent lower previsions on finite spaces. *Fuzzy Sets and Systems*, 160:1286–1307, 2009. (Cited on page 53)

462 E. Miranda, I. Couso, and P. Gil. Extreme points of credal sets generated by 2-alternating capacities. *International Journal of Approximate Reasoning*, 33:95–115, 2003. (Cited on page 81)

463 E. Miranda, I. Couso, and P. Gil. A random set characterization of possibility measures. *Information Sciences*, 168:51–75, 2004. (Cited on page 92)

464 E. Miranda, I. Couso, and P. Gil. Random sets as imprecise random variables. *Journal of Mathematical Analysis and Applications*, 307:32–47, 2005. (Cited on pages 83 and 99)

465 E. Miranda and G. de Cooman. Epistemic independence in numerical possibility theory. *International Journal of Approximate Reasoning*, 103:23–42, 1999. (Cited on page 108)

466 E. Miranda and G. de Cooman. Coherence and independence on non-linear spaces. Technical Report TR05/10, Technical Reports on Statistics and Decision Sciences, Rey Juan Carlos University, 2005. Downloadable at http://bellman.ciencias.uniovi.es/~emiranda/ (Cited on pages 45 and 49)

467 E. Miranda and G. de Cooman. Marginal extension in the theory of coherent lower previsions. *International Journal of Approximate Reasoning*, 46:188–225, 2007. (Cited on page 49)

468 E. Miranda and M. Zaffalon. Coherence graphs. *Artificial Intelligence*, 173:104–144, 2009. (Cited on pages 45 and 47)

469 E. Miranda and M. Zaffalon. Conditional models: Coherence and inference through sequences of joint mass functions. *Journal of Statistical Planning and Inference*, 140:1805–1833, 2010. (Cited on pages 45 and 53)

470 E. Miranda and M. Zaffalon. Notes on desirability and conditional lower previsions. *Annals of Mathematics and Artificial Intelligence*, 60:251–309, 2010. (Cited on page 39)

471 E. Miranda and M. Zaffalon. Conglomerable coherence. *International Journal of Approximate Reasoning*, 54:1322–1350, 2013. (Cited on pages 47 and 52)

472 E. Miranda, M. Zaffalon, and G. de Cooman. Conglomerable natural extension. *International Journal of Approximate Reasoning*, 53:1200–1227, 2012. (Cited on pages 51 and 52)

473 K. Miyabe and A. Takemura. Convergence of random series and the rate of convergence of the strong law of large numbers in game-theoretic probability. *Stochastic Processes and their Applications*, 122:1–30, 2012. (Cited on page 133)

474 K. Miyabe and A. Takemura. The law of the iterated logarithm in game-theoretic probability with quadratic and stronger hedges, 2012. arXiv:1208.5088. (Cited on page 127)

475 M. Modarres. *Risk Analysis in Engineering. Techniques, Tools, and Trends*. Taylor & Francis, Boca Raton, 2006. (Cited on page 303)

476 D. Moens and D. Vandepitte. A survey of non-probabilistic uncertainty treatment in finite element analysis. *Computer Methods in Applied Mechanics and Engineering*, 194:1527–1555, 2005. (Cited on page 303)

477 F. Molinari. Partial identification of probability distributions with misclassified data. *Journal of Econometrics*, 144:81–117, 2008. (Cited on page 185)

478 F. Molinari. Missing treatments. *Journal of Business & Economic Statistics*, 28:82–95, 2010. (Cited on page 187)

479 B. Möller. Fuzzy-Modellierung in der Baustatik. *Bauingenieur*, 72:75–84, 1997. (Cited on page 303)

480 B. Möller and M. Beer. *Fuzzy Randomness. Uncertainty in Civil Engineering and Computational Mechanics*. Springer, Berlin, 2004. (Cited on page 303)

481 B. Möller, M. Beer, W. Graf, and A. Hoffmann. Possibility theory based safety assessment. *Computer-Aided Civil & Infrastructure Engineering*, 14:81–91, 1999. (Cited on pages 294 and 303)

482 S. Moral. Epistemic irrelevance on sets of desirable gambles. *Annals of Mathematics in Artificial Intelligence*, 45:197–214, 2005. (Cited on pages 11, 26, and 60)

483 R. Muhanna and R. Mullen. Formulation of fuzzy finite-element methods for solid mechanics problems. *Computer-Aided Civil & Infrastructure Engineering*, 14:107–117, 1999. (Cited on page 303)

484 R. Muhanna, H. Zhang, and R. Mullen. Interval finite elements as a basis for generalized models of uncertainty in engineering mechanics. *Reliable Computing*, 13:173–194, 2007. (Cited on page 303)

485 D. Mundici. The logic of Ulam's game with lies. In C. Bicchieri, editor, *Knowledge, Belief and Strategic Interaction*, pages 275–284. Cambridge University Press, Cambridge, UK, 1998. (Cited on page 111)

486 T. Murofushi and M. Sugeno. An interpretation of fuzzy measures and the Choquet integral as an integral with respect to a fuzzy measure. *Fuzzy Sets and Systems*, 29:201–227, 1989. (Cited on page 80)

487 A. Murphy and E. Epstein. Verification of probabilistic predictions: A brief review. *Journal of Applied Meteorology*, 6:748–755, 1967. (Cited on page 117)

488 K. Murphy, Y. Weiss, and M. Jordan. Loopy belief propagation for approximate inference: An empirical study. In K. Laskey and H. Prade, editors, *UAI '99: Proceedings of the Fifteenth Conference on Uncertainty in Artificial Intelligence*, pages 467–475, San Francisco, CA, 1999. Morgan Kaufmann. (Cited on page 224)

489 R. Nau. Indeterminate probabilities on finite sets. *The Annals of Statistics*, 20:1737–1767, 1992. (Cited on page 29)

490 R. Nau. The shape of incomplete preferences. *The Annals of Statistics*, 34:2430–2448, 2006. (Cited on pages 26 and 290)

491 K. Nehring. Imprecise probabilistic beliefs as a context for decision-making under ambiguity. *Journal of Economic Theory*, 144:1054–1091, 2009. (Cited on page 113)

492 A. Neumaier. Clouds, fuzzy sets and probability intervals. *Reliable Computing*, 10:249–272, 2004. (Cited on pages 87, 294, and 303)

493 H. Nguyen, V. Kreinovich, B. Wu, and G. Xiang. *Computing Statistics under Interval and Fuzzy Uncertainty: Applications to Computer Science and Engineering*. Springer, Berlin, 2011. (Cited on pages 182 and 187)

494 E. Nikolaidis, D. Ghiocel, and S. Singhal, editors. *Engineering Design Reliability Handbook*, Boca Raton, 2005. CRC. (Cited on page 304)

495 A. Nilim and L. El Ghaoui. Robust control of Markov decision processes with uncertain transition matrices. *Operations Research*, 53:780–798, 2005. (Cited on page 264)

496 R. Noubiap and W. Seidel. An algorithm for calculating Γ-minimax decision rules under generalized moment conditions. *Annals of Statistics*, 29:1094–1116, 2001. (Cited on page 165)

497 M. Oberguggenberger and W. Fellin. Reliability bounds through random sets: Nonparametric methods and geotechnical applications. *Computers & Structures*, 86:1093–1101, 2008. (Cited on pages 299 and 303)

498 M. Oberguggenberger, J. King, and B. Schmelzer. Classical and imprecise probability methods for sensitivity analysis in engineering: A case study. *International Journal of Approximate Reasoning*, 50:680–693, 2009. (Cited on pages 87, 300, 303, and 312)

499 W. Oberkampf and J. Helton. Investigation of Evidence Theory for Engineering Applications. Technical Report AIAA 2001–1569, American Institute of Aeronautics and Astronautics, Reston VA, 2002. (Cited on pages 294 and 303)

500 A. O'Hagan, C. Buck, A. Daneshkhah, J. Eiser, P. Garthwaite, D. Jenkinson, J. Oakley, and T. Rakow. *Uncertain Judgements: Eliciting Experts' Probabilities*. Wiley, Chichester, 2006. (Cited on pages 326 and 328)

501 K. Ovaere, G. Deschrijver, and E. Kerre. Application of fuzzy decision making to the fuzzy finite element method. *Fuzzy Information and Engineering*, 11:27–36, 2009. (Cited on page 303)

502 A. Papamarcou and T. Fine. A note on undominated lower probabilities. *The Annals of Probability*, 14:710–723, 1986. (Cited on page 81)

503 A. Papamarcou and T. Fine. Unstable collectives and envelopes of probability measures. *The Annals of Probability*, 19:893–906, 1991. (Cited on page 146)

504 Y. Pawitan. *In All Likelihood: Statistical Modelling and Inference Using Likelihood*. Oxford University Press, Oxford, 2001. (Cited on pages 179 and 239)

505 J. Pearl. *Probabilistic Reasoning in Intelligent Systems: Networks of Plausible Inference*. Morgan Kaufmann, San Mateo, CA, 1988. (Cited on pages 217, 220, and 223)

506 J. Pearl. Reasoning with belief functions: An analysis of compatibility. *International Journal of Approximate Reasoning*, 4:363–389, 1990. (Cited on page 99)

507 R. Pelessoni and P. Vicig. Convex imprecise previsions. *Reliable Computing*, 9:465–485, 2003. (Cited on pages 283, 286, 287, and 289)

508 R. Pelessoni and P. Vicig. Imprecise previsions for risk measurement. *International Journal of Uncertainty, Fuzziness and Knowledge-Based Systems*, 11:393–412, 2003. (Cited on pages 283, 284, 285, and 289)

509 R. Pelessoni and P. Vicig. Uncertainty modelling and conditioning with convex imprecise previsions. *International Journal of Approximate Reasoning*, 39:297–319, 2005. (Cited on pages 29, 283, 286, and 289)

510 R. Pelessoni and P. Vicig. Williams coherence and beyond. *International Journal of Approximate Reasoning*, 50:612–626, 2009. (Cited on page 54)

511 R. Pelessoni, P. Vicig, and M. Zaffalon. Inference and risk measurement with the Pari-Mutuel model. *International Journal of Approximate Reasoning*, 51:1145–1158, 2010. (Cited on pages 90, 283, and 284)

512 L. Pericchi and P. Walley. Robust Bayesian credible intervals and prior ignorance. *International Statistical Review*, 59:1–23, 1991. (Cited on pages 77 and 164)

513 G. Peschl. *Reliability Analysis in Geotechnics with the Random Set Finite Element Method*. Technische Universität Graz, 2005. (Cited on page 303)

514 A. Piatti, A. Antonucci, and M. Zaffalon. Building knowledge-based systems by credal networks: A tutorial. In A. Baswell, editor, *Advances in Mathematics Research*, volume 11, pages 227–279. Nova Science, New York, 2010. (Cited on page 229)

515 F. Pichon, D. Dubois, and T. Denoeux. Relevance and truthfulness in information correction and fusion. *International Journal of Approximate Reasoning*, 53:159–175, 2012. (Cited on page 109)

516 M. Ponomareva and E. Tamer. Misspecification in moment inequality models: back to moment equalities? *The Econometrics Journal*, 14:186–203, 2011. (Cited on page 187)

517 U. Pötter. *Statistical Models of Incomplete Data and Their Use in Social Sciences*, Ruhr-Universität Bochum, 2008. Habilitation thesis. (Cited on page 182)

518 H. Prautzsch, W. Boehm, and M. Paluszny. *Bézier and B-Spline Techniques*. Springer, Berlin, 2002. (Cited on page 76)

519 E. Quaeghebeur. *Learning from Samples Using Coherent Lower Previsions*. PhD thesis, Ghent University (available from http://hdl.handle.net/1854/LU-495650), 2009. (Cited on pages 26, 155, and 161)

520 E. Quaeghebeur. Completely monotone outer approximations of lower probabilities on finite possibility spaces. In S. Li, X. Wang, Y. Okazaki, J. Kawabe, T. Murofushi, and L. Guan, editors, *Nonlinear Mathematics for Uncertainty and its Applications*, volume 100, pages 169–178. Springer, Berlin, 2011. (Cited on page 84)

521 E. Quaeghebeur and G. de Cooman. Imprecise probability models for inference in exponential families. In F. Cozman, R. Nau, and T. Seidenfeld, editors, *ISIPTA '05. Proceedings of the Fourth International Symposium on Imprecise Probabilities and Their Applications*, pages 287–296, Pittsburgh, PA, 2005. SIPTA. (Cited on pages 155, 156, 161, 162, and 164)

522 J. Quiggin. *Generalized Expected Utility Theory: The Rank Dependent Model*. Kluwer, Boston, MA, 1993. (Cited on page 324)

523 J. Quinlan. Induction of decision trees. *Machine Learning*, 1:81–106, 1986. (Cited on page 242)

524 R. Rackwitz. Reviewing probabilistic soils modelling. *Computers and Geotechnics*, 26:199–223, 2000. (Cited on pages 296 and 297)

525 A. Raftery, M. Tanner, and M. Wells. Statistics in the year 2000: Vignettes. *Journal of the American Statistical Association*, 95:281, 2000. (Cited on page 139)

526 H. Raiffa and R. Schlaifer. *Applied Statistical Decision Theory*. Harvard University Press, Cambridge, MA, 1961. (Cited on pages 197 and 199)

527 F. Ramsey. Truth and probability. In H. Kyburg Jr. and H. Smokler, editors, *Studies in Subjective Probability*, pages 23–52. Krieger, New York, 2nd edition, 1998. (Cited on page 112)

528 R. Regan, F. Mosteller, and C. Youtz. Quantitative meanings of verbal probability expressions. *Journal of Applied Psychology*, 74:433–442, 1990. (Cited on pages 320, 321, and 328)

529 L. Rego and T. Fine. Estimation of chaotic probabilities. In F. Cozman, R. Nau, and T. Seidenfeld, editors, *ISIPTA'05: Proceedings of the Fourth International Symposium on Imprecise Probabilities and Their Applications*, pages 297–305, Pittsburgh, PA, 2005. SIPTA. (Cited on pages 175 and 188)

530 G. Regoli. Comparative probability orderings, 1999. http://www.sipta.org/documentation/comparative_prob/regoli.pdf. (Cited on page 91)

531 N. Rescher and R. Manor. On inference from inconsistent premises. *Theory and Decision*, 1:179–219, 1970. (Cited on page 110)

532 J. Reunanen. Overfitting in making comparisons between variable selection methods. *Journal of Machine Learning Research*, 3:1371–1382, 2003. (Cited on page 252)

533 H. Rieder. Least favorable pairs for special capacities. *The Annals of Statistics*, 5:909–921, 1977. (Cited on page 171)

534 H. Rieder. A finite-sample minimax regression estimator. *Statistics*, 20:211–221, 1989. (Cited on page 173)

535 H. Rieder. *Robust Asymptotic Statistics*. Springer, New York, 1994. (Cited on page 174)

536 H. Rieder, M. Kohl, and P. Ruckdeschel. The costs of not knowing the radius. *Statistical Methods and Applications*, 7:13–40, 2008. (Cited on page 174)

537 S. Rinderknecht, M. Borsuk, and P. Reichert. Eliciting density ratio classes. *International Journal of Approximate Reasoning*, 52:792–804, 2011. (Cited on page 164)

538 D. Ríos Insua and F. Ruggeri, editors. *Robust Bayesian Analysis*. Springer, Berlin, 2000. (Cited on pages 38, 144, and 154)

539 G. Rohwer and U. Pötter. *Grundzüge der sozialwissenschaftlichen Statistik*. Juventa, Weinheim, 2001. (Cited on page 182)

540 T. Ross. *Fuzzy Logic with Engineering Applications*. Wiley, New Delhi, 2nd edition, 2008. (Cited on page 304)

541 T. Ross, J. Booker, and W. Parkinson, editors. *Fuzzy Logic and Probability Applications. Bridging the Gap*, Philadelphia, PA/Alexandria, VA, 2002. Society for Industrial Mathematics/Americal Statistical Association. (Cited on page 304)

542 P. Rousseeuw. Maxbias curve. In S. Kotz, C. Read, N. Balakrishnan, and B. Vidakovic, editors, *Encyclopedia of Statistical Sciences*, volume 7, pages 4622–4625. Wiley, 2nd edition, 2004. (Cited on page 173)

543 D. Rubin. Inference and missing data. *Biometrika*, 63:581–592, 1976. (Cited on page 236)

544 D. Rubin. Multiple imputation after 18+ years. *Journal of the American Statistical Association*, 91:473–489, 1996. (Cited on page 181)

545 E. Rubio, J. Hall, and M. Anderson. Uncertainty analysis in a slope hydrology and stability model using probabilistic and imprecise information. *Computers and Geotechnics*, 31:529–536, 2004. (Cited on page 303)

546 P. Ruckdeschel. A motivation for $1/\sqrt{(n)}$- shrinking neighborhoods. *Metrika*, 63:295–307, 2006. (Cited on page 174)

547 B. Rüger. *Test- und Schätztheorie. Band I: Grundlagen*. Oldenbourg, München, 1999. (Cited on page 158)

548 F. Ruggeri, D. Ríos Insua, and J. Martín. Robust Bayesian analysis. In D. Dey and C. Rao, editors, *Handbook of Statistics. Bayesian Thinking: Modeling and Computation*, volume 25, pages 623–667. Elsevier, Amsterdam, 2005. (Cited on pages 154, 164, and 187)

549 J. Russo and K. Kolzow. Where is the fault in fault trees? *Journal of Experimental Psychology: Human Perception Performance*, 20:17–32, 1994. (Cited on page 326)

550 J. Russo and P. Schoemaker. Managing overconfidence. *Sloan Management Review*, 33:7–17, 1992. (Cited on pages 318, 325, and 328)

551 A. Sadrolhefazi and T. Fine. Finite-dimensional distributions and tail behavior in stationary interval-valued probability models. *The Annals of Statistics*, 22:1840–1870, 1994. (Cited on page 146)

552 U. Saint-Mont. *Statistik im Forschungsprozess. Eine Philosophie der Statistik als Baustein einer integrativen Wissenschaftstheorie*. Physica, Heidelberg, 2011. (Cited on page 187)

553 S. Sandri, D. Dubois, and H. Kalfsbeek. Elicitation, assessment and pooling of expert judgments using possibility theory. *IEEE Transactions on Fuzzy Systems*, 3:313–335, 1995. (Cited on page 86)

554 J. Satia and R. Lave Jr. Markovian decision processes with uncertain transition probabilities. *Operations Research*, 21:728–740, 1973. (Cited on pages 194 and 264)

555 L. Savage. *The Foundations of Statistics*. Wiley, New York, 1954. (Cited on pages 94 and 204)

556 C. Schenk and G. Schuëller. *Uncertainty Assessment of Large Finite Element Systems*. Springer, Berlin, 2005. (Cited on page 303)

557 M. Schervish, T. Seidenfeld, and J. Kadane. The fundamental theorem of prevision and asset pricing. *International Journal of Approximate Reasoning*, 49:148–158, 2008. (Cited on page 290)

558 A. Schied. Optimal investments for robust utility functionals in complete market models. *Mathematics of Operations Research*, 30:750–764, 2005. (Cited on page 171)

559 B. Schmelzer. On solutions to stochastic differential equations with parameters modeled by random sets. *International Journal of Approximate Reasoning*, 51:1159–1171, 2010. (Cited on pages 277, 301, and 303)

560 B. Schmelzer. Set-valued assessments of solutions to stochastic differential equations with random set parameters. *Journal of Mathematical Analysis and Applications*, 400:425–438, 2013. (Cited on page 303)

561 B. Schmelzer, M. Oberguggenberger, and C. Adam. Efficiency of tuned mass dampers with uncertain parameters on the performance of structures under stochastic excitation. *Journal of Risk and Reliability*, 224:297–308, 2010. (Cited on page 303)

562 C. Schneider. Dempster's combination rule and sufficient statistics. *Communications in Statistics – Theory and Methods*, 18:3841–3850, 1989. (Cited on page 179)

563 C. Schneider. Das Fiduzialargument von Fisher. Versuch einer Rekonstruktion (Fisher's fiducial argument. An attempt of reconstruction). In H. Rinne, B. Rüger, and H. Strecker, editors, *Grundlagen der Statistik und ihre Anwendungen. Festschrift für Kurt Weichselberger*, pages 87–99. Physica, Heidelberg, 1995. (Cited on page 179)

564 G. Schollmeyer and T. Augustin. On sharp identification regions for regression under interval data. Technical Report 128 (available from http://epub.ub.uni-muenchen.de/15231/), Department of Statistics, LMU Munich, 2013. (Cited on page 187)

565 G. Schuëller. On the treatment of uncertainties in structural mechanics and analysis. *Computers & Structures*, 85:235–243, 2007. (Cited on page 296)

566 G. Schuëller and H. Pradlwarter. Uncertainty analysis of complex structural systems. *International Journal for Numerical Methods in Engineering*, 80:881–913, 2009. (Cited on page 303)

567 H. Schweiger and G. Peschl. Reliability analysis in geotechnics with the random set finite element method. *Computers and Geotechnics*, 32:422–435, 2005. (Cited on page 303)

568 H. Schweiger, G. Peschl, and R. Pöttler. Application of the random set finite element method for analysing tunnel excavation. *Georisk*, 1:43–56, 2007. (Cited on page 303)

569 T. Seidenfeld. Decision theory without 'independence' or without 'ordering': What is the difference? *Economics and Philosophy*, 4:267–290, 1988. (Cited on page 201)

570 T. Seidenfeld. A contrast between two decision rules for use with (convex) sets of probabilities: Γ-maximin versus E-admissibility. *Synthese*, 140:69–88, 2004. (Cited on pages 193, 196, 198, and 200)

571 T. Seidenfeld, M. Schervish, and J. Kadane. Decisions without ordering. In W. Sieg, editor, *Acting and Reflecting: The Interdisciplinary Turn in Philosophy*, volume 211, pages 143–170. Kluwer, Dordrecht, 1990. Reprinted in [385]. (Cited on page 26)

572 T. Seidenfeld, M. Schervish, and J. Kadane. A representation of partially ordered preferences. *The Annals of Statistics*, 23:2168–2217, 1995. Reprinted in [385]. (Cited on page 26)

573 T. Seidenfeld and L. Wasserman. Dilation for sets of probabilities. *The Annals of Statistics*, 21:1139–1154, 1993. (Cited on pages 104 and 227)

574 K. Sentz and S. Ferson. Combination of evidence in Dempster-Shafer theory. Technical Report SAND2002-0835 (available from http://www.sandia.gov/epistemic/Reports/SAND2002-0835.pdf), Sandia National Laboratories, Albuquerque, 2002. (Cited on page 303)

575 G. Shackle. *Decision, Order and Time in Human Affairs*. Cambridge University Press, Cambridge, UK, 1961. (Cited on pages 94 and 100)

576 G. Shafer. *A Mathematical Theory of Evidence*. Princeton University Press, Princeton, NJ, 1976. (Cited on pages 84, 86, 90, 94, 98, 106, 242, and xiv)

577 G. Shafer. Allocations of probability. *Annals of Probability*, 7:827–839, 1979. (Cited on page 84)

578 G. Shafer. Bayes's two arguments for the rule of conditioning. *Annals of Statistics*, 10:1075–1089, 1982. (Cited on page 133)

579 G. Shafer. A betting interpretation for probabilities and Dempster-Shafer degrees of belief. *International Journal of Approximate Reasoning*, 52:127–136, 2011. (Cited on pages 117, 133, and 134)

580 G. Shafer, P. Gillett, and R. Scherl. A new understanding of subjective probability and its generalization to lower and upper prevision. *International Journal of Approximate Reasoning*, 33:1–49, 2003. (Cited on page 133)

581 G. Shafer, A. Shen, N. Vereshchagin, and V. Vovk. Test martingales, Bayes factors, and p-values. *Statistical Science*, 26:84–101, 2011. (Cited on page 133)

582 G. Shafer and V. Vovk. *Probability and Finance: It's Only a Game!* Wiley, New York, 2001. (Cited on pages 61, 117, 121, 123, 127, 128, 134, 261, 278, 289, and xiv)

583 G. Shafer and V. Vovk. The sources of Kolmogorov's *Grundbegriffe*. *Statistical Science*, 21:70–98, 2006. (Cited on page 122)

584 G. Shafer, V. Vovk, and R. Chychyla. How to base probability theory on perfect information games. *Bulletin of the European Association for Theoretical Computer Science*, 100:115–148, February 2010. (Cited on pages 132 and 134)

585 G. Shafer, V. Vovk, and A. Takemura. Lévy's zero-one law in game-theoretic probability. *Journal of Theoretical Probability*, 25:1–24, 2012. (Cited on page 128)

586 C. Shannon. A mathematical theory of communication. *The Bell System Technical Journal*, 27:379–423, 1948. (Cited on page 243)

587 R. Shirota, F. Cozman, F. Trevizan, and C. de Campos. Multilinear and integer programming for Markov decision processes with imprecise probabilities. In G. de Cooman, J. Vejnarová, and M. Zaffalon, editors, *ISIPTA '07: Proceedings of the Fifth International Symposium on Imprecise Probability: Theories and Applications*, pages 395–404, Prague, 2007. Action M Agency for SIPTA. (Cited on page 228)

588 R. Shirota, D. Kikuti, and F. Cozman. Solving decision trees with imprecise probabilities through linear programming. In T. Augustin, F. Coolen, S. Moral, and M. Troffaes, editors, *ISIPTA '09: Proceedings of the Sixth International Symposium on Imprecise Probability: Theories and Applications*, Durham, UK, 2009. SIPTA. (Cited on pages 199 and 228)

589 N. Shyamalkumar. Likelihood robustness. In D. Ríos Insua and F. Ruggeri, editors, *Robust Bayesian Analysis*, pages 127–143. Springer, New York, 2000. (Cited on page 165)

590 Z. Siddique. Partially identified treatment effects under imperfect compliance: The case of domestic violence. *Journal of the American Statistical Association*, 108:504–513, 2013. (Cited on page 187)

591 R. Simpson. The specific meanings of certain terms indicating differing degrees of frequency. *Quarterly Journal of Speech*, 30:328–330, 1944. (Cited on page 320)

592 R. Simpson. Stability in meanings for quantitative terms: A comparison over 20 years. *Quarterly Journal of Speech*, 49:146–151, 1963. (Cited on page 320)

593 D. Škulj. Generalized conditioning in neighbourhood models. In F. Cozman, R. Nau, and T. Seidenfeld, editors, *ISIPTA '05: Proceedings of the Fourth International Symposium on Imprecise Probabilities and Their Applications*, pages 322–331, Pittsburgh, PA, 2005. SIPTA. (Cited on page 147)

594 D. Škulj. Jeffrey's conditioning rule in neighbourhood models. *International Journal of Approximate Reasoning*, 42:192–211, 2006. (Cited on page 147)

595 D. Škulj. Discrete time Markov chains with interval probabilities. *International Journal Approximate Reasoning*, 50:1314–1329, 2009. (Cited on pages 264, 265, 269, and 272)

596 D. Škulj. The use of Markov operators to constructing generalised probabilities. *International Journal of Approximate Reasoning*, 52:1392–1408, 2011. (Cited on page 77)

597 D. Škulj and R. Hable. Coefficients of ergodicity for imprecise Markov chains. In T. Augustin, F. P. A. Coolen, M. Serafin, and M.C.M. Troffaes, editors, *ISIPTA '09: Proceedings of the Sixth International Symposium on Imprecise Probability: Theories and Applications*, pages 377–386. Durham, UK, 2009. SIPTA (Cited on pages 272 and 277)

598 P. Smets. Possibilistic inference from statistical data. In A. Ballester, D. Cardus, and E. Trillas, editors, *Second World Conference on Mathematics at the Service of Man*, pages 611–613. Universidad Politecnica de Las Palmas, Spain, 1982. (Cited on page 101)

599 P. Smets. The canonical decomposition of a weighted belief. In *IJCAI '95: Proceedings of the Fourteenth International Joint Conference on Artificial Intelligence*, pages 1896–1901. Morgan Kaufmann, 1995. (Cited on page 111)

600 P. Smets. Belief functions on real numbers. *International Journal of Approximate Reasoning*, 40:181–223, 2005. (Cited on page 84)

601 P. Smets and R. Kennes. The transferable belief model. *Artificial Intelligence*, 66:191–234, 1994. (Cited on pages 98, 106, and 107)

602 C. Smith. Consistency in statistical inference and decision. *Journal of the Royal Statistical Society, Series B*, 23:1–37, 1961. (Cited on pages 26 and xiv)

603 M. Smithson. Scale construction from a decisional viewpoint. *Minds and Machines*, 16:339–364, 2006. (Cited on page 319)

604 M. Smithson, T. Hatori, and M. Gurr. Additivity and Sample Space Ambiguity. Unpublished Manuscript, The Australian National University, Canberra, 2010. (Cited on page 326)

605 M. Smithson, E. Merkle, and J. Verkuilen. Beta regression finite mixture models of polarization and priming. *Journal of Educational and Behavioral Statistics*, 36:804–831, 2011. (Cited on pages 323 and 324)

606 M. Smithson and C. Segale. Partition priming in judgments of imprecise probabilities. *Journal of Statistical Theory and Practice*, 3:169–182, 2009. Reprinted in [146], pages 139-151. (Cited on pages 321, 323, 327, and 328)

607 M. Smithson and J. Verkuilen. A better lemon-squeezer? Maximum likelihood regression with beta-distributed dependent variables. *Psychological Methods*, 11:54–71, 2006. (Cited on pages 321 and 322)

608 I. Sobol. Sensitivity analysis for nonlinear mathematical models. *Mathematical Modeling & Computational Experiment*, 1:407–414, 1993. (Cited on page 300)

609 C. Soize. A comprehensive overview of a non-parametric probabilistic approach of model uncertainties for predictive models in structural dynamics. *Journal of Sound and Vibration*, 288:623–652, 2005. (Cited on page 303)

610 C. Soize. Random matrix theory for modeling uncertainties in computational mechanics. *Computer Methods in Applied Mechanics and Engineering*, 194:1333–1366, 2005. (Cited on page 303)

611 J. Soll and J. Klayman. Overconfidence in interval estimates. *Journal of Experimental Psychology: Learning, Memory, and Cognition*, 30:299–314, 2004. (Cited on page 325)

612 H. Steadman, E. Silver, J. Monahan, P. Appelbaum, P. Rorbbins, E. Mulvey, T. Grisso, L. Roth, and S. Banks. A classification tree approach to the development of actuarial violence risk assessment tools. *Law and Human Behavior*, 24:83–100, 2000. (Cited on page 320)

613 S. Stigler. The changing history of robustness. *American Statistician*, 64:277–281, 2010. (Cited on page 187)

614 D. Stone and R. Johnson. A study of words indicating frequency. *Journal of Educational Psychology*, 50:224–227, 1959. (Cited on page 320)

615 J. Stoye. Partial identification and robust treatment choice: An application to young offenders. *Journal of Statistical Theory and Practice*, 3:239–254, 2009. Reprinted in [146], pages 173-188. (Cited on pages 180 and 187)

616 J. Stoye. Partial identification of spread parameters. *Quantitative Economics*, 1:323–357, 2010. (Cited on pages 183 and 186)

617 T. Strat. Continuous belief functions for evidential reasoning. In R. Brachman, editor, *Proceedings of the Fourth National Conference on Artificial Intelligence*, pages 308–313, 1984. AAAI Press. (Cited on page 84)

618 M. Studeny. Semigraphoid and structures of probabilistic conditional independence. *Annals of Mathematics and Artificial Intelligence*, 21:71–98, 1997. (Cited on page 77)

619 K. Takeuchi, M. Kumon, and A. Takemura. A new formulation of asset trading games in continuous time with essential forcing of variation exponent. *Bernoulli*, 15:1243–1258, 2009. (Cited on page 133)

620 E. Tamer. Partial identification in econometrics. *Annual Review of Economics*, 2:167–195, 2010. (Cited on page 187)

621 F. Tonon. Using random set theory to propagate epistemic uncertainty through a mechanical system. *Reliability Engineering and System Safety*, 85:169–181, 2004. (Cited on page 303)

622 F. Tonon, A. Bernardini, and A. Mammino. Determination of parameters range in rock engineering by means of random set theory. *Reliability Engineering and System Safety*, 70:241–261, 2000. (Cited on pages 294 and 303)

623 F. Tonon, A. Bernardini, and A. Mammino. Reliability analysis of rock mass response by means of random set theory. *Reliability Engineering and System Safety*, 70:263–282, 2000. (Cited on pages 294 and 303)

624 F. Tonon and S. Chen. Inclusion properties for random relations under the hypotheses of stochastic independence and non-interactivity. *International Journal of General Systems*, 34:615–624, 2005. (Cited on page 109)

625 F. Tonon and C. Pettit. Toward a definition and understanding of correlation for variables constrained by random relations. In G. Augusti, G. Schuëller, and M. Ciampoli, editors, *ICOSSAR '05: International Conference on Structural Safety and Reliability of Engineering Systems and Structures*, pages 1741–1745. Rotterdam, 2005. Millpress. (Cited on page 299)

626 F. Trevizan, F. Cozman, and L. de Barros. Mixed probabilistic and nondeterministic factored planning through Markov decision processes with set-valued transitions. In *Workshop on a Reality Check for Planning and Scheduling Under Uncertainty at the Eighteenth International Conference on Automated Planning and Scheduling (ICAPS)*, AAAI, 2008. (Cited on page 228)

627 M. Troffaes. `improb`: A python module for working with imprecise probabilities. http://packages.python.org/improb/. (Cited on page 329)

628 M. Troffaes. Learning and optimal control of imprecise Markov decision processes by dynamic programming using the imprecise Dirichlet model. In M. Lopéz-Díaz, M. Gil, P. Grzegorzewski, O. Hyrniewicz, and J. Lawry, editors, *Soft Methodology and Random Information Systems*, pages 141–148, Berlin, 2004. Springer. (Cited on page 228)

629 M. Troffaes. Conditional lower previsions for unbounded random quantities. In J. Lawry, E. Miranda, A. Bugarin, S. Li, M. Gil, P. Grzegiorzewski, and O. Hryniewicz, editors, *Soft Methods in Integrated Uncertainty Modelling*, volume 37, pages 201–209. Springer, Berlin, 2006. (Cited on pages 29 and 53)

630 M. Troffaes. Decision making under uncertainty using imprecise probabilities. *International Journal of Approximate Reasoning*, 45:17–29, 2007. (Cited on page 196)

631 M. Troffaes and F. Coolen. Applying the imprecise Dirichlet model in cases with partial observations and dependencies in failure data. *International Journal of Approximate Reasoning*, 50:257–268, 2009. (Cited on pages 306 and 311)

632 M. Troffaes and G. de Cooman. Extensions of coherent lower previsions to unbounded random variables. In B. Bouchon-Meunier, L. Foulloy, and R. Yager, editors, *Intelligent Systems for Information Processing: From Representation to Applications*, pages 277–288. North-Holland, Amsterdam, 2003. (Cited on page 29)

633 M. Troffaes and S. Destercke. Probability boxes on totally preordered spaces for multivariate modelling. *International Journal of Approximate Reasoning*, 52:767–791, 2011. (Cited on page 88)

634 M. Troffaes, N. Huntley, and R. Shirota. Sequential decision processes under act-state independence with arbitrary choice functions. In E. Hüllermeier, R. Kruse, and F. Hoffmann, editors, *IPMU '10: Information Processing and Management of Uncertainty in Knowledge-Based Systems. Theory and Methods*, volume 80, pages 98–107. Berlin, 2010. Springer. (Cited on pages 199 and 228)

635 M. Troffaes, E. Miranda, and S. Destercke. On the connection between probability boxes and possibility measures. *Information Sciences*, 224:88–108, 2013. (Cited on page 92)

636 G. Tsoumakas and I. Vlahavas. Random k-labelsets: An ensemble method for multilabel classification. In J. Kok, R. López de Mántaras, S. Matwin, D. Mladenic, and A. Skowron, editors, *ECML '07: Proceedings of the 18th European conference on Machine Learning*, volume 4701, pages 406–417, Berlin, 2007. Springer. (Cited on page 250)

637 A. Tversky and D. Kahneman. Advances in prospect theory: Cumulative representation of uncertainty. *Journal of Risk and Uncertainty*, 5:297–323, 1992. (Cited on pages 324 and 328)

638 A. Tversky and D. Koehler. Support theory: A nonextensional representation of subjective probability. *Psychological Review*, 101:547–567, 1994. (Cited on pages 321 and 326)

639 L. Utkin. Probabilities of judgements provided by unknown experts by using the imprecise Dirichlet model. *Risk, Decision and Policy*, 9:391–400, 2004. (Cited on page 311)

640 L. Utkin. Extensions of belief functions and possibility distributions by using the imprecise Dirichlet model. *Fuzzy Sets and Systems*, 154:413–431, 2005. (Cited on pages 183 and 311)

641 L. Utkin. Cautious analysis of project risks by interval-valued initial data. *International Journal of Uncertainty, Fuzziness and Knowledge-Based Systems*, 14:663–685, 2006. (Cited on pages 183 and 290)

642 L. Utkin. A method for processing the unreliable expert judgments about parameters of probability distributions. *European Journal of Operational Research*, 175:385–398, 2006. (Cited on page 160)

643 L. Utkin. Fuzzy one-class classification model using contamination neighborhoods. *Advances in Fuzzy Systems*, 2012:1–10, 2012. (Cited on page 188)

644 L. Utkin and T. Augustin. Powerful algorithms for decision making under partial prior information and general ambiguity attitudes. In F. Cozman, R. Nau, and T. Seidenfeld, editors, *ISIPTA '05, Proceedings of the Fourth International Symposium on Imprecise Probabilities and Their Applications*, pages 349–358. Pittsburgh, PA, 2005. SIPTA. (Cited on pages 335 and 337)

645 L. Utkin and T. Augustin. Decision making under imperfect measurement using the imprecise Dirichlet model. *International Journal of Approximate Reasoning*, 44:322–338, 2007. (Cited on page 183)

646 L. Utkin and F. Coolen. Imprecise reliability: An introductory overview. In G. Levitin, editor, *Computational Intelligence in Reliability Engineering, Volume 2: New Metaheuristics, Neural and Fuzzy Techniques in Reliability*, volume 40, pages 261–306. Springer, Berlin, 2007. (Cited on page 303)

647 L. Utkin and S. Gurov. Imprecise reliability for the new lifetime distribution classes. *Journal of Statistical Planning and Inference*, 105:215–232, 2002. (Cited on page 311)

648 L. Utkin and I. Kozine. Computing the reliability of complex systems. In G. de Cooman, T. Fine, and T. Seidenfeld, editors, *ISIPTA '01: Proceedings of the Second International Symposium on Imprecise Probabilities and their Applications*, pages 324–331, Maastricht, 2001. Shaker. (Cited on pages 294 and 303)

649 L. Utkin and I. Kozine. Different faces of the natural extension. In G. de Cooman, T. Fine, and T. Seidenfeld, editors, *ISIPTA '01: Proceedings of the Second International Symposium on Imprecise Probabilities and Their Applications*, pages 316–323, Maastricht, 2001. Shaker. (Cited on page 54)

650 L. Utkin and I. Kozine. Stress-strength reliability models under incomplete information. *International Journal of General Systems*, 31:549–568, 2002. (Cited on pages 294 and 303)

651 L. Utkin and I. Kozine. On new cautious structural reliability models in the framework of imprecise probabilities. *Structural Safety*, 32:411–416, 2010. (Cited on page 160)

652 L. Utkin, S. Zatenko, and F. Coolen. New interval Bayesian models for software reliability based on non-homogeneous Poisson processes. *Automation and Remote Control*, 71:935–944, 2010. (Cited on page 160)

653 A. van Heerwarden and R. Kaas. The dutch premium principle. *Insurance, Mathematics and Economics*, 14:129–133, 1994. (Cited on page 287)

654 S. Vansteelandt, E. Goetghebeur, M. Kenward, and G. Molenberghs. Ignorance and uncertainty regions as inferential tools in a sensitivity analysis. *Statistica Sinica*, 16:953, 2006. (Cited on pages 182 and 183)

655 J. Verkuilen and M. Smithson. Mixed and mixture regression models for continuous bounded responses using the beta distribution. *Journal of Educational and Behavioral Statistics*, 37:82–113, 2012. (Cited on page 322)

656 P. Vicig. Epistemic independence for imprecise probabilities. *International Journal of Approximate Reasoning*, 24:235–250, 2000. (Cited on page 63)

657 P. Vicig. Financial risk measurement with imprecise probabilities. *International Journal of Approximate Reasoning*, 49:159–174, 2008. (Cited on pages 284, 288, and 289)

658 P. Vicig. A gambler's gain prospects with coherent imprecise previsions. In E. Hüllermeier, R. Kruse, and F. Hoffmann, editors, *Information Processing and Management of Uncertainty in Knowledge-Based Systems*, pages 50–59. Springer, Heidelberg, 2010. (Cited on pages 281 and 289)

659 P. Vicig, M. Zaffalon, and F. Cozman. Notes on 'Notes on conditional previsions'. *International Journal of Approximate Reasoning*, 44:358–365, 2007. (Cited on page 53)

660 J. von Neumann and O. Morgenstern. *Theory of Games and Economic Behavior*. Princeton University Press, Princeton, NJ, 1944. (Cited on page 26)

661 V. Vovk. A logic of probability, with application to the foundations of statistics (with discussion). *Journal of the Royal Statistical Society, Series B*, 55:317–351, 1993. (Cited on page 117)

662 V. Vovk. Continuous-time trading and the emergence of volatility. *Electronic Communications in Probability*, 13:319–324, 2008. (Cited on page 133)

663 V. Vovk. Rough paths in idealized financial markets. *Lithuanian Mathematical Journal*, 51:274–285, 2011. (Cited on page 132)

664 V. Vovk. Continuous-time trading and the emergence of probability. *Finance and Stochastics*, 16:561–609, 2012. (Cited on pages 124 and 133)

665 V. Vovk. Kolmogorov's strong law of large numbers in game-theoretic probability: Reality's side, 2013. arXiv:1304.1074. (Cited on page 133)

666 V. Vovk and G. Shafer. The game-theoretic capital asset pricing model. *International Journal of Approximate Reasoning*, 49:175–197, 2008. (Cited on page 289)

667 V. Vovk and A. Shen. Prequential randomness and probability. *Theoretical Computer Science*, 411:2632–2646, 2010. (Cited on page 132)

668 V. Vovk, A. Takemura, and G. Shafer. Defensive forecasting. In R. Cowell and G. Zoubin, editors, *AISTATS '05: Proceedings of the Tenth International Workshop on Artificial Intelligence and Statistics*, pages 365–372. Society for Artificial Intelligence and Statistics, Bridgetown, (available at http://www.gatsby.ucl.ac.uk/aistats/), 2005. (Cited on page 133)

669 P. Walley. Coherent lower (and upper) probabilities. Technical report, University of Warwick, Coventry, 1981. Statistics Research Report. (Cited on page 105)

670 P. Walley. The elicitation and aggregation of beliefs. Technical report, University of Warwick, Coventry, 1982. (Cited on page 110 and 111)

671 P. Walley. Belief function representations of statistical evidence. *The Annals of Statistics*, 15:1439–1465, 1987. (Cited on pages 106 and 112)

672 P. Walley. *Statistical Reasoning with Imprecise Probabilities*. Chapman & Hall, London, 1991. (Cited on pages 11, 18, 22, 26, 28, 33, 34, 35, 36, 37, 38, 39, 40, 44, 47, 48, 49, 51, 52, 53, 54, 63, 71, 72, 77, 81, 89, 91, 143, 149, 154, 156, 159, 161, 162, 163, 165, 186, 187, 196, 209, 221, 231, 233, 242, 243, 244, 246, 263, 283, 285, 289, 326, 335, and xiv)

673 P. Walley. Inferences from multinomial data: Learning about a bag of marbles. *Journal of the Royal Statistical Society, Series B*, 58:3–34, 1996. (Cited on pages 52, 155, 156, 159, 160, 177, 231, 233, 245, 306, and 310)

674 P. Walley. Towards a unified theory of imprecise probability. *International Journal of Approximate Reasoning*, 24:125–148, 2000. (Cited on pages 26, 80, 115, and 132)

675 P. Walley. Reconciling frequentist properties with the likelihood principle. *Journal of Statistical Planning and Inference*, 105:35–65, 2002. (Cited on page 179)

676 P. Walley and G. de Cooman. Coherence of rules for defining conditional possibility. *International Journal of Approximate Reasoning*, 21:63–107, 1999. (Cited on page 107)

677 P. Walley and T. Fine. Varieties of modal (classificatory) and comparative probability. *Synthese*, 41:321–374, 1979. (Cited on page 112)

678 P. Walley and T. Fine. Towards a frequentist theory of upper and lower probability. *Annals of Statistics*, 10:741–761, 1982. (Cited on pages 143, 146, 149, 165, 170, and xiv)

679 P. Walley, R. Pelessoni, and P. Vicig. Direct algorithms for checking consistency and making inferences from conditional probability assessments. *Journal of Statistical Planning and Inference*, 126:119–151, 2004. (Cited on pages 47, 53, 75, 330, and 331)

680 A. Wallner. Bi-elastic neighbourhood models. In J.-M. Bernard, T. Seidenfeld, and M. Zaffalon, editors, *ISIPTA '03: Proceedings of the Third International Symposium on Imprecise Probabilities and Their Applications (Lugano)*, pages 593–607, Waterloo, 2003. Carleton Scientific. (Cited on page 147)

681 T. Wallsten, D. Budescu, and I. Erev. Understanding and using linguistic uncertainties. *Acta Psychologica*, 68:39–52, 1988. (Cited on page 320)

682 T. Wallsten, D. Budescu, A. Rappoport, R. Zwick, and B. Forsyth. Measuring the vague meanings of probability terms. *Journal of Experimental Psychology: General*, 115:348–365, 1986. (Cited on pages 320, 321, and 328)

683 G. Walter. A technical note on the Dirichlet-multinomial model: The Dirichlet distribution as the canonically constructed conjugate prior. Technical Report 131 (available from http://epub.ub.uni-muenchen.de/14068/), Department of Statistics, LMU Munich, 2012. (Cited on page 153)

684 G. Walter and T. Augustin. Imprecision and prior-data conflict in generalized Bayesian inference. *Journal of Statistical Theory and Practice*, 3:255–271, 2009. Reprinted in [146], pages 107–123. (Cited on pages 151, 155, 156, 161, 162, and 164)

685 G. Walter, T. Augustin, and F. Coolen. On prior-data conflict in predictive Bernoulli inferences. In F. Coolen, G. de Cooman, T. Fetz, and M. Oberguggenberger, editors, *ISIPTA '11: Proceedings of the Seventh International Symposium on Imprecise Probability: Theories and Applications*, pages 391–400, Innsbruck, 2011. SIPTA. (Cited on pages c07S51b01-bib-0312, 157, and 164)

686 G. Walter, T. Augustin, and A. Peters. Linear regression analysis under sets of conjugate priors. In G. de Cooman, J. Vejnarova, and M. Zaffalon, editors, *ISIPTA '07: Proceedings of the Fifh International Symposium on Imprecise Probability: Theories and Applications*, pages 445–455, Prague, 2007. Action M Agency for SIPTA. (Cited on pages 164 and 188)

687 L. Wasserman. *All of Statistics: A Concise Course in Statistical Inference*. Springer, New York, 2004. (Cited on page 187)

688 L. Wasserman, M. Lavine, and R. Wolpert. Linearization of Bayesian robustness problems. *Journal of Statistical Planning and Inference*, 37:307–316, 1993. (Cited on page 165)

689 K. Weichselberger. Axiomatic foundations of the theory of interval probability. In V. Mammitzsch and H. Schneeweiss, editors, *Proceedings of the 2nd Gauss Symposium*, (Conference B) pages 47–64. de Gruyter, Berlin, 1995. (Cited on page 54)

690 K. Weichselberger. Stichproben und Intervallwahrscheinlichkeit. *ifo studien*, 41:653–676, 1995. (Cited on page 147)

691 K. Weichselberger. The theory of interval probability as a unifying model for uncertainty. *International Journal of Approximate Reasoning*, 24:149–170, 2000. (Cited on pages 54 and 264)

692 K. Weichselberger. *Elementare Grundbegriffe einer allgemeineren Wahrscheinlichkeitsrechnung I: Intervallwahrscheinlichkeit als umfassendes Konzept*. Physica, Heidelberg, 2001. (Cited on pages 54, 143, 159, 179, 264, 285, and xiv)

693 K. Weichselberger. The logical concept of probability and statistical inference. In F. Cozman, R. Nau, and T. Seidenfeld, editors, *ISIPTA '05: Proceedings of the Fourth International Symposium on Imprecise Probabilities and Their Applications*, pages 396–405, Pittsburgh, PA, 2005. SIPTA. (Cited on pages 55 and 179)

694 K. Weichselberger. *Elementare Grundbegriffe einer allgemeineren Wahrscheinlichkeitsrechnung II. Symmetrische Wahrscheinlichkeitstheorie*. (In preparation), 2014. (Cited on page 179)

695 K. Weichselberger and T. Augustin. Analysing Ellsberg's paradox by means of interval-probabilty. In R. Galata and H. Küchenhoff, editors, *Econometrics in Theory and Practice. Festschrift for Hans Schneeweiß*, pages 291–304. Physica, Heidelberg, 1998. (Cited on page 204)

696 K. Weichselberger and T. Augustin. On the symbiosis of two concepts of conditional interval probability. In J. Bernard, T. Seidenfeld, and M. Zaffalon, editors, *ISIPTA '03: Proceedings of the Third International Symposium on Imprecise Probabilities and Their Applications*, pages 608–630, Waterloo, 2003. Carleton Scientific. (Cited on page 55)

697 K. Weichselberger and S. Pöhlmann. *A Methodology for Uncertainty in Knowledge-based Systems*, volume 419. Springer Heidelberg, 1990. (Cited on pages 82 and 84)

698 M. Welsh, M. Lee, and S. Begg. More-Or-Less Elicitation (MOLE): Testing a heuristic elicitation method. In B. Love, K. McRae, and V. Sloutsky, editors, *CogSci '08: Proceedings of the 30th Annual Conference of the Cognitive Science Society*, pages 493–498. Cognitive Science Society, Washington, D.C., 2008. (Cited on page 327)

699 C. White and H. Eldeib. Markov decision processes with imprecise transition probabilities. *Operations Research*, 42:739–749, 1994. (Cited on page 264)

700 P. Whittle. *Probability via Expectation*. Springer, New York, 4th edition, 2000. (Cited on pages 34, 259, and 261)

701 P. Williams. Indeterminate probabilities. In M. Przelecki, K. Szaniawski, and R. Wójcicki, editors, *Formal Methods in the Methodology of Empirical Sciences: Proceedings of the Conference for Formal Methods in the Methodology of Empirical Sciences*, pages 229–246, Warsaw, 1974. Reidel, Dordrecht, Holland / Boston, MA; Ossolineum, Wrocław, Poland. (Cited on pages 26 and 53)

702 P. Williams. Coherence, strict coherence and zero probabilities. In *DLMPS '75: Proceedings of the Fifth International Congress of Logic, Methodology and Philosophy of Science*, volume VI, pages 29–33, 1975. (Cited on pages 26, 53, and xiv)

703 P. Williams. Notes on conditional previsions. *International Journal of Approximate Reasoning*, 44:366–383, 2007. (Cited on pages 26, 28, 53, 54, 289, 330, and 331)

704 N. Wilson. Algorithms for Dempster-Shafer theory. In D. Gabbay, P. Smets, J. Kohlas, and S. Moral, editors, *Handbook of Defeasible Reasoning and Uncertainty Management Systems*, volume 5, pages 421–475. Kluwer, Dordrecht, 2000. (Cited on page 84)

705 N. Wilson and S. Moral. A logical view of probability. In A. Cohn, editor, *ECAI '94: Proceedings of the 11th European Conference on Artificial Intelligence*, pages 386–390. Wiley, 1994. (Cited on page 26)

706 A. Winman, P. Hansson, and P. Juslin. Subjective probability intervals: How to cure overconfidence by interval evaluation. *Journal of Experimental Psychology: Learning, Memory, and Cognition*, 30:1167–1175, 2004. (Cited on pages 322, 325, 327, and 328)

707 I. Witten and E. Frank. *Data Mining: Practical Machine Learning Tools and Techniques*. Morgan Kaufmann, 2005. (Cited on page 251)

708 H. Witting. *Mathematische Statistik I, Parametrische Verfahren bei festem Stichprobenumfang*. Teubner, Stuttgart, 1985. (Cited on pages 138 and 170)

709 S. Wong, Y. Yao, P. Bollmann, and H. Burger. Axiomatization of qualitative belief structures. *IEEE Transactions on Systems, Man, and Cybernetics*, 21:259–276, 1991. (Cited on page 113)

710 Xi. Wu, V. Kumar, J. Quinlan, J. Ghosh, Q. Yang, H. Motoda, G. McLachlan, A. Ng, B. Liu, P. Yu, Z.-H. Zhou, M. Steinbach, D. Hand, and D. Steinberg. Top 10 algorithms in data mining. *Knowledge and Information Systems*, 14:1–37, 2008. (Cited on page 231)

711 R. Yager. Entropy and specificity in a mathematical theory of evidence. *International Journal of General Systems*, 9:249–260, 1983. (Cited on page 242)

712 I. Yaniv and D. Foster. Graininess of judgment under uncertainty: An accuracy-informativeness tradeoff. *Journal of Experimental Psychology: General*, 124:424–432, 1995. (Cited on page 325)

713 I. Yaniv and D. Foster. Precision and accuracy of judgmental estimation. *Journal of Behavioral Decision Making*, 10:21–32, 1997. (Cited on pages 325 and 328)

714 L. Zadeh. Fuzzy sets as a basis for a theory of possibility. *Fuzzy Sets and Systems*, 1:3–28, 1978. (Cited on pages 94, 100, and 109)

715 M. Zaffalon. Statistical inference of the naive credal classifier. In G. de Cooman, T. Fine, and T. Seidenfeld, editors, *ISIPTA '01: Proceedings of the Second International Symposium on Imprecise Probabilities and Their Applications*, pages 384–393, Maastricht, 2001. Shaker. (Cited on pages 231, 234, and 235)

716 M. Zaffalon. The naive credal classifier. *Journal of Statistical Planning and Inference*, 105:5–21, 2002. (Cited on page 90)

717 M. Zaffalon. Credible classification for environmental problems. *Environmental Modelling & Software*, 20:1003–1012, 2005. (Cited on pages 160 and 235)

718 M. Zaffalon, G. Corani, and D. Mauá. Utility-based accuracy measures to empirically evaluate credal classifiers. In F. Coolen, G. de Cooman, T. Fetz, and M. Oberguggenberger, editors, *ISIPTA '11: Proceedings of the Seventh International Symposium on Imprecise Probability: Theories and Applications*, pages 401–410, Innsbruck, 2011. SIPTA. (Cited on pages 250 and 251)

719 M. Zaffalon, G. Corani, and D. Mauá. Evaluating credal classifiers by utility-discounted predictive accuracy. *International Journal of Approximate Reasoning*, 53:1282–1301, 2012. (Cited on page 251)

720 M. Zaffalon and E. Fagiuoli. Tree-based credal networks for classification. *Reliable Computing*, 9:487–509, 2003. (Cited on page 241)

721 M. Zaffalon and E. Miranda. Conservative inference rule for uncertain reasoning under incompleteness. *Journal of Artificial Intelligence Research*, 34:757–821, 2009. (Cited on pages 65, 182, 222, and 236)

722 M. Zaffalon and E. Miranda. Probability and time. *Artificial Intelligence*, 198:1–51, 2013. (Cited on pages 42 and 52)

723 M. Zaffalon, K. Wesnes, and O. Petrini. Reliable diagnoses of dementia by the naive credal classifier inferred from incomplete cognitive data. *Artificial Intelligence in Medicine*, 29:61–79, 2003. (Cited on pages 160 and 235)

724 H. Zhang, R. Mullen, and R. Muhanna. Finite element structural analysis using imprecise probabilities based on p-box representation. In M. Beer, R. Muhanna, and R. Mullen, editors, *REC '10: 4th International Workshop on Reliable Engineering Computing. Robust Design – Coping with Hazards, Risk and Uncertainty*, pages 211–225, Singapore, 2010. Research Publishing Services. (Cited on page 303)

725 Z. Zhang. Profile likelihood and incomplete data. *International Statistical Review*, 78:102–116, 2010. (Cited on page 187)

Author index

Introduction to Imprecise Probabilities, First Edition.
Edited by Thomas Augustin, Frank P. A. Coolen, Gert de Cooman and Matthias C. M. Troffaes.

Subject index

Introduction to Imprecise Probabilities, First Edition.
Edited by Thomas Augustin, Frank P. A. Coolen, Gert de Cooman and Matthias C. M. Troffaes.

WILEY SERIES IN PROBABILITY AND STATISTICS
ESTABLISHED BY WALTER A. SHEWHART AND SAMUEL S. WILKS

The ***Wiley Series in Probability and Statistics*** is well established and authoritative. It covers many topics of current research interest in both pure and applied statistics and probability theory. Written by leading statisticians and institutions, the titles span both state-of-the-art developments in the field and classical methods.

Reflecting the wide range of current research in statistics, the series encompasses applied, methodological and theoretical statistics, ranging from applications and new techniques made possible by advances in computerized practice to rigorous treatment of theoretical approaches.

This series provides essential and invaluable reading for all statisticians, whether in academia, industry, government, or research.

† ABRAHAM and LEDOLTER · Statistical Methods for Forecasting
AGRESTI · Analysis of Ordinal Categorical Data, *Second Edition*
AGRESTI · An Introduction to Categorical Data Analysis, *Second Edition*
AGRESTI · Categorical Data Analysis, *Third Edition*
ALSTON, MENGERSEN and PETTITT (editors) · Case Studies in Bayesian Statistical Modelling and Analysis
ALTMAN, GILL, and McDONALD · Numerical Issues in Statistical Computing for the Social Scientist
AMARATUNGA and CABRERA · Exploration and Analysis of DNA Microarray and Protein Array Data
ANDĚL · Mathematics of Chance
ANDERSON · An Introduction to Multivariate Statistical Analysis, *Third Edition*
* ANDERSON · The Statistical Analysis of Time Series
ANDERSON, AUQUIER, HAUCK, OAKES, VANDAELE, and WEISBERG · Statistical Methods for Comparative Studies
ANDERSON and LOYNES · The Teaching of Practical Statistics
ARMITAGE and DAVID (editors) · Advances in Biometry
ARNOLD, BALAKRISHNAN, and NAGARAJA · Records
* ARTHANARI and DODGE · Mathematical Programming in Statistics
AUGUSTIN, COOLEN, DE COOMAN and TROFFAES (editors) · Introduction to Imprecise Probabilities
* BAILEY · The Elements of Stochastic Processes with Applications to the Natural Sciences
BAJORSKI · Statistics for Imaging, Optics, and Photonics
BALAKRISHNAN and KOUTRAS · Runs and Scans with Applications
BALAKRISHNAN and NG · Precedence-Type Tests and Applications

*Now available in a lower priced paperback edition in the Wiley Classics Library.
†Now available in a lower priced paperback edition in the Wiley–Interscience Paperback Series.

BARNETT · Comparative Statistical Inference, *Third Edition*

BARNETT · Environmental Statistics

BARNETT and LEWIS · Outliers in Statistical Data, *Third Edition*

BARTHOLOMEW, KNOTT, and MOUSTAKI · Latent Variable Models and Factor Analysis: A Unified Approach, *Third Edition*

BARTOSZYNSKI and NIEWIADOMSKA-BUGAJ · Probability and Statistical Inference, *Second Edition*

BASILEVSKY · Statistical Factor Analysis and Related Methods: Theory and Applications

BATES and WATTS · Nonlinear Regression Analysis and Its Applications

BECHHOFER, SANTNER, and GOLDSMAN · Design and Analysis of Experiments for Statistical Selection, Screening, and Multiple Comparisons

BEIRLANT, GOEGEBEUR, SEGERS, TEUGELS, and DE WAAL · Statistics of Extremes: Theory and Applications

BELSLEY · Conditioning Diagnostics: Collinearity and Weak Data in Regression

† BELSLEY, KUH, and WELSCH · Regression Diagnostics: Identifying Influential Data and Sources of Collinearity

BENDAT and PIERSOL · Random Data: Analysis and Measurement Procedures, *Fourth Edition*

BERNARDO and SMITH · Bayesian Theory

BHAT and MILLER · Elements of Applied Stochastic Processes, *Third Edition*

BHATTACHARYA and WAYMIRE · Stochastic Processes with Applications

BIEMER, GROVES, LYBERG, MATHIOWETZ, and SUDMAN · Measurement Errors in Surveys

BILLINGSLEY · Convergence of Probability Measures, *Second Edition*

BILLINGSLEY · Probability and Measure, *Anniversary Edition*

BIRKES and DODGE · Alternative Methods of Regression

BISGAARD and KULAHCI · Time Series Analysis and Forecasting by Example

BISWAS, DATTA, FINE, and SEGAL · Statistical Advances in the Biomedical Sciences: Clinical Trials, Epidemiology, Survival Analysis, and Bioinformatics

BLISCHKE and MURTHY (editors) · Case Studies in Reliability and Maintenance

BLISCHKE and MURTHY · Reliability: Modeling, Prediction, and Optimization

BLOOMFIELD · Fourier Analysis of Time Series: An Introduction, *Second Edition*

BOLLEN · Structural Equations with Latent Variables

BOLLEN and CURRAN · Latent Curve Models: A Structural Equation Perspective

BOROVKOV · Ergodicity and Stability of Stochastic Processes

BOSQ and BLANKE · Inference and Prediction in Large Dimensions

BOULEAU · Numerical Methods for Stochastic Processes

* BOX and TIAO · Bayesian Inference in Statistical Analysis

BOX · Improving Almost Anything, *Revised Edition*

*Now available in a lower priced paperback edition in the Wiley Classics Library.

†Now available in a lower priced paperback edition in the Wiley–Interscience Paperback Series.

* BOX and DRAPER · Evolutionary Operation: A Statistical Method for Process Improvement

BOX and DRAPER · Response Surfaces, Mixtures, and Ridge Analyses, *Second Edition*

BOX, HUNTER, and HUNTER · Statistics for Experimenters: Design, Innovation, and Discovery, *Second Editon*

BOX, JENKINS, and REINSEL · Time Series Analysis: Forcasting and Control, *Fourth Edition*

BOX, LUCEÑO, and PANIAGUA-QUIÑONES · Statistical Control by Monitoring and Adjustment, *Second Edition*

* BROWN and HOLLANDER · Statistics: A Biomedical Introduction

CAIROLI and DALANG · Sequential Stochastic Optimization

CASTILLO, HADI, BALAKRISHNAN, and SARABIA · Extreme Value and Related Models with Applications in Engineering and Science

CHAN · Time Series: Applications to Finance with R and S-Plus®, *Second Edition*

CHARALAMBIDES · Combinatorial Methods in Discrete Distributions

CHATTERJEE and HADI · Regression Analysis by Example, *Fourth Edition*

CHATTERJEE and HADI · Sensitivity Analysis in Linear Regression

CHERNICK · Bootstrap Methods: A Guide for Practitioners and Researchers, *Second Edition*

CHERNICK and FRIIS · Introductory Biostatistics for the Health Sciences

CHILÈS and DELFINER · Geostatistics: Modeling Spatial Uncertainty, *Second Edition*

CHIU, STOYAN, KENDALL and MECKE · Stochastic Geometry and its Applications, *Third Edition*

CHOW and LIU · Design and Analysis of Clinical Trials: Concepts and Methodologies, *Third Edition*

CLARKE · Linear Models: The Theory and Application of Analysis of Variance

CLARKE and DISNEY · Probability and Random Processes: A First Course with Applications, *Second Edition*

* COCHRAN and COX · Experimental Designs, *Second Edition*

COLLINS and LANZA · Latent Class and Latent Transition Analysis: With Applications in the Social, Behavioral, and Health Sciences

CONGDON · Applied Bayesian Modelling

CONGDON · Bayesian Models for Categorical Data

CONGDON · Bayesian Statistical Modelling, *Second Edition*

CONOVER · Practical Nonparametric Statistics, *Third Edition*

COOK · Regression Graphics

COOK and WEISBERG · An Introduction to Regression Graphics

COOK and WEISBERG · Applied Regression Including Computing and Graphics

CORNELL · A Primer on Experiments with Mixtures

*Now available in a lower priced paperback edition in the Wiley Classics Library.

†Now available in a lower priced paperback edition in the Wiley–Interscience Paperback Series.

CORNELL · Experiments with Mixtures, Designs, Models, and the Analysis of Mixture Data, *Third Edition*

COX · A Handbook of Introductory Statistical Methods

CRESSIE · Statistics for Spatial Data, *Revised Edition*

CRESSIE and WIKLE · Statistics for Spatio-Temporal Data

CSÖRGÖ and HORVÁTH · Limit Theorems in Change Point Analysis

DAGPUNAR · Simulation and Monte Carlo: With Applications in Finance and MCMC

DANIEL · Applications of Statistics to Industrial Experimentation

DANIEL · Biostatistics: A Foundation for Analysis in the Health Sciences, *Eighth Edition*

* DANIEL · Fitting Equations to Data: Computer Analysis of Multifactor Data, *Second Edition*

DASU and JOHNSON · Exploratory Data Mining and Data Cleaning

DAVID and NAGARAJA · Order Statistics, *Third Edition*

DAVINO, FURNO and VISTOCCO · Quantile Regression: Theory and Applications

* DEGROOT, FIENBERG, and KADANE · Statistics and the Law

DEL CASTILLO · Statistical Process Adjustment for Quality Control

DEMARIS · Regression with Social Data: Modeling Continuous and Limited Response Variables

DEMIDENKO · Mixed Models: Theory and Applications

DENISON, HOLMES, MALLICK and SMITH · Bayesian Methods for Nonlinear Classification and Regression

DETTE and STUDDEN · The Theory of Canonical Moments with Applications in Statistics, Probability, and Analysis

DEY and MUKERJEE · Fractional Factorial Plans

DE ROCQUIGNY · Modelling Under Risk and Uncertainty: An Introduction to Statistical, Phenomenological and Computational Models

DILLON and GOLDSTEIN · Multivariate Analysis: Methods and Applications

* DODGE and ROMIG · Sampling Inspection Tables, *Second Edition*

* DOOB · Stochastic Processes

DOWDY, WEARDEN, and CHILKO · Statistics for Research, *Third Edition*

DRAPER and SMITH · Applied Regression Analysis, *Third Edition*

DRYDEN and MARDIA · Statistical Shape Analysis

DUDEWICZ and MISHRA · Modern Mathematical Statistics

DUNN and CLARK · Basic Statistics: A Primer for the Biomedical Sciences, *Fourth Edition*

DUPUIS and ELLIS · A Weak Convergence Approach to the Theory of Large Deviations

EDLER and KITSOS · Recent Advances in Quantitative Methods in Cancer and Human Health Risk Assessment

* ELANDT-JOHNSON and JOHNSON · Survival Models and Data Analysis

*Now available in a lower priced paperback edition in the Wiley Classics Library.

†Now available in a lower priced paperback edition in the Wiley–Interscience Paperback Series.

ENDERS · Applied Econometric Time Series, *Third Edition*

† ETHIER and KURTZ · Markov Processes: Characterization and Convergence

EVANS, HASTINGS, and PEACOCK · Statistical Distributions, *Third Edition*

EVERITT, LANDAU, LEESE, and STAHL · Cluster Analysis, *Fifth Edition*

FEDERER and KING · Variations on Split Plot and Split Block Experiment Designs

FELLER · An Introduction to Probability Theory and Its Applications, Volume I, *Third Edition*, Revised; Volume II, *Second Edition*

FITZMAURICE, LAIRD, and WARE · Applied Longitudinal Analysis, *Second Edition*

* FLEISS · The Design and Analysis of Clinical Experiments

FLEISS · Statistical Methods for Rates and Proportions, *Third Edition*

† FLEMING and HARRINGTON · Counting Processes and Survival Analysis

FUJIKOSHI, ULYANOV, and SHIMIZU · Multivariate Statistics: High-Dimensional and Large-Sample Approximations

FULLER · Introduction to Statistical Time Series, *Second Edition*

† FULLER · Measurement Error Models

GALLANT · Nonlinear Statistical Models

GEISSER · Modes of Parametric Statistical Inference

GELMAN and MENG · Applied Bayesian Modeling and Causal Inference from Incomplete-Data Perspectives

GEWEKE · Contemporary Bayesian Econometrics and Statistics

GHOSH, MUKHOPADHYAY, and SEN · Sequential Estimation

GIESBRECHT and GUMPERTZ · Planning, Construction, and Statistical Analysis of Comparative Experiments

GIFI · Nonlinear Multivariate Analysis

GIVENS and HOETING · Computational Statistics

GLASSERMAN and YAO · Monotone Structure in Discrete-Event Systems

GNANADESIKAN · Methods for Statistical Data Analysis of Multivariate Observations, *Second Edition*

GOLDSTEIN · Multilevel Statistical Models, *Fourth Edition*

GOLDSTEIN and LEWIS · Assessment: Problems, Development, and Statistical Issues

GOLDSTEIN and WOOFF · Bayes Linear Statistics

GRAHAM · Markov Chains: Analytic and Monte Carlo Computations

GREENWOOD and NIKULIN · A Guide to Chi-Squared Testing

GROSS, SHORTLE, THOMPSON, and HARRIS · Fundamentals of Queueing Theory, *Fourth Edition*

GROSS, SHORTLE, THOMPSON, and HARRIS · Solutions Manual to Accompany Fundamentals of Queueing Theory, *Fourth Edition*

* HAHN and SHAPIRO · Statistical Models in Engineering

HAHN and MEEKER · Statistical Intervals: A Guide for Practitioners

HALD · A History of Probability and Statistics and their Applications Before 1750

*Now available in a lower priced paperback edition in the Wiley Classics Library.

†Now available in a lower priced paperback edition in the Wiley–Interscience Paperback Series.

† HAMPEL · Robust Statistics: The Approach Based on Influence Functions
HARTUNG, KNAPP, and SINHA · Statistical Meta-Analysis with Applications
HEIBERGER · Computation for the Analysis of Designed Experiments
HEDAYAT and SINHA · Design and Inference in Finite Population Sampling
HEDEKER and GIBBONS · Longitudinal Data Analysis
HELLER · MACSYMA for Statisticians
HERITIER, CANTONI, COPT, and VICTORIA-FESER · Robust Methods in Biostatistics
HINKELMANN and KEMPTHORNE · Design and Analysis of Experiments, Volume 1: Introduction to Experimental Design, *Second Edition*
HINKELMANN and KEMPTHORNE · Design and Analysis of Experiments, Volume 2: Advanced Experimental Design
HINKELMANN (editor) · Design and Analysis of Experiments, Volume 3: Special Designs and Applications
* HOAGLIN, MOSTELLER, and TUKEY · Fundamentals of Exploratory Analysis of Variance
* HOAGLIN, MOSTELLER, and TUKEY · Exploring Data Tables, Trends and Shapes
* HOAGLIN, MOSTELLER, and TUKEY · Understanding Robust and Exploratory Data Analysis
HOCHBERG and TAMHANE · Multiple Comparison Procedures
HOCKING · Methods and Applications of Linear Models: Regression and the Analysis of Variance, *Third Edition*
HOEL · Introduction to Mathematical Statistics, *Fifth Edition*
HOGG and KLUGMAN · Loss Distributions
HOLLANDER and WOLFE · Nonparametric Statistical Methods, *Second Edition*
HOSMER and LEMESHOW · Applied Logistic Regression, *Second Edition*
HOSMER, LEMESHOW, and MAY · Applied Survival Analysis: Regression Modeling of Time-to-Event Data, *Second Edition*
HUBER · Data Analysis: What Can Be Learned From the Past 50 Years
HUBER · Robust Statistics
† HUBER and RONCHETTI · Robust Statistics, *Second Edition*
HUBERTY · Applied Discriminant Analysis, *Second Edition*
HUBERTY and OLEJNIK · Applied MANOVA and Discriminant Analysis, *Second Edition*
HUITEMA · The Analysis of Covariance and Alternatives: Statistical Methods for Experiments, Quasi-Experiments, and Single-Case Studies, *Second Edition*
HUNT and KENNEDY · Financial Derivatives in Theory and Practice, *Revised Edition*
HURD and MIAMEE · Periodically Correlated Random Sequences: Spectral Theory and Practice
HUSKOVA, BERAN, and DUPAC · Collected Works of Jaroslav Hajek – with Commentary
HUZURBAZAR · Flowgraph Models for Multistate Time-to-Event Data

*Now available in a lower priced paperback edition in the Wiley Classics Library.
†Now available in a lower priced paperback edition in the Wiley–Interscience Paperback Series.

INSUA, RUGGERI and WIPER · Bayesian Analysis of Stochastic Process Models

JACKMAN · Bayesian Analysis for the Social Sciences

† JACKSON · A User's Guide to Principle Components

JOHN · Statistical Methods in Engineering and Quality Assurance

JOHNSON · Multivariate Statistical Simulation

JOHNSON and BALAKRISHNAN · Advances in the Theory and Practice of Statistics: A Volume in Honor of Samuel Kotz

JOHNSON, KEMP, and KOTZ · Univariate Discrete Distributions, *Third Edition*

JOHNSON and KOTZ (editors) · Leading Personalities in Statistical Sciences: From the Seventeenth Century to the Present

JOHNSON, KOTZ, and BALAKRISHNAN · Continuous Univariate Distributions, Volume 1, *Second Edition*

JOHNSON, KOTZ, and BALAKRISHNAN · Continuous Univariate Distributions, Volume 2, *Second Edition*

JOHNSON, KOTZ, and BALAKRISHNAN · Discrete Multivariate Distributions

JUDGE, GRIFFITHS, HILL, LÜTKEPOHL, and LEE · The Theory and Practice of Econometrics, *Second Edition*

JUREK and MASON · Operator-Limit Distributions in Probability Theory

KADANE · Bayesian Methods and Ethics in a Clinical Trial Design

KADANE AND SCHUM · A Probabilistic Analysis of the Sacco and Vanzetti Evidence

KALBFLEISCH and PRENTICE · The Statistical Analysis of Failure Time Data, *Second Edition*

KARIYA and KURATA · Generalized Least Squares

KASS and VOS · Geometrical Foundations of Asymptotic Inference

† KAUFMAN and ROUSSEEUW · Finding Groups in Data: An Introduction to Cluster Analysis

KEDEM and FOKIANOS · Regression Models for Time Series Analysis

KENDALL, BARDEN, CARNE, and LE · Shape and Shape Theory

KHURI · Advanced Calculus with Applications in Statistics, *Second Edition*

KHURI, MATHEW, and SINHA · Statistical Tests for Mixed Linear Models

* KISH · Statistical Design for Research

KLEIBER and KOTZ · Statistical Size Distributions in Economics and Actuarial Sciences

KLEMELÄ · Smoothing of Multivariate Data: Density Estimation and Visualization

KLUGMAN, PANJER, and WILLMOT · Loss Models: From Data to Decisions, *Third Edition*

KLUGMAN, PANJER, and WILLMOT · Solutions Manual to Accompany Loss Models: From Data to Decisions, *Third Edition*

KOSKI and NOBLE · Bayesian Networks: An Introduction

KOTZ, BALAKRISHNAN, and JOHNSON · Continuous Multivariate Distributions, Volume 1, *Second Edition*

*Now available in a lower priced paperback edition in the Wiley Classics Library.

†Now available in a lower priced paperback edition in the Wiley–Interscience Paperback Series.

KOTZ and JOHNSON (editors) · Encyclopedia of Statistical Sciences: Volumes 1 to 9 with Index

KOTZ and JOHNSON (editors) · Encyclopedia of Statistical Sciences: Supplement Volume

KOTZ, READ, and BANKS (editors) · Encyclopedia of Statistical Sciences: Update Volume 1

KOTZ, READ, and BANKS (editors) · Encyclopedia of Statistical Sciences: Update Volume 2

KOWALSKI and TU · Modern Applied U-Statistics

KRISHNAMOORTHY and MATHEW · Statistical Tolerance Regions: Theory, Applications, and Computation

KROESE, TAIMRE, and BOTEV · Handbook of Monte Carlo Methods

KROONENBERG · Applied Multiway Data Analysis

KULINSKAYA, MORGENTHALER, and STAUDTE · Meta Analysis: A Guide to Calibrating and Combining Statistical Evidence

KULKARNI and HARMAN · An Elementary Introduction to Statistical Learning Theory

KUROWICKA and COOKE · Uncertainty Analysis with High Dimensional Dependence Modelling

KVAM and VIDAKOVIC · Nonparametric Statistics with Applications to Science and Engineering

LACHIN · Biostatistical Methods: The Assessment of Relative Risks, *Second Edition*

LAD · Operational Subjective Statistical Methods: A Mathematical, Philosophical, and Historical Introduction

LAMPERTI · Probability: A Survey of the Mathematical Theory, *Second Edition*

LAWLESS · Statistical Models and Methods for Lifetime Data, *Second Edition*

LAWSON · Statistical Methods in Spatial Epidemiology, *Second Edition*

LE · Applied Categorical Data Analysis, *Second Edition*

LE · Applied Survival Analysis

LEE · Structural Equation Modeling: A Bayesian Approach

LEE and WANG · Statistical Methods for Survival Data Analysis, *Third Edition*

LEPAGE and BILLARD · Exploring the Limits of Bootstrap

LESSLER and KALSBEEK · Nonsampling Errors in Surveys

LEYLAND and GOLDSTEIN (editors) · Multilevel Modelling of Health Statistics

LIAO · Statistical Group Comparison

LIN · Introductory Stochastic Analysis for Finance and Insurance

LITTLE and RUBIN · Statistical Analysis with Missing Data, *Second Edition*

LLOYD · The Statistical Analysis of Categorical Data

LOWEN and TEICH · Fractal-Based Point Processes

MAGNUS and NEUDECKER · Matrix Differential Calculus with Applications in Statistics and Econometrics, *Revised Edition*

MALLER and ZHOU · Survival Analysis with Long Term Survivors

MARCHETTE · Random Graphs for Statistical Pattern Recognition

MARDIA and JUPP · Directional Statistics

MARKOVICH · Nonparametric Analysis of Univariate Heavy-Tailed Data: Research and Practice

MARONNA, MARTIN and YOHAI · Robust Statistics: Theory and Methods

MASON, GUNST, and HESS · Statistical Design and Analysis of Experiments with Applications to Engineering and Science, *Second Edition*

McCULLOCH, SEARLE, and NEUHAUS · Generalized, Linear, and Mixed Models, *Second Edition*

McFADDEN · Management of Data in Clinical Trials, *Second Edition*

* McLACHLAN · Discriminant Analysis and Statistical Pattern Recognition

McLACHLAN, DO, and AMBROISE · Analyzing Microarray Gene Expression Data

McLACHLAN and KRISHNAN · The EM Algorithm and Extensions, *Second Edition*

McLACHLAN and PEEL · Finite Mixture Models

McNEIL · Epidemiological Research Methods

MEEKER and ESCOBAR · Statistical Methods for Reliability Data

MEERSCHAERT and SCHEFFLER · Limit Distributions for Sums of Independent Random Vectors: Heavy Tails in Theory and Practice

MENGERSEN, ROBERT, and TITTERINGTON · Mixtures: Estimation and Applications

MICKEY, DUNN, and CLARK · Applied Statistics: Analysis of Variance and Regression, *Third Edition*

* MILLER · Survival Analysis, *Second Edition*

MONTGOMERY, JENNINGS, and KULAHCI · Introduction to Time Series Analysis and Forecasting

MONTGOMERY, PECK, and VINING · Introduction to Linear Regression Analysis, *Fifth Edition*

MORGENTHALER and TUKEY · Configural Polysampling: A Route to Practical Robustness

MUIRHEAD · Aspects of Multivariate Statistical Theory

MULLER and STOYAN · Comparison Methods for Stochastic Models and Risks

MURTHY, XIE, and JIANG · Weibull Models

MYERS, MONTGOMERY, and ANDERSON-COOK · Response Surface Methodology: Process and Product Optimization Using Designed Experiments, *Third Edition*

MYERS, MONTGOMERY, VINING, and ROBINSON · Generalized Linear Models. With Applications in Engineering and the Sciences, *Second Edition*

NATVIG · Multistate Systems Reliability Theory With Applications

† NELSON · Accelerated Testing, Statistical Models, Test Plans, and Data Analyses

† NELSON · Applied Life Data Analysis

NEWMAN · Biostatistical Methods in Epidemiology

NG, TAIN, and TANG · Dirichlet Theory: Theory, Methods and Applications

OKABE, BOOTS, SUGIHARA, and CHIU · Spatial Tesselations: Concepts and Applications of Voronoi Diagrams, *Second Edition*

OLIVER and SMITH · Influence Diagrams, Belief Nets and Decision Analysis

*Now available in a lower priced paperback edition in the Wiley Classics Library.

†Now available in a lower priced paperback edition in the Wiley–Interscience Paperback Series.

PALTA · Quantitative Methods in Population Health: Extensions of Ordinary Regressions

PANJER · Operational Risk: Modeling and Analytics

PANKRATZ · Forecasting with Dynamic Regression Models

PANKRATZ · Forecasting with Univariate Box-Jenkins Models: Concepts and Cases

PARDOUX · Markov Processes and Applications: Algorithms, Networks, Genome and Finance

PARMIGIANI and INOUE · Decision Theory: Principles and Approaches

* PARZEN · Modern Probability Theory and Its Applications

PEÑA, TIAO, and TSAY · A Course in Time Series Analysis

PESARIN and SALMASO · Permutation Tests for Complex Data: Applications and Software

PIANTADOSI · Clinical Trials: A Methodologic Perspective, *Second Edition*

POURAHMADI · Foundations of Time Series Analysis and Prediction Theory

POURAHMADI · High-Dimensional Covariance Estimation

POWELL · Approximate Dynamic Programming: Solving the Curses of Dimensionality, *Second Edition*

POWELL and RYZHOV · Optimal Learning

PRESS · Subjective and Objective Bayesian Statistics, *Second Edition*

PRESS and TANUR · The Subjectivity of Scientists and the Bayesian Approach

PURI, VILAPLANA, and WERTZ · New Perspectives in Theoretical and Applied Statistics

† PUTERMAN · Markov Decision Processes: Discrete Stochastic Dynamic Programming

QIU · Image Processing and Jump Regression Analysis

* RAO · Linear Statistical Inference and Its Applications, *Second Edition*

RAO · Statistical Inference for Fractional Diffusion Processes

RAUSAND and HØYLAND · System Reliability Theory: Models, Statistical Methods, and Applications, *Second Edition*

RAYNER, THAS, and BEST · Smooth Tests of Goodnes of Fit: Using R, *Second Edition*

RENCHER and SCHAALJE · Linear Models in Statistics, *Second Edition*

RENCHER and CHRISTENSEN · Methods of Multivariate Analysis, *Third Edition*

RENCHER · Multivariate Statistical Inference with Applications

RIGDON and BASU · Statistical Methods for the Reliability of Repairable Systems

* RIPLEY · Spatial Statistics

* RIPLEY · Stochastic Simulation

ROHATGI and SALEH · An Introduction to Probability and Statistics, *Second Edition*

ROLSKI, SCHMIDLI, SCHMIDT, and TEUGELS · Stochastic Processes for Insurance and Finance

*Now available in a lower priced paperback edition in the Wiley Classics Library.

†Now available in a lower priced paperback edition in the Wiley–Interscience Paperback Series.

ROSENBERGER and LACHIN · Randomization in Clinical Trials: Theory and Practice

ROSSI, ALLENBY, and McCULLOCH · Bayesian Statistics and Marketing

† ROUSSEEUW and LEROY · Robust Regression and Outlier Detection

ROYSTON and SAUERBREI · Multivariate Model Building: A Pragmatic Approach to Regression Analysis Based on Fractional Polynomials for Modeling Continuous Variables

* RUBIN · Multiple Imputation for Nonresponse in Surveys

RUBINSTEIN and KROESE · Simulation and the Monte Carlo Method, *Second Edition*

RUBINSTEIN and MELAMED · Modern Simulation and Modeling

RUBINSTEIN, RIDDER, and VAISMAN · Fast Sequential Monte Carlo Methods for Counting and Optimization

RYAN · Modern Engineering Statistics

RYAN · Modern Experimental Design

RYAN · Modern Regression Methods, *Second Edition*

RYAN · Sample Size Determination and Power

RYAN · Statistical Methods for Quality Improvement, *Third Edition*

SALEH · Theory of Preliminary Test and Stein-Type Estimation with Applications

SALTELLI, CHAN, and SCOTT (editors) · Sensitivity Analysis

SCHERER · Batch Effects and Noise in Microarray Experiments: Sources and Solutions

* SCHEFFE · The Analysis of Variance

SCHIMEK · Smoothing and Regression: Approaches, Computation, and Application

SCHOTT · Matrix Analysis for Statistics, *Second Edition*

SCHOUTENS · Levy Processes in Finance: Pricing Financial Derivatives

SCOTT · Multivariate Density Estimation: Theory, Practice, and Visualization

* SEARLE · Linear Models

† SEARLE · Linear Models for Unbalanced Data

† SEARLE · Matrix Algebra Useful for Statistics

† SEARLE, CASELLA, and McCULLOCH · Variance Components

SEARLE and WILLETT · Matrix Algebra for Applied Economics

SEBER · A Matrix Handbook For Statisticians

† SEBER · Multivariate Observations

SEBER and LEE · Linear Regression Analysis, *Second Edition*

† SEBER and WILD · Nonlinear Regression

SENNOTT · Stochastic Dynamic Programming and the Control of Queueing Systems

* SERFLING · Approximation Theorems of Mathematical Statistics

SHAFER and VOVK · Probability and Finance: It's Only a Game!

SHERMAN · Spatial Statistics and Spatio-Temporal Data: Covariance Functions and Directional Properties

*Now available in a lower priced paperback edition in the Wiley Classics Library.

†Now available in a lower priced paperback edition in the Wiley–Interscience Paperback Series.

SILVAPULLE and SEN · Constrained Statistical Inference: Inequality, Order, and Shape Restrictions

SINGPURWALLA · Reliability and Risk: A Bayesian Perspective

SMALL and McLEISH · Hilbert Space Methods in Probability and Statistical Inference

SRIVASTAVA · Methods of Multivariate Statistics

STAPLETON · Linear Statistical Models, *Second Edition*

STAPLETON · Models for Probability and Statistical Inference: Theory and Applications

STAUDTE and SHEATHER · Robust Estimation and Testing

STOYAN · Counterexamples in Probability, *Second Edition*

STOYAN and STOYAN · Fractals, Random Shapes and Point Fields: Methods of Geometrical Statistics

STREET and BURGESS · The Construction of Optimal Stated Choice Experiments: Theory and Methods

STYAN · The Collected Papers of T. W. Anderson: 1943–1985

SUTTON, ABRAMS, JONES, SHELDON, and SONG · Methods for Meta-Analysis in Medical Research

TAKEZAWA · Introduction to Nonparametric Regression

TAMHANE · Statistical Analysis of Designed Experiments: Theory and Applications

TANAKA · Time Series Analysis: Nonstationary and Noninvertible Distribution Theory

THOMPSON · Empirical Model Building: Data, Models, and Reality, *Second Edition*

THOMPSON · Sampling, *Third Edition*

THOMPSON · Simulation: A Modeler's Approach

THOMPSON and SEBER · Adaptive Sampling

THOMPSON, WILLIAMS, and FINDLAY · Models for Investors in Real World Markets

TIERNEY · LISP-STAT: An Object-Oriented Environment for Statistical Computing and Dynamic Graphics

TROFFAES and DE COOMAN · Lower Previsions

TSAY · Analysis of Financial Time Series, *Third Edition*

TSAY · An Introduction to Analysis of Financial Data with R

TSAY · Multivariate Time Series Analysis: With R and Financial Applications

UPTON and FINGLETON · Spatial Data Analysis by Example, Volume II: Categorical and Directional Data

† VAN BELLE · Statistical Rules of Thumb, *Second Edition*

VAN BELLE, FISHER, HEAGERTY, and LUMLEY · Biostatistics: A Methodology for the Health Sciences, *Second Edition*

VESTRUP · The Theory of Measures and Integration

VIDAKOVIC · Statistical Modeling by Wavelets

*Now available in a lower priced paperback edition in the Wiley Classics Library.

†Now available in a lower priced paperback edition in the Wiley–Interscience Paperback Series.

VIERTL · Statistical Methods for Fuzzy Data

VINOD and REAGLE · Preparing for the Worst: Incorporating Downside Risk in Stock Market Investments

WALLER and GOTWAY · Applied Spatial Statistics for Public Health Data

WANG and WANG · Structural Equation Modeling: Applications Using M*plus*

WEISBERG · Applied Linear Regression, *Fourth Edition*

WEISBERG · Bias and Causation: Models and Judgment for Valid Comparisons

WELSH · Aspects of Statistical Inference

WESTFALL and YOUNG · Resampling-Based Multiple Testing: Examples and Methods for *p*-Value Adjustment

* WHITTAKER · Graphical Models in Applied Multivariate Statistics

WINKER · Optimization Heuristics in Economics: Applications of Threshold Accepting

WOODWORTH · Biostatistics: A Bayesian Introduction

WOOLSON and CLARKE · Statistical Methods for the Analysis of Biomedical Data, *Second Edition*

WU and HAMADA · Experiments: Planning, Analysis, and Parameter Design Optimization, *Second Edition*

WU and ZHANG · Nonparametric Regression Methods for Longitudinal Data Analysis

YAKIR · Extremes in Random Fields

YIN · Clinical Trial Design: Bayesian and Frequentist Adaptive Methods

YOUNG, VALERO-MORA, and FRIENDLY · Visual Statistics: Seeing Data with Dynamic Interactive Graphics

ZACKS · Stage-Wise Adaptive Designs

* ZELLNER · An Introduction to Bayesian Inference in Econometrics

ZELTERMAN · Discrete Distributions – Applications in the Health Sciences

ZHOU, OBUCHOWSKI, and MCCLISH · Statistical Methods in Diagnostic Medicine, *Second Edition*

*Now available in a lower priced paperback edition in the Wiley Classics Library.

†Now available in a lower priced paperback edition in the Wiley–Interscience Paperback Series.

www.ingramcontent.com/pod-product-compliance
Lightning Source LLC
Chambersburg PA
CBHW082045280125
20788CB00044B/46
9780470973813